# MATERIAL and ENERGY
# BALANCES
## for
## ENGINEERS and ENVIRONMENTALISTS

### Second Edition

# Advances in Chemical and Process Engineering

ISSN: 2045-0036

**Series Editor:** Stephen M Richardson *(Imperial College London UK)*

---

*Published*

Advances in Chemical and Process Engineering – Vol. 3

# MATERIAL and ENERGY BALANCES

## for

## ENGINEERS and ENVIRONMENTALISTS

### Second Edition

## Colin Oloman

University of British Columbia, Canada

 **World Scientific**

NEW JERSEY · LONDON · SINGAPORE · BEIJING · SHANGHAI · HONG KONG · TAIPEI · CHENNAI · TOKYO

*Published by*

World Scientific Publishing Europe Ltd.

57 Shelton Street, Covent Garden, London WC2H 9HE

*Head office:* 5 Toh Tuck Link, Singapore 596224

*USA office:* 27 Warren Street, Suite 401-402, Hackensack, NJ 07601

**Library of Congress Cataloging-in-Publication Data**
Names: Oloman, Colin, author.
Title: Material and energy balances for engineers and environmentalists /
    Colin Oloman, University of British Columbia, Canada.
Description: Second edition. | London ; Hackensack, NJ : World Scientific Publishing Europe Ltd., [2023] |
    Series: Advances in chemical and process engineering, 2045-0036 ; vol. 3 |
    Includes bibliographical references and index.
Identifiers: LCCN 2022040916 | ISBN 9781800613102 (hardcover) | ISBN 9781800613249 (paperback) |
    ISBN 9781800613119 (ebook for institutions) | ISBN 9781800613126 (ebook for individuals)
Subjects: LCSH: Chemical processes. | Chemical engineering--Mathematics. | Mathematical optimization. |
    Nonlinear programming. | Conservation laws (Physics)
Classification: LCC TP155.7 .O46 2023 | DDC 660--dc23/eng/20221006
LC record available at https://lccn.loc.gov/2022040916

**British Library Cataloguing-in-Publication Data**
A catalogue record for this book is available from the British Library.

For any available supplementary material, please visit
https://www.worldscientific.com/worldscibooks/10.1142/Q0387#t=suppl

Desk Editors: Balasubramanian Shanmugam/Steven Patt

Typeset by Stallion Press
Email: enquiries@stallionpress.com

Printed in Singapore

# About the Author

 **Colin Oloman** (BE, M.A.Sc., P.Eng) is professor emeritus in the Department of Chemical and Biological Engineering at the University of British Columbia. Oloman's career encompasses 45 years of engineering in industry, commercial R&D, and academia, including 25 years teaching undergraduate courses in Material and Energy Balances, Process Design, Process Synthesis, and Electrochemical Engineering. He is the author of two books and author/co-author or inventor/co-inventor of over 100 scientific papers, patents, and proprietary reports on aspects of electrochemical and thermochemical process technology. A graduate of the University of Sydney and University of British Columbia, he resides in Vancouver, B.C.

# Acknowledgements

Thanks to the undergraduate students in the Department of Chemical and Biological Engineering at the University of British Columbia for motivating me to write this book, and to my wife, Mab, for her invaluable assistance in bringing it to fruition.

My gratitude also goes to colleagues, students and friends who reviewed early drafts of the text and made helpful suggestions for its improvement, notably Bruce Bowen, Dusko Posarac, Chad Bennington, Joel Bert, Sheldon Duff, Paul Watkinson, Karyn Ho, Alvaro Reyes and Bob Harvey.

# Contents

| Symbol | Meaning | Typical Units | |
|--------|---------|---------------|---|
| A, B, C | constants in the Antoine equation | –, K, K | |
| $A_{hex}$ | area of heat transfer surface | $m^2$ | |
| $A_i$ | interface area | $m^2$ | |
| a, b, c, d | empirical constants | various units | |
| ã | acceleration | $ms^{-2}$ | $[m/s^2]$ |
| $a$ | activity of a species | — | |
| $C^{\#}$ | number of components in a system | — | |
| $C_p$ | heat capacity at constant pressure | $kJ.kmol^{-1}.K^{-1}$ | [kJ/kmol.K] |
| $C_v$ | heat capacity at constant volume | $kJ.kmol^{-1}.K^{-1}$ | [kJ/kmol.K] |
| $C_{p,m}$ | mean heat capacity at constant pressure | $kJ.kmol^{-1}.K^{-1}$ | [kJ/kmol.K] |
| $C_{v,m}$ | mean heat capacity at constant volume | $kJ.kmol^{-1}.K^{-1}$ | [kJ/kmol.K] |
| c | velocity of light in vacuum | $ms^{-1}$ | [m/s] |
| D | liquid–liquid distribution coefficient | – | |
| D of F | degrees of freedom of an M&E balance problem | – | |
| d | diameter of dispersed bubbles, drops or particles | m | |
| E | energy | kJ | |
| $E_i$ | interfacial energy | kJ | |
| $E_p$ | potential energy | kJ | |
| $E_k$ | kinetic energy | kJ | |
| $E_v$ | electric voltage | V | |
| e | volume fraction of dispersed phase | – | |
| $e_E$ | energy balance closure | – | |
| $e_m$ | mass balance closure | – | |
| F | Faraday's number | $kCkmol^{-1}$ | [kC/kmol] |
| $F^{\#}$ | number of degrees of freedom of the intensive properties that define the thermodynamic state of a system | – | |
| $F_e$ | rate of energy degradation by friction of flowing fluid | kW | |
| f | force | kN | |
| $G^o$ | standard free energy of reaction | $kJ.kmol^{-1}$ | [kJ/kmol] |
| GEP | gross economic potential | $US.y^{-1}$ | [$US/y] |
| g | gravitational constant | $m.s^{-2}$ | $[m/s^2]$ |
| H | Enthalpy, w.r.t. a reference state (condition) | kJ | |
| Ḣ | Stream enthalpy flow w.r.t. compounds at reference state | kW | [kJ/s] |
| Ḣ* | Stream enthalpy flow w.r.t elements at reference state | kW | [kJ/s] |
| Ḧ | Stream enthalpy flow w.r.t. elements or compounds at reference states | kW | [kJ/s] |

| | | | |
|---|---|---|---|
| $H_{mix}$ | enthalpy (heat) of mixing | kJ/kmol | kJkmol$^{-1}$ |
| h | specific enthalpy, w.r.t. compounds at reference state | kJ.kmol$^{-1}$ | [kJ/kmol] |
| h* | specific enthalpy, w.r.t. elements at reference state | kJ.kmol$^{-1}$ | [kJ/kmol] |
| $h^{o}_{f,Tref}$ | standard enthalpy (heat) of formation at $T_{ref}$ | kJ.kmol$^{-1}$ | [kJ/kmol] |
| $h^{o}_{c}$ | standard enthalpy (heat) of combustion | kJ.kmol$^{-1}$ | [kJ/kmol] |
| $h_{c,gross}$ | gross heat of combustion (–ve higher heating value) | kJ.kmol$^{-1}$ | [kJ/kmol] |
| $h_{c,net}$ | heat of combustion (–ve lower heating value) | kJ.kmol$^{-1}$ | [kJ/kmol] |
| $h_m$ | specific enthalpy (heat) of fusion | kJ.kmol$^{-1}$ | [kJ/kmol] |
| $h_p$ | specific enthalpy (heat) of a phase change | kJ.kmol$^{-1}$ | [kJ/kmol] |
| $h_v$ | specific enthalpy (heat) of vaporisation | kJ.kmol$^{-1}$ | [kJ/kmol] |
| $h_{rxn}$ | specific enthalpy (heat) of reaction | kJ.kmol$^{-1}$ | [kJ/kmol] |
| I | number of process streams | — | |
| I′ | current | kA | |
| J | number of process species | — | |
| K | number of process units | — | |
| $K_A$ | adsorption constant | kPa$^{-1}$ | [1/kPa] |
| $K_{DIV}$ | divider flow ratio | – | |
| $K_{eq}$ | reaction equilibrium constant | – | |
| $K_H$ | Henry's constant | kPa(abs) | kPa(a) |
| k | chemical reaction rate constant | s$^{-1}$ | [1/s] |
| L′ | vertical height of fluid column | m | |
| L | number of chemical reactions | – | |
| M | molar mass (molecular weight) | kg.kmol$^{-1}$ | [kg/kmol] |
| $M_m$ | mean molar mass (molecular weight) | kg.kmol$^{-1}$ | [kg/kmol] |
| m | mass | kg | |
| ṁ | mass flow rate | kg.s$^{-1}$ | [kg/s] |
| m̄ | total stream mass flow rate | kg.s$^{-1}$ | [kg/s] |
| NEP | net economic potential | $US.y$^{-1}$ | [$US/y] |
| n | number of moles | kmol | |
| ṅ | mole flow rate | kmol.s$^{-1}$ | [kmol/s], [kmol/h] |
| n̄ | total stream mole flow rate | kmol.s$^{-1}$ | [kmol/s], [kmol/h] |
| $n_w$ | amount of water from combustion reaction | kmol(kmolfuel)$^{-1}$ | [kmol/kmolfuel] |
| P | total pressure | kPa(abs) | [kPa(a)] |
| $P_c$ | critical pressure | kPa(abs) | [kPa(a)] |
| $\Delta P$ | pressure difference | kPa | |
| p | partial pressure | kPa(abs) | [kPa(a)] |
| p* | vapour pressure | kPa(abs) | [kPa(a)] |
| $p_w$ | partial pressure of water vapour | kPa(abs) | [kPa(a)] |

| Symbol | Description | Units | Alt units |
|---|---|---|---|
| $p_w^*$ | vapour pressure of water | kPa(abs) | [kPa(a)] |
| Q | net heat input to the system | kJ | |
| $\dot{Q}$ | rate of net heat transfer into the system | kW | [kJ/s], [kJ/h] |
| R | universal gas constant | $kJ.kmol^{-1}.K^{-1}$ | [kJ/kmol.K] |
| ROI | return on investment | $\%.y^{-1}$ | [%/y] |
| r | ratio of heat capacities $= C_p/C_v$ | – | |
| rh | relative humidity | – | |
| SG | specific gravity | – | |
| S | selectivity | – | |
| S′ | solubility | $kg.kg^{-1}$ | [kg/kg] |
| s | split fraction | – | |
| T | temperature | K | |
| $T_B$ | bubble-point temperature | K | |
| $T_c$ | critical temperature | K | |
| $T_D$ | dew-point temperature | K | |
| $T_{ref}$ | reference temperature | K | |
| $\Delta T$ | temperature difference | K | |
| t | time | s, h | |
| U | internal energy, w.r.t. a reference state | kJ | |
| $U_{hex}$ | heat transfer coefficient | $kW.m^{-2}.K^{-1}$ | [kW/m². K] |
| u | specific internal energy, w.r.t. a reference state | $kJ.kmol^{-1}$ | [kJ/kmol] |
| $\tilde{u}$ | velocity | $m.s^{-1}$ | [m/s] |
| V | volume | $m^3$ | |
| $\dot{V}$ | volume flow rate | $m^3.s^{-1}$ | [m³/s] |
| $V_R$ | reactor volume | $m^3$ | |
| v | specific volume | $m^3.kg^{-1}$ | [m³/kg] |
| $v_m$ | molar volume | $m^3.kmol^{-1}$ | [m³/kmol] |
| v′ | partial volume | $m^3$ | |
| $\tilde{v}$ | volume fraction | – | |
| W | net work output from (i.e. work done by) system | kJ | |
| $\dot{W}$ | rate of net work transfer out of system | kW | [kJ/h] |
| w | mass fraction | – | |
| $w_q$ | quality | – | |
| X | conversion of reactant | – | |
| $X_{act}$ | actual conversion of reactant | – | |
| $X_{eq}$ | equilibrium conversion of reactant | – | |
| x | mole fraction (in ideal mixture gas phase) | – | |
| Y | yield | – | |
| y | mole fraction (in gas phase) | – | |
| Z | value of h, u or v in a gas-liquid mixture | various | |
| z | compressibility factor | – | |

## Greek Letters

| | | | |
|---|---|---|---|
| $\alpha, \beta, \chi, \delta$ | reaction stoichiometric coefficients | – | |
| $\nu$ | general reaction stoichiometric coefficient | – | |
| $\varepsilon$ | extent of reaction for the specified reaction | kmol | |
| $\theta$ | coefficient of thermal expansion | $K^{-1}$ | [1/K] |
| $\sigma$ | coefficient of compressibility | $kPa^{-1}$ | [1/kPa] |
| $\rho$ | density | $kg.m^{-3}$ | [kg/m$^3$] |
| $\rho_m$ | density of mixture | $kg.m^{-3}$ | [kg/m$^3$] |
| $\rho_{ref}$ | density of reference substance | $kg.m^{-3}$ | [kg/m$^3$] |
| $\Pi^{\#}$ | number of phases in a system at equilibrium | – | |
| $\Sigma$ | sum of values | – | |
| $\gamma$ | surface tension | $kN.m^{-1}$ | [kN/m] |
| $\Gamma$ | surface concentration | $kmol.m^{-2}$ | [kmol/m$^2$] |

## Subscripts

| | |
|---|---|
| g | gas phase |
| l | liquid phase |
| s | solid phase |
| i | process stream counter |
| j | process species counter |
| k | process unit counter |
| in | input to a system |
| out | output from a system |

## Superscripts

| | |
|---|---|
| o | standard state |

– = dimensionless

## Abbreviations

| | | Typical Units |
|---|---|---|
| ACC | (final amount – initial amount) of a specified quantity in a system | kg, kmol, kJ |
| IN | input of a specified quantity to a system | kg, kmol, kJ |
| OUT | output of a specified quantity from a system | kg, kmol, kJ |
| GEN | generation of a specified quantity in a system | kg, kmol, kJ |
| CON | consumption of a specified quantity in a system | kg, kmol, kJ |
| RateACC | rate of change of ACC, w.r.t. time | kg/s, kmol/s, kW, kg/h, kmol/h, kJ/h |
| Rate IN | rate of change of IN, w.r.t. time | kg/s, kmol/s, kW, kg/h, kmol/h, kJ/h |
| Rate OUT | rate of change of OUT, w.r.t. time | kg/s, kmol/s, kW, kg/h, kmol/h, kJ/h |
| Rate GEN | rate of change of GEN, w.r.t. time | kg/s, kmol/s, kW, kg/h, kmol/h, kJ/h |
| Rate CON | rate of change of CON, w.r.t. time | kg/s, kmol/s, kW, kg/h, kmol/h, kJ/h |

**Arrays:** Array dimensions "i", "j" and "k" designate respectively the streams, species and process units in equations, flowsheets and stream tables.

| | |
|---|---|
| ACC = IN − OUT + GEN − CON | General balance |
| Rate ACC = Rate IN − Rate OUT + Rate GEN − Rate CON | Differential general balance |
| Abs pressure = Atm pressure + gauge pressure | Superatmospheric |
| Abs pressure = Atm pressure − vacuum | Subatmospheric |
| $P = \rho g L'$ | Column of fluid |
| $K = 273.15 + {}^\circ C$ | Kelvin |
| ${}^\circ R = 459.67 + {}^\circ F$ | Rankine |
| ${}^\circ F = 32 + 1.8\,{}^\circ C$ | Fahrenheit |
| $p(j) = y(j)P$ | Partial pressure |
| $v'(j) = y(j)V$ | Partial volume |
| $\Sigma[p(j)] = P$ | Dalton's law |
| $\Sigma[v'(j)] = V$ | Amagat's law |
| $F^{\#} = C^{\#} - P^{\#} + 2$ | Phase rule |
| $PV = nRT$ | Ideal gas law |
| $PV = znRT$ | Non-ideal gas |
| $v = v_o(1 + \beta(T - T_o) - \delta(P - P_o))$ | Liquid or solid |
| $\rho = m/V$ | Density |
| $\rho = MP/RT$ | Ideal gas density |
| $M_m = \Sigma[M(j)y(j)]$ | Mean molar mass |
| $\rho_m = 1/\Sigma\,[w(j)/\rho(j)]$ | Ideal mix density |
| $SG = \rho/\rho_{ref}$ | Specific gravity |
| $V_m = Mv = M/\rho$ | Molar volume |
| $w(j) = m(j)/\Sigma[m(j)]$ | Mass fraction |
| $x(j) = n(j)/\Sigma[n(j)]$ | Mole fraction |
| $\Sigma[w(j)] = 1 \quad \Sigma[x(j)] = 1 \quad \Sigma[y(j)] = 1$ | Stream composition |
| $x(j) = w(j)/M(j)/\Sigma[w(j)/M(j)]$ | Mass to mole fraction |
| $w(j) = x(j)M(j)/\Sigma[x(j)M(j)]$ | Mole to mass fraction |
| Mass % = weight % = wt % = 100w(j) | Mass percent |
| Mole % = 100x(j) | Mole percent |
| $p^{*}(j) = \exp[A(j) - B(j)/(T + C(j))]$ | Antoine equation |
| $\ln(p_1^{*}/p_2^{*}) = (h_v/R)(1/T_2 - 1/T_1)$ | Clausius-Clapeyron equation |
| $p(j) = x(j)p^{*}(j)$ | Raoult's law |
| $p(j) = K_H x(j)$ | Henry's law |
| $1 = \Sigma[x(j)p^{*}(j)]/P$  (ideal mixture liquid phase) | Bubble-point |
| $1 = P\Sigma[y(j)/p^{*}(j)]$  (ideal mixture gas phase) | Dew-point |
| $\Sigma[v(j)M(j)] = 0$ | Stoichiometry |
| $E_i = \gamma A_i$ | Surface energy |

$$X(j) = \frac{\text{amount of } j \text{ converted to all products}}{\text{amount of } j \text{ introduced to reaction}}$$

Conversion

$$\varepsilon(l) = \frac{\text{amount of species } j \text{ converted in reaction } l}{\text{stoichiometric coefficient of } j \text{ in reaction } l}$$

Extent of reaction

$$S(j, q) = \frac{\text{amount of } j \text{ converted to product } q}{\text{amount of } j \text{ converted to all products}}$$

Selectivity

$$Y(j, q) = \frac{\text{amount of } j \text{ converted to product } q}{\text{amount of } j \text{ introduced to reaction}}$$

Yield

$Y(j, q) = X(j)S(j, q)$                                                                                Yield

$K_{eq} = (a_C^{\chi} a_D \delta)/(a_C^{\alpha} a_D^{\beta})$                                          Equilibrium constant

$a(j) \approx x(j)$                                                                                                Activity (solids)

$a(j) \approx [J]/[1 \text{ kmol/m}^3]$                                                                    Activity (liquids)

$a(j) \approx p(j)/101.3 \text{ kPa}$                                                                         Activity (gases)

$X_{act}/X_{eq} < 1$                                                                                            Conversion

$C_p = C_v + R$                                                                                                 Heat capacity

$C_p = a + bT + cT^2 + dT^3$                                                                            Heat capacity

$$C_{p,m} = \int_{T_{ref}}^{T} C_p \, dt/(T - T_{ref})$$

[No phase change, no reaction]                 Mean heat capacity

$H = U + PV$                                                                                                    Enthalpy

$h_{f,Tref}(\text{gas}) = h_{f,Tref}(\text{solid}) + h_{m,Tref} + h_{v,Tref}$        Heat of formation

$h_{f,Tref}(\text{gas}) = h_{f,Tref}(\text{liquid}) + h_{v,Tref}$                          Heat of formation

$h_{rxn} = \Sigma(v(j)h_f(j)) \text{ products} - \Sigma(v(j)h_f(j)) \text{ reactants}$     Heat of reaction

$A_i = (6e/d)V$                                                                                                 Interface area

$Z_m = w_q Z_g + (1 - w_q)Z_l \quad [Z = h, u \text{ or } v]$                                Quality

$rh = p_w/p^*_w$                                                                                                Relative humidity

GEP = Value of products − Value of feeds                                         Gross economic potential

NEP = GEP − cost (utilities + labour + maintenance + interest)     Net economic potential

ROI = Annual NEP/Capital cost                                                          Return on investment

$s(i, j) = \dot{n}(i, j)/\dot{n}(l, j) = \dot{m}(i, j)/\dot{m}(l, j)$                       Split-fraction

$0 = \Sigma[\bar{m}(i)in] - \Sigma[\bar{m}(i)out]$                                              Steady-state mass balance

$e_M = 100 \, \Sigma[\bar{m}(i)out]/ \Sigma[\bar{m}(i)in]$                               Mass balance closure

D of F = no. of unknowns − no. of independent equations             Specification

$E_{final} - E_{initial} = Q - W$                          [Closed system]       Energy balance

$E_{final} - E_{initial} = E_{input} - E_{output} + Q - W - (PV_{out} - PV_{in})$     [Open system]       Energy balance

$dE/dt = \dot{E}_{input} - \dot{E}_{output} \, \dot{Q} - \dot{W} - (P\dot{V}_{out} - P\dot{V}_{in})$     [Open system]       Energy balance

| Equation | Condition | Description |
|---|---|---|
| $dE/dt = 0 = [\dot{H} + \dot{E}_k + \dot{E}_p]_{in} - [\dot{H} + \dot{E}_k + \dot{E}_p]_{out} + \dot{Q} - \dot{W}$ | [Open system] | Energy balance (steady-state) |
| $0 = \dot{H}_{input} - \dot{H}_{ouput} + \dot{Q} - \dot{W}$ | [Open system] | Enthalpy balance (steady-state) |
| $\Gamma(j) = aK_A p(j)/(1 + K_A p(j))$ | | Adsorption isotherm |
| $D(j) = x(j)_A / x(j)_B$ | | Distribution coefficient |
| $S'(j) = m(j)/m_s$ | | Solubility |
| Separation efficiency of component $j = \dfrac{\text{amount of } j \text{ in specified outlet stream}}{\text{amount of } j \text{ in the inlet stream}}$ | | Separator |
| $\dot{m}(1,j)/\overline{m}(1) = \dot{m}(2,j)/\overline{m}(2) = \dot{m}(3,j)/\overline{m}(3),$ etc. | | Divider composition |
| $K_{DIV}(i) = \overline{n}(i)/\overline{n}(1) = \overline{m}(i)/\overline{m}(1)$ | | Divider ratio |
| $C_v = [\partial u/\partial T]_v$ | [No phase change, no reaction] | Heat capacity constant volume |
| $C_p = [\partial h/\partial T]_p$ | [No phase change, no reaction] | Heat capacity constant pressure |
| $\dot{H}^*(i) = \Sigma[\dot{n}(i,j)h^*(j)]$ | [Stream enthalpy] | Ideal mixture |
| $\dot{H}^*(i) = \Sigma[\dot{n}(i,j)h^*(j)] + \dot{H}_{mix}$ | [Stream enthalpy] | Non-ideal mixture |
| $h^* = \displaystyle\int_{T_{ref}}^{T(i)} C_p dT + h^{\circ}_{f,Tref}$ | [Respect the phase] | Specific enthalpy, w.r.t. elements |
| $h^* \approx C_{p,m}[T(i) - T_{ref}] + h^{\circ}_{f,Tref}$ | [Respect the phase] | Specific enthalpy, w.r.t. elements |
| $h^* = \displaystyle\int_{T_{ref}}^{T(i)} C_p dT + h_{p,Tref}$ | [Respect the phase] | Specific enthalpy, w.r.t. compounds |
| $h \approx C_{p,m}[T(i) - T_{ref}] + h_{p,Tref}$ | [Respect the phase] | Specific enthalpy, w.r.t. compounds |
| $\dot{Q} = U_{hex}A_{hex}\,\Delta T$ | [Thermal duty] | Heat exchanger |
| $\dot{W} = -(P1\dot{V}_1)(r/(r-1))[(P_2/P_1)^{((r-1)/r)} - 1]$ | [Power] | Adiabatic gas compressor |
| $T_2 = T_1[P_2/P_1]^{((r-1)/r)}$ | [Temperature] | Adiabatic gas compressor |
| $\dot{W} = -(P1\dot{V}_1)[\ln(P_2/P_1)]$ | [Power] | Isothermal compressor |
| $\dot{W} = -\dot{V}_1(P_2/P_1)$ | [Power] | Liquid pump |
| $\dot{W} = E_v I'$ | [Power] | Battery, fuel-cell, resistor |
| $e_E = 100\,[\Sigma\dot{H}^*(i)\text{out} + \dot{W}]/[\Sigma\dot{H}^*(i)\text{in} + \dot{Q}]$ | | Energy balance closure |
| $T_{out} = T_{ref} + [(\dot{H}^*(i)\text{in} + \dot{Q} - \dot{W})/\dot{n}(\text{in}) - h^{\circ}_{f,Tref}]/C_{p,m}$ | [No phase change] | Heat exchanger |
| $0 = \Sigma[\dot{H}(i)]\text{in} - \Sigma[\dot{H}(i)]\text{out} + \dot{Q} - \dot{W} - \dot{n}_{fuel}\,h_c$ | | Combustion |

$h_{c,net} = h_{c,\,gross} + n_w h_{v,H2O}$      Heat of combustion

$d(m)/dt = (\dot{m})_{in} - (\dot{m})_{out} + (\dot{m})_{gen} - (\dot{m})_{con}$    [Unsteady-state]    Mass balance

$d(n)/dt = (\dot{n})_{in} - (\dot{n})_{out} + (\dot{n})_{gen} - (\dot{n})_{con}$    [Unsteady-state]    Mole balance

$dU/dt = \dot{H}_{in} - \dot{H}_{out} + \dot{Q} - \dot{W}$    [Unsteady-state]    Enthalpy balance

$0 = 0.5[(\tilde{u}_{out})^2 - (\tilde{u}_{in})^2] + g[L'_{out} - L'_{in}] + [P_{out} - P_{in}]/\rho$    Bernoulli equation

$0 = 0.5(u_{out})^2[\dot{E}_k + \dot{E}_p]_{in} - [\dot{E}_k + \dot{E}_p]_{out} - \int_{Pin}^{Pout} \dot{V}_{dp} - \dot{F}_e - \dot{W}$    Mechanical energy balance

| | |
|---|---|
| **Absorption** | Transfer of species between two bulk phases. |
| **Acceleration** | Rate of change of velocity with respect to time. |
| **Adiabatic** | Zero transfer of heat across the system envelope. |
| **Accumulation** | Change in the amount of a material and/or energy inside the system envelope. |
| **Adsorption** | Concentration of species at an interface. |
| **Atom** | Smallest particle of an element, identifiable as that element by its nucleus. |
| **Atom balance** | Material balance in which the quantity is atoms or moles of specified elements. |
| **Batch** | Operation in a closed system. |
| **Blower** | Gas compressor operating up to ca. 400 kPa(abs). |
| **Boiler** | Apparatus in which liquid is vaporised (boiled), usually by transfer of heat from combustion. |
| **Bubble-point** | Condition at which first bubbles of vapour form in a liquid. |
| **Capital cost** | Total cost for the design and installation of a plant. |
| **Carbon cycle** | Geochemical cycle of carbon through carbon dioxide, bio-mass and carbonates. |
| **Catalyst** | Substance that speeds a reaction without itself suffering a net conversion. |
| **Closed system** | System with zero transfer of material across the boundary. |
| **Closure** | Ratio of output to input of mass or energy for an open system. Steady-state closure = 1. |
| **Colloid** | Mixture with super-molecular particles or drops below ca. 1 micron dispersed in liquid. |
| **Combustion** | Thermochemical reaction of a fuel with an oxidant (the oxidant is usually oxygen from air). |
| **Component** | Constituent of a mixture identified as a phase or as a species. |
| **Compressibility factor** | Multiplier (range approx. 0.2 to 4) that adapts the ideal gas law for non-ideal gases. |
| **Compressor** | Apparatus used to pump gas (to super-atmospheric pressure). |
| **Concentration** | Quantitative measure of the composition of a mixture. |
| **Condensation** | Change of phase from vapour (gas) to liquid or solid. |
| **Condensed phase** | Solid or liquid phase (usually related to a corresponding vapour). |
| **Continuous** | Operation in an open system. |
| **Controlled mass** | Closed system. Zero transfer of material across system boundary. |
| **Controlled volume** | Open system. Finite transfer of material across the system boundary. |

| | |
|---|---|
| **Convergence** | Tending to a single condition, usually the solution of a non-linear problem. |
| **Conversion** | Fraction of a reactant species consumed by chemical reaction. |
| **Critical pressure** | Pressure required to liquefy a substance at its critical temperature. |
| **Critical temperature** | Temperature above which a substance cannot be liquefied at any pressure. |
| **Degrees of freedom** | Number of variables that can be independently manipulated to influence a system. |
| **Density** | Ratio of mass to volume. |
| **Dew-point** | Condition at which first drops of liquid form from a vapour. |
| **Differential equation** | Relation between the rates of change of variables with respect to other variables. |
| **Dispersion** | Mixture in which one (or more) phase is dispersed as bubbles, drops or particles. |
| **Divider** | Process unit that divides an input stream to multiple outputs of the same composition. |
| **Dry-bulb** | Dry thermometer bulb. |
| **Electrolysis** | Decomposition of a substance at complementary electrodes by an electric current (usually DC). |
| **Emulsion** | Mixture of liquids with a dispersed liquid phase (usually in drops < ca. 1 mm). |
| **Energy** | Measure of the potential to do work. |
| **Energy balance** | Quantitative account of amounts of energy within and across a system. |
| **Enthalpy** | Internal energy plus pressure–volume energy. |
| **Enthalpy balance** | Energy balance with the stream energy replaced by stream enthalpy. |
| **Entropy** | Capacity factor for isothermally unavailable energy. Measure of system disorder. |
| **Equation of state** | Mathematical relation between intensive variables that uniquely fix the thermodynamic state of a system. A single-phase, single-component system requires 3 intensive variables — typically P, T and v. |
| **Equilibrium** | Condition of balance. Zero gradient in chemical potential. |
| **Evaporation** | Change of phase from solid or liquid to vapour. |
| **Extensive property** | A property that depends on the amount of material. |
| **Extent of reaction** | Ratio of moles of a species consumed or produced in a chemical reaction to its stoichiometric coefficient. |

| | |
|---|---|
| **Fan** | Apparatus used to pump gas (usually within about 5 kPa of atmospheric pressure). |
| **Figure of merit** | Quantitative measure of the value of an object, design, process, etc. |
| **1st law of thermodynamics** | Mathematical statement of the conservation of energy in non-nuclear processes. |
| **Flash split** | Separation of a mixture into liquid and vapour by dropping the pressure. |
| **Flowsheet** | Diagram of process units and their functions, with their interconnecting process streams. |
| **Flow work** | Net energy needed as pressure–volume work to move material (usually a fluid) in and out of a system. |
| **Fluid** | Gas or liquid. Easily deformed when subject to differential pressure. |
| **Foam** | Mixture of gas with liquid, in which the continuous liquid phase exists as bubble films. |
| **Force** | Product of mass with acceleration. |
| **Fuel cell** | Spontaneous electrochemical cell that generates DC electrical power with catalytic electrodes. |
| **Furnace** | Apparatus in which a fuel is burned (usually in air) to generate heat. |
| **Gas** | A compressible state of matter that cannot hold its shape without support. Material in the gas phase above its critical temperature. |
| **Gauge pressure** | Pressure measured relative to the local atmospheric pressure. |
| **Greenhouse effect** | Heating of Earth's atmosphere due to absorption of IR radiation by greenhouse gases. |
| **Greenhouse gases** | Polyatomic gases (e.g. $CO_2$, $H_2O$, $CH_4$) that absorb IR radiation more effectively than air ($N_2$ and $O_2$). |
| **Gross economic potential** | Value of products minus value of feeds. |
| **Heat** | Energy transfer to a system that is driven by a temperature difference. |
| **Heat capacity** | Specific heat input to raise the temperature of a substance without changing its phase. |
| **Heat exchanger** | Process unit that transfers heat from one input stream to a second input stream. |
| **Heat of combustion**[1] | Heat input required to carry out a combustion reaction under isothermal conditions. |

---

[1] The heats (a.k.a. enthalpies) of combustion, formation, fusion, mixing, phase change and reaction involve the condition of constant pressure, which is usually one standard atmosphere.

| | |
|---|---|
| **Heat of formation**[1] | Heat input required to form a compound from its elements under isothermal conditions. |
| **Heat of fusion**[1] | Heat input to melt a substance under isothermal conditions. |
| **Heat of mixing**[1] | Heat input required to form a mixture from its components under isothermal conditions. |
| **Heat of phase change**[1] | Heat input required to change the phase of a substance under isothermal conditions. |
| **Heat of reaction**[1] | Heat input required to carry out a (chemical) reaction under isothermal conditions. |
| **Heat of vaporisation**[1] | Heat input to vaporise a substance under isothermal conditions. |
| **Ideal gas** | Gas with zero interaction between molecules. |
| **Ideal mixture** | Mixture with zero interaction between components (at the molecular level). |
| **Independent equation** | Equation not obtained by combination of other equations in the set. |
| **Infinite dilution** | Concentration corresponding to an infinite ratio of solvent to solute. |
| **Intensive property** | A property that is independent of the amount of material. |
| **Interaction** | Condition where the level of one variable affects the response of another variable(s). |
| **Isothermal** | Constant temperature. |
| **Iteration** | Repeating a calculation with recycled values of key variables, usually to convergence. |
| **Irreversible reaction** | Reaction that goes in only the forward (L to R) direction. |
| **Kilomole** | 1000 moles. Molar mass expressed in kilograms. |
| **Kinetic energy** | Energy of a mass due to its motion (velocity). |
| **Latent heat** | Heat input to change the phase of matter without changing composition or temperature. |
| **Linear equation** | Equation in which each independent variable appears in a separate term and with exponent = 1. |
| **Liquid** | A (relatively) incompressible state of matter that cannot hold its shape without support. |
| **Mass** | Measure of the quantity of material. |
| **Mass fraction** | Ratio of mass of a component to the total mass of a mixture. |
| **Material** | Matter, physical substance, stuff. |
| **Material balance** | Quantitative account of amounts of material within and across a system boundary. |
| **Mixer** | Process unit that combines multiple input streams to a single output stream. |

| | |
|---|---|
| **Mixture** | Combination of components in a single or multi-phase system without chemical reaction. |
| **Model** | Representation of a physical system in equations that predict the system behaviour. |
| **Molarity** | Measure of concentration expressed as kilomole solute per cubic meter solution. |
| **Mole** | Amount of material in Avogadro's number (6.023E23) of "molecules". |
| **Mole balance** | Material balance in which the quantity is moles of specified substance(s). |
| **Mole fraction** | Ratio of moles of a component to the total moles of a mixture. |
| **Momentum** | Product of mass and velocity. |
| **Net economic potential** | Gross economic potential minus operating costs. |
| **Non-condensable** | Gas above its critical temperature. |
| **Non-ideal mixture** | Mixture in which the components interact (properties of the mixture are not a linear combination of those of the components). |
| **Non-linear** | Relation between variables that is not linear (e.g. involving products, exponents, etc.). |
| **Non-linear equation** | Equation in which one or more independent variables are multiplied by other variables, and/or exponentiated and/or are arguments in algebraic functions (e.g. exp, log, tan, etc.). |
| **Non-process element** | Element that is not necessary (i.e. is superfluous) for operation of a process. |
| **Normal boiling point** | Boiling point temperature of a pure substance under 101.3 kPa(abs) pressure. |
| **Normality** | Measure on concentration expressed as kiloequivalents solute per cubic meter solution. |
| **N.T.P.** | Normal temperature and pressure [various, but usually ca. 293 K, 101.3 kPa(abs)]. |
| **Open system** | System with finite transfer of material across the boundary. |
| **Optimisation** | Finding conditions for the best (usually a maximum or minimum) value of an objective. |
| **Path function** | Value dependent on the path between two states. [Heat and work are both path functions] |
| **Partial pressure** | Pressure exerted by a component of a gas mixture if it alone were at the temperature and volume of the mixture. |

| | |
|---|---|
| **Partial volume** | Volume occupied by a component of a gas mixture if it alone were present at the total pressure and temperature of the mixture. |
| **Phase** | State of matter identified as solid, liquid or gas. |
| **Phase split** | Distribution of species between the phases in a multi-phase system. |
| **Photovoltaic cell** | Device for converting radiant energy (e.g. sunlight) to electricity. |
| **Potential energy** | Energy of a mass due to its position in a potential field (e.g. gravitational). |
| **Power** | Rate of transfer of energy with respect to time. |
| **Pressure** | Force per unit area. |
| **Process** | Operation that changes the condition of material or energy. |
| **Process rate** | Specific rate w.r.t. time, determined by intrinsic kinetics and/or mass or energy transfer. |
| **Process unit** | Apparatus that operates on material or energy input(s) to produce modified output(s). |
| **Pump** | Process unit that changes the pressure of a process stream. |
| **Purge** | Removing undesired substances by bleeding material from recycle loops. |
| **Quantity** | Property that can be measured. |
| **Quality** | Mass fraction of vapour in vapour–liquid mixture. |
| **Reactor** | Process unit that transforms molecular or atomic species by chemical or nuclear reaction. |
| **Reaction rate** | Rate of consumption of a reactant or generation of a product with respect to time. |
| **Recycle** | Return to a previous stage of a cyclic process (e.g. of material, from downstream to upstream process units, via a "recycle loop"). |
| **Relaxation** | Procedure used to speed convergence of iterative calculations. |
| **Return on investment** | Ratio of net economic potential to capital cost. |
| **Reversible reaction** | Reaction that can go in both the forward (L to R) and reverse (R to L) directions. |
| **R.T.P.** | Room temperature and pressure. [ca. 293 K, 101.3 kPa(abs)] |
| **Rule of thumb** | Approximation (based on experience) adequate for practical calculations in most cases. |
| **Selectivity** | Fraction of a reactant converted to a specified product. |
| **Sensible heat** | Heat to change the temperature of matter without changing composition or phase. |

| | |
|---|---|
| **Separator** | Process unit that separates an input stream to multiple output streams of different composition. |
| **Sequential modular** | Method of calculating material and/or energy balances taking process units in sequence. |
| **Shaft work** | Work transfer from a system, excluding pressure–volume work (flow work) on/by material entering/leaving the system. |
| **Simultaneous equations** | A set of independent equations embracing multiple variables. |
| **Sink** | Consumption term in the general balance equation. |
| **Slurry** | Mixture in which particles of solid exceeding about 1 micron diameter are dispersed in a liquid. |
| **Solid** | State of matter that retains its shape without support. |
| **Solubility** | Maximum concentration of a solute in its solution (under specified conditions). |
| **Solute** | Constituent of a solution dissolved in the solvent. |
| **Solution** | Mixture in which the components are (uniformly) distributed on a molecular scale. |
| **Solvent** | Basic constituent of a solution, usually present in the largest molar amount. |
| **Source** | Generation term in the general balance equation. |
| **Sparge** | Blow gas into a liquid (usually through an orifice to produce small bubbles). |
| **Species** | Constituent identified by a specific atomic or molecular configuration. |
| **Specification** | Determination of the degrees of freedom of a system (esp. in process design). |
| **Specific gravity** | Ratio of density of substance to density of a reference substance (usually liquid water or air). |
| **Specific heat** | Ratio of heat capacity of substance to heat capacity of liquid water. |
| **Standard state** | Equilibrium state of a substance under 101.3 kPa(abs) pressure (and usually 298 K). |
| **State** | Condition of a material that fixes its thermodynamic properties. |
| **State function** | Property depending on the state of a substance and independent of the path to that state. |
| **Steady-state** | Condition of a system with zero rate of accumulation. Invariant with respect to time. |
| **Steam table** | Table of the thermodynamic properties of water in its gas and liquid phases. |
| **Stoichiometry** | Quantitative relation between substances consumed and generated in chemical reaction. |

**Stream table**       Tabular quantitative summary of a material and energy balance on an open system.

**Strip**             Remove material by desorption or redistribution into a carrier phase.

**S.T.P.**            Standard temperature and pressure [273.15 K, 101.3 kPa(abs)].

**Surface energy**    Excess internal energy of a surface relative to an equal amount of bulk material.

**Sublimation**       Change of phase from solid to vapour.

**System**            Region of space defined by a (real or imaginary) closed envelope (envelope = system boundary).

**Temperature**       Measure of potential to transfer heat.

**Threshold effect**  Sudden change in behaviour of a system when process variable(s) cross a critical value.

**Transient**         Varying over time. Unsteady-state.

**Triple-point**      Condition of a pure substance at which the solid, liquid and gas co-exist at equilibrium.

**Turbine**           Rotating machine that generates energy from pressure drop in a flowing fluid.

**Unsteady-state**    Condition of a system in which rate of accumulation is not zero. Variant with respect to time. Transient.

**Vacuum**            Sub-atmospheric pressure (absolute vacuum = zero absolute pressure).

**Vacuum pump**       Apparatus used to pump gas from a system under vacuum.

**Vaporisation**      Change of phase from solid or liquid to gas (a.k.a. vapour).

**Vapour**            Material in the gas phase below its critical temperature (i.e. condensable).

**Vapour pressure**   Pressure exerted by a solid or liquid in equilibrium with its own vapour.

**Velocity**          Rate of change of distance with respect to time.

**Volume fraction**   Ratio of partial volume of a component to the total volume of a mixture.

**Weight**            Force exerted on a mass by a gravitational field. [Weight = (mass)(acceleration of gravity) = mg]

**Wet-bulb**          Thermometer bulb wetted by liquid (usually wrapped in a wetted porous material).

**Work**              Force times distance. Energy transfer from a system, other than as heat or internal energy of transferred material.

**Yield**             Product of reactant conversion and the corresponding selectivity for a given species. Fraction of reactant converted to a specific product (usually the desired product).

## An Introduction

A material and energy balance is essentially a quantitative account of the redistribution of material and/or energy that occurs when anything happens. This basic tool of process engineering can be used to solve many practical problems. The list below shows that such problems range from those of interest to engineers and scientists in their daily work (see *Refs. 3,4*) to those of relevance to all people concerned with the sustainability of life on planet Earth (see *Refs. 1,2,5 and 6*).

## USES OF MATERIAL AND ENERGY BALANCES

- Modelling and design of industrial chemical[1] processes
- Life-cycle analysis and industrial ecology
- Development of medical technology
- Modelling the global environment, geochemical cycles and climate change
- Management of resources (energy, food and water) for the population of planet Earth
- Design for space stations and interplanetary travel

This text focuses on the use of material and energy balances in the modelling and design of industrial chemical processes, but includes examples that illustrate the application of material and energy balances to chemical processes in other fields, such as those listed above.

---

[1] The term "chemical process" in this context means any change resulting in the redistribution of material and/or energy in a system, and embraces processes that involve *biochemical, electrochemical, photochemical, physicochemical or thermochemical phenomena*. As used in this text the term "chemical process" excludes changes that involve the inter-conversion of mass and energy via nuclear reactions.

## PREREQUISITES

The principles involved in material and energy balance calculations are those that you probably learned in high school and/or the first year of university. These principles include, for example:

- The law of conservation
- Dimensions and units of physical quantities
- Chemical reactions and stoichiometry
- The ideal gas law
- Multi-phase equilibrium
- The first law of thermodynamics
- Basic thermochemistry and thermophysics
- Simple algebra and a little calculus
- Computer calculations with spreadsheets

These basics are reviewed in Chapters 1 and 2 then combined and extended in Chapters 3 to 7 to a formalised procedure for treating chemical processes, which is a key skill in the practice of chemical, biochemical, electrochemical and environmental engineering.

## DEALING WITH COMPLEX PROBLEMS

Material and energy balances provide a tool for students in many fields to grasp and to quantify the factors that determine the behaviour of both man-made and natural systems. Chemical processes (as defined broadly above) are complex and typically involve many interacting factors in non-linear relationships. In this respect, the material and energy balance techniques used by engineers to model industrial processes can be used as both an educational and a predictive tool to deal with complex natural phenomena such as, the global geochemical cycles, population dynamics and the thermal "greenhouse" effect on planet Earth.

The constraints of material and energy in natural systems appear poorly understood by our politicians, as well as by many economists and practitioners of the social sciences (see *Refs. 6,7*). Perhaps this text will help to provide the needed conceptual basis for dealing with the problems of material and energy flow that are basic to the progress and sustainability of our civilisation.

## FURTHER READING

[1]  A. Kopecky, *The Environmentalist's Dilemma. Promise and Peril in an Age of Climate Crisis*, ECW Press, 2021.

[2] Intergovernmental Panel on Climate Change, *Sixth Assessment Report — The Physical Science Basis*, IPCC, 2022.

[3] A. Vesilind, L. Heine and S. Morgan, *Introduction to Environmental Engineering*, Thomson Engineering, 2010.

[4] K. Solen and J. Harb, *Introduction to Chemical Engineering*, Wiley, 2011.

[5] A. Ford, *Modeling the Environment*, Island Press, 2009.

[6] R. Wright, *A Short History of Progress*, Anansi, 2019.

[7] T. Homer-Dixon, *The Ingenuity Gap*, Alfred Knofp, 2000.

# CHAPTER ONE

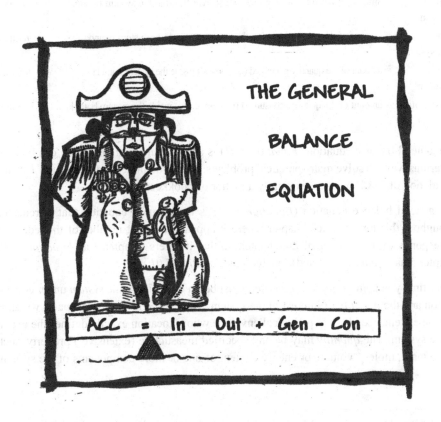

THE GENERAL

BALANCE

EQUATION

ACC = In – Out + Gen – Con

## THE GENERAL BALANCE

All material and energy (M&E) balance calculations are based on our experience that matter and energy may change its form, but it cannot appear from nor disappear to nothing. This observation is expressed mathematically in *Equation 1.01*, the *General Balance Equation.*

For a defined *system* and a specified *quantity*:

| **Accumulation** | = **Input** | − **Output** | + **Generation** | − **Consumption** | |
|---|---|---|---|---|---|
| **in system** | **to system** | **from system** | **in system** | **in system** | *Equation 1.01* |

where:

Accumulation = [final amount of the quantity − initial amount of the quantity] inside the system boundary.
in system

Input = amount of the quantity entering the system through the system boundary.          (input)
to system

Output = amount of the quantity leaving the system through the system boundary.          (output)
from system

Generation = amount of the quantity generated (i.e. formed) inside the system boundary.          (source)
in system

Consumption = amount of the quantity consumed (i.e. converted) inside the system boundary.          (sink)
in system

The general balance equation (*Equation 1.01*) is a powerful equation, which can be used in various ways to solve many practical problems. Once you understand *Equation 1.01* the calculation of M&E balances is simply a matter of bookkeeping.

The general balance equation (*Equation 1.01*) is the primary equation that is repeated throughout this text. In each chapter where it appears, the first number of this equation corresponds to the number of the chapter, so that it enters Chapter 4 as *Equation 4.01*, Chapter 5 as *Equation 5.01* and Chapter 7 as *Equation 7.01*.

Every time you apply *Equation 1.01*, you must begin by defining the *system* under consideration and the *quantity* of interest in the system. The *system* is a physical space, which is completely enclosed by a hypothetical envelope whose location exactly defines the extent of the system. The *quantity* may be any specified measurable (extensive[1]) property, such as the mass, moles,[2] volume or energy content of one or more components of the system.

---

[1] An extensive property is a property whose value is proportional to the amount of material.

[2] A *mole* is the amount of a substance containing the same number of *elementary particles* as there are atoms in 0.012 kg of carbon 12 (i.e. Avogadro's number = 6.028E23 particles). The number of moles in a given quantity of a molecular species (or element) is its mass divided by its molar mass, i.e. $n = m/M$. For molecular and ionic compounds in chemical processes (e.g. $H_2O$, $H_2$, $O_2$, $CH_4$, NaCl) the *elementary particles* are molecules. For unassociated elements and for ions the *elementary particles* are atoms (e.g. C, $Na^+$, $Cl^-$) or charged groups of atoms (e.g. $NH_4^+$, $ClO^-$). Refer to a basic chemistry text for more comprehensive information of atoms, ions, molecules and chemical bonding.

When *Equation 1.01* is applied to energy balances, the quantity also includes energy transfer across the system envelope as heat and/or work (see Chapter 5).

When you use the general balance equation it is best to begin with a conceptual diagram of the system, which clearly shows the complete system boundary plus the inputs and outputs of the system. *Figure 1.01* is an example of such a diagram. Note that:

- *Figure 1.01* is a two-dimensional representation of a real three-dimensional system
- The system has a closed boundary
- The system may have singular or multiple inputs and/or outputs

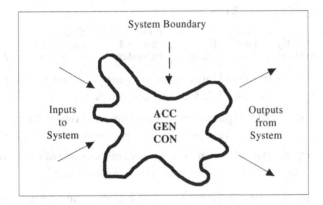

Figure 1.01.   Conceptual diagram of a system.

## CONSERVED AND CHANGING QUANTITIES

The general balance equation (*Equation 1.01*) applies to any extensive quantity. However, when you use *Equation 1.01*, it is helpful to distinguish between quantities that are conserved and quantities that are not conserved. A conserved quantity is a quantity whose total amount is maintained constant and is understood to obey the *principle of conservation*, which states that such a quantity can be neither created nor destroyed. Alternatively, a conserved quantity is one whose total amount remains constant in an isolated system,[3] regardless of what changes occur inside the system.

For quantities that are conserved, the generation and consumption terms in *Equation 1.01* are both zero, whereas for quantities that are not conserved, either or both of the generation and consumption terms is not zero.

---

[3] An isolated system is a hypothetical system that has zero interaction with its surroundings, i.e. zero transfer of material, heat, work, radiation, etc. across the boundary.

## MATERIAL AND ENERGY BALANCES

In general, material and energy are equivalent through *Equation 1.02*.

$$E = mc^2$$

*Equation 1.02*

where:

E = energy      J
m = mass (a.k.a. the rest mass)      kg
c = velocity of light in vacuum = 3E8      $ms^{-1}$

In a system where material and energy are inter-converted, *Equation 1.01* can thus be written for the quantity ($mc^2 + E$) as:

$$\begin{array}{ccccccc}
\text{Acc}_2 & = & \text{In}_2 & - & \text{Out}_2 & + & \text{Gen}_2 & - & \text{Con}_2 \\
(mc^2 + E) & & (mc^2 + E) & & (mc^2 + E) & & (mc^2 + E) & & (mc^2 + E) \\
\text{in system} & & \text{to system} & & \text{from system} & & \text{in system} & & \text{in system}
\end{array}$$

*Equation 1.03*

The sum ($mc^2 + E$) is a conserved quantity, so the generation and consumption terms in *Equation 1.03* are zero. *Equation 1.03* thus reduces to *Equation 1.04*.

**Accumulation of ($mc^2 + E$) = Input of ($mc^2 + E$) – Output of ($mc^2 + E$)**
**in system                to system               from system**      *Equation 1.04*

*Equation 1.04* is a mathematical statement of the general principle of conservation of matter and energy that applies to all systems within human knowledge.[4]

Excluding processes in nuclear engineering, nuclear physics, nuclear weapons and radiochemistry, systems on Earth do not involve significant inter-conversion of mass and energy. In such non-nuclear systems, individual atomic species (i.e. elements) as well as the total mass and total energy are *independently* conserved quantities.

For individual atomic species in non-nuclear systems, *Equation 1.01* then reduces to *Equation 1.05*.

**Accumulation of = Input of atoms – Output of atoms**
**atoms in system      to system          from system**      *Equation 1.05*

Also, *Equation 1.04* can be written as two separate and independent equations, one for total mass (*Equation 1.06*) and another for total energy (*Equation 1.07*).

**Accumulation of = Input of mass – Output of mass**
**mass in system      to system          from system**      *Equation 1.06*

**Accumulation      = Input of energy – Output of energy**
**of energy in system    to system          from system**      *Equation 1.07*

*Equation 1.05*, for the conservation of elements   ≡  ATOM BALANCE
*Equation 1.06*, for the conservation of total mass   ≡  MASS BALANCE
*Equation 1.07*, for the conservation of total energy ≡  ENERGY BALANCE

---

[4] Modern cosmology may have more to say about the principle of conservation (see *Ref. 5*).

The word "total" is used here to ensure that *Equations 1.06* and *1.07* are not applied to the mass or energy associated with only part of the system (e.g. a single species or form of energy) — which may not be a conserved quantity.

*Equations 1.05, 1.06* and *1.07* are **special cases of *Equation 1.01*** that apply to conserved quantities. The strength of *Equation 1.01* is that it can also be applied to quantities that are not conserved. The generation and consumption terms of *Equation 1.01* then account for changes in the amounts of non-conserved quantities in the system (see *Ref. 6*).

Some of the quantities that can be the subject of *Equation 1.01*, applied to non-nuclear processes, are:

Elements     : Conserved — *Equation 1.05* — Atom balance
Total mass   : Conserved — *Equation 1.06* — Total mass balance
Total energy : Conserved — *Equation 1.07* — Total energy balance
Momentum  : Conserved — Momentum balance
Moles        : Conserved in systems without chemical reaction.
               Not necessarily conserved in systems with chemical reaction
Enthalpy     : Not necessarily conserved
Entropy      : Not necessarily conserved
Volume       : Not necessarily conserved

*Examples 1.01* and *1.02* show how *Equation 1.01* is applied respectively to a conserved quantity and to a non-conserved quantity.

### EXAMPLE 1.01    *Material balance on water in a reservoir.*

A water reservoir initially contains 100 tonne of water. Over a period of time 60 tonne of water flow into the reservoir, 20 tonne of water flow out of the reservoir and 3 tonne of water are lost from the reservoir by evaporation. Water is not involved in any chemical reactions in the reservoir.

### Problem:

What is the amount of water in the reservoir at the end of the period of time? [Answer in tonnes of water]

### Solution:
Define the system    = the reservoir
Specify the quantity = mass of water [conserved quantity]

Write the balance equation:

ACC of water  = IN of water  − OUT of water  + GEN of water  − CON of water
in system         to system        from system      in system           in system

Interpret the terms:

Accumulation of water = final mass of water − initial mass of water = unknown = X
in system                        in system                     in system

Input of water to system        = 60 tonne
Output of water from system     = 20 (flow) + 3 (evaporation) = 23 tonne
Generation of water in system   = 0 (water is conserved) tonne
Consumption of water in system  = 0 (water is conserved) tonne

Substitute values for each term into the general balance equation

$$X = 60 - (20 + 3) + 0 - 0$$

Solve the balance equation for the unknown.

$$X = 37 \text{ tonne}$$

Final mass of water in system

$$= \text{Initial mass of water in system} + X = 100 + 37 = \textbf{137 tonne}$$


### EXAMPLE 1.02    Material balance on the population of a country.

A country had a population of 10 million people in 1900 CE. Over the period from 1900 to 2000 CE, 4 million people immigrated into the country, 2 million people emigrated from the country, 5 million people were born in the country and 3 million people died in the country.


### Problem:[5]

What is the population of the country in the year 2000 CE?


### Solution:

Define the system    = country
Specify the quantity = number of people
                        [a non-conserved quantity]

Immigration — 1 → | Country |
                  | 5E6 births |  — 2 → Emigration
                  | 3E6 deaths |

---

[5] This text uses scientific notation for large numbers, i.e. aEb ≡ (a) $(10^6)$.

Write the balance equation:

ACC of people = IN of people – OUT of people + GEN of people – CON of people
in system                  to system         from system       in system                in system

Interpret the terms:

Accumulation = Final no. of people in system – Initial no. of people in system = unknown = X

Input of people to system            = immigration = 4 million people
Output of people from system      = emigration   = 2 million people
Generation of people in system    = born            = 5 million people
Consumption of people in system = died            = 3 million people

Substitute values into the general balance equation

$$X = 4 - 2 + 5 - 3 = 4 \text{ million people}$$

Final number of people in 2000 CE:

Initial no. of people in 1900 + X = 10 + 4 = **14 million people**

*Example 1.03* shows a slightly more difficult case of a non-conserved quantity involving chemical reaction stoichiometry.

## EXAMPLE 1.03   Material balance on carbon dioxide from an internal combustion engine.

An automobile driven by an internal combustion engine burns 10 kmol of gasoline[6] consisting of 100% octane ($C_8H_{18}$) and converts it completely to carbon dioxide and water vapour by a combustion reaction, whose stoichiometric equation is:

$$2C_8H_{18} + 25O_2 \rightarrow 16\ CO_2 + 18H_2O \qquad\qquad\qquad Reaction\ 1$$

All $CO_2$ and $H_2O$ produced in the reaction is discharged to the atmosphere via the engine's exhaust pipe (i.e. zero accumulation).

Assume $CO_2$ content of input combustion air = zero

**Problem:** What is the amount of carbon dioxide discharged to the atmosphere from 10 kmol octane? [Answer in kg of $CO_2$]

**Solution:** First note that the amount (mass or moles) of $CO_2$ is not a conserved quantity. Chemical reactions involve the consumption of reactant species and the

---

[6] kmol ≡ kilomole ≡ molar mass (molecular weight) expressed as kilograms.
1 kmol $C_8H_{18}$ = (12.01)(8) + (1.008)(18) = 114.22 kg = 162 litre = 43 US gallons, i.e. about 3 tanks of gasoline (a.k.a. petrol) for a family car.

generation of product species, so that no species appearing in the stoichiometric equation is conserved when a reaction occurs.

Define the system    = automobile internal combustion engine

Specify the quantity = mol (kmol) of $CO_2$ [mole quantities give the simplest calculations for chemical reactions]

(Gasoline + air) in ⟶ | IC Engine | ⟶ Exhaust gas out

Write the balance equation:

ACC of $CO_2$  = IN of $CO_2$  − OUT of $CO_2$  + GEN of $CO_2$  − CON of $CO_2$
in system         to system     from system     in system         in system

Interpret the terms:

Accumulation of $CO_2$ in system = 0          kmol (all $CO_2$ is discharged)
Input of $CO_2$ to system       = 0          kmol (zero $CO_2$ enters engine)
Output of $CO_2$ from system    = unknown = X kmol
Generation of $CO_2$ in system  = (16/2)(consumption of $C_8H_{18}$ in system) = (16/2)(10 kmol)
                                = 80          kmol (from stoichiometry of *Reaction 1*)
Consumption of $CO_2$ in system = 0          kmol ($CO_2$ is a product, not a reactant)

Substitute values for each term into the general balance equation.

   $0 = 0 - X + 80 - 0$

Solve the balance equation for the unknown "X".

X = 80 kmol $CO_2$ discharged to atmosphere
   ≡ (80 kmol)(((1)(12.01) + (2)(16.00)) kg/kmol) = **3521 kg $CO_2$**

## FORMS OF THE GENERAL BALANCE EQUATION

There are several forms of the general balance equation (*Equation 1.01*), with each form suited to a specific class of problem.

First, for convenience, *Equation 1.01* is usually written in shorthand as:

   **ACC = IN − OUT + GEN − CON**                    (see *Equation 1.01*)

where:
   ACC = Accumulation of specified quantity in system = (Final amount − Initial amount)
        of quantity in system.
   IN   = Input of specified quantity to system         (input)
   OUT = Output of specified quantity from system       (output)

GEN = Generation of specified quantity in system       (source)
CON = Consumption of specified quantity in system    (sink)

As written above *Equation 1.01* is an *integral balance* suited to deal with the *integrated (total) amounts* of the specified quantity in each of the five terms over a period of time from the initial to the final state of the system. The integral balance equation can be used for any system with any process but is limited to seeing only the total changes occurring between the initial and final states of the system. Calculations with the integral balance require the solution of algebraic equations.

When *Equation 1.01* is differentiated with respect to (w.r.t.) time it becomes a *differential balance* equation (*Equation 1.08*), with each term representing a RATE of change of the specified quantity with respect to time, i.e.

**Rate ACC = Rate IN – Rate OUT + Rate GEN – Rate CON**          *Equation 1.08*

where:

Rate ACC  = Rate of accumulation of specified quantity in system,       w.r.t. time

Rate IN     = Rate of input of specified quantity to system,                   w.r.t. time

Rate OUT  = Rate of output of specified quantity from system,             w.r.t. time

Rate GEN  = Rate of generation of specified quantity in system,           w.r.t. time

Rate CON  = Rate of consumption of specified quantity in system,       w.r.t. time

*Equation 1.08* is suited to deal with both closed systems and open systems involving unsteady-state processes, as well as with open systems involving steady-state processes. These terms are defined below in "Class of Problem".

The differential balance equation is more versatile than the integral balance equation because it can trace changes in the system over time. Application of the differential balance to unsteady-state processes requires knowledge of calculus to solve differential equations. For steady-state processes, the differential equation(s) can be simplified and solved as algebraic equation(s).

## CLASS OF PROBLEM

Practical problems are classified according to the type of system and the nature of the process occurring in the system, as follows:

**Closed system**            Zero <u>material</u>* is transferred in or out of the system [during the time period of interest].
[Controlled mass]       i.e. in the material balance equation:      IN = OUT = 0
A process occurring in a closed system is called a BATCH process.

**Open system**              Material is transferred in and/or out of the system.
[Controlled volume]    i.e. in the material balance equation:      IN $\neq$ 0 and/or OUT $\neq$ 0

A process occurring in an open system is called a CONTINUOUS process.

**Steady-state process**    A process in which all conditions are invariant with time. i.e. at steady-state: Rate ACC = 0    for all quantities.

**Unsteady-state process** A process in which one or more conditions vary with time [these are *transient* conditions], i.e. at unsteady-state:

Rate ACC ≠ 0    for one or more quantities.

*\*Energy can be transferred in and/or out of both closed and open systems.*

From these generalisations we can write balance equations for special cases that occur frequently in practical problems.

Differential material[7] balance
on a closed system:      **Rate ACC = Rate GEN – Rate CON**    *Equation 1.09*

Differential total mass balance
on a closed system:      **Rate ACC = 0**      *Equation 1.10*

Differential material balance on
an open system at steady-state:    **0 = Rate IN – Rate OUT + Rate GEN – Rate CON**
     *Equation 1.11*

Differential total mass balance
on an open system:      **Rate ACC = Rate IN – Rate OUT**    *Equation 1.12*

Differential total mass balance
on open system at steady-state:    **0 = Rate IN – Rate OUT**      *Equation 1.13*

Differential total energy balance
on a closed or an open system:    **Rate ACC = Rate IN – Rate OUT**    *Equation 1.14*

Differential total energy balance
at steady-state:      **0 = Rate IN – Rate OUT**      *Equation 1.15*

*Equations 1.09* to *1.15* can be applied either as differential balances or as integral balances with the terms integrated over time.

A powerful feature of *Equation 1.01* is your freedom to define both the *system* and the *quantity* to suit the problem at hand. For example, if you are working on global warming, the *system* could be a raindrop, a tree, a cloud, an ocean or the whole Earth with its atmosphere, while the *quantity* may be the mass (or moles) of carbon dioxide, the mass (or moles) water vapour or the energy content associated with the system. If you are working

---

[7]Note that "material" and "mass" are not synonymous terms. Material may change its form, but the total mass is constant (in non-nuclear processes).

on a slow release drug, the *system* could be a single cell, an organ or the whole human body, while the *quantity* may be the mass of the drug or its metabolic product, etc.

For an industrial chemical process (the focus of this text[8]) the *system* could be defined as, for example:

- a single molecule
- a microelement of fluid
- a bubble of gas, drop of liquid or particle of solid
- a section of a process unit (e.g. a tube in a heat exchanger, a plate in a distillation column, etc.)
- a complete process unit (e.g. a heat exchanger, an evaporator, a reactor, etc.)
- part of a process plant including several process units and piping
- a complete process plant
- a complete process plant plus the surrounding environment

In an industrial chemical process the *quantity* of interest may be, for example:

- total mass or total moles of all species
- mass or moles of any single species
- mass or atoms of any single element
- total energy

Other quantities that may be used in process calculations are, for example:

- total mass plus energy of all species (nuclear processes)
- one form of energy (e.g. internal energy, kinetic energy, potential energy)
- volume of one or more components
- enthalpy
- entropy [not treated in this text]
- momentum [not treated in this text]

The choice of both the *system* and the *quantity* are major decisions that you must make when beginning an M&E balance calculation. These choices are usually obvious but sometimes they are not. You should always give these choices careful consideration because they will determine the degree of difficulty you encounter in solving the problem.

*Examples 1.04* and *1.05* show that selection of the *system* and the *quantity* can affect the ease of solution of M&E balance problems. *Example 1.04* is a multi-stage balance problem that can be solved either by stage-wise calculations or by an overall balance. Solution A shows the stage-wise approach while solution B shows the more efficient solution obtained by defining an overall system that includes all the process steps inside a single envelope. In *Example 1.05* you can see how a relatively complex multi-stage problem involving

---

[8] The techniques described here for an industrial chemical process can be applied to any system in which you are interested.

chemical reactions can be reduced to one overall balance on the quantity of a single element (carbon). Such atom balances (on one or more elements) are handy for resolving many material balance problems. However there are other more complex problems represented in Chapter 4 whose solution requires stage-wise balances on each chemical species over every process unit.

### EXAMPLE 1.04    Material balance on a two-stage separation process.

A 100 kg initial mixture of oil and water contains 10 kg oil + 90 kg water. The mixture is separated in two steps.[9]

**Step 1.** 8 kg oil with 2 kg water is removed by decantation.
**Step 2.** 1 kg oil with 70 kg water is removed by evaporation.

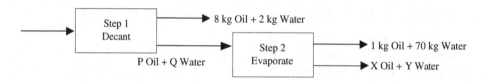

**Problem:** Find the amounts of oil and water in the final mixture (i.e. X kg oil + Y kg water).
**Solution A:** [Stepwise balances]

**Step 1:** Define the system      = Decanter (an open system)
            Specify the quantities = Mass of oil, mass of water

Write the integral balance equation: **ACC = IN – OUT + GEN – CON**

| **For the oil:** | **For the water:** |
|---|---|

ACC = Final oil in system – Initial oil in   ACC = Final water in system – Initial water in
           system = 0                                        system = 0
IN    = 10      kg                              IN    = 90      kg
OUT  = 8 + P    kg                             OUT  = 2 + Q    kg
GEN  = 0        kg [Oil is conserved]   GEN  = 0        kg [Water is conserved]
CON  = 0        kg                              CON  = 0        kg

Mass balance on oil:   ACC = 0 = 10 – (8 + P) + 0 – 0  P = 2 kg oil
Mass balance on water: ACC = 0 = 90 – (2 + Q) + 0 – 0  Q = 88 kg water

**Step 2:** Define the system      = Evaporator (an open system)
            Specify the quantities = Mass of oil, mass of water

Write the integral balance equation:   **ACC = IN – OUT + GEN – CON**

---

[9]Assumes the decanter and the evaporator are empty before and after the operation; i.e. accumulation = 0.

**For the oil:**
ACC = Final oil in system – Initial oil
    in system = 0
IN   = 2            kg
OUT = 1 + X        kg
GEN = 0            kg  [Oil is conserved]
CON = 0            kg

**For the water:**
ACC = Final water in system – Initial
    water in system = 0
IN    = 88          kg
OUT = 70 + Y kg
GEN = 0            kg [Water is conserved]
CON = 0            kg

Mass balance on oil:    ACC = 0 = 2 – (1 + X) + 0 – 0    **X = 1 kg oil**
Mass balance on water: ACC = 0 = 88 – (70 + Y) + 0 – 0   **Y = 18 kg water**

***Solution B:*** [Overall balance]
        Define the system    = Decanter + evaporator (i.e. overall process)
        Specify the quantities = Mass of oil, mass of water

Write the integral balance equation:    **ACC = IN – OUT + GEN – CON**

**For the oil:**
ACC = Final oil in system – Initial oil
    in system = 0
IN    = 10            kg
OUT = 8 + 1 + X kg
GEN = 0            kg [Oil is conserved]
CON = 0            kg

**For the water:**
ACC = Final water in system – Initial water
    in system = 0
IN    = 90            kg
OUT = 2 + 70 + Y    kg
GEN = 0            kg [Water is conserved]
CON = 0            kg

Mass balance on oil:    ACC = 0 = 10 – (8 + 1 + X) + 0 – 0    **X = 1 kg oil**
Mass balance on water:   ACC = 0 = 90 – (2 + 70 + Y) + 0 – 0    **Y = 18 kg water**

*Example 1.04* shows that (when the intermediate conditions are not required) the <u>overall</u> <u>balance</u> gives the most efficient solution.

**EXAMPLE 1.05    *Material balance on carbon dioxide from a direct methanol fuel cell.***

Methanol ($CH_3OH$) is used as the fuel in an automobile engine based on a direct methanol fuel cell. The fuel supply for one tank of methanol is produced from 10 kmol[10] of natural gas methane ($CH_4$) by the following process:

1.  Eight kmol of $CH_4$ reacts completely with water (steam) to produce CO and $H_2$ by the stoichiometry of *Reaction 1*.

    $$CH_4 + H_2O \rightarrow CO + 3H_2$$                                                 *Reaction 1*

2.  Two kmol of $CH_4$ is burned in *Reaction 2* to supply heat to drive *Reaction 1*.

    $$CH_4 + 2O_2 \rightarrow CO_2 + 2H_2O$$    ($CO_2$ to atmosphere)                *Reaction 2*

3.  One-third of the $H_2$ from *Reaction 1* is converted to water by *Reaction 3*.

    $$2H_2 + O_2 \rightarrow 2H_2O$$                                                       *Reaction 3*

---

[10] 10 kmol $CH_4$ ≡ (10 kmol)(16.04 kg/kmol) = 160.4 kg $CH_4$.

4.  The remaining CO and $H_2$ are converted to $CH_3OH$ by *Reaction 4*.

    $CO + 2H_2 \rightarrow CH_3OH$                                                           *Reaction 4*

5.  The methanol is subsequently completely converted to $CO_2$ and $H_2O$ by the net *Reaction 5*, which occurs in the fuel cell to produce power for the automobile.

    $CH_3OH + 1.5O_2 \rightarrow CO_2 + 2H_2O$ ($CO_2$ to atmosphere)                         *Reaction 5*

***Problem***: Find the total amount of $CO_2$ released to the atmosphere from these operations, starting with the 10 kmol of $CH_4$. Assume zero accumulation of material in all the process steps. [kg $CO_2$]

***Solution***: We can define each of the 5 reaction steps as a separate system and trace the production of $CO_2$ through the sequence of operations, or we can define any combination of steps as the system (provided it is enclosed by a complete envelope). In this case it is most efficient to define the system to include all 5 reaction steps and carry out an overall atom balance on the open system.

Define the system    = the overall process of 5 reaction steps [closed envelope, shown by broken line].

Specify the quantity = moles (kmol) of the element carbon [contained in various compounds].

Write the integral balance equation:    **ACC = IN – OUT + GEN – CON**

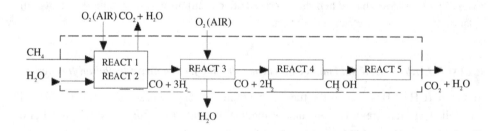

ACC = Final C in system – Initial C in system = 0 (i.e. zero accumulation)
IN    = (10 kmol $CH_4$)(1 kmol C/kmol $CH_4$) = 10 kmol C
OUT = (X kmol $CO_2$)(1 kmol C/kmol $CO_2$) = X kmol C
GEN = 0 [carbon is conserved]
CON = 0 [carbon is conserved]

0 = 10 – X + 0 – 0
X = 10 kmol $CO_2$ ≡ (10 kmol)(44 kg/kmol) = **440 kg $CO_2$ total to atmosphere**

*Example 1.06* shows the solution of a material balance using a differential balance on an open system and also illustrates how a problem involving chemical reaction can be solved either by mole balances from the reaction stoichiometry or by atom balances on selected elements. You should note here that for a single reaction the atom balance method is simpler than the mole balance because the atom balance does not need to use the reaction stoichiometric equation. In more complex cases, such as those involving reactions with incomplete conversion and/or where the specified elements appear in several products, the mole balance is often the preferred method of solution.

### EXAMPLE 1.06 Material balance on water vapour from a jet engine.

Water vapour trails from jet planes are known to initiate cloud formation and are one suspected cause of global climate change.

A jet engine burns kerosene fuel ($C_{14}H_{30}$) at the steady rate of 1584 kg/h.

The fuel undergoes complete combustion to $CO_2$ and $H_2O$.

**Problem:**    Calculate the flow of water vapour in the engine exhaust stream.        [kg/h]

**Solution A:**  (Mole balance) Define the system = jet engine (an open system)
                        Specify the quantity = moles (kmol) of water

Write the reaction stoichiometry:

$$2C_{14}H_{30} + 43O_2 \rightarrow 28CO_2 + 30H_2O \hspace{3cm} \textit{Reaction 1}$$

Let X = flow of water in engine exhaust.        [kmol/h]

Write the differential material balance:

**Rate ACC = Rate IN – Rate OUT + Rate GEN – Rate CON**

Rate ACC = 0 (steady-state)
Rate IN    = 0
Rate OUT = X
Rate GEN = (15 kmol $H_2O$/kmol $C_{14}H_{30}$)(1584 kg $C_{14}H_{30}$/h)/(198 kg $C_{14}H_{30}$/kmol $C_{14}H_{30}$)
              = 120 kmol/h $H_2O$
Rate CON = 0

0 = 0 – X + 120 – 0   X = 120 kmol $H_2O$/h = (120 kmol/h)(18 kg/kmol) = **2160 kg/h $H_2O$**

**NOTE:** Nitrogen ($N_2$) in the air is assumed to pass unchanged through the combustion process. In reality a small fraction of the nitrogen is converted to nitrogen oxides, NO and $NO_2$.

**Solution B**: (Atom balance) Define the system = jet engine
Specify the quantity = moles (kmol) of hydrogen atoms, i.e.
hydrogen as $H_1$

Let X = flow of water in engine exhaust.     [kmol/h]

Write the differential material balance:

**Rate ACC = Rate IN − Rate OUT + Rate GEN − Rate CON**

Rate ACC = 0 (steady-state)
Rate IN  = (1584 kg/h $C_{14}H_{30}$)/(198 kg $C_{14}H_{30}$/kmol $C_{14}H_{30}$)
$\times$ (30 kmol $H_1$/kmol $C_{14}H_{30}$)                                 = 240 kmol/h $H_1$
Rate OUT = (X kmol/h $H_2O$)(2 kmol $H_1$/kmol $H_2O$)                 = 2X kmol/h $H_1$
Rate GEN = Rate CON = 0 (atoms are conserved)

0 = 240 − 2X + 0 − 0
X = 120 kmol $H_2O$/h = (120 kmol/h)(18 kg/kmol) = **2160 kg/h $H_2O$**

Differential balances on closed systems and unsteady-state open systems require solution by calculus and are not considered until Chapter 7.

## SUMMARY

[1]  The general balance equation (GBE) is the key to material and energy (M&E) balance problems.
The GBE can be used in integral or in differential form.
For a defined *system* and a specified *quantity*:
Integral form of GBE
**ACC = IN − OUT + GEN − CON**                                    (see *Equation 1.01*)
Differential form of GBE
**Rate ACC = Rate IN − Rate OUT + Rate GEN − Rate CON**   (see *Equation 1.08*)

[2]  Always begin work with the GBE by defining the system (a closed envelope) and specifying the quantity of interest. Take care here to avoid ambiguity. *Ambiguity is the mother of confusion.*

[3]  The choice of both *system* and *quantity* in the GBE determines the ease of solution of M&E balance problems. The *system* can range from a sub-micron feature to a whole planet and beyond, or from a single unit in a multi-stage sequence to the overall operation. The *quantity* may be any specified measurable (extensive) property, such as the mass, moles, volume or energy of one or more components of the system.

[4]  The GBE applies to both conserved and non-conserved quantities.
For conserved quantities:      GEN = CON      = 0
For non-conserved quantities:  GEN and/or CON $\neq$ 0

[5]  In non-nuclear processes the individual atomic species (elements) as well as both total mass and total energy are independently conserved quantities, for which the GBE's simplify to:

Atom (element) balance:

**Atoms ACC = Atoms IN – Atoms OUT**                 (see *Equation 1.05*)

Total mass balance:

**Mass ACC = Mass IN – Mass OUT**                      (see *Equation 1.06*)

Total energy balance:

**Energy ACC = Energy IN – Energy OUT**              (see *Equation 1.07*)

[6]  M&E balance problems are broadly classified by system and process as:

Closed system:        Material IN = Material OUT = 0    [Controlled mass]
Open system:         Material IN and/or Material OUT ≠ 0  [Controlled volume]
Batch process:       Process occurring in a closed system.
Continuous process:   Process occurring in an open system.
Steady-state process:  Invariant with time.
                      Rate ACC = 0 for all quantities
Unsteady-state process: Variant with time.
                      Rate ACC ≠ 0 for one or more quantities

[7]  Energy may be transferred in and/or out of both closed and open systems.

[8]  The calculation of an M&E balance can sometimes be simplified by choosing a conserved quantity and/or by making an overall balance on a multi-unit sequence. Always check these options before starting a more detailed analysis.

[9]  *Examples 1.01* to *1.06* are simple material balance problems that illustrate the concepts of the GBE. Subsequent chapters will show how the GBE is used to solve more complex material and energy problems, with emphasis on practical chemical processing.

## FURTHER READING

[1]  K. Solen and J. Harb, *Introduction to Chemical Engineering*, Wiley, 2011.
[2]  A. Vesilind, L. Heine and S. Morgan, *Introduction to Environmental Engineering*, Thomson Engineering, 2010.
[3]  R. Felder, R. Rousseau and L. Bullard, *Elementary Principles of Chemical Processes*, Wiley, 2018.

MATERIAL TRANSFER ACROSS A SYSTEM BOUNDARY

[4]  J. Verret, R. Barghout and R. Qiao, *Foundations of Chemical and Biological Engineering*, Victoria, BC Campus, 2020. (Creative Commons).

[5]  S. Hawking, *The Theory of Everything — The Origin and Fate of the Universe*, Millennium, 2002.

[6]  R. K. Sinnot, Chemical Engineering Design, Butterworth-Heinmann, Oxford, 1999.

[7]  S. I. Sandler, Chemical and Engineering Thermodynamics, John Wiley & Sons, New York, 1999.

# CHAPTER TWO

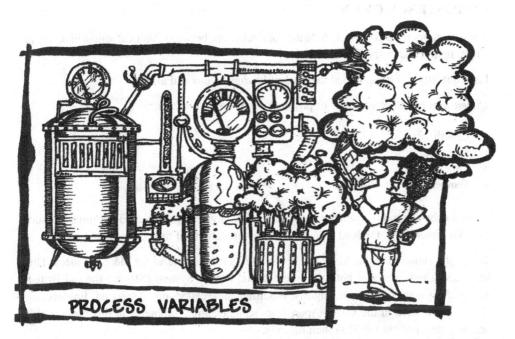

PROCESS VARIABLES

AND THEIR RELATIONSHIPS

# UNITS AND CALCULATIONS

Material and energy balances are essentially mathematical models that are used to predict the performance of chemical processes. The calculation of M&E balances involves setting up and solving equations related to subjects such as equations of state, chemical stoichiometry, thermochemistry, phase equilibria, etc. in the context of the general balance equation (*Equation 1.01*). These calculations may range from simple arithmetic to complex manipulations with algebraic and differential equations.

This chapter reviews some concepts, conventions and jargon that will help you to navigate M&E balance problems and keep your calculations honest.

# CONSISTENCY OF UNITS

The equations used for M&E balances are relations between *physical quantities,* all of which require exact specification.

Each quantity can be specified in terms of its *dimensions* and *units.*

*Dimensions* are basic concepts of physical quantities, which are:

- Length
- Mass
- Time
- Temperature

*Units* are the commonly used means of measuring physical quantities. Basic units measure the four dimensions. Derived units measure combinations of the basic units.

*Table 2.01* summarises the basic and derived units that are useful for M&E balances. This table also shows units in both the International Metric System [SI] and the old British Imperial System (otherwise slightly modified and called the American Engineering System, now used in the USA). Apart from some example problems in this chapter, this text will deal only in the SI system of units.

You should check that the units "balance" at every step of a calculation, using the rules of *Table 2.02*, as shown in *Examples 2.01* and *2.02*.

Always specify the units of your final result.

### Table 2.01.   Units used in material and energy balance calculations.

| Quantity | SI Unit | Symbol | Combination | Imperial/American Unit | Symbol | Combination |
|---|---|---|---|---|---|---|
| **Basic Units** | | | | | | |
| Length | metre | m | – | foot | ft | – |
| Mass | kilogram | kg | – | pound (mass)[a] | lb | – |
| Time | second | s | – | second | s | – |
| Thermodynamic temp. | Kelvin | K | | Rankine | °R | – |
| Current | Ampere | A | | Ampere | A | – |
| Amount of substance | mole | mol | – | pound mole | lbmol | |
| Force | derived unit, see below | – | – | pound (force) | $lb_f$ | – |
| **Derived Units** | | | | | | |
| Force | Newton | N | $kg.m.s^{-2}$ | basic unit, see above | $lb_f$ | basic unit |
| Energy | Joule | J | $kg.m^2.s^2$ | British thermal unit | BTU | $778\ ft.lb_f$ |
| Power | Watt = $J.s^{-1}$ | W | $kg.m^2.s^{-3}$ | horsepower | HP | $550\ ft.lb\ .s^{-1}$ |
| Pressure | Pascal = $Nm^{-2}$ | Pa | $kg.m^{-1}.s^{-2}$ | pound per square inch | psi | $lb_f.ft^{-2}/144$ |
| Electric charge | Coulomb | C | A.s | Coulomb | C | A.s |
| Frequency | Hertz (cycle/s) | Hz | $s^{-1}$ | cycle/s | Hz | $s^{-1}$ |
| **Commonly Used Compound Units** | | | | | | |
| Customary temp. | Celsius | °C | K –273.15 | Fahrenheit | °F | °R – 459.67 |
| Volume | cubic metre | – | $m^3$ | cubic foot | – | $ft^3$ |
| Density | kilogram per cubic metre – | $kg.m^{-3}$ | pound per cubic foot | – | lb.ft–3 | Density |
| Velocity | metre per second | – | $m.s^{-1}$ | foot per second | – | $ft.s^{-1}$ |
| Acceleration | metre per sec. squared | – | $m.s^{-2}$ | foot per sec. squared | – | $ft.s^{-2}$ |
| Heat capacity | Joule per mole. Kelvin = kJ per kmol.Kelvin | – | $kg.m^2.s^{-2}.mol^{-1}.K^{-1}$ | BTU per pound. mole Fahrenheit | – | $778\ ft.lb\ lbmol^{-1}F^{-1}$ |
| Heat of formation | Joule per mole = kJ/kmol | – | $kg.m^2.s^{-2}.mol^{-1}$ | BTU per pound mole | – | $778\ ft.lb\ .lbmol^{-1}$ |
| Amount of substance | kilomole | kmol | 1000 mol | ton[b] mole | tmol | 2240 lbmol |
| Molar mass | – | – | $kg.kmol^{-1}$ | – | – | $lb.lbmol^{-1}$ |
| Energy | kiloJoule | kJ | 1000 $kg.m^2.s^{-2}$ | – | – | – |
| Power | kiloWatt = $kJ.s^{-1}$ | kW | 1000 $kg.m^2.s^{-3}$ | – | – | – |
| Pressure | kiloPascal = $kN.m^{-2}$ | kPa | 1000 $kg.m^{-1}.s^{-2}$ | – | – | – |

*In M&E balance calculations a quantity is meaningless without its units. Every time you write an equation you should specify the units of each variable in that equation. You should then check the equation for <u>dimensional consistency</u> by ensuring that each term in the equation has the same units as those terms to which it is added, subtracted or equated.*

*All correct equations are dimensionally consistent, and any equation that is not dimensionally consistent is incorrect. All dimensionally consistent equations are not necessarily correct, but when you write an equation from memory there is a good chance that if it is dimensionally consistent then it is correct. A check of dimensional consistency is a simple method to eliminate common errors in your calculations.*

[a]The pound mass is sometimes written as $lb_m$.

[b]One "long ton" = 2240 lb, one "short ton" = 2000 lb, one "metric" ton = 2204.6 lb = 1000 kg.

**Table 2.02.   Rules for manipulating units.**

1.  Treat units like algebraic symbols.
2.  Add, subtract or equate terms only when they have the same units as each other.
3.  Identical units can be cancelled from numerator and denominator in products and dividends.
4.  Logarithm, exponential and trigonometric functions have dimensionless arguments and yield dimensionless results.

For example:

| | |
|---|---|
| $X = A + B - C$ | The terms A, B, C and X all have the same units. |
| $X = AB$ | The term AB has the same units as X. |
| $X = AB + CDE$ | The terms AB, CDE and X all have the same units. |
| $X = \ln(Y)$ | X and Y are both dimensionless.[1] |
| $X = \exp(Y)$ | X and Y are both dimensionless. |
| $X = (2A(\exp(B)) + C(\log(D)))/(P - RS)$ | B and D are dimensionless. |
| | Units of A = units of C. |
| | Units of P = units of RS. |
| | Units of X = units of A/P, etc. |

BALANCING ▲ THE UNITS

## USE OF UNITS TO CHECK EQUATIONS

### EXAMPLE 2.01   Checking the dimensional consistency of equations.

The relation between the force "f" required to move an object of mass "m" with an acceleration "ã" is one of the equations (a), (b) and (c) below:

| | | | | |
|---|---|---|---|---|
| (a) | $f = m + ã$ | where: | f  = force | N |
| (b) | $f = m/ã$ | | m = mass | kg |
| (c) | $f = mã$ | | ã  = acceleration | $ms^{-2}$ |

---

[1] Some equations apparently call for you to exponentiate or take the logarithm of a dimensional variable [e.g. log (p*) in the Antoine equation (*Equation 2.32*)]. In such cases the argument of the function should strictly be made dimensionless by division with unit value of the dimensioned variable.

**Problem:** Which of these three equations is dimensionally consistent?
**Solution:** Substitute the units for every variable in each equation.

| LHS | RHS | |
|-----|-----|-----|
| (a) $kg.m.s^{-2} = kg + ms^{-2}$ | | Inconsistent units. Incorrect equation. |
| (b) $kg.m.s^{-2} = kg/m.s^{-2} = kg.m^{-1}.s^{2}$ | | Inconsistent units. Incorrect equation. |
| (c) $kg.m.s^{-2} = kg.m.s^{-2}$ | | Consistent units. Correct equation. |

Equation (c) is dimensionally consistent (you will probably recognise it as Newton's second law of motion). Note that if the RHS of equation (c) was multiplied by a dimensionless constant (e.g. $f = 2m\ddot{a}$), then it will be dimensionally consistent but incorrect. A check on dimensional consistency does not guarantee that an equation is correct but it does eliminate many common mistakes.

### EXAMPLE 2.02   Finding the units for variables in a simple equation.

The pressure at the base of a vertical column of fluid is given by the (correct) equation:

$$P = \rho g L'$$

where:
  $P$ = pressure
  $\rho$ = fluid density          $kg.m^{-3}$
  $g$ = gravitation constant      $ms^{-2}$
  $L'$ = height of column of fluid   $m$

**Problem:** What are the units of P?
**Solution:** Substitute units for each variable.

$$P = (kg.m^{-3})\,(m.s^{-2})\,(m)$$

Cancel $m^2$ from numerator and denominator.

$$P = kg.m^{-1}.s^{-2} = kg.m.s^{-2}\,m^{-2} = Nm^{-2} = Pa \text{ (i.e. Pascal)}$$

### EXAMPLE 2.03   Finding the units of variables in a complicated equation.

The temperature of the fluid leaving a continuous flow heat exchanger is given by the (correct) equation:

$$T_2 = \frac{\{z + \exp[(U_{hex}A_{hex}/(m_cC_c))(z+1)]\}T_1 + \{\exp[(U_{hex}A_{hex}/(m_cC_c))(z+1)]-1\}zT_3}{(z+1)\exp[(U_{hex}A_{hex}/(m_cC_c))(z+1)]}$$

where:

$A_{hex}$ = heat transfer area $\quad$ m$^2$ $\qquad$ $T_2$ = hot fluid outlet temperature ?

$C_c$ = cold fluid heat capacity kJ.kg$^{-1}$.K$^{-2}$ $\quad$ T = hot fluid inlet temperature ?

$C_h$ = hot fluid heat capacity kJ.kg$^{-1}$.K$^{-1}$ $\quad$ $T_3$ = cold fluid inlet temperature K

$m_c$ = cold fluid flow rate $\qquad$ ? $\qquad$ $U_{hex}$ = heat transfer coefficient $\quad$ ?

$m_h$ = hot fluid flow rate $\qquad$ kg.s$^{-1}$ $\qquad$ z = $m_c C_c/(m_c C_c)$ $\qquad$ ?

**Problem**: What are the units of z, $T_1$, $T_2$, $m_c$ and U?

**Solution**:

| | |
|---|---|
| Units of z | = dimensionless |
| Units of $T_1$ | = units of $T_2$ = units of $T_3$ = K |
| Units of $(m_c C_c/(m_h C_h))$ | = units of z = dimensionless |
| Units of $m_c$ | = units of $(m_h C_c/C_c)$ |
| | = (kg.s$^{-1}$)(kJ.kg$^{-1}$.K$^{-1}$)/(kJ.kg$^{-1}$.K$^{-1}$) |
| | = kg.s$^{-1}$ |
| Units of $(UA_{hex}/(m_c C_c))$ | = dimensionless |
| Units of U | = units of $(m_c C_c/A_{hex})$ |
| | = kg.s$^{-1}$. kJ.kg$^{-1}$. K$^{-1}$/m$^2$ |
| | = kJ.s$^{-1}$.m$^{-2}$.K$^{-1}$ |
| | = kW.m$^{-2}$.K$^{-1}$ $\equiv$ kW/(m$^2$.K) |

# CONVERSION OF UNITS

Sometimes you need to convert a value in the SI system to its equivalent in the Imperial/American system, or vice-versa. An easy way to make such a conversion is with a table of conversion factors. *Table 2.03* lists some common conversion factors. Extensive tables of unit conversion factors are in the handbooks and texts listed at the end of this chapter.

You can also work out any unit conversions from the equivalents in *Table 2.03*, as shown in *Example 2.04*.

**Table 2.03.  Unit conversion factors.**

| Quantity | Value | Units | | Value | Units |
|---|---|---|---|---|---|
| Mass | 1 | kg | | 2.2046 | lb$_m$ |
| Length | 1 | m | | 3.2808 | ft |
| Temperature | 1 | K | | 1.8 | R° |
| Amount of substance | 1 | kmol | *Equivalent to:* | 2.2046 | lbmol |
| Energy | 1 | kJ | | 0.9486 | BTU |
| Power | 1 | kW | | 1.341 | HP |
| Force | 1 | kN | | 224.8 | lb$_f$ |
| Pressure | 1 | kPa | | 0.145 | psi |
| Volume | 1 | m$^3$ | | 264.17 | US gall |
| | | | | 220.83 | Imp gall |

## EXAMPLE 2.04    *Converting units from SI to Imperial/American.*

***Problem A***: Convert a density of 5.000 kg.m$^{-3}$ to its equivalent in units of lb$_m$ ft$^{-3}$.
***Solution***:    Substitute conversion factors from *Table 2.03*.

$$5 \text{ kg.m}^{-3} \equiv 5 \text{ kg } (2.2 \text{ lb}_m/\text{kg})/\text{m } (3.28 \text{ ft/m})^3$$

Cancel the SI units from numerator and denominator of the RHS, and do the arithmetic.

$$5 \text{ kg.m}^{-3} \equiv 0.312 \text{ lb}_m \text{ ft}^{-3} \text{ (lb} \equiv \text{pound mass)}$$

***Problem B***: Convert a heat transfer coefficient of 1.00 kW.m$^{-2}$.K$^{-1}$ to its equivalent in units
of BTU.hr$^{-1}$.ft$^{-2}$.°F$^{-1}$.
***Solution***:    Substitute conversion factors from *Table 2.03*.

$$1\text{kW.m}^{-2}\text{K}^{-1} \equiv \frac{1\text{kJ}(0.9486\,\text{BTU/kJ})}{(\text{s}(1/3600\,\text{s/h}))(\text{m}(3.28\text{ft/m}))^2\,(\text{K}(1.8°\,\text{F/K}))}$$

Cancel SI units from numerator and denominator and do the arithmetic.

$$1.00 \text{ kW.m}^{-2}.\text{K}^{-1} \equiv 176 \text{ BTU.h}^{-1}.\text{ft}^{-2}.°\text{F}^{-1}$$

## SIGNIFICANT FIGURES IN CALCULATIONS

The number of significant figures in a value is the number of digits in the argument when expressed in scientific notation, i.e. x.xxxx Ey ≡ x.xxxx (10$^y$) shows the 5 significant figures "x.xxxx". Electronic calculators and computers produce up to 10 significant figures but you will rarely do an engineering calculation with a result justified to more than 5 significant figures. Often, 1 to 3 significant figures are satisfactory. Table 2.04 shows the standard prefixes and abbreviations for numbers ranging from 1E9 to 1E-9.

The final results of your calculations should be given with a number of significant figures corresponding to the accuracy of measurements and calculations. Extra significant figures are misleading and inappropriate. They also clutter your work and waste the time of others who have to read it. When you use a calculator or a computer, a good procedure is to carry extra significant figures to the end of the calculation, then round off the final "displayed" result(s) to an appropriate number of significant figures.

In deciding the number of significant figures you should consider the sources of your data, the equations used to process them and the needs of your audience, as illustrated in *Example 2.05 A, B* and *C*.

***EXAMPLE 2.05   Specifying significant figures from measurements and calculations.***

***Problem A:*** In a laboratory test the volume of 2.81 ± 0.005 kg of nitrogen gas at 300.0 ± 0.1 K and 122 ± 1 kPa(abs) was measured as 2.05 ± 0.005 $m^3$. Calculate the value of the universal gas constant.

***Solution:***   By the ideal gas law, PV = nRT, solve for R and substitute values.

R = PV/nT = (122 ± 1) kPa * (2.05 ± 0.005) $m^3$/[((2.81 ± 0.005)/28.013) kmol * (300 ± 0.1) K]

= 8.310855635 ± (1/122 + 0.005/2.05 + 0.005/2.81 + 0.1/300)(8.310855635)

= 8.310855635 ± 0.105950432

= **8.31** ± 0.11 kJ.$kmol^{-1}$.$K^{-1}$

***Problem B:*** The capital cost of a wood pulp mill can be estimated by the correlation:
C = 17E6 $(P)^{0.6}$        ± 20%

where:

   C = capital cost; accurate to ± 20%;           $US (2000 CE)

   P = pulp production rate                        metric ton per day

Calculate the capital cost of a pulp mill producing 2000 metric ton per day of pulp.

***Solution:***   Substitute P = 2000 ton/day
                C = 17E6 $(P)^{0.6}$ = 17E6 $(2000)^{0.6}$ = 1,625,799,249.64 ( ± 306E6)
                i.e. C = **1.6E9 $US** (2000 CE)

***Problem C:*** The population of a country in year 2000 CE is 1 billion people, with a projected growth rate of 4% per year. Calculate the population of the country in the year 2020.

***Solution:***   Pop = 1E9 $(1+ 0.04)^{20}$ = 2,191,123,143 people i.e. Population in year 2020 CE = **2.2 billion people**

**Table 2.04.   Prefixes in SI system.**

| Value | Prefix | Abbreviation |
|-------|--------|--------------|
| 1E9   | giga   | G            |
| 1E6   | mega   | M            |
| 1E3   | kilo   | k            |
| 1E-3  | milli  | m            |
| 1E-6  | micro  | μ            |
| 1E-9  | nano   | n            |

*Note*: Numbers in this text are expressed as integers with decimals, e.g. "x.xx" or in scientific format, i.e. x.xxEy ≡ (x.xx)($10^y$).

## PROCESS VARIABLES

### PRESSURE

Pressure is the normal force per unit area. The basic unit of pressure in the SI system is the Pascal (Pa), but since this is a small quantity, ordinary pressures are usually measured in kiloPascals (kPa). Some measuring devices, such as the barometer, display the *absolute pressure*, but others, such as the open ended manometer and the Bourdon gauge display the *gauge pressure*, which is the pressure relative to the local atmospheric pressure. The relation between these scales is given in *Equations 2.01* and *2.02*.

For super-atmospheric pressures: **Absolute = atmospheric + gauge**        *Equation 2.01*
                                                     **pressure     pressure          pressure**

For sub-atmospheric pressures:    **Absolute = atmospheric − vacuum**        *Equation 2.02*
                                                     **pressure     pressure**

Many calculations with pressure require the absolute pressure (e.g. the gas law PV = nRT). When a *pressure difference* (ΔP) is required either the absolute or the relative scales can be used consistently to calculate ΔP. *Example 2.06* shows calculations with absolute and relative pressures.

The atmospheric pressure features in many problems with pressure. Equivalents between standard atmospheric pressure and other common measures of pressure are given in *Table 2.05*. The pressure equivalent to a column of fluid comes from *Equation 2.03*.

$$P = \rho g\ L'$$        *Equation 2.03*

where:

| | | |
|---|---|---|
| P | = pressure | Pa |
| ρ | = fluid density | $kg.m^{-3}$ |
| g | = gravitational acceleration | $m.s^{-2}$ |
| L′ | = vertical height of fluid column | m |

**Table 2.05.   Absolute pressure equivalents.**

| Standard atmosphere | Bar | kPa | psi* | mm (mercury at 0°C) | m (water at 4°C) |
|---|---|---|---|---|---|
| 1.000 | 0.989 | 101.3 | 14.696 | 760.0 | 10.333 |

*psi ≡ pounds (force) per square inch (see *Table 2.01*).
psia = psi (absolute)              psig = psi (gauge)

To avoid ambiguity you should always indicate whether pressures are absolute or relative, e.g. 448 kPa(abs), 50 psig, etc. Also note that atmospheric pressure varies with place and time and is rarely exactly one standard atmosphere. For accurate work the local atmospheric pressure should be established, for example, by a barometer measurement. In solving problems where the atmospheric pressure is not given it is usual to assume the value of one standard atmosphere. However, this assumption will give a substantial error for locations far above sea level, e.g. the atmospheric pressure near the top of Mt. Everest is about 30 kPa(abs).

## TEMPERATURE

The temperature of a body is a measure of its potential to transfer heat to other bodies. There are two temperature scales in both the SI and the Imperial/American systems. The relations between these scales are given in *Equations 2.04, 2.05* and *2.06*.

|  | SI | Imperial/ American |
|---|---|---|
| Thermodynamic (absolute) temperature | Kelvin (K) | Rankine (°R) |
| Customary (relative) temperature | Celsius (°C) | Fahrenheit (°F) |

$$K = 273.15 + °C \quad \textit{Equation 2.04}$$
$$°R = 459.67 + °F \quad \textit{Equation 2.05}$$
$$°F = 32 + 1.8\,°C \quad \textit{Equation 2.06}$$

Common temperature measurement devices, such as thermometers, thermocouples and thermistors usually display temperature on the customary scale. However, many calculations require the use of temperature on the thermodynamic scale (e.g. the gas law $PV = nRT$). When a temperature *difference* ($\Delta T$) is required, either the customary or the thermodynamic scale can be used consistently to calculate $\Delta T$. *Example 2.06* shows calculations with absolute and relative pressures and temperatures.

**EXAMPLE 2.06    *Using pressure and temperature measurements for process calculations.***

*Figure 2.01* shows a pipe with a set of pressure and temperature gauges used to measure a continuous steady flow of nitrogen gas, which is simultaneously being heated from an external heat source.

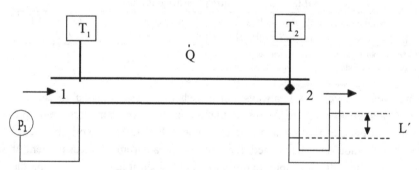

**Figure 2.01.    Gas flow measurement.**

The flow of nitrogen is approximated by:[2]    $\dot{m} \approx A(\rho_{av}\Delta P)^{0.5}$
The rate of heat transfer to the flowing nitrogen is given by:    $\dot{Q} = \dot{m}C_p\Delta T)$
where:
   A  = cross section area of tube = $0.002\ m^2$
   $C_p$ = heat capacity of nitrogen = $1.05\ kJ.kg^{-1}K^{-1}$
   L′ = difference in mercury levels in the open manometer = 105 mm Hg

---

[2] This equation applies only for a narrow range of conditions.

$\dot{m}$ = mass flow rate of nitrogen

$\rho_{av}$ = density of nitrogen at the average temperature and pressure

$\Delta P = P_1 - P_2$ = pressure drop along the pipe

$\dot{Q}$ = rate of heat transfer into the flowing nitrogen

$\Delta T = T_2 - T_1$ = temperature rise along the pipe

$P_1$ = inlet nitrogen pressure measured on a Bourdon gauge = 20 kPa(gauge)

$T_1$ = inlet nitrogen temperature measured by a thermometer = 20°C

$T_2$ = outlet nitrogen temperature measured by a thermocouple = 158°F

**Problem**: Calculate the velocity of the nitrogen gas at the pipe outlet (point 2) and the rate of heat transfer to the nitrogen.

**Solution**:

Note 1   Units are not specified for several of the variables, but they can be found by dimensional consistency.

$N_2$ outlet pressure     = $P_2$ = (105 mm Hg (gauge))(101.3 kPa/760 mm Hg)
                                = 14.0 kPa(gauge)

Average pressure        = $(P_1 + P_2) / 2$ = (20 kPa(gauge) + 14 kPa(gauge)) / 2
                                = 17 kPa(gauge)

$N_2$ outlet temperature = (158°F – 32°F) / 1.8 (°F/°C) = 70°C

Average temperature   = $(T_1 + T_2) / 2$ = (20°C + 70°C) / 2 = 45°C

Note 2   The outlet values calculated above are the relative pressure and temperature.

Ideal gas density (*Equation 2.14*) = (M/R)(P/T) where M = 28.01 kg/kmol,
                                       R = 8.314 kJ/(kmol K)

Density of $N_2$ at average P and T = (28.01 kg/kmol/8.314 kJ/(kmol.K))
                                       (17 + 101.3) kPa(abs)/(45 + 273) K
                                       = 1.25 kg/m$^3$

Note 3   The P and T used in the ideal gas equation must be absolute values.

Atmospheric pressure is taken from *Table 2.05* as 101.3 kPa(abs).

Pressure difference = $\Delta P$ = (20 kPa(gauge) – 14 kPa(gauge)) = (121.3 kPa(abs)
                             – 115.3 kPa(abs)) = 6 kPa

Flow of $N_2$ = $\dot{m}$ = $A(\rho_{av} \Delta P)^{0.5}$ = (0.002 m$^2$)(1.25 kg/m$^3$)(1000 Pa/kPa)(6 kPa)$^{0.5}$
                       = 0.173 kg/s

Note 4   The pressure difference $\Delta P$ can be calculated either from both relative pressures or both absolute pressures.

Note 5   From *Table 2.01* Pa = kg/(m.s$^2$), thus the units of $\dot{m}$ = $m^2(kg/m^3)(kg/(m.s^2))^{0.5}$
         = kg/s

Volumetric flow of $N_2$ at outlet = $\dot{V} = \dot{n}\,RT_2/P_2 = (\dot{m}/M)RT_2/P_2$

Ideal gas, *Equation 2.13*

$$= (0.173 \text{ kg/s/28.01 kg/kmol})(8.314 \text{ kJ/(kmol.K)}(70 + 273)\text{K}/(14 + 101.3)$$
$$\text{kPa(abs)} = 0.153 \text{ m}^3/\text{s}$$

Note 6  The P and T used in the ideal gas equation must be absolute values.

Velocity of $N_2$ at outlet    $= \dot{V}/A = (0.158\,\text{m}^3/\text{s})/(0.002\,\text{m}^2) = \textbf{76 m/s}$

Temperature difference    $= \Delta T = (70°C - 20°C) = 50C° = (343\text{ K} - 293\text{ K})$
$= 50 \text{ K}$

Rate of heat transfer into $N_2 = \dot{Q} = \dot{m}C_p\Delta T = (0.173 \text{ kg/s})(1.05 \text{ kJ/(kg.K)})(50\text{ K})$
$= 9.1 \text{ kJ/s} = \textbf{9.1 kW}$

Note 7  The temperature difference $\Delta$ T can be calculated either from both relative temperatures or both absolute temperatures.

## PARTIAL PRESSURE AND PARTIAL VOLUME

The *partial pressure* of a component of a gas mixture is the pressure that component would exert if it alone occupied the full volume of the mixture at the temperature of the mixture.

The *partial volume* of a component of a gas mixture is the volume that component would occupy if it alone were present at the total pressure and temperature of the mixture.

The partial pressure and partial volume of a component of an ideal gas mixture are both proportional to its mole fraction, i.e.:

$p(j) = y(j)P$     (fixed temperature)                        *Equation 2.07*

$v'(j) = y(j)V$     (fixed temperature)                        *Equation 2.08*

Also, since $\Sigma$ [y(j)] = 1, then:

$\Sigma$ [p(j)] = P     (Dalton's law of partial pressures)             *Equation 2.09*

$\Sigma$ [v'(j)] = V     (Amagat's law of partial volumes)          *Equation 2.10*

where:

p(j) = partial pressure of component    "j"    kPa(abs)
v'(j) = partial volume of component    "j"    $m^3$
y(j) = mole fraction of component    "j" in ideal gas mixture    –
P    = total pressure of mixture    kPa(abs)
V    = total volume of mixture    $m^3$

## PHASE

Many substances can exist as one or more of the phases: solid (S), liquid (L) or gas (G). The phase of a substance at equilibrium depends on its conditions of temperature and pressure. For example, the phase diagram of water in *Figure 2.02* shows that at equilibrium water is:

- a single phase solid at 200 K,150 kPa(abs)           region (S)
- a single phase liquid at 300 K,150 kPa(abs)          region (L)
- a single phase gas at 600 K,150 kPa(abs)             region (G)
- a two phase solid–liquid along the line between      (L) and (S)
- a two phase solid–gas along line between             (S) and (G)
- a two phase liquid–gas along line between            (L) and (G)
- a three phase solid–liquid–gas at the *triple-point*
  The triple-point is the unique condition at which the solid, liquid and gas co-exist at equilibrium.

A vapour is a gas below its critical temperature and in this condition can be condensed by a sufficient increase in the pressure.

Note that the phase relations of *Figure 2.02* apply only at equilibrium. It is possible to obtain water (and many other substances) in non-equilibrium states, corresponding to conditions such as supercooled liquid water (e.g. P = 101.3 kPa(abs), T = 272 K), super-heated liquid water (e.g. P = 101.3 kPa(abs), T = 374 K) and supercooled water vapour (e.g. P = 101.3 kPa(abs), T = 372 K). These non-equilibrium states are thermodynamically unstable but do exist under some circumstances.

The physical properties of any substance depend on its phase, so the description of any process stream should include the stream phase(s).

Depending on the circumstances, determining the phase of an unseen material can be easy, or it can be difficult. In the case of pure substances (i.e. a single molecular species), when you know the temperature and pressure you can look up the phase in a table of thermodynamic properties or in a phase diagram (see *Refs. 1–6*). Alternatively you can find the phase from equations representing the inter-phase equilibria as shown in *Example 2.07*.

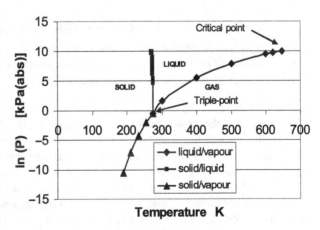

Figure 2.02.    Phase diagram for water.

*EXAMPLE 2.07    Finding the phase of water by using the Antoine equation.*

The *vapour pressure*[3] of pure water is given by:

**p\* = exp[16.5362 – 3985.44/(T – 38.9974 )]**                     *Equation 2.11*

where:
   p\*  = vapour pressure      kPa(abs)
   T   = temperature          K

*Problem*:  Is pure water a liquid or a gas at 374 K, 100 kPa(abs)?
*Solution*:  Substitute T = 374 K into *Equation 2.11*.
          p\* = exp[16.5362 – 3985.44 / (374 – 38.9974)] = 104 kPa(abs)

Since p\* at 374 K is greater than 100 kPa(abs) [i.e. 374 K, 100 kPa lies below the G/L equilibrium line] then water in its equilibrium state at 374 K, 100 kPa(abs) is in the GAS phase.

In this case the gas phase is a *vapour* because the temperature is below the *critical temperature*[4] of water, which is 647 K, and the water (gas) can be condensed to water (liquid) by increasing the pressure.

### Phase rule

When you are trying to determine the phase of a material it is sometimes helpful to use the phase rule, as shown in *Example 2.08*.

   $F^{\#} = C^{\#} – \Pi^{\#} + 2$ (Phase rule — applies to non-reactive systems)          *Equation 2.12*

where:
   $F^{\#}$  = number of thermodynamic degrees of freedom of the system
        = number of *intensive variables*[5] that must be fixed to fully-specify the state of each
          phase in the system
          When the values of $F^{\#}$ intensive variables are fixed, then the values of all other
          intensive variables are automatically and uniquely set *for each phase* in the system
   $C^{\#}$  = number of *components* ≡ number of independent chemical species, in the system
   $\Pi^{\#}$  = number of phases in the system at equilibrium

---

[3] *Vapour pressure* is the pressure exerted by a substance when its solid or its liquid phase is in contact and at equilibrium with its vapour (e.g. in a closed container) at a specified temperature.

[4] The *critical temperature* of a substance is the temperature above which the substance cannot be liquefied by increasing the pressure.

[5] *Intensive variables* are the properties of each phase that are independent of the amount of material, e.g. P, T, v, u, h, x. Note that for a multi-phase system at equilibrium the values of P and T are uniform throughout the phases, but the values of v, u, h, x, etc. are different in each phase.

The phase rule shown above does not apply when the system involves chemical reactions. In reactive systems the number of components $C^\#$ in *Equation 2.12* is reduced by the number of independent reactions that occur in the system at equilibrium.

### EXAMPLE 2.08    Using the phase rule to find the phase of water in a multi-phase system.

***Problem A***: How many intensive variables are needed to define the equilibrium state of pure liquid water?

**Solution:**    From the phase rule *Equation 2.12*:
$$F^\# = C^\# - \Pi^\# + 2 = 1 \text{ component} - 1 \text{ phase} + 2 = \underline{\textbf{2 intensive variables}}$$

Values of any two intensive variables (e.g. [T, P] or [T, v] or [P, h], etc.) will fully-specify the state of pure liquid water.

***Problem B***: How many intensive variables are needed to define the equilibrium state of each phase of a mixture of liquid water and gas water?

**Solution:**    From the phase rule *Equation 2.12*:
$$F^\# = C^\# - \Pi^\# + 2 = 1 \text{ component} - 2 \text{ phases} + 2 = \underline{\textbf{1 intensive variable}}$$

The value of any one intensive variable (e.g. P or T or v or h, etc.) will fully-specify the states of the liquid and of the gas.

Note however that the value of one more system variable, such as the specific volume or specific enthalpy, of the 2 phase mixture is needed to fix the distribution of mass between the two phases (i.e. to fix the phase split).

***Problem C***: The equilibrium state of a mixture of ethanol and water is determined by fixing ONLY the temperature of the mixture and the mole fraction of ethanol in one phase. How many phases are present in the system at equilibrium?

**Solution:**    From the phase rule *Equation 2.12*: $\Pi^\# = C^\# + 2 - F^\#$

$C^\# = 2$ (i.e. ethanol, water)
$F^\# = 2$ (i.e. T, x)
$\Pi^\# = 2 + 2 - 2 = \underline{\textbf{2 phases}}$

The phase relations for mixtures of substances are outlined later in this chapter under "Multi-Species, Multi-Phase Equilibria".

## EQUATIONS OF STATE

An *equation of state*[6] is a relation between the specific volume of a substance, its pressure and its temperature (i.e. v, P and T).

### IDEAL GAS

An *ideal gas* is a hypothetical gas in which the individual molecules do not interact with each other. No real gas is truly ideal, but many gases are nearly ideal and can be modelled by the ideal gas *equation of state*, which is called the "ideal gas law":

$$PV = nRT \text{ (Ideal gas law)} \qquad\qquad\qquad\qquad Equation\ 2.13$$

where:

| | | |
|---|---|---|
| P | = pressure in gas | kPa(abs) |
| n | = m / M = amount of gas | kmol |
| T | = temperature of gas K | |
| m | = mass of gas | kg |
| V | = volume of gas | $m^3$ |
| R | = universal gas constant = 8.314 | $kJ.kmol^{-1}.K^{-1}$ |
| M | = molar mass (molecular weight) of gas | $kg.kmol^{-1}$ |

As a rule of thumb, the ideal gas law predicts values within 1% of reality when RT/P > 5 $m^3$/kmol for diatomic gases, or RT/P > 20 $m^3$/kmol for other polyatomic gases. The deviation of a real gas from the ideal gas law increases as the conditions approach the *critical state* of the gas, which is defined by its *critical temperature* ($T_c$) and *critical pressure* ($P_c$).[7] *Table 2.06* lists the critical temperature and pressure of some common substances. Deviations from ideality are measured by the gas compressibility factor "z", which ranges in value from about 0.2 to 4, depending on the proximity to the critical state. To account for compressibility, the ideal gas equation is modified to *Equation 2.14*.

$$PV = znRT \qquad\qquad\qquad\qquad Equation\ 2.14$$

Calculations with non-ideal gases involve the use of the *reduced temperature* ($T_r = T/T_c$) and *reduced pressure* (Pr = P/Pc) in relatively complex non-linear equations of state that are beyond the scope of this text (see *Refs. 4, 5*). As a rough guide, if Tr ≥ 2 and Pr < 10 the compressibility factor ranges from about 0.9 to 1.2, so the gas can be modelled as an ideal gas with accuracy better than +/–20%.

---

[6] Specifically, a volumetric equation of state. Equations of state for single-phase, single-component systems may relate any three intensive variables.

[7] The *critical temperature* ($T_c$) is the temperature above which the gas cannot be liquefied by increasing the pressure. The *critical pressure* ($P_c$) is the minimum pressure required to liquefy the gas at its critical temperature.

From *Table 2.06* you can see that around normal atmospheric conditions (i.e. 293 K, 101 kPa(abs)), hydrogen, oxygen, nitrogen, air and methane behave effectively as ideal gases. Butane and water are liquids at 293 K, 101 kPa(abs), but in multi-component gas systems above 273 K their vapours approximate the ideal gas law if the partial pressure is below about 10% of the critical pressure (i.e. $P_r < 0.1$).

**Table 2.06. Critical temperature and pressure of common substances.**

| Substance | $T_c$ K | $P_c$ kPa(abs) |
|---|---|---|
| Hydrogen (H$_2$) | 33.3 | 1297 |
| Oxygen (O$_2$) | 154.4 | 5035 |
| Nitrogen (N$_2$) | 126.2 | 3394 |
| Methane (CH$_4$) | 190.7 | 4653 |
| Carbon dioxide (CO$_2$) | 304.2 | 7385 |
| n-Butane (C$_4$H$_{10}$) | 425.2 | 3796 |
| Water (H$_2$O) | 647.3 | 22110 |

*EXAMPLE 2.09   Using the ideal gas law to calculate a gas volume.*

**Problem:** Dry air[8] contains 21 vol% O$_2$ ($M = 32.00$ kg.kmol$^{-1}$) + 79 vol% N$_2$ ($M = 28.01$ kg.kmol$^{-1}$). Calculate the volume occupied by 5.00 kg of dry air at 400 K, 500 kPa(gauge). Assume ideal gas behaviour.

**Solution:** By the ideal gas law:

**PV = nRT**                                           (see *Equation 2.13*)

Molar mass of dry air (see *Equation 2.18*):
$M_m = \Sigma ( M(j).y(j)) = (32.00)(0.21) + (28.01)(0.79)$     $= 28.85$ kg.kmol$^{-1}$

Amount of dry air:
$n = m / M_{av} = 5$ kg $/ 28.85$ kg.kmol$^{-1}$     $= 0.173$ kmol

Absolute pressure (see *Equation 2.02*):
$P$ = Atmospheric pressure + gauge pressure = 101.3 kPa + 500 kPa(gauge)
= 601.3 kPa(abs)

Solve *Equation 2.13* for V and substitute values.
$V = nRT/P = (0.173$ kmol$)(8.314$ kJ.kmol$^{-1}$.K$^{-1})(400$ K$)/(601.3$ kPa(abs)$) = \underline{\mathbf{0.959}}$ m$^3$

**NOTE:** The ideal gas law does not apply to liquids or solids.

**LIQUIDS AND SOLIDS**

For most purposes the pressure–temperature–volume relation for liquids and solids can be approximated by *Equation 2.15*.

$$v \cong v_o(1 + \theta (T - T_o) - \sigma(P - P_o))$$                     *Equation 2.15*

---
[8] More accurately the composition of dry air is: N$_2$ = 78.08, O$_2$ = 20.95, Ar = 0.93, CO$_2$ = 0.04 vol%.

where:

   $v$   = specific volume                                                    $m^3/kg$

   $v_o$  = specific volume at $P = P_o, T = T_o$                          $m^3/kg$

   $\theta$   = coefficient of thermal expansion of the material = $1/v_o$ $(dv/dT)$    $K^{-1}$

   $\sigma$   = coefficient of compressibility of the material = $-1/v$ $(dv/dP)$    $kPa^{-1}$

Both $\theta$ and $\sigma$ are slightly affected by temperature and pressure, and this fact should be accounted for in accurate calculations, especially when T and/or P change over a wide range.

## DENSITY

Density is the ratio of mass to volume.

   $\rho = m/V$                                                       *Equation 2.16*

where:

   $\rho$ = density    m = mass    V = volume

The density of a material depends on its phase, the temperature and the pressure. *Table 2.07* shows values of density for some earthly materials, which ranges from about 0.09 $kg.m^{-3}$ for hydrogen gas at room conditions to 22,000 $kg.m^{-3}$ for solid osmium.

**Table 2.07.  Density and specific volume of some materials.**

| Material | Phase | M | Pressure | Temperature | Density | Specific Volume |
|----------|-------|---|----------|-------------|---------|-----------------|
| | | $kg.kmol^{-1}$ | kPa(abs) | K | $kg.m^{-3}$ | $m^3.kg^{-1}$ |
| Hydrogen | gas | 2 | 101 | 273 | 0.089 | 11.2 |
| Air | gas | 28.8 | 101 | 273 | 1.28 | 0.781 |
| Chlorine | gas | 71 | 101 | 273 | 3.16 | 0.316 |
| Pentane | liquid | 72 | 101 | 291 | 630 | 0.00159 |
| Water | liquid | 18 | 101 | 277 | 1000 | 0.00100 |
| Bromine | liquid | 253 | 101 | 293 | 3119 | 0.000321 |
| Mercury | liquid | 201 | 101 | 293 | 13550 | 0.0000738 |
| Octadecane | solid | 254 | 101 | 298 | 775 | 0.00129 |
| Water | solid | 18 | 101 | 273 | 920 | 0.00109 |
| Iron | solid | 56 | 101 | 293 | 7860 | 0.000127 |
| Osmium | solid | 190 | 101 | 293 | 22500 | 0.0000444 |

Variations in temperature and pressure have a small effect on the density of solids and liquids, as measured respectively by the coefficient of thermal expansion and the coefficient of compressibility. For example, for liquid water at 298 K:

Coefficient of thermal expansion   =  $(1/v)(dv/dT)$   = $d(\ln(v))/dT$   = 257E-6 $K^{-1}$

Coefficient of compressibility        =  $-(1/v)(dv/dP)$  = $-d(\ln(v))/dP$  = 45E-8 $kPa^{-1}$

The density of gases is strongly affected by variation in temperature and pressure. For ideal gases the density can be calculated from *Equation 2.17*.

$$\rho = MP/RT \qquad\qquad\qquad\qquad\qquad\qquad\qquad\qquad\text{\textit{Equation 2.17}}$$

where:

| | | |
|---|---|---|
| $\rho$ | = density of ideal gas or gas mixture | $kg.m^{-3}$ |
| $M$ | = molar mass of gas or gas mixture | $kg.kmol^{-1}$ |
| $P$ | = pressure | $kPa(abs)$ |
| $R$ | = gas constant = 8.314 | $kJ.kmol^{-1}.K^{-1}$ |
| $T$ | = temperature | $K$ |

For a gas mixture the molar mass is the mean value defined by *Equation 2.18*

$$M_m = \Sigma(M(j)y(j)) \qquad\qquad\qquad\qquad\qquad\qquad\text{\textit{Equation 2.18}}$$

where:

| | | |
|---|---|---|
| $M_m$ | = mean molar mass ("molecular weight") of gas mixture | $kg.kmol^{-1}$ |
| $M(j)$ | = molar mass of component "j" | $kg.kmol^{-1}$ |
| $y(j)$ | = mole fraction of component "j" | – |

The density of pure solids, liquids and non-ideal gases are relatively difficult to calculate (see *Ref. 5*) and are best obtained by measurement or from tabulations such as those in *Refs. 1, 2* and *3*. For *ideal mixtures*[9] of substances (solids, liquids or gases) the density of the mixture can be calculated from the individual component densities by *Equation 2.19*.

$$\rho_m = 1/\Sigma\ (w(j)/\rho(j)) \qquad\qquad\qquad\qquad\qquad\text{\textit{Equation 2.19}}$$

where:

| | | |
|---|---|---|
| $\rho_m$ | = density of ideal mixture | $kg.m^{-3}$ |
| $w(j)$ | = mass fraction of component "j" | – |
| $\rho(j)$ | = density of component "j" | $kg.m^{-3}$ |

**Table 2.08.  Density of sulphuric acid — water mixtures at 293 K, 101 kPa(abs).**

| $H_2SO_4$ conc.<br>wt% | Actual density<br>$kg.m^{-3}$ | Density from Eqn 2.19<br>$kg.m^{-3}$ |
|---|---|---|
| 0 | 998 | 998 |
| 10 | 1066 | 1048 |
| 50 | 1395 | 1249 |
| 90 | 1814 | 1691 |
| 100 | 1831 | 1831 |

For *non-ideal mixtures* of substances the density of the mixture is only approximated by *Equation 2.19* and is best obtained by measurement or from tabulations in the literature. For example, *Table 2.08* shows a comparison of the actual density of sulphuric acid — water liquid mixtures with values calculated by the ideal mixture *Equation 2.19*.

---

[9] In an *ideal mixture* the partial volumes of the components can be added to give the total volume of the mixture. This is not the case for a *non-ideal* mixture.

## SPECIFIC GRAVITY

Specific gravity is the ratio of the density of a material to the density of a reference substance.

$$\mathbf{SG} = \rho/\rho_{ref}$$                                    *Equation 2.20*

where:
  SG = specific gravity of material            –
  $\rho$   = density of material                $kg.m^{-3}$
  $\rho_{ref}$ = density of reference substance      $kg.m^{-3}$

For solids and liquids the reference substance is conventionally liquid water at 277 K, 101 kPa(abs), with a density of 1000 $kg.m^{-3}$. For gases the reference substance is usually air at 293 K, 101 kPa(abs), with density 1.20 $kg.m^{-3}$. Due to this potential ambiguity you should check the reference substance when you use values of specific gravity.

## SPECIFIC VOLUME

Specific volume is the ratio of volume to mass, i.e. the inverse of density.

$$\mathbf{v = V/m = 1/\rho}$$                                *Equation 2.21*

where:
  m  = mass                    kg
  v   = specific volume          $m^3.kg^{-1}$
  V  = volume                  $m^3$

## MOLAR VOLUME

Molar volume is the volume of one mole of a substance and equal to the product of the specific volume and molar mass.

$$\mathbf{v_m = Mv = M/\rho}$$                               *Equation 2.22*

where:
  $v_m$ = molar volume            $m^3.kmol^{-1}$
  M  = molar mass              $kg.kmol^{-1}$
  $\rho$   = density               $kg.m^{-3}$

## CONCENTRATION

Concentration is the measure of the relative amount of a component in a mixture (i.e. the mixture composition). *Table 2.09* summarises the common ways of expressing concentration. With so many different ways to express concentration there is much potential for ambiguity. For a given component in a given mixture, each of the measures in *Table 2.09* will usually give a different value, so it is important to specify the measure when giving values of concentration.

Table 2.09.   Common measures of concentration.

| Name | Definition | Symbol | Units |
|------|-----------|--------|-------|
| Mass fraction | Mass of component/total mass of mixture | w | dimensionless |
| Mole fraction | Moles of component/total moles of mixture | x,y | dimensionless |
| Volume fraction | Partial volume of component/total volume of mixture | $\tilde{v}$ | dimensionless |
| Molarity | Moles of component/volume of mixture | M | $mol.l^{-1}$ ($kmol.m^{-3}$) |
| Molality | Moles of component/mass of solvent | *m* | $mol.kg^{-1}$ |
| Parts per million | 1E6 w(j)<br>1E6 $\tilde{v}$(j) | ppm(wt)<br>ppm(vol) | dimensionless |
| Parts per billion | 1E9 w(j)<br>1E9 $\tilde{v}$(j) | ppb(wt)<br>ppb(vol) | dimensionless |
| Mass per volume | Mass of component/volume of mixture | – | $kg.m^{-3}$ |
| Mass per mass | Mass of component/mass of solvent | – | dimensionless |
| Activity | Function of chemical potential (see *Refs. 5–6*) | *a* | dimensionless |

The mass fraction and mole fraction are well defined measures of concentration commonly used in process engineering.

$$w(j) = m(j) / \Sigma \, m(j) \hspace{4cm} \textit{Equation 2.23}$$

$$x(j) = n(j) / \Sigma \, n(j) = (m(j) / M(j)) / \Sigma \, (m(j) / M(j)) \hspace{2cm} \textit{Equation 2.24}$$

where:
  m(j)  = mass of component "j"                                kg
  M(j)  = molar mass of component "j"                          $kg.kmol^{-1}$
  w(j)  = mass fraction (i.e. weight fraction) of component "j"   –
  x(j)  = mole fraction of component "j"                        –

Note that:[10]   $\Sigma \, (w(j)) = 1$ $\hspace{5cm}$ *Equation 2.25*

$\hspace{2.5cm} \Sigma \, (x(j)) = 1$ and $\Sigma \, (y(j)) = 1$ $\hspace{3cm}$ *Equation 2.26*

Inter-conversion of mass fraction with mole fraction in a given mixture can be done by *Equations 2.27* and *2.28*, as shown in *Example 2.10 A* and *B*.

To convert mass fraction to mole fraction:

$$x(j) = (w(j) / M(j))/\Sigma \, (w(j) / M(j)) \hspace{3cm} \textit{Equation 2.27}$$

To convert mole fraction to mass fraction:

$$w(j) = (x(j) \, M(j))/\Sigma \, (x(j) \, M(j)) \hspace{3cm} \textit{Equation 2.28}$$

Fractional concentrations are often expressed as percentages, where:

$$\textbf{Mass \% = weight \% = wt\% = 100 \, w(j)} \hspace{2cm} \textit{Equation 2.29}$$

$$\textbf{Mole \% = 100 \, x(j)} \hspace{4cm} \textit{Equation 2.30}$$

---

[10] Mole fraction is usually given the symbol "x" in solid and liquid mixtures, "y" in gas mixtures.

In the case of ideal gas mixtures the mole fraction and volume fraction have the same value, i.e.

$\tilde{v}(j) = y(j) = p(j) / P$     (ideal gas mixtures)                    *Equation 2.31*

where:
$\tilde{v}(j)$ = partial volume fraction of component j in gas mixture          –
y(j) = mole fraction of component j in gas mixture                   –
p(j) = partial pressure of component j                              kPa(abs)
P    = total pressure                                              kPa(abs)

*Equation 2.31* applies approximately to non-ideal gas mixtures, but does not apply to solid or liquid mixtures.

### EXAMPLE 2.10   *Converting between mass fraction and mole fraction compositions.*

**Problem A**: Gunpowder is a mixture of sulphur (29 wt%), carbon (14 wt%) and potassium nitrate. Find the mole fraction of potassium nitrate ($KNO_3$) in the gunpowder mixture.

**Solution**:   Define the components.

| Component | S | C | $KNO_3$ |
|---|---|---|---|
| Number (j) | 1 | 2 | 3 |
| w(j) | 0.29 | 0.14 | w(3) |
| M(j) | 32 | 12 | 101 |

From *Equation 2.25*   w(3) = 1 – (0.29 + 0.14) = 0.57
From *Equation 2.27*   x(3) = w(3)/M(3)/Σ (w(j)/M(j)) = 0.57/101 / (0.29/32 + 0.14/12 + 0.57/101) = **0.20**

**Problem B**: Air (gas) is a mixture of oxygen (21 vol%) and nitrogen. Find the mass fraction of nitrogen in air.

**Solution**:   Define the components.

| Component | $O_2$ | $N_2$ |
|---|---|---|
| Number | 1 | 2 |
| y(j) | 0.21 | y(2) |
| M(j) | 32 | 28 |

From *Equation 2.26*   y(2) = 1 – 0.21 = 0.79
From *Equation 2.28*   w(2) = (y(j) M(j)) / Σ (y(j) M(j)) = (0.79)(28)/(0.21*32 + 0.79*28) = **0.77**

Other measures of concentration are used in various situations. For example, low concentrations are conventionally given as *ppm(wt)* in liquids and solids and as *ppm(vol)* in gases. Concentrations of toxic dusts in air are given as micrograms per cubic metre and solubilities of solids in liquids are commonly listed as grams of solute per 100 grams of solvent. The concentrations of solutions used in the chemical laboratory are usually given as molarity (M) or *normality*[11] (N), whereas thermodynamic calculations involving solutions use the *molality*, or more generally the dimensionless *activity*.

## MULTI-SPECIES, MULTI-PHASE EQUILIBRIA

When two (or more) phases are in contact they tend to an equilibrium in which the concentration of each species in one phase is related to its concentration in the other phase(s). The equilibrium relations for the distribution of species between phases are used in M&E balances to calculate the composition of process streams associated with multi-phase systems.

The equilibrium distribution of species between the phases of multi-phase systems is strongly temperature dependent, and the temperature dependence is usually non-linear. Some of the relations commonly used to model equilibrium in multi-phase systems are as follows:

### GAS–LIQUID SYSTEMS

#### *Vapour pressure*

The vapour pressure of a substance is its partial pressure in the gas phase in equilibrium with a condensed phase (solid or liquid) of the pure substance. The vapour pressure of a substance is

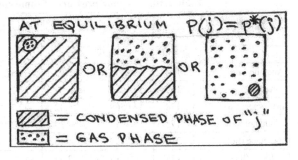

determined by the nature of the substance and the temperature. Whenever a pure (single component) condensed substance is in equilibrium with a gas the partial pressure of that substance in the gas phase will be its vapour pressure at the specified temperature. [Note that this condition is modified (cf. the Kelvin equation) as the radius of curvature of the condensed phase becomes very small (e.g. < 1 micron), due to the effect of surface energy in such systems.] Vapour pressures of pure substances are tabulated and correlated with temperature in sources such as *Refs. 1–5* of this chapter.

Vapour pressures of many substances can be estimated by empirical *equations*[12] such as the Antoine equation:

$$p^*(j) = \exp[A(j) - B(j) / (T + C(j))] \qquad \text{(Antoine equation)} \qquad Equation\ 2.32$$

---

[11] The normality (N) is the gram equivalents of solute per litre of solution, where the solute equivalent weight is defined relative to the type of reaction in which the solute is used, for example, an acid/base neutralisation or an oxidation/reduction reaction.

[12] Vapour pressure can also be estimated (less accurately) by the Clausius-Clapeyron equation: $\ln(p_1^*/p_2^*) = (h_v/R)(1/T_2 - 1/T_1)$.

where:

| | | |
|---|---|---|
| p*(j) | = vapour pressure of pure substance j | kPa(abs) |
| T | = temperature | K |
| A(j), B(j), C(j) | = Antoine constants for substance j | –, K, K |

The Antoine equation defines a line that is the locus of the unique set of points where a pure liquid and its vapour can co-exist at equilibrium. *Table 2.10* shows values for the Antoine constants and the vapour pressure of some common substances.

The vapour pressures of solids are much lower than those of liquids, typically less than 0.1 kPa(abs) at temperatures below the melting point. However some solids such as camphor, iodine and naphthalene do exert substantial vapour pressure (e.g. up to 10 kPa(abs)) at temperatures below the melting point.

The phase relations for mixtures of substances (i.e. multi-species systems) range from simple to complex. In mixtures of mutually insoluble compounds (e.g. L/L dispersions or emulsions) each liquid exerts its own vapour pressure independent of others.

Table 2.10.   **Vapour pressure of some common substances.**

| Condensed phase | $T_m$ | Antoine constants[#] | | | Temperature K | | | |
|---|---|---|---|---|---|---|---|---|
| | | A | B | C | 250 | 300 | 500 | 3000 |
| | K | – | K | K | Vapour pressure kPa(abs) | | | |
| Acetone (liq) | 178 | 14.7171 | 2975.95 | –34.5228 | 2.5 | 33.4 | 4121 | – |
| Ethanol (liq) | 159 | 16.1952 | 3423.53 | –55.7152 | 0.24 | 8.9 | 4863 | – |
| n-Hexane (liq) | 178 | 14.0568 | 2825.42 | –42.7089 | 1.53 | 21.7 | 2639 | – |
| Methanol (liq) | 175 | 16.4948 | 3593.39 | –35.2249 | 0.79 | 18.6 | 6395 | – |
| n-Octane (liq) | 216 | 14.2368 | 3304.16 | –55.2278 | 0.07 | 2.1 | 905 | – |
| Toluene (liq) | 178 | 14.2515 | 3242.38 | –47.1806 | 0.18 | 4.2 | 1201 | – |
| Water (liq) | 273 | 16.5362 | 3985.44 | –38.9974 | – | 3.5 | 2673 | – |
| Mercury (liq) | 234 | – | – | – | 1.8E-6 | 3E-4 | 5.2 | – |
| Water (solid) | 273 | – | – | – | 0.08 | – | – | – |
| Iodine (solid) | 387 | – | – | – | – | 0.07 | – | – |
| Tungsten (solid) | 3655 | – | – | – | – | – | – | 9E-6 |
| $T_m$ = melting point temperature. # p* = kPa(abs), T = K | | | | | | | | |

When liquids are mutually soluble they form homogeneous single-phase liquid systems, classified as either *ideal* or *non-ideal* mixtures. An *ideal* mixture is a mixture in which the molecules of component species do not interact with molecules of other component species. A *non-ideal* mixture involves interactions between component species that affect their chemical activity.

The vapour/liquid equilibrium for homogeneous *ideal mixtures* in the *condensed phase*[13] can be described by linear relations such as Raoult's law:

**p(j) = x (j) p*(j)**      (Raoult's law)                              *Equation 2.33*

---

[13] The *condensed phase* is the liquid or solid phase in contact with the gas.

where:
  p(j)  = partial pressure of pure substance j in the gas phase          kPa(abs)
  p*(j)= vapour pressure of pure substance j at the specified          kPa(abs)
          temperature
  x(j)  = mole fraction of substance j in ideal liquid phase mixture at    –
          equilibrium with the gas phase

Many multi-component systems of practical interest involve homogeneous non-ideal mixtures. Over small concentration ranges (e.g. x(j) < 0.05) the phase equilibria of non-ideal mixtures can be approximated by linear relations, such as Henry's law for liquid/gas systems:[14]

$$p(j) = K_H \, x(j) = f\,(T) \qquad \text{(Henry's law)} \qquad\qquad \textit{Equation 2.34}$$

where:
  $K_H(j)$ = Henry's constant for solute substance j (in a specified
          solvent at a specified temperature)                          kPa
  x(j)   = mole fraction of substance j in (non-ideal) liquid
          phase mixture at equilibrium with the gas phase              –

Values of Henry's constant are tabulated in various sources (see *Refs. 1 and 4*) and some of these values are given in *Table 2.11*.

**EXAMPLE 2.11    *Using Raoult's and Henry's laws to calculate composition in a gas/liquid systems.***

A closed vessel contains hydrogen gas plus an ideal liquid mixture of 40 mol% n-hexane + 60 mol% n-octane in equilibrium with the gas phase at 400 K, with a total pressure P = 500 kPa(abs). Henry's constant for $H_2$ dissolution in the liquid mixture = 1E6 kPa.

**Problem:**
Calculate the composition of the gas phase in the vessel and the concentration of $H_2$ in the liquid phase [mol%.].

**Table 2.11. Henry's constant for some liquid–gas systems.**

| Solvent | Solute | Temperature K | | |
|---------|--------|-----|-----|-----|
| | | 298 | 333 | 363 |
| | | Henry's constant kPa | | |
| Water(l) | $O_2$ | 4.4E6 | 6.5E6 | 7.1E6 |
| Water(l) | $N_2$ | 8.7E6 | 12.1E6 | 12.7E6 |
| Water(l) | $H_2$ | 7.1E6 | 7.7E6 | 7.5E6 |
| Benzene(l) | $H_2$ | 0.37E6 | – | – |
| Benzene(l) | $CO_2$ | 0.01E6 | – | – |

**Solution:**
Identify the components as: hydrogen: j = A, n-hexane: j = B, n-octane: j = C
Take Antoine constants from *Table 2.10*:
Vapour pressure n-hexane at 400 K = p*(B) = exp(14.0568 – 2825.42/(400 – 42.7089)) = 468 kPa(abs)
Vapour pressure n-octane at 400 K = p*(C) = exp(14.2368 – 3304.16/(400 – 55.2278)) = 105 kPa(abs)

---

[14] Henry's law is paricularly useful to calculate the solubility of *non-condensable gases* in liquids.

Assume negligible effect of the hydrogen content of the liquid on the partial pressures of n-hexane and n-octane in the gas. [The low concentration of $H_2$ in the liquid (0.025 mol%) justifies the assumption.]

Then by Raoult's law (*Equation 2.33*):

$p(B) = x(B)p^*(B) = (0.4)(468 \text{ kPa}) = 187 \text{ kPa(abs)}$
$p(C) = x(C)p^*(C) = (0.6)(105 \text{ kPa}) = 63 \text{ kPa(abs)}$

By Dalton's law (see *Equation 2.09*): $p(A) = P - (p(B) + p(C)) = 500 - (187 + 63) =$
$$250 \text{ kPa(abs)}$$

By Henry's law (see *Equation 2.34*): $p(A) = KH\, x(A)$
$$x(A) = p(A)/KH = 250 \text{ kPa}/1E6 \text{ kPa} = 2.5E-4$$

| Component | Liquid phase | mol% | Gas phase | mol% |
|---|---|---|---|---|
| $H_2$ | – | 0.03 | 250 kPa/500 kPa = 0.50 | 50 |
| n-hexane | – | 39.99 | 187 kPa/500 kPa = 0.37 | 37 |
| n-octane | – | 59.98 | 63 kPa/500 kPa = 0.13 | 13 |

Non-ideal mixtures in general show complex non-linear behaviour that is dealt with by sophisticated thermodynamic calculations beyond the scope of this text (see *Refs. 4, 5*). Empirical data are tabulated for common non-ideal mixtures (see *Ref. 1*) and will be used in some examples in this text.

### Bubble-point and dew-point

The *bubble-point* is the condition of pressure and temperature at which a liquid mixture just begins to "boil" (i.e. first bubbles of vapour form). If the pressure is decreased at a fixed temperature then the pressure at which the first bubbles of vapour form is the *bubble-point pressure* at that temperature. If the temperature is increased at a fixed pressure then the temperature at which the first bubbles of vapour form is the *bubble- point temperature*, at that pressure.

For an ideal liquid mixture[15] the bubble point condition can be calculated from *Equation 2.35*:

$$1 = \Sigma\, [x(j)p^*(j)] / P \qquad\qquad\qquad \textit{Equation 2.35}$$

where:

| | | |
|---|---|---|
| P | = total pressure | kPa(abs) |
| $p^*(j)$ | $= f(T_B)$ = vapour pressure of component "j" | kPa(abs) |
| $T_B$ | = bubble-point temperature | K |
| $x(j)$ | = mole fraction of component "j" in the liquid mixture | – |

The *dew-point* is the condition of pressure and temperature at which a vapour mixture forms the first drops of liquid (i.e. dew drops). If the pressure is raised at a fixed temperature then the pressure at which liquid drops first form is the *dew-point pressure*. If the temperature is lowered at a fixed pressure then the temperature at which liquid drops first form is the *dew-point temperature*. When the condensate is an ideal liquid mixture[15] the dew-point condition can be calculated from *Equation 2.36*:

---

[15] *Equations 2.35* and *2.36* come from Raoult's law, with respectively: $\Sigma\,[y(j)] = 1$ and $\Sigma\,[x(j)] = 1$.

$$1 = P(\Sigma \ [y(j)/p^*(j)]) \qquad\qquad\qquad\qquad\qquad \textit{Equation 2.36}$$

where:

| | | |
|---|---|---|
| P | = total pressure | kPa(abs) |
| y(j) | = mole fraction of component "j" in the vapour mixture | – |
| p*(j) | = f ($T_D$) = vapour pressure of component "j" | kPa(abs) |
| $T_D$ | = dew-point temperature | K |

In *Equations 2.35* and *2.36*, if the total pressure is fixed, the Antoine equation (*Equation 2.32*) can be inserted as the vapour pressure function f($T_B$) or f($T_D$) and the resulting non-linear equations solved for $T_B$ or $T_D$.

When some components of the liquid mixture have very low vapour pressures (relative to the others) the bubble-point calculation is simplified by setting their vapour pressures as zero in *Equation 2.35*. For example, the bubble point of a solution of sugar (sucrose) in water is determined only by the mole fraction of water in the solution. Similarly, when some components of a gas are non-condensable[16] or have a very high vapour pressure (relative to the others) then the dew-point calculation is simplified by setting their vapour pressures at infinity in *Equation 2.36*. For example, in the case of humid air the vapour pressures of $N_2$ and $O_2$ are set at infinity and *Equation 2.36* becomes:

$$1 = P(y_w/p_w^*) \qquad\qquad\qquad\qquad\qquad\qquad \textit{Equation 2.37}$$

where:

| | | |
|---|---|---|
| P | = total pressure | kPa(abs) |
| $y_w$ | = mole fraction of water vapour in humid air | – |
| $p_w^*$ | = f ($T_D$) = vapour pressure of water | kPa(abs) |

*Equation 2.37* is also useful to find the dew-point for combustion product gas mixtures of $N_2$, $O_2$, $CO_2$ and $H_2O$.

**NOTE**: For water mixed with "non-condensable" gases, *Equation 2.37* is easily "solved" by looking up the vapour pressure of water in a *steam table* (*Table 2.20*).

---

[16] Non-condensable gases are those above their critical temperature.

***EXAMPLE 2.12***    *Calculating the bubble-point and dew-point conditions of ideal mixtures.*

***Problem***: Calculate

A.   The bubble-point temperature of a mixture of 20 mol% n-hexane + 80 mol% n-octane under a total pressure = 300 kPa(abs).

B.   The dew-point pressure of a combustion exhaust gas with 10 vol% $H_2O$, 15 vol% $CO_2$, 73 vol% $N_2$, 2 vol% $O_2$ at 400 K.

***Solution***:

A.   Substitute values to the bubble-point *Equation 2.35*: $1 = \Sigma [x(j)p^*(j)] / P$
$300 = (0.20) \exp(14.0568 - 2825.42 / (T_B - 42.7089)) + (0.8)\exp(14.2368 - 3304.16 / (T_B - 55.2278))$
Solve this non-linear equation for $T_B$ (e.g. by Excel "Solver") to give: $T_B = \underline{\textbf{435 K}}$

B.   Substitute values into the dew-point *Equation 2.36*: $1 = P(\Sigma [y(j) / p^*(j)])$
Note that for each component $CO_2$, $N_2$ and $O_2$      $p^*(j) \gg p^*(H_2O)$
                                      [In fact $CO_2$, $N_2$ and $O_2$ are *non-condensable* at 400 K]
Then: $1 \approx P_D [0.1/\exp(16.5362 - 3985.44/(400 - 38.9974)) + 0.15/\infty + 0.73/\infty + 0.02/\infty]$
Solve this equation to give: $P_D = \underline{\textbf{2440 kPa(abs)}}$

Note that the same value of $P_D$ is obtained by examination of the saturated liquid/vapour data in the steam table (*Table 2.20*). A similar approach can be used to find the dew-point temperature under a given pressure.

## GAS–SOLID SYSTEMS

### *Surface concentration*

Gases are adsorbed on solid surfaces to form mono- or multi-molecular layer adsorbate films. The equilibrium between a species adsorbed on a solid surface and its gas is represented by an "adsorption isotherm" such as:

**$\Gamma = aK_Ap(j)/(1 + K_Ap(j))$**
(Langmuir adsorption isotherm (fixed temperature))                        *Equation 2.38*

where:

| | | |
|---|---|---|
| a | = empirical constant (relates surface coverage to surface concentration) | $kmol.m^{-2}$ |
| $\Gamma$ | = equilibrium surface concentration of the adsorbed species (i.e. the adsorbate) | $kmol.m^{-2}$ |
| $K_A$ | = empirical adsorption constant | $kPa^{-1}$ |
| p(j) | = equilibrium partial pressure of the adsorbate in the gas phase | kPa(abs) |

Adsorption isotherms are used, for example, in calculation of the distribution of adsorbed species between gases or liquids and microporous solids (e.g. activated carbon, silica gel, molecular sieves) that have specific surface areas of the order 10E3 to 1000E3 $m^2$/kg. These adsorption equilibria are important in the design of processes such as solid catalysed gas phase reactions (e.g. synthesis of ammonia) and gas separations (e.g. pressure swing adsorption to remove nitrogen from air).

## LIQUID–LIQUID SYSTEMS

### *Distribution coefficient*

Liquid–liquid (L–L) systems typically consist of two mutually insoluble liquids (like oil and water) in contact with each other. The distribution of a solute species between the two liquids at equilibrium is determined by the value of the distribution coefficient.

The distribution coefficient (also called the "partition coefficient") is defined as:

$$D(j) = x(j)_A/x(j)_B = f(T) \hspace{3cm} \text{\textit{Equation 2.39}}$$

where:

  $D(j)$  = liquid–liquid distribution coefficient for species j                 –
  $x(j)_A$ = mole fraction of solute species j in liquid A at equilibrium       –
  $x(j)_B$ = mole fraction of solute species j in liquid B at equilibrium       –

Values of the distribution coefficient for many systems are tabulated in several sources (see *Ref. 1*) and *Table 2.12* gives values of the distribution coefficients for some common L–L systems.

**Table 2.12.   Distribution coefficients for some common systems.**

| Solute | Liquid A | Liquid B | Temperature K | | | | | |
|--------|----------|----------|------|------|------|------|------|------|
| | | | 273 | 298 | 303 | 313 | 333 | 343 |
| | | | Distribution coefficient (dimensionless) | | | | | |
| Acetic acid | benzene | water | – | 0.033 | 0.098 | 0.102 | 0.064 | – |
| Ethanol | ethyl acetate | water | 0.026 | 0.50 | – | – | – | 0.455 |
| Methanol | n-butanol | water | 0.600 | – | 0.510 | – | 0.682 | – |
| Toluene | aniline | n-heptane | 0.577 | – | – | 0.425 | – | – |

Liquid–liquid (L–L) distribution calculations are used in the design of liquid–liquid separation (a.k.a. liquid extraction or solvent extraction) processes, such as the recovery of penicillin from fermentation liquor (*Fig. 3.01*), the separation of aromatics (e.g. benzene and toluene) from petroleum and the reprocessing of spent fuel from nuclear reactors to recover unused uranium, plutonium and other isotopes.

## LIQUID–SOLID SYSTEMS

### *Solubility*

The solubility of a material (the solute) is the amount of that material that is dissolved in a saturated solution with a specified solvent. Solubility values vary over a wide range and are expressed in several ways. One common way of expressing solubility is given in *Equation 2.40.*

$$S'(j) = m(j) / m_s = f(T)$$                                   *Equation 2.40*

where:

| | | |
|---|---|---|
| $S'(j)$ | = solubility of solute species j in solvent | kg solute $(kg.solvent)^{-1}$ |
| $m(j)$ | = mass of solute in the saturated solution | kg |
| $m_s$ | = mass of solvent in the saturated solution | kg |

Solubility data are tabulated in sources such as *Refs. 1, 2. Table 2.13* gives values of solubility for some common materials in water. The solubility of gases in liquids (at fixed pressure) generally decreases with increasing temperature (cf. *Table 2.11*). The solubility of most solids increases with temperature, although a few solids (e.g. $Ca(OH)_2$) show a negative temperature coefficient of solubility. Note that these tabulated solubility values are for *saturated* solutions under equilibrium conditions at the specified temperature. By suppressing crystal nucleation it is possible to obtain supersaturated solutions in which the concentration of solute exceeds the tabulated solubility value and such solutions are fairly common in practical processes.

Table 2.13.   Solubility of some common materials in water.

| Solute | Solvent | Temperature K | | | | | |
|---|---|---|---|---|---|---|---|
| | | 273 | 293 | 313 | 333 | 353 | 373 |
| | | Solubility kg solute. $(kg\ solvent)^{-1}$ | | | | | |
| $O_2$ (1atm) | Water | 70E-6 | 44E-6 | 33E-6 | 28E-6 | 26E-6 | 25E-6 |
| $CO_2$ (1atm) | Water | 3.35E-3 | 1.69E-3 | 0.97E-3 | 0.58E-3 | – | 0.00 |
| $NH_3$ (1atm) | Water | 0.88 | 0.53 | 0.32 | 0.18 | – | – |
| $CaCO_3$ | Water | – | 12E-6 | – | – | – | 20E-6 |
| $Ca(OH)_2$ | Water | 1.85E-3 | 1.65E-3 | 1.41E-3 | 1.16E-3 | 0.94E-3 | 0.77E-3 |
| $CaCl_2$ | Water | 0.595 | 0.745 | – | 1.37 | 1.47 | 1.59 |
| NaCl | Water | 0.357 | 0.360 | 0.366 | 0.373 | 0.384 | 0.398 |

## CHEMICAL EQUATIONS AND STOICHIOMETRY

Chemical equations represent the transformation of species in a chemical reaction. They are conventionally written with the reactants (initial species) on the left hand side (LHS) and the products (final species) on the right hand side (RHS):

$$\alpha A + \beta B \rightarrow \chi C + \delta D$$                         *Reaction 2.01*

In *Reaction 2.01* the reactants are species A and B, the products are species C and D. The numbers $\alpha$, $\beta$, $\chi$ and $\delta$ are the stoichiometric coefficients of the reaction. *Reaction 2.01* shows that:

α molecules of species A reacts with β molecules of species B to generate χ molecules of species C plus δ molecules of species D.

In terms of mole[17] quantities *Reaction 2.01* means that:
α moles A + β moles B react to generate χ moles + δ moles D

In terms of mass quantities *Reaction 2.01* means that:

α M(A) kg of A + β M(B) kg of B react to generate χ M(C) kg of C + δ M(D) kg of D

where M(j) = molar mass (molecular weight) of species j      kg kmol$^{-1}$

Chemical reactions transform molecules, but the atoms that constitute the molecules are conserved. A properly written chemical equation must thus be *balanced* with respect to its atoms, i.e. for each element:

Number of atoms on LHS = Number of atoms on RHS

Also, the conservation of mass requires that for a balanced chemical reaction:

$\quad$ **a M(A) + β M(B) = χ M(C) + δ M(D)** $\hspace{4cm}$ *Equation 2.41*

or more generally:

$\quad$ **Σ V(j)M(j) = 0** $\hspace{6cm}$ *Equation 2.42*

where:
$\quad$ v(j) = stoichiometric coefficient of species j, with a (–) sign for reactants and a
$\quad\quad$ (+) sign for products.


**EXAMPLE 2.13   *Calculating reaction stoichiometry for the oxidation of ammonia.***

**Problem**: Calculate the mass of $NH_3$ consumed, the mass of NO produced and the mass of
$\quad\quad\quad\quad$ water produced when 320 kg of $O_2$ is consumed by the reaction:

$\quad$ $4NH_3 + 5O_2 \rightarrow 4NO + 6H_2O$ $\hspace{5cm}$ *Reaction 1*

**Solution**:
NH3 consumed:

$\quad$ = (320 kg $O_2$ / 32 kg $O_2$/kmol $O_2$)(4 kmol $NH_3$/5 kmol $O_2$)(17 kg NH3 / kmol $NH_3$)
$\quad$ = **136 kg NH3**

---

[17] A *mole* is the amount of a substance containing the same number of *elementary particles* as there are atoms in 0.012 kg of carbon 12. (i.e. Avogadro's number = 6.028E23 particles.) The number of moles in a given quantity of a molecular species (or element) is its mass divided by its molar mass, i.e. n = m / M. For molecular and ionic compounds in chemical processes (e.g. $H_2O$, $H_2$, $O_2$, $CH_4$, NaCl) the *elementary particles* are molecules. For unassociated elements and for ions the elementary particles are atoms (e.g. C, $Na^+$, $Cl^-$) or charged groups of atoms (e.g. $NH_4^+$, $ClO^-$). Refer to a basic chemistry text for more comprehensive information of atoms, ions, molecules and chemical bonding.

NO produced:

= (320 kg $O_2$/ 32 kg $O_2$/kmol $O_2$)(4 kmol NO/5 kmol O2)(30 kg NO / kmol NO)

= **240 kg NO**

H2O produced:

= (320 kg $O_2$/ 32 kg $O_2$/kmol $O_2$)(6 kmol $H_2O$/5 kmol $O_2$)(18 kg $H_2O$ / kmol $H_2O$)

= **216 kg $H_2O$**

Check the overall mass balance:

Mass INITIAL = 320 kg $O_2$ + 136 kg $NH_3$  = 456 kg

Mass FINAL   = 240 kg NO + 216 kg $H_2O$ = 456 kg          OK

*NOTE*: In this example the molar mass values are rounded for simplicity. The more accurate values are:

$O_2$ = 32.000      $NH_3$ = 17.030      NO = 30.006      $H_2O$ = 18.016  kg/kmol

## LIMITING REACTANT

When reactants are mixed in ratios corresponding to their stoichiometric coefficients they are said to be in *stoichiometric proportions*. If reactants are not mixed in *stoichiometric proportions* then the reactant present in the smallest relative amount will determine the maximum *extent of reaction*. That reactant is called the *limiting reactant* and the remaining reactants are called *excess reactants*. *Example 2.14* shows how to find the limiting reactant.

## EXAMPLE 2.14    Calculating the limiting reactant in the combustion of ethane.

A batch of 6 kmol of $C_2H_6$ is burned with 14 kmol of $O_2$ and the limiting reactant is completely consumed by the reaction:

$2C_2H_6(g) + 7O_2(g) \rightarrow 4CO_2(g) + 6H_2O(g)$                          *Reaction 1*

**Problem**: A. What is the limiting reactant?

            B. Calculate the composition of the reaction product mixture.      [mole%.]

**Solution**:

A.   Stoichiometric $O_2$ required for 6 kmol $C_2H_6(g)$ = (6 kmol $C_2H_6(g)$)(7 kmol $O_2$/2 kmol $C_2H_6(g)$) = 21 kmol $O_2$

     $O_2$ supplied = 14 kmol $O_2$ < 21 kmol $O_2$ required            i.e. **limiting reactant = $O_2$**

B.   By mole balances        ACC = IN – OUT + GEN – CON        IN – OUT = 0
     on the closed system:

|                   |                                                                 |                              | mol%       |
|-------------------|-----------------------------------------------------------------|------------------------------|------------|
| Balance on $O_2$  | ACC = Final – Initial = Final – 14 =<br>GEN – CON = 0 – 14       | Final $O_2$ = 0 kmol         | **0.0**    |
| Balance on $C_2H_6$ | ACC = Final – Initial = Final – 6 =<br>GEN – CON = 0 – (14)(2/7) | Final $C_2H_6$ = 2 kmol      | **9.1**    |
| Balance on $CO_2$ | ACC = Final – Initial = Final – 0 =<br>GEN – CON = (14)(4/7) – 0 | Final $CO_2$ = 8 kmol        | **36.4**   |
| Balance on $H_2O$ | ACC = Final – Initial = Final – 0 =<br>GEN – CON = (14)(6/7) – 0 | Final $H_2O$ = 12 kmol       | **54.5**   |
|                   |                                                                 | Total         = 22 kmol      | **100.0**  |

Check the overall mass balance:

Mass INITIAL = 6 kmol $C_2H_6$(30 kg/kmol) + 14 kmol $O_2$(32 kg/kmol)       = **628 kg**

Mass FINAL   = 2 kmol $C_2H_6$(30 kg/kmol) + 0 kmol $O_2$(32 kg/kmol)

   + 8 kmol $CO_2$(44 kg/kmol) + 12 kg $H_2O$ (18 kg/kmol)   = **628 kg**

## CONVERSION, EXTENT OF REACTION, SELECTIVITY AND YIELD

In real chemical processes the progress of a reaction depends on the chemical equilibria and reaction rates. Reactions may not go to completion and reactants may engage in secondary reactions to give undesired products. The terms "conversion", "extent of reaction", "selectivity" and "yield" are used to specify the progress of the reaction. You will see various definitions of some of these terms in different places, so be clear on the definitions when you use them.

*Conversion* $\mathbf{X(j)} = \dfrac{\textbf{amount of reactant species j converted to all products}}{\textbf{amount of reactant species j introduced to the reaction}}$

*Equation 2.43*

*Extent of reaction* $\mathbf{\varepsilon(\ell)} = \dfrac{\textbf{amount of species j converted in a specified reaction } \ell}{\textbf{stoichiometric coefficient of species j in the specified reaction } \ell}$

*Equation 2.44*

*Selectivity* $\mathbf{S(j,q)} = \dfrac{\textbf{amount of reactant species j converted to product species q}}{\textbf{amount of reactant species j converted to all products}}$

*Equation 2.45*

*Yield* $\mathbf{Y(j,q)} = \dfrac{\textbf{amount of reactant species j converted to product species q}}{\textbf{amount of reactant species j introduced to the reaction}}$

*Equation 2.46*

where:

X(j)   = conversion of reactant species j                                             dimensionless

$\varepsilon(\ell)$   = extent of reaction for the specified reaction $\ell$
        (independent of species)                                                      kmol

S(j, q) = selectivity for product species q from reactant species j        dimensionless

Y(j, q) = yield of product species q from reactant species j                dimensionless

Note that by the above definitions the conversion must be defined relative to a specific reactant and the extent of reaction refers to a specific reaction, while the selectivity and yield must all be defined relative to a specific reactant and a specific product. Then $X(j) \leq 1$, $S(j, q) \leq 1$, $Y(j, q) \leq 1$ and

**Y(j, q) = X(j) S(j, q)**                                               *Equation 2.47*

### EXAMPLE 2.15    Conversion, extent of reaction, selectivity and yield from simultaneous reactions.

Hydrogen ($H_2$) can react with oxygen ($O_2$) by *Reaction 1* to produce hydrogen peroxide ($H_2O_2$) or by *Reaction 2* to produce water ($H_2O$).

$H_2 + O_2 \rightarrow H_2O_2$                                            *Reaction 1*

$2H_2 + O_2 \rightarrow 2H_2O$                                           *Reaction 2*

*Reactions 1* and *2* are competitive reactions that occur in parallel when hydrogen reacts with oxygen on a special catalyst. There are no other reactions.

The initial reaction mixture (before reaction) contains: 5 kmol $H_2$ + 3 kmol $O_2$
The final reaction mixture (after reaction) contains: 1 kmol $H_2O_2$ + 2 kmol $H_2O$
                                                     + unreacted $H_2$ and $O_2$

**Problem**: Calculate the conversion of $H_2$, the extent of *Reaction 2*, the selectivity for $H_2O_2$ from $H_2$ and the yield of $H_2O_2$ from $H_2$.

**Solution**: Note that the initial reaction mixture does not have stoichiometric proportions of $H_2$ and $O_2$ for either *Reaction 1* or *Reaction 2*.

$H_2$ converted in *Reaction 1* = (1 kmol $H_2O_2$)(1 kmol $H_2$/kmol $H_2O_2$)     = 1 kmol
$H_2$ converted in *Reaction 2* = (2 kmol $H_2O$)(1 kmol $H_2$/kmol $H_2O$)         = 2 kmol
                                            Total $H_2$ converted          = 3 kmol

$O_2$ converted in *Reaction 1* = (1 kmol $H_2O_2$)(1 kmol $O_2$/kmol $H_2O_2$)     = 1 kmol
$O_2$ converted in *Reaction 2* = (2 kmol $H_2O$)(0.5 kmol $O_2$/kmol $H_2O$)       = 1 kmol
                                            Total $O_2$ converted          = 2 kmol

Conversion of $H_2$   = $H_2$ converted/$H_2$ initial = 3 kmol/5 kmol = 0.60          ≡ **60%**

Extent of *Reaction 2* = moles $H_2$ converted in *Reaction 2*/stoich. coeff. of $H_2$
                in *Reaction 2* = 2 kmol/2                                    = **1 kmol**

Selectivity for $H_2O_2$ from $H_2$   = $H_2$ converted to $H_2O_2$/Total $H_2$ converted
                    =1 kmol/3 kmol = 0.33                                  ≡ **33%**

Yield of $H_2O_2$ from $H_2$ = $H_2$ converted to $H_2O_2$/$H_2$ initial = 1 kmol/5 kmol = 0.20   ≡ **20%**

or

Yield of $H_2O_2$ from $H_2$ = (Conversion of $H_2$)(Selectivity for $H_2O_2$ from $H_2$)
                    = (0.60)(0.33) = 0.20                                   ≡ **20%**

### EQUILIBRIUM CONVERSION AND ACTUAL CONVERSION

Many chemical reactions are *reversible*[18] and *tend to* an equilibrium state defined by the temperature dependent equilibrium constant.

$$\alpha\, A + \beta\, B \leftrightarrow \chi\, C + \delta\, D \qquad\qquad\qquad Reaction\ 2.01$$

$$\mathbf{K_{eq} = (a_C^\chi\, a_D^\delta)/(a_A^\alpha/a_B^\beta) = f(T)}\ \text{(reaction equilibrium constant)} \qquad Equation\ 2.48$$

where:
   $K_{eq}$  = reaction equilibrium constant = $\exp(-\Delta G_T^\circ/RT)$     – (dimensionless)
   $a(j)$ = activity of species j in the reaction mixture *at equilibrium*   – (dimensionless)
   $\Delta G_T^\circ$ = f(T) = standard free energy change for the reaction at
                temperature T                                            kJ.kmol$^{-1}$
   R   = gas constant                                              kJ.(kmol.K)$^{-1}$
   T   = reaction temperature                                        K

For most calculations the activities in *Equation 2.48* can be approximated as follows:

Reactions in the solid phase:  $a(j) \approx x(j)$                        *Equation 2.49*
Reactions in the liquid phase: $a(j) \approx [J]/(1\ \mathbf{kmol.m^{-3}})$                *Equation 2.50*
Reactions in the gas phase:    $a(j) \approx p(j)/(101.3\ \mathbf{kPa})$                *Equation 2.51*

where:
   [J]   = concentration of species j            kmol.m$^{-3}$
   p(j)  = partial pressure of species j          kPa(abs)
   x(j)  = mole fraction of species j              –

---

[18] A *reversible* reaction is one that can go in both directions, shown by the two-way arrow ↔. Strictly all chemical reactions are reversible, but for many the equilibrium conversion is near 100% (i.e. >99%) so these are usually described as irreversible reactions and indicated by a one-way arrow →.

Note that:

A.  The denominators in *Equations 2.50* and *2.51* refer to the species' *standard state*[19] and have the effect of making the activity a dimensionless number. Consequently, the equilibrium constant is also a dimensionless number.
B.  The equilibrium constant is a strong function of temperature but is nearly independent of pressure.
C.  For endothermic reactions $\Delta G_T^\circ$ decreases and $K_{eq}$ increases with rising temperature. For exothermic reactions $\Delta G_T^\circ$ increases and $K_{eq}$ decreases with rising temperature.

The *equilibrium conversion* is the hypothetical conversion that would be obtained if the reaction reached equilibrium at a specified temperature. However all reactions proceed at finite net rates that tend to zero as the mixture approaches equilibrium, and thus can never reach true equilibrium. The *actual conversion* is the conversion obtained in reality, as determined by the reaction rate and residence time in the reactor. The actual conversion is always less than the equilibrium conversion, i.e.:

$$X_{act}/X_{eq} < 1 \hspace{4cm} \textit{Equation 2.52}$$

where:
$X_{act}$ = actual conversion
$X_{eq}$ = equilibrium conversion

In practice, chemical reactions can be roughly divided into four categories:

1.  **High equilibrium constant + high[20] rate**   e.g. acid-base neutralisation at 293 K, actual conversion of limiting reactant exceeds 99% within milliseconds.
2.  **High equilibrium constant + low rate**   e.g. fermentation of glucose at 303 K, actual conversion of limiting reactant reaches 99% in several days.
3.  **Low equilibrium constant + high rate**   e.g. ammonia synthesis at 700 K, $X_{eq} \approx 0.2$, actual conversion of limiting reactant exceeds 99% of $X_{eq}$ within seconds.
4.  **Low equilibrium constant + low rate**   e.g. esterification at 293 K, $X_{eq} \approx 0.5$, actual conversion of limiting reactant reaches 99% of $X_{eq}$ in several minutes.

The rate of reaction nearly always increases with temperature (there are very few exceptions) and as a rule of thumb the reaction rate roughly doubles for each 10 K rise in temperature.

---

[19] The *standard state* of a species is its normal state at 101.3 kPa(abs) and (usually) 298.15 K [some texts define the standard pressure as 1 bar = 100.0 kPa(abs), but this difference has negligible effect in practical calculations].

[20] The terms "fast rate" and "slow rate", often seen in technical writing, are grammatically incorrect. You should not use them.

The calculation of actual conversion in a chemical reaction requires a knowledge of reaction kinetics plus heat and mass transfer dynamics that is generally beyond the scope of this text (see *Refs. 1, 10, 12*). A simple case showing the comparison between equilibrium conversion and actual conversion in a batch reactor is given here in *Example 2.16*. *Example 7.03* takes a more detailed look at the time dependent conversion in a batch reactor.

### EXAMPLE 2.16   Comparing the equilibrium conversion with the actual conversion in a chemical reaction.

The reversible liquid phase *Reaction 2.07* is carried out at 400 K in an isothermal batch reactor with constant volume.

$$A(l) \leftrightarrow B(l) \text{ The reactor initially contains a solution of A with zero B} \qquad Reaction\ 2.07$$

The equilibrium constant for *Reaction 2.07* at 400 K is $K_{eq} = 5.0$ and the reaction rates[21] at 400 K are:

Forward rate = $d[A]/dt = -k_1[A]$
Reverse rate = $d[B]/dt = k_2[B]$

where:

| | | |
|---|---|---|
| $k_1$ | = forward reaction rate constant = 0.03 | $s^{-1}$ |
| $k_2$ | = reverse reaction rate constant = $k_1/K_{eq}$ = 0.006 | $s^{-1}$ |
| $[A],[A]_0, [A]_{eq}$ | = concentration of A at time t, at time zero, at equilibrium | $kmol.m^{-3}$ |
| $[B],[B]_0, [B]_{eq}$ | = concentration of B at time t, at time zero, at equilibrium | $kmol.m^{-3}$ |
| t | = time | s |

Note that *at equilibrium* (t = ∞): Forward rate = Reverse rate.Thus: $K_{eq} \approx [B]_{eq}/[A]_{eq} = k_1/k_2$

***Problem***: Calculate:
A.  $X_{eq}$ = the equilibrium conversion of A                                    [%]
B.  $X_{act}$ = the actual conversion of A at reaction time = 10 seconds        [%]

***Solution***:
A.  The *equilibrium conversion* at 400 K depends only on the value of $K_{eq}$ at 400 K.

  Define the system   = contents of batch reactor (a closed system)
  Specify the quantity = (i) total mass (ii) moles of A

---

[21] Reaction rate is the rate of consumption of a reactant or generation of a product with respect to time. This rate depends on the instantaneous concentration(s) of reactive species, the temperature and the effect of any catalyst that may be present.

(i)  Integral balance on total mass:

ACC = Final − Initial = IN − OUT + GEN − CON

$V_R(M_A[A] + M_B[B]) - V_R M_A[A]_o = 0 - 0 + 0 - 0$ [mass is conserved]

Since $M_A = M_B$      $[A] + [B] - [A]_o = 0$                                [1]

$M_A, M_B$ = molar mass of A, B kg.kmol$^{-1}$          $V_R$ = reaction volume m$^3$
                                                        (specified as constant)

(ii)  Integral balance on moles of A:

$V_R([A]_{eq} - [A]_o) = 0 - 0 + 0 - X_{eq}V_R[A]_o$          $X_{eq} = ([A]_o - [A]_{eq})/[A]_o$     [2]

From equation [1]:     $[B]_{eq} = [A]_o - [A]_{eq} = X_{eq}[A]_o$

Equilibrium constant: $K_{eq} = [B]_{eq}/[A]_{eq} = X_{eq}[A]_o/(1 - X_{eq})[A]_o = X_{eq}/(1 - X_{eq})$     [3]

(*Equations 2.48 and 2.50*)     Substitute $K_{eq} = 5.0$     Solve equation [3] for: $X_{eq}$

= 0.83 ≡ **83%**

B.  The *actual conversion* at 400 K depends on the reaction rate and the residence time:

Define the system     = contents of batch reactor (a closed system)

Specify the quantity = amount (moles) of A

Differential balance on moles of A:

Rate ACC = Rate IN − Rate OUT + Rate GEN − Rate CON

$V_R d[A]/dt = 0 - 0 + V_R k_2[B] - V_R k_1[A]$

Substitute equation [1]:

$d[A]/dt = -[A]_o d(X_{act})/dt = k_2[B] - k_1[A] = k_2[A]_o X_{act} - k_1[A]_o(1 - X_{act})$

where:

$X_{act} = ([A]_o - [A])/[A]_o$

$dX_{act}/dt = k_1 - (k_1 + k_2)X_{act}$                                       [4]

Substitute values: $dX_{act}/dt = 0.03 - (0.036)X_{act}$     Limits: $X_{act} = 0$ at t = 0

Integrate and solve for $X_{act}$

$X_{act} = 0.83[1 - \exp(-0.036\,t)]$                                          [5]

When t = 10 s $X_{act} = 0.25 \equiv$ **25%**

Note that $X_{act} \rightarrow X_{eq}$ only when t $\rightarrow \infty$ (infinity), i.e. complete equilibrium cannot be reached in a finite time!

## THERMOCHEMISTRY

Each molecule of any species contains chemical energy. When a chemical reaction occurs this chemical energy is redistributed among the reactant and product species and partially converted to heat. The study of such thermal effects in chemical systems is called "thermochemistry". Thermochemistry provides several measures that are useful in the calculation of energy balances.[22]

---

[22] Many sources quote these values in units of kJ.mol$^{-1}$. To preserve dimensional consistency this text uses units of kJ.kmol$^{-1}$ (see *Table 2.01*).

## Table 2.14.   Thermochemical values.*

| Substance | Phase | $h^o_{f,298K}$ | $H^o_{c,298K}$ | $C^o_{p,298K}$ | $h_{m,Tm}$ | $h_{v,Tb}$ | $h_{v,298K}$ | $T_c$ |
|-----------|-------|------|------|------|------|------|------|------|
| | | kJ.kmol$^{-1}$ | kJ.kmol$^{-1}$ | kJ.kmol$^{-1}$.K$^{-1}$ | kJ.kmol$^{-1}$ | kJ.kmol$^{-1}$ | kJ.kmol$^{-1}$ | K |
| | | | [Gross] | | | | | |
| $H_2$ | G | 0.0 | −285.8E3 | 28.8 | 117 | 903 | – | 33.3 |
| $O_2$ | G | 0.0 | 0.0 | 29.4 | 443 | 6.8E3 | – | 154.4 |
| $H_2O$ | G | −241.8E3 | 0.0 | 33.6 | – | – | – | 647.4 |
| $H_2O$ | L | −285.8E3 | 0.0 | 75.3 | – | 40.6E3 | 44.0E3 | 647.4 |
| $H_2O$ | S | −291.8E3 | 0.0 | 37 | 6.00E3 | – | – | 647.4 |
| $CO_2$ | G | −393.5E3 | 0.0 | 37.1 | 8.36E3 | 25.2E3 | – | 304.2 |
| $CH_4$ | G | −74.8E3 | −889.5E3 | 35.3 | 936 | 8.1E3 | – | 190.7 |
| $CH_3OH$ | L | −238.7E3 | −726.6E3 | 81.6 | 3.16E3 | 34.8E3 | 37.8E3 | 513.2 |
| $C_6H_6$ | L | +49E3 | −3264E3 | 136 | 9.91E3 | 30.7E3 | 33.8E3 | 562.6 |

*The superscript "o" means at the standard state pressure of 101.3 kPa(abs). $T_c$ = critical temperature.

### HEAT (I.E. ENTHALPY) OF FORMATION

The heat of formation of a compound ($h_f$) is the heat absorbed by a reaction mixture (i.e. the heat <u>input</u> to the mixture) when that compound is formed from its elements, at a specified condition of temperature, pressure and phase. Heat of formation values for many compounds are tabulated in sources such as *Refs. 1,2,3,9*. By convention these tables give the heat of formation of each compound in the *standard state* at 298.15 K, 101.3 kPa(abs). The effect of pressure on the heat of formation is small, but the effect of temperature is large — so the temperature is specified in the subscript, e.g. $h_{f,298k}$. Values of the heat of formation of some common substances are given in *Table 2.14*.

Some sources designate the heats (i.e. enthalpies) of combustion, formation, reaction, etc. with a "Δ", e.g. $\Delta h^o_{f,298k}$, to indicate that the quantity is a change in enthalpy relative to a reference state. That symbolism is not used in this text.

### HEAT (I.E. ENTHALPY) OF REACTION

The heat of reaction ($h_{rxn}$) is the heat absorbed by a reaction mixture (i.e. the heat <u>input</u> to the mixture) when the reaction occurs, with complete conversion of the reactants in stoichiometric proportions, at a specified condition of temperature, pressure and phase.

$$\mathbf{h_{rxn} = \Sigma \ (v(j)h_f(j)) \ products - \Sigma \ (v(j)h_f(j)) \ reactants} \qquad \textit{Equation 2.53}$$

where:

| | | |
|---|---|---|
| $h_{rxn}$ | = heat of reaction | kJ.kmol$^{-1}$ |
| $h_f(j)$ | = heat of formation of species j at the specified conditions | kJ.kmol$^{-1}$ |
| $v(j)$ | = stoichiometric coefficient of species j (in the balanced reaction) | – |

$v(j)$ is taken as positive for reactants and for products in *Equation 2.53*.

Note that the units of $h_{rxn}$ can cause confusion. When the units are given as kJ per kmol the question is: "Per kmol of what?" One way to resolve this ambiguity is to state $h_{rxn}$ per kmol of a specific reactant or product, as shown in *Example 2.17*.

### EXAMPLE 2.17   Calculating the standard heat of reaction in the oxidation of hydrogen by oxygen.

**Problem**: Calculate the standard heat of reaction for *Reaction 2.01*.
Note that the phase of each species is shown in brackets.

$$2H_2(g) + O_2(g) \rightarrow 2H_2O(g) \qquad\qquad\qquad Reaction\ 2.01$$

**Solution**: By *Equation 2.53*, $\mathbf{h_{rxn} = \Sigma\ (v\ (j)h_f(j))}$ **products**
$$- \Sigma\ (v(j)h_f(j))\ \textbf{reactants} \qquad Equation\ 2.53$$

For the reaction as written, with data from *Table 2.14*:

$$h_{rxn,\ 298k} = (2)(-242E3\ kJ.kmol^{-1}) - [(2)(0\ kJ.kmol^{-1})$$
$$+ (1)(0\ kJ.kmol^{-1})] \qquad\qquad = \underline{-484E3\ kJ.(kmol\ rxn)^{-1}}$$

For a specific reactant (e.g. $H_2$):

$$h_{rxn,\ 298k} = -484E3\ kJ.(kmol\ rxn)^{-1}/(2\ kmol\ H_2/kmol\ rxn) \quad = \underline{-242E3\ kJ.(kmol\ H_2)^{-1}}$$

The algebraic sign of $h_{rxn}$ shows whether the reaction is *endothermic* or *exothermic*.
An *endothermic* reaction ($h_{rxn}$ = positive) requires heat <u>input</u> to the reaction mixture to keep the temperature at 298 K as the reaction occurs.
An *exothermic* reaction ($h_{rxn}$ = negative) requires heat <u>output</u> from the reaction mixture to keep the temperature at 298 K as the reaction occurs.

### HEAT (I.E. ENTHALPY) OF COMBUSTION

The heat of combustion of a fuel ($h_c$) is the heat of reaction for the *complete combustion* of the fuel in oxygen (or air), at specified conditions of temperature, pressure and phase. For hydrocarbon fuels *complete combustion* means conversion of all carbon to $CO_2$ and all hydrogen to $H_2O$, in a reaction such as *Example 1.03*. When fuels contain other elements such as sulphur (S) and nitrogen (N), their combustion products should be specified, for example, as $SO_2$ and $N_2$.

The *standard heat of combustion* is the heat of combustion at standard state 298 K, 101.3 kPa(abs) which is tabulated in sources such as *Refs. 1, 2, 9*. These tables usually give the *gross heat of combustion* as kJ (or kCal, where 1 kCal = 4.18 kJ) per mole of fuel.

The *gross heat of combustion* (–ve higher heating value) is the heat of reaction obtained when the product water is a liquid. The *net heat of combustion* (–ve lower heating value) is the heat of reaction obtained when the product water is a gas. The difference between the gross and net heats of combustion corresponds to the heat of vaporisation of the product water.

$$h_{c,net} = h_{c,gross} + n_w h_v$$                                            *Equation 2.54*

where:

| | | |
|---|---|---|
| $h_{c,gross}$ | = gross heat of combustion | kJ.kmol$^{-1}$ |
| $h_{c,net}$ | = net heat of combustion | kJ.kmol$^{-1}$ |
| $n_w$ | = amount of water produced in the reaction | kmol(kmol fuel)$^{-1}$ |
| $h_v$ | = heat of vaporisation of water | kJ.kmol$^{-1}$ |

*NOTE*: The combustion of fuels is always an exothermic process, but beware that some sources (e.g. the *Chemical Engineers' Handbook*) reverse the sign of $h_c$ and tabulate it with a positive value.

## HEATS (ENTHALPIES) OF PHASE CHANGES

The energy inputs required to melt and to vaporise a substance are called respectively the (*latent*) *heat of fusion* and the (*latent*) *heat of vaporisation*.

*Heat of fusion (melting)* ($h_m$) is the heat input required to change the phase of a substance from solid to liquid at specified conditions of temperature and pressure.

*Heat of vaporisation* ($h_v$) is the heat input required to change the phase of a substance from a condensed phase (solid or liquid) to gas in equilibrium with the condensed phase at specified conditions of temperature and pressure. Unless otherwise stated the condensed phase is usually a liquid.

Values for the heat of fusion and heat of vaporisation are tabulated in sources such as *Refs. 1, 2, 9*. These values are usually given, respectively, at the *normal melting point*[23] and *normal boiling point* of the substance.

Heat of fusion is affected little by changes in temperature, but heat of vaporisation decreases with increasing temperature according to *Equation 2.55*, and becomes zero at the critical temperature.

$$h_{v,T} = h_{v,298K} + \int_{298K}^{T} (C_{p,g} - C_{p,l})dT$$                    *Equation 2.55*

where:

| | | |
|---|---|---|
| $h_{v,T}$ | = heat (enthalpy) of vaporisation at T | kJ.kmol$^{-1}$ |
| $h_{v,298K}$ | = heat (enthalpy) of vaporisation at 298 | kJ.kmol$^{-1}$ |
| $C_{p,g}$ | = heat capacity of gas | kJ.kmol$^{-1}$.K$^{-1}$ |
| $C_{p,l}$ | = heat capacity of liquid | kJ.kmol$^{-1}$.K$^{-1}$ |

---

[23] The *normal* melting and boiling points are the melting and boiling temperatures under a pressure of 1 standard atmosphere, i.e. 101.3kPa(abs).

Note that as a rule: $C_{p,g} < C_{p,l}$ and $h_v$ decreases as T increases, as exemplified for the case of water in *Figure 2.03*.

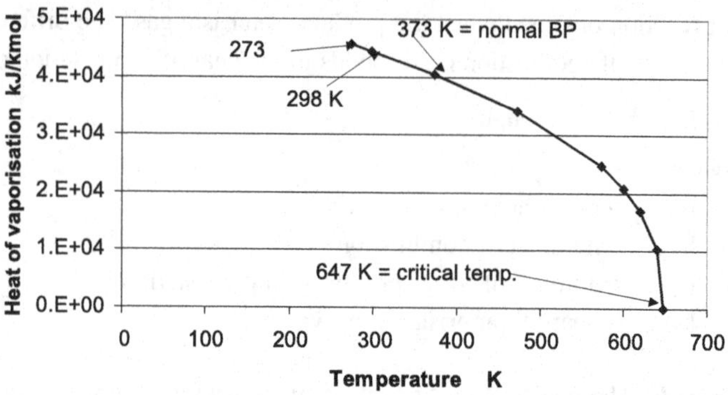

**Figure 2.03.   Effect of temperature on the heat of vaporisation of water.**

## HEAT CAPACITY[24]

The heat capacity of a substance is effectively the heat input required to raise the temperature of a specified amount of that substance by 1 degree, *without a change of composition or of phase*.

Heat capacity is defined more precisely by a pair of *partial differential equations*.[25]

$C_v = [\partial u/\partial T]_v$    (no phase change, no reaction)                    *Equation 2.56*

$C_p = [\partial h/\partial T]_p$    (no phase change, no reaction)                    *Equation 2.57*

where:

  $C_v$ = heat capacity at constant volume         $kJ.kmol^{-1}.K^{-1}$
  $C_p$ = heat capacity at constant pressure       $kJ.kmol^{-1}.K^{-1}$
  u  = specific *internal energy*                     $kJ.kmol^{-1}$
  h  = specific *enthalpy*                           $kJ.kmol^{-1}$

For a given substance in the solid or liquid phase, $C_v$ and $C_p$ have nearly the same value (i.e. $C_p \cong C_v$), but in the gas phase $C_v$ differs from $C_p$, according to *Equation 2.58*.

$C_p = C_v + R$     (Ideal gas)                                    *Equation 2.58*

where:

  R = universal gas constant = 8.314 $kJ.kmol^{-1}.K^{-1}$

In energy balance calculations involving gases, $C_v$ is used to calculate energy in closed systems and for the accumulation in open systems, whereas $C_p$ is used for the flowing streams in open (continuous) systems.

The heat capacity of each substance changes with its temperature, pressure and phase. The effect of pressure is small (except for gases near the critical state) but changes of temperature and phase have a substantial effect on the heat capacity, as shown in *Figure 2.04*.

---

[24] Heat capacity cannot be defined over a phase change because a phase change involves the transfer of heat to a substance without a change in its temperature.

[25] A *partial differential equation* represents the rate of change of a variable (x) with respect to another variable (y) while other variables that affect (x) are held constant.

Some points you should note about *Figure 2.04* are as follows:

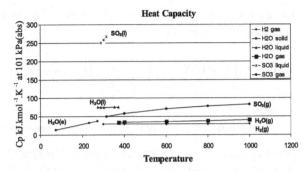

- Heat capacity is zero at 0 K.
- The discontinuity of heat capacity values at each phase change.
- The liquid heat capacity of a substance is usually higher than its gas heat capacity at the normal boiling point. Hydrogen (BP = 20 K) is an exception to this rule.

**Figure 2.04.    Changes of heat capacity with temperature (K) and phase at 101.3 kPa(abs).**

- The heat capacity of each gas increases with increasing temperature.
- The rate of increase of the gas heat capacity with temperature rises with the number of atoms in the gas molecule, (e.g. He < $H_2$ < $H_2O$ < $SO_3$ < $C_2H_6$).
- The heat capacity is undefined at the critical state.[26]

For a substance where the phase is fixed (no phase change) the heat capacity can be calculated as a polynomial function of temperature, such as *Equation 2.59*.

$$C_p = a + bT + cT^2 + dT^3 \qquad\qquad Equation\ 2.59$$

where:

T = temperature        K        a, b, c, d = empirical constants specific to each substance [appropriate units]

*Table 2.15* lists values of a, b, c and d for some common gases.

**Table 2.15.    Polynomial coefficients for the heat capacity of gases at 101.3 kPa(abs).**

| Gas | a | b | c | d | Range | $C_p$ at 298 K |
|-----|-----|-----|-----|-----|-----|-----|
| | | | | | K | kJ kmol$^{-1}$ K$^{-1}$ |
| $H_2$ | 28.79 | –0.092E-3 | 0.879E-6 | 0.544E-9 | 298–1500 | 28.9 |
| $N_2$ | 29.58 | –5.52E-3 | 13.85E-6 | –5.27E-9 | 298–1500 | 29.0 |
| $O_2$ | 26.02 | 11.3E-3 | –1.55E-6 | 0.921E-9 | 298–1500 | 29.3 |
| $H_2O$ | 33.89 | –3.01E-3 | 15.19E-6 | –4.85E-9 | 298–1500 | 34.2 |
| $CO_2$ | 21.51 | 64.4E-3 | –41.6E-6 | 10.1E-9 | 298–1500 | 37.3 |
| $CH_4$ | 21.09 | 39.0E-3 | 37.1E-6 | –22.5E-9 | 298–1500 | 35.7 |

$$C_p = a + bT + cT^2 + dT^3 \qquad kJ.kmol^{-1}K^{-1} \qquad T = K$$

---

[26] The critical temperatures and critical pressures of $H_2$, $SO_2$ and $H_2O$ are respectively 33 K, 1.3E3 kPa(abs), 431 K, 7.88E3 kPa(abs) and 647 K, 22.1E3 kPa(abs). These substances do not approach the critical state in *Figure 2.04*.

For energy balance calculations it is often satisfactory to use the *mean heat capacity* defined by *Equation 2.60*.

$$C_{p,m} = \int_{T_{ref}}^{T} C_p dT/(T - T_{ref})$$     (no phase change, no reaction)     *Equation 2.60*

where:

$C_{p,m}$ = mean heat capacity over the temperature range $T_{ref}$ to T     $kJ.kmol^{-1}.K^{-1}$

Some values of mean heat capacity are listed in *Table 2.16*.

Be careful when you use the heat capacity that the value applies to the correct phase, and it is not taken across a phase change.

**Table 2.16.   Mean heat capacities of gases at 101.3 kPa(abs) relative to 298 K.**

| Temperature K | $H_2(g)$ | $N_2(g)$ | $O_2(g)$ | $H_2O(g)$ | $CO_2(g)$ | $CH_4(g)$ | $C_2H_6(g)$ | $SO_2(g)$ |
|---|---|---|---|---|---|---|---|---|
| | $C_{p,m}$ $kJ.kmol^{-1}.K^{-1}$ | | | | | | | |
| 298 | 28.8 | 29.1 | 29.4 | 33.8 | 37.2 | 35.8 | 52.8 | 39.9 |
| 573 | 29.2 | 29.4 | 30.5 | 34.4 | 42.3 | 43.1 | 70.0 | 44.4 |
| 873 | 29.3 | 30.2 | 31.9 | 36.3 | 46.2 | 51.3 | 86.1 | 47.9 |
| 1173 | 29.6 | 31.1 | 32.9 | 38.1 | 49.1 | 58.7 | 99.1 | 50.2 |
| 1473 | 30.2 | 31.9 | 33.8 | 39.8 | 51.3 | 64.8 | 109.4 | 51.8 |
| 1773 | 30.7 | 32.6 | 34.3 | 41.4 | 53.1 | – | – | – |
| 2070 | 31.2 | 33.2 | 34.9 | 42.8 | 54.1 | – | – | – |
| 2373 | 31.7 | 33.6 | 35.4 | 44.0 | 55.1 | – | – | – |

### INTERNAL ENERGY

Internal energy (U) is a "catch-all" term that includes all the energy in a substance from the intra-atomic, inter-atomic and inter-molecular forces that give it coherence.[27] Internal energy does not include *potential energy* due to the position or *kinetic energy* due to the motion of a mass of the substance.

The internal energy of a substance is a conceptual *state function*[28] whose absolute value cannot be measured or calculated. When the internal energy is used in energy balance calculations, its value is a relative value defined with respect to a *reference state* whose internal energy is arbitrarily set at zero.

---

[27] The concept of internal energy originated with J. W. Gibbs (1839–1903) before the study of nuclear energy.

[28] A *state function* is a quantity whose value depends only on the state of a substance and is independent of the route taken to reach that state.

**ENTHALPY**

The enthalpy of a substance is a state function defined by *Equation 2.61*.

$$H = U + PV$$                                                                                                    *Equation 2.61*

where:

| | | | | |
|---|---|---|---|---|
| H = enthalpy | kJ | P = pressure | kPa(abs) | |
| U = internal energy | kJ | V = volume | $m^3$ | |

In energy balance calculations, enthalpy must be defined relative to a specified *reference state*. The enthalpy of pure substances is given in *thermodynamic tables* that are available in sources such as *Refs. 1* and *2*, but when you use such tables you should be aware of the *reference state*, as discussed below.

A convenient *reference state* for energy balance calculations on chemical processes is the chemical elements in their standard states at 101.3 kPa(abs), 298.15 K. With this basis the specific enthalpy of a pure substance can be *estimated* by *Equation 2.62*.

You can see from *Equation 2.62* why enthalpy is sometimes called the "heat content" of a substance.

$$h* \approx \int_{T_{ref}}^{T} C_p .dT + h^{\circ}_{f,T_{ref}} \approx C_{p,m}(T - T_{ref}) + h^{\circ}_{f,T_{ref}} \qquad \text{[Respect the phase]} \qquad \textit{Equation 2.62}$$

where:

| | | |
|---|---|---|
| h* | = specific enthalpy of the substance at temperature T in phase Π w.r.t. *elements* at the standard state | $kJ.kmol^{-1}$ |
| $C_p$ | = heat capacity at constant pressure of the substance in phase Π | $kJ.kmol^{-1}.K^{-1}$ |
| $h^{\circ}_{f,T_{ref}}$ | = enthalpy of formation of the substance at $T_{ref}$ in phase Π | $kJ.kmol^{-1}$ |
| T | = temperature | K |
| $T_{ref}$ | = reference temperature (usually 298 K) | K |

Note that *Equations 2.62* and *2.65* are approximate because they do not account for the (usually small) effect of pressure on enthalpy.

The value of $h^{\circ}_{f,T_{ref}}$ in *Equation 2.62* includes the latent heat of any phase changes required to convert the substance from its standard state at $T_{ref}$ to the phase Π at $T_{ref}$. For example, if the substance is a solid at its standard state but the phase Π is a gas, then:

$$h^{\circ}_{f,T_{ref}}(\textbf{gas}) = h^{\circ}_{f,T_{ref}}(\textbf{solid}) + h_{m,Tref} + h_{v,Tref} \qquad \textit{Equation 2.63}$$

Or if the substance is a liquid in its standard state, but phase Π is a gas, then:

$$h^{\circ}_{f,T_{ref}}(\textbf{gas}) = h^{\circ}_{f,T_{ref}}(\textbf{solid}) + h_{v,Tref} \qquad \textit{Equation 2.64}$$

where:

$h^{\circ}_{f,Tref}$ (gas)   = heat of formation of substance in gas phase at $T_{ref}$          kJ.kmol$^{-1}$

$h^{\circ}_{f,Tref}$ (liquid) = heat of formation of substance in liquid phase at $T_{ref}$         kJ.kmol$^{-1}$

$h^{\circ}_{f,Tref}$ (solid)  = heat of formation of substance in solid phase at $T_{ref}$          kJ.kmol$^{-1}$

$h_{m,Tref}$            = heat of fusion of substance at $T_{ref}$                       kJ.kmol$^{-1}$

$h_{v,Tref}$            = heat of vaporisation of substance at $T_{ref}$               kJ.kmol$^{-1}$

In general the enthalpy defined by *Equation 2.62* can be considered in three parts:

- Sensible heat $= \int C_p.dT = C_{p,m}(T-T_{ref})$ required to change temperature at constant phase
- Latent heat $= h_m$ and/or $h_v$ required to change phase at constant temperature
- Heat of formation $= h_f$ required to form a compound (at specified phase) from its elements

The specific enthalpy (h\*) defined by *Equation 2.62* is called the "total specific enthalpy" to differentiate it from the "specific enthalpy" (h) found in many sources of *thermodynamic data*. The specific enthalpy reported in thermodynamic tables and charts, such as the steam table (*Table 2.20*), enthalpy-pressure diagrams and psychometric charts (*Figure 2.07*) *does not include heat of formation*. This specific enthalpy "h" is measured relative to *compounds* at a reference state and can be *estimated*[29] by *Equation 2.65*.

$$h \approx \int_{T_{ref}}^{T} C_p.dT + h_{p,Tref} \approx C_{p,m}(T-T_{ref}) + h_{p,Tref} \quad \text{[Respect the phase]} \qquad \textit{Equation 2.65}$$

where:

h      = specific enthalpy of the substance at temperature T
         in phase Π w.r.t. *compounds* at the reference state       kJ.kmol$^{-1}$

$C_p$     = heat capacity at constant pressure of the substance in phase Π   kJ.kmol$^{-1}$.K$^{-1}$

$h_{p,Tref}$ = latent heat of any phase changes to convert the substance
         from its reference state at $T_{ref}$ to the phase Π at $T_{ref}$.        kJ.kmol$^{-1}$

T      = temperature                                          K

$T_{ref}$    = reference temperature                                K

*Figure 2.05* shows the relation between the total specific enthalpy and temperature for a pure substance as it passes through phase changes at constant pressure. You should examine *Figure 2.05* to see how the total specific enthalpy (h\*) can be calculated in several ways, by summing the enthalpy changes along different paths,[30] to give the same final result as that of *Equation 2.62*. In *Figure 2.05* you can also see that shifting the reference state (baseline) from the elements to the compound would give the specific enthalpy (h) of *Equation 2.65*.

---

[29] The values of "h\*" and "h" from *Equations 2.62* and *2.65* are not exact because they do not account for the effect of pressure on $C_p$ and $h_p$.

[30] Since enthalpy is a state function a change in enthalpy depends only on the initial and final states and is independent of the path between those states (this condition is the basis for Hess's Law).

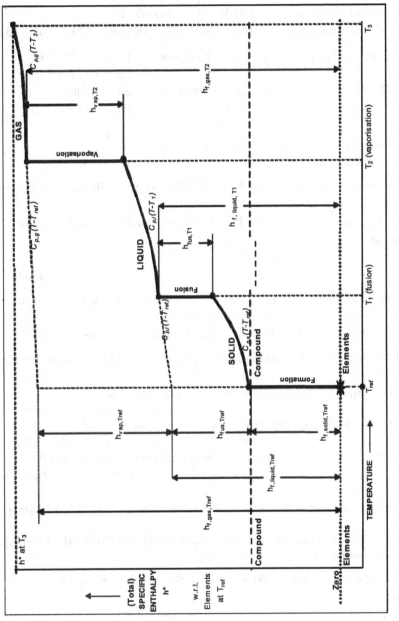

**Figure 2.05. Thermodynamic values ($h$, $h_f$, $h_v$, $h_m$, $C_p$, etc.) at constant pressure.**

*Graphical interpretation of the specific enthalpy ($h*$) for a single pure substance at a constant pressure.*

*Reference condition = elements in standard state at reference temperature $T_{ref}$*

*NOTE*: For simplicity the enthalpy of formation is shown here as positive, although in many compounds the heat of formation is negative.

***EXAMPLE 2.18    Finding the enthalpy of water by calculation and from the steam table.***

**Problem:**   Calculate the total specific enthalpy of pure water at:

      (A) 200 kPa(abs), 400 K     (B) 3000 kPa(abs), 600 K

      (i) By *Equation 2.62*.      (ii) From steam tables (*Table 2.20*, see page 75)

Data:  For $H_2O$ (l)    $h^\circ_{f,298k}$    = −285.8E3 kJ/kmol    $h_{v,298k}$ = 44E3 kJ.kmol$^{-1}$

                                                               (at 101.3 kPa(abs))

      For $H_2O$ (g)    $C_{p,m}$ ref 298 K = 35.2 kJ.kmol$^{-1}$.K$^{-1}$    36.3 kJ.kmol$^{-1}$.K$^{-1}$ 600 K

                                       400 K                    (at 101.3 kPa(abs))

**Solution:**

A.  Define the reference state for total specific enthalpy (h*)
    = elements ($H_2$ and $O_2$) at 101.3 kPa(abs), 298.15 K, i.e. standard state.

(i)  The vapour pressure of water at 400 K is 246 kPa(abs), so at 400 K, 200 kPa(abs)
    water is a gas.

*Equation 2.64:* $h_{f,Tref}$ (gas) = $h_{f,Tref}$ (liquid) + $h_{v,Tref}$ = −285.8E3 + 44E3 = −241.8E3 kJ.kmol$^{-1}$

*Equation 2.62:* $h^* = C_{p,m}(T − T_{ref}) + h_{f,Tref}$ = (35.2 kJ.kmol$^{-1}$.K$^{-1}$)(400 − 298)K
        + (−241.8E3 kJ.kmol$^{-1}$)

                                                         = **−238.2E3 kJ.kmol$^{-1}$**

The steam table gives more accurate values of the sensible and latent heats than do *Equations 2.62* and *2.64*, but these enthalpy values are for a reference condition of liquid water at 0.61 kPa(abs), 273.16 K and do not include the heat of formation of water.

(ii)  To find the enthalpy w.r.t. the elements at 101.3 kPa(abs), 298.15 K consider a process
     path as follows:

    1.  $H_2O$(l) is formed from $H_2$ and $O_2$ at 101.3 kPa(abs), 298.15 K.
                                      $\Delta h$ = −285.8E3 kJ.kmol$^{-1}$ (*Table 2.14*)
    2.  $H_2O$(l) at 101.3 kPa(abs), 298.15 K is cooled to 273.16K at 0.6 kPa(abs).
                                        $\Delta h$ = −1.86E3 kJ.kmol$^{-1}$ (*Table 2.20*)
    3.  $H_2O$(l) at 0.6 kPa(abs), 273.16 K is vaporised and heated to 400 K, 200 kPa(abs)
                                        $\Delta h$ = + 48.96E3 kJ.kmol$^{-1}$ (*Table 2.20*)

Then

Total specific enthalpy of water at 200 kPa(abs), 400 K = h* = Sum   = **−238.7E3 kJ.kmol$^{-1}$**

B.  Define the reference state for total enthalpy = elements ($H_2$ and $O_2$) at 101.3 kPa(abs),
    298.15 K, i.e. standard state.

(i)  The vapour pressure of water at 600 K is 12330 kPa(abs), so at 600K, 3000 kPs(abs)
    water is a gas.

*Equation 2.64:* $h_{f,Tref}$ (gas) = $h_{f,Tref}$ (liquid) + $h_{v,Tref}$ = −285.8E3 + 44E3 = −241.8E3 kJ.kmol$^{-1}$

*Equation 2.62:* h*    = $C_{p,m}(T − T_{ref}) + h_{f,Tref}$ = (36.3 kJ.kmol$^{-1}$.K$^{-1}$)(600 − 298)K
        + (−241.8E3 kJ.kmol$^{-1}$)

                                                          = **−230.8E3 kJ.kmol$^{-1}$**

(ii) To find the enthalpy w.r.t. the elements at 101.3 kPa(abs), 298.15 K consider a process path as follows:

1.  $H_2O(l)$ is formed from $H_2$ and $O_2$ at 101.3 kPa(abs), 298.15 K
$$\Delta h = -285.8E3 \text{ kJ.kmol}^{-1} \text{ } (Table \text{ } 2.13)$$

2.  $H_2O(l)$ at 101.3 kPa(abs), 298.15 K is cooled to 273.16 K at 0.6 kPa(abs).
$$\Delta h = -1.86E3 \text{ kJ.kmol}^{-1} \text{ } (Table \text{ } 2.20)$$

3.  $H_2O(l)$ at 0.6 kPa(abs), 273.16 K is vaporised and heated to 3000 kPa(abs), 600 K
$$\Delta h = +\underline{55.07E3} \text{ kJ.kmol}^{-1} \text{ } (Table \text{ } 2.20)$$

Then

Total specific enthalpy of water at 3000 kPa(abs), 600 K = h* = Sum = **−232.6E3 kJ/kmol**
The difference between the values in (i) and (ii) is mostly due to the variation of $C_p$ of the gas with pressure, which is not accounted for in (i).

## POTENTIAL ENERGY AND KINETIC ENERGY

Potential energy is the energy contained in a mass of material due to its position in a force field, such as an electric, gravitational or magnetic field. The most common of these is the gravitational potential energy, calculated by *Equation 2.66*.

$$E_p = (1E\text{-}3)mgL' \qquad\qquad\qquad\qquad Equation \text{ } 2.66$$

where:

| | | |
|---|---|---|
| $E_p$ | = potential energy | kJ |
| m | = mass of material | kg |
| g | = gravitational constant (= 9.81 m.s$^{-2}$ on Earth) | m.s$^{-2}$ |
| L' | = elevation of the mass above a reference level | m |

Kinetic energy is the energy contained in a mass due to its motion and is calculated by *Equation 2.67*.

$$E_k = (1E\text{-}3)0.5 \text{ } m\tilde{u}^2 \qquad\qquad\qquad\qquad Equation \text{ } 2.67$$

where:

| | | |
|---|---|---|
| $E_k$ | = kinetic energy | kJ |
| m | = mass of material | kg |
| $\tilde{u}$ | = velocity of the mass | m.s$^{-1}$ |

## SURFACE ENERGY

Due to the existence of unbalanced cohesive forces at phase boundaries, material within about one nanometer (1E-9 m) of an interface has a slightly higher specific internal energy than the same material in a bulk phase. The corresponding excess surface energy of a multi-phase system is calculated by *Equation 2.68*.

$$E_i = \gamma A_i \qquad\qquad\qquad\qquad Equation \text{ } 2.68$$

where:

$E_i$   = excess surface energy                                              kJ
$\gamma$   = interfacial tension                                              kN m$^{-1}$
$A_i$   = interfacial area (i.e. area of phase boundaries)                   m$^2$

For multi-phase systems in which the disperse phase exists as mono-disperse spherical bubbles, drops or particles the interfacial area is given by *Equation 2.69.*

$$A_i = (6e/d)V \qquad\qquad\qquad\qquad\qquad\qquad\qquad\qquad \textit{Equation 2.69}$$

where:

$A_i$   = interfacial area (i.e. area of phase boundaries)                   m$^2$
e   = volume fraction of the disperse phase                              –
d   = diameter of dispersed bubble, drop or particle                     m
V   = total volume of system                                             m$^3$

Surface energy is an "exotic" energy term that is negligible in most chemical processes (e.g. for water/air at 298 K, $\gamma = 72\text{E-}6$ kN.m$^{-1}$) but can be significant in systems with high specific surface area, such as in *colloidal dispersions*[31] (e.g. emulsions, foams, mists).

## MIXING

When two (or more) pure substances are mixed together several things can happen. If the substances are, and remain in, different phases the mixture is called a *dispersion.* Dispersions are classified into various types, such as:

- Aerosol          = solid or liquid in gas dispersion    (e.g. smog, smoke, mist)
- Conglomerate or = solid in solid dispersion            (e.g. nut chocolate, fibreglass)
  Composite
- Emulsion         = liquid in liquid dispersion          (e.g. mayonnaise, milk, oily water)
- Foam             = gas in liquid or solid dispersion    (e.g. soap lather, foamed rubber)
- Slurry or        = solid in liquid dispersion           (e.g. muddy water, yoghurt)
  Suspension

If the substances dissolve to form a single phase the mixture becomes a *solution.* Solutions are classified into three phases:

- Solid solution       (e.g. metal alloys)
- Liquid solution      (e.g. sea water, gasoline, whiskey)
- Gas solution — usually called a *gas mixture* — because all gases are mutually soluble (e.g. air).

The mixed substances may engage in chemical reaction(s) that transform their molecules. Chemical reactions are classified as:

---

[31] A colloidal dispersion is a multi-phase mixture in which the dispersed phase has bubbles, drops or particles with diameter in the range (approximately) 1 to 5000 nm. Most ions and molecules (except large polymers) have diameters below about 1 nm.

- Homogeneous reaction = a reaction that occurs in a single phase.
- Heterogeneous reaction = a reaction that occurs across a phase boundary.

In reality nearly all substances have a degree of mutual solubility that varies from very low (e.g. parts per million (wt) for mineral oil in water) to infinity (e.g. ethanol in water) — so the separate phases of a dispersion nearly always contain species from the other phases in solution. Often, the degree of mutual solubility is so low that it can be ignored, but sometimes even a very low solubility is important, for example, when dealing with catalyst poisons in chemical reactors or toxic materials in the environment.

You must be careful because the word "mixture" is ambiguous. In scientific and engineering jargon, unless a reaction is specified, a *mixture* implies that no chemical reactions occur between the components. The terms *ideal mixture* and *non-ideal mixture* generally mean multi-species, single phase solutions, whereas, for example, a *mixture* of oil and water usually means a two-phase oil-water *dispersion* and a *mixture* of sand and water is a two-phase *slurry* of sand in water.

### HEAT OF MIXING

When two (or more) pure substances are mixed to form a solution (see above), the resulting mixture can be classified as ideal or non-ideal.

- **Ideal mixture**   No interactions between molecules (or atoms) of the individual components.
- **Non-ideal mixture**   Interactions between molecules (or atoms) of the individual components.

The formation of a non-ideal mixture at constant pressure and temperature involves an input of an amount of heat called the *heat of mixing*. When the mixture is formed by dissolving a gas or a solid in a liquid the heat of mixing is called the *heat of solution*.

For substances that are soluble in water the heat of mixing is added to the heat of formation of the solute to give the heat of formation of the solute at *infinite dilution*,[32] which is shown in thermodynamic tables as $h_f$ (aq). *Table 2.17* lists values of the heat of mixing with corresponding heat of formation for sulphuric acid in water and *Table 2.18* gives some values of heat of solution in water. Heats of mixing are tabulated in sources such as *Refs. 1* and *2*.

**Table 2.17. Heat of mixing sulphuric acid with water at 298 K.**

| Amount of water | Heat of mixing | $h^{\circ}_{f,298K}$ |
|---|---|---|
| moles $H_2O$/mole $H_2SO_4$ | kJ.kmol$^{-1}$ $H_2SO_4$ | kJ.kmol$^{-1}$ $H_2SO_4$ |
| 0 | 0.00 | −811E3 |
| 1 | −28E3 | −839E3 |
| 2 | −42E3 | −853E3 |
| 10 | −67E3 | −878E3 |
| Infinity | −96E3 | −907E3* |

*Heat of formation of $H_2SO_4$ at infinite dilution*

**NOTE:** *The heat of formation of $H_2SO_4$(aq) does not include the heat of formation of water.*

---

[32] *Infinite dilution* means the presence of a large excess (hypothetically infinite) of solvent.

**Table 2.18.   Heat of solution of some solids in water at 291 K.**

| Substance | | NaOH | NaClO₃ | KNO₃ | Citric Acid | Sucrose | Quinone |
|---|---|---|---|---|---|---|---|
| Heat of solution at infinite dilution | kJ.kmol⁻¹ | −43E3 | +22E3 | +36E3 | +23E3 | +5.5E3 | +17E3 |

Values of the heat of mixing can be positive (+) or negative (−). A positive value means that the mixing process is endothermic, so if no heat is supplied the system temperature will drop as the mixture is formed. This occurs, for example, when $KNO_3$ dissolves in water. If the mixing process is exothermic ($H_{mix}$ is negative) and no heat is removed the system temperature will rise as the mixture is formed. This occurs when NaOH is dissolved or $H_2SO_4$ is mixed into water. The addition of water to cold sulphuric acid can release enough heat to vaporise the water and cause an explosion.

In some cases (e.g. sulphuric acid-water) the heat of mixing is substantial and should be included in the enthalpy values used for energy balance calculations. Enthalpy-concentration diagrams, which are described below, contain data for heat of mixing.

## THERMODYNAMIC TABLES AND REFERENCE STATES

Many sources, [e.g. *Refs. 1–5*] record the thermodynamic properties of pure substances. The information may be in the form of diagrams, tables or equations that correlate the properties with temperature and/or pressure. *Table 2.14* lists some general thermodynamic data for a few substances and *Table 2.20* presents a condensed form of a thermodynamic table for water commonly called the *steam table*. Other sources of thermodynamic data are enthalpy-concentration diagrams, enthalpy-pressure diagrams, enthalpy-entropy diagrams (called Mollier diagrams) and psychrometric charts.

The thermodynamic literature uses several different reference states that create ambiguity and confusion among students. *Table 2.19* lists common reference conditions with their usual context. When you use thermodynamic data for energy balance calculations you should always be aware of the reference conditions and ensure that they cancel from the balance equation (see Chapter 5).

**Table 2.19.   Conventional reference conditions for thermodynamic data.**

| Name | Condition | Application |
|---|---|---|
| Standard temperature and pressure [STP] | 101.3 kPa(abs), 273.15 K | Gas law |
| Normal temperature and pressure [NTP] | 101.3 kPa(abs), 293 K | Gas metering (used in industry) |
| Triple-point of water | Liquid water 0.611 kPa(abs), 273.16 K | Steam tables |
| Standard state (elements) | Element at 101.3* kPa(abs), 298.15 K | Heat of formation, reactive energy balances |
| Standard state (compounds) | Compound at 101.3* kPa(abs), 298.15 K | Heat of mixing, non-reactive energy balances |
| Absolute zero temperature | Compounds at 0 K | Mollier diagrams |
| Miscellaneous conditions | Compounds at 233 K | Refrigerants |
| Ice point | Compounds at 101.3 kPa(abs) 273.15K | Enthalpy-concentration diagrams. |

*Some sources use 1 bar = 100 kPa(abs) as the standard state pressure.*

The reference condition for both enthalpy and internal energy values in the steam table is the liquid compound water at its triple-point. The enthalpy and internal energy values in the steam table *do not include the heat of formation of water*.

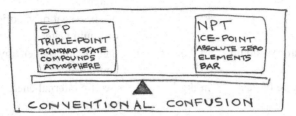

**Table 2.20.   Condensed version of the steam table. Properties of pure water.**

*Reference condition for "h" and "u" is liquid water at 0.611 kPa(abs), 273.16 K.*

| T | P | $v_l$ | $v_g$ | $h_l$ | $h_g$ | $h_v$ | $u_l$ | $u_g$ |
|---|---|---|---|---|---|---|---|---|
| K | kPa(abs) | m³.kg⁻¹ | m³.kg⁻¹ | kJ.kg⁻¹ | kJ.kg⁻¹ | kJ.kg⁻¹ | kJ.kg⁻¹ | kJ.kg⁻¹ |
| *Subcooled liquid water* | | | | | | | | |
| 273.15 | 5000 | 0.000998 | – | 5.1 | – | – | 0.07 | – |
| 273.15 | 50000 | 0.000977 | – | 49.4 | – | – | 0.52 | – |
| 400 | 500 | 0.001067 | – | 532.8 | – | – | 532.3 | – |
| 400 | 50000 | 0.001040 | – | 567.3 | – | – | 515.3 | – |
| 500 | 3000 | 0.001202 | – | 975.7 | – | – | 972.1 | – |
| 500 | 50000 | 0.001150 | – | 991.1 | – | – | 933.6 | – |
| 600 | 14000 | 0.001524 | – | 1499.4 | – | – | 1478.1 | – |
| 600 | 50000 | 0.001359 | – | 1454.0 | – | – | 1386.1 | – |
| *Saturated liquid-vapour water* | | | | | | | | |
| ᵀ273.16 | 0.611* | 0.001000 | 206.1 | 0.0 | 2500.9 | 2500.9 | 0.0 | 2375.7 |
| 300 | 3.536* | 0.001004 | 39.10 | 111.7 | 2550.1 | 2438.4 | 112.5 | 2412.5 |
| 350 | 41.66* | 0.001027 | 3.844 | 322.5 | 2637.9 | 2315.4 | 321.7 | 2478.4 |
| 400 | 245.6* | 0.001067 | 0.7308 | 532.7 | 2715.6 | 2182.9 | 532.6 | 2536.2 |
| 450 | 931.5* | 0.001123 | 0.2080 | 749.0 | 2774.9 | 2025.9 | 748.2 | 2579.8 |
| 500 | 2637* | 0.001202 | 0.07585 | 975.6 | 2803.1 | 1827.5 | 972.3 | 2601.7 |
| 550 | 6122* | 0.001322 | 0.03179 | 1219.9 | 2782.6 | 1562.7 | 1212.3 | 2589.3 |
| 600 | 12330* | 0.001540 | 0.01375 | 1504.6 | 2677.1 | 1172.5 | 1486.9 | 2511.6 |
| ᶜ647.3 | 21830* | 0.003155 | 0.003155 | 2098.8 | 2098.8 | 0.0 | 2037.3 | 2037.3 |
| *Superheated gas water* | | | | | | | | |
| 600 | 1.0 | – | 276.9 | – | 3130.1 | – | – | 2853.2 |
| 1100 | 1.0 | – | 507.7 | – | 4221.7 | – | – | 3714.1 |
| 600 | 101.3 | – | 2.727 | – | 3128.0 | – | – | 2851.6 |
| 1100 | 101.3 | – | 5.009 | – | 4221.3 | – | – | 3713.8 |
| 600 | 3000 | – | 0.08626 | – | 3059.6 | – | – | 2891.4 |
| 1100 | 3000 | – | 0.1684 | – | 4209.4 | – | – | 3704.7 |
| 600 | 10000 | – | 0.02008 | – | 2818.3 | – | – | 2617.5 |
| 1100 | 10000 | – | 0.04992 | – | 4180.3 | – | – | 3681.2 |

T = Triple-point of water (solid/liquid/vapour)      C = Critical point of water

*Vapour pressure over pure liquid water (with zero surface curvature.)[33]

Values between points in the full version of the steam table are usually obtained by linear interpolation, i.e. calculated assuming the points are connected by a straight line.

e.g. For $T_1 < T < T_2$, $h_T = h_{T1} + (h_{T2} - h_{T1})((T - T_1)/(T_2 - T_1))$

---

[33] The effect of curvature on vapour pressure becomes significant when the radius of curvature of the condensed phase is small (e.g. <1 micron cf. the Kelvin equation).

The columns of *Table 2.20* are interpreted as follows:

| | | | |
|---|---|---|---|
| T = temperature | K | $h_g$ = specific enthalpy of gas water | kJ.kg$^{-1}$ |
| P = pressure | kPa(abs) | $h_v = h_g - h_l$ = heat of vaporisation of liquid water | kJ.kg$^{-1}$ |
| $v_l$ = specific volume of liquid water | m$^3$.kg$^{-1}$ | $u_l$ = specific internal energy of liquid water | kJ.kg$^{-1}$ |
| $v_g$ = specific volume of gas water | m$^3$.kg$^{-1}$ | $u_g$= specific internal energy of gas water | kJ.kg$^{-1}$ |
| $h_l$ = specific enthalpy of liquid water | kJ.kg$^{-1}$ | | |

The line obtained by plotting P versus T from the liquid–vapour section of *Table 2.20* is the liquid/vapour (L/V) equilibrium line for water, shown in *Figure 2.02*. When the pressure and temperature of pure water are on the L/V line then water at this condition can exist as two phases (liquid and vapour) in equilibrium and is said to be *saturated*. The mass fraction of vapour in the saturated liquid/vapour mixture is called the *quality* of the mixture. The relation between quality and specific enthalpy (h), internal energy (u), volume (v) or entropy (s) of the two-phase mixture is:

$$Z_m = w_q Z_g + (1 - w_q)Z_l \hspace{4cm} Equation\ 2.70$$

where:

$Z_m$ = value of h, u, v or s for the mixture (mass basis)
$Z_l$ = value of h, u, v or s for the liquid phase alone
$Z_g$ = value of h, u, v or s for the gas (vapour) phase alone
$w_q$ = quality = mass fraction vapour in mixture

If the temperature is below the saturation temperature at a given pressure, then pure water can exist <u>at equilibrium</u> only as a liquid and is said to be *subcooled*. If the temperature exceeds the saturation temperature at a given pressure then pure water can exist <u>at equilibrium</u> only as a gas and is said to be *superheated*. As an exercise you should consider the L/V equilibrium of water described above in terms of the *phase rule*.

The thermodynamic properties of subcooled liquid water depend mainly on the temperature and are only slightly affected by pressure, whereas the properties of superheated gas water depend on both pressure and temperature. For most purposes the properties of subcooled liquid water can be taken as those of the liquid in the saturated liquid–vapour table, at the specified temperature. Thermodynamic data for superheated gas water are tabulated in a separate part of the steam table and should be used for calculations involving superheated conditions (see *Table 2.20*).

Note that enthalpy in the steam table is based on a reference condition of liquid water at its triple-point and does not include the heat of formation of water. This value of enthalpy is fine for energy balance calculations involving only physical changes to water (e.g. compression, expansion, evaporation or heating) but is not sufficient when water is involved in chemical reactions. If you use the steam table as a source of enthalpy data in energy balances where water is a reactant or a reaction product then you must add the heat of formation and consider the implications of the different reference conditions for $h_l$, $h_g$ and $h_f$; or otherwise account for the heat of reaction, which is usually a major term in the energy balance.[34]

### EXAMPLE 2.19   *Using the steam table to find the phase, enthalpy and volume of water.*

**Problem A:** 1 kg of pure water at 1000 kPa(abs), 400K is heated to 101.3 kPa (abs), 1100 K. Specify the phase at each condition then calculate the change of volume and enthalpy for the process.

*Solution*:
At 1000 kPa(abs), 400 K pure water is below its saturation temperature, i.e. subcooled liquid.
PHASE = liquid     $v_{in} = 0.001061$ m³/kg,     $h_{in} = 533.1$ kJ/kg [From the full steam table]
(NOTE: the similarity to saturated liquid at 400 K)

At 101.3 kPa(abs), 1100 K pure water is above its saturation temperature, i.e. superheated gas.
PHASE = gas,     $v_{fin} = 5.009$ m³ /kg,     $h_{fin} = 4221.3$ kJ/kg
Change of volume = final volume – initial volume     $= v_{fin} - v_{in} = 5.009 - 0.001061$
$$= \underline{\mathbf{5.008 \ m^3}}$$

Change in enthalpy = final enthalpy – initial enthalpy $= h_{fin} - h_{in} = 4221.3 - 533.1$
$$= \underline{\mathbf{3687.2 \ kJ}}$$

**Problem B:** 1 kg of pure water at 500 K has a volume of 0.06 m³ at equilibrium. Specify the phase and the pressure then calculate the quality and the specific enthalpy.

*Solution*:
From *Table 2.20* at 500 K:   $v_l \leq 0.001202$ m³/kg, and $v_g \geq 0.07585$ m³/kg
Since 500 K > 273.16 K (triple-point), then water with $v = 0.06$ m³/kg must be a mixture of liquid and gas.
PHASE = saturated (liquid + gas), Pressure = saturation pressure = 2637 kPa(abs)

Let $w_q$ = quality
Then by *Equation 2.70* $v = (1 - w_q)v_l + w_q v_g$     $0.06 = 0.001202 (1 - w_q) + 0.07585 w_q$
$$w_q = \underline{\mathbf{0.79}}$$
Specific enthalpy = $h = (1 - w_q)h_l + w_q h_g = (1 - 0.79) (975.6 \ kJ/kg) + (0.79) (2803.1 \ kJ/kg)$
$$h = \underline{\mathbf{2419.3 \ kJ.kg^{-1}}}$$

---

[34] See Chapter 5 for more on energy balance calculations.

***Problem C***: A gas mixture of 2 kg $H_2$ + 16 kg $O_2$ at 101.3 kPa(abs), 298.15 K is reacted to give to 1 kmol of water at 10000 kPa(abs),1100 K.

Calculate the change of enthalpy using *Equation 2.62* with *Tables 2.14* and *2.20*.

***Solution***:

Water is involved in a chemical reaction, so the steam table alone is not sufficient to calculate the enthalpy change. From *Table 2.20*, at 10000 kPa(abs), 1100 K water is a superheated gas.

Define the reference condition = elements ($H_2$ and $O_2$) in standard state at 101.3 kPa(abs), 298.15 K.

Initial specific enthalpy by *Equation 2.62*: $h_{in}^* = C_{p,m}(T - T_{ref}) + h_{f,T_{ref}}^o$

For $H_2$:   $h_{in,H_2}^*$ = (1 kmol) (28.8 kJ/(kmol.K) (298 − 298) K + 0 kJ.kmol$^{-1}$))      = 0 kJ

For $O_2$:   $h_{in,O_2}^*$ = (0.5 kmol)((29.4 kJ/(kmol.K))(298 − 298) K + 0 kJ.kmol$^{-1}$))   = 0 kJ

Total    = $h_{in}^*$                = $h_{in,H_2}^* + h_{in,O_2}^*$         = 0 + 0                 = 0 kJ

Final specific enthalpy = $h_{fin}^*$ = [$h_g$ at 10000 kPa,1100K − $h_l$ at 101.3 kPa,298K] + $h_f$, liquid water at 101.3 kPa,298K

Values of $h_g$ and $h_l$ are from the steam table (*Table 2.20*), whose enthalpy values include the sensible heat and latent heat effects but *exclude the heat of formation*. Value of $h_f$ comes from the thermodynamic property *Table 2.14*.

    $h_{fin}^*$   = 18 kg/kmol [4180.3 kJ/kg − 105 kJ/kg] + (−285.8E3 kJ/kmol)

         = <u>**−212.4E3 kJ/kmol $H_2O$**</u>


# THERMODYNAMIC DIAGRAMS

Thermodynamic diagrams present complex thermodynamic data for materials over wide ranges of conditions. Such charts are used to simplify difficult thermodynamic calculations. The list of thermodynamic diagrams includes:

- Enthalpy-concentration diagrams.
- Psychrometric diagrams (properties of gas/vapour mixtures)
- Enthalpy-pressure diagrams
- Entropy-temperature diagrams
- Enthalpy-entropy diagrams (called Mollier diagrams)

As a rule the reference conditions for thermodynamic charts are the pure compounds at some specified pressure and temperature. Thus the enthalpies found from these charts *do not include heat of formation*.

### ENTHALPY-CONCENTRATION DIAGRAMS

Enthalpy-concentration diagrams show thermodynamic data for multi-component non-ideal systems. Such systems are fairly common in practical processes and cause difficulties in calculations because the properties of the mixture are not a simple combination of the properties of the pure components. Enthalpy–concentration diagrams give a simple method to account for the *heat of mixing* that results when two pure components are combined to form a non-ideal mixture.

*Figure 2.06* shows the enthalpy–concentration diagram for sulphuric acid–water mixtures. The reference condition for "enthalpy" in enthalpy–concentration

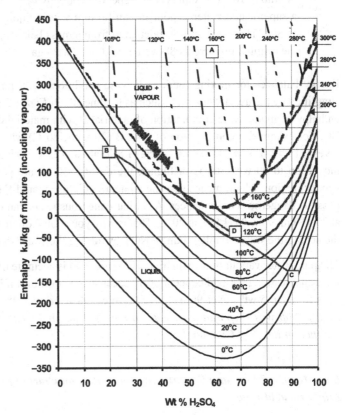

**Figure 2.06. Enthalpy–concentration diagram for H$_2$SO$_4$–H$_2$O mixtures at 101.3 kPa(abs).**

and similar mixing diagrams is the pure component compounds at a specified pressure and temperature. This mixture enthalpy *does not include the heat of formation for any of the components*. If you use a mixture enthalpy (from an enthalpy–concentration diagram) in an energy balance where the components are involved in chemical reactions you should add appropriate heats of formation, and beware of different reference conditions. *Example 2.20* demonstrates a basic mixing calculation using *Figure 2.06*.

*NOTE*: The gas phase of the [liquid + vapour] region in *Figure 2.06* is taken as pure
superheated water vapour. For example, a mixture with 60 wt% $H_2SO_4$ at 101.3
kPa(abs), 160°C (point A) has a specific enthalpy of ca. 375 kJ/kg and consists of
a superheated water vapour in equilibrium with a liquid solution of ca. 68 wt %
$H_2SO_4$. The phase split is calculated from a mass balance, which in this case gives
the original mixture as 88 wt% liquid + 12 wt% gas.

Energy balances for mixing (under adiabatic conditions with no chemical reactions) are
simplified by the use of "tie-lines" such as the line BC in *Figure 2.06*. The composition and
enthalpy of an adiabatic combination of material B with material C lies on the straight line
BC. For example, a combination of 1 kg B (20 wt%, 80°C) with 2 kg C (90 wt%, 20°C)
gives 3 kg of mixture D with 67 wt% $H_2SO_4$ at ca. 125°C. Observe that if point C were
moved up to 90 wt%, 60°C or higher, the tie-line BC would pass through the [liquid +
vapour] region, which means that the adiabatic mixture would flash into two phases with
possible dangerous consequences. A similar dangerous result can occur from mixing water
at 20°C with 98 wt% $H_2SO_4$ at 20°C. That is why you should dilute sulphuric acid by adding
the acid to the water (while mixing), NOT by adding the water to the acid.

The pure water vapour approximation of *Figure 2.06* is accurate below 90 wt% $H_2SO_4$, but
less reliable at higher acid concentration because the equilibrium $H_2SO_4$ content of the gas
rises from ca. 5 wt% over 90 wt% $H_2SO_4$ to ca. 98 wt% over 98 wt% $H_2SO_4$ (see *Ref. 1*).
*Figure 2.06 was prepared from data in the International Critical Tables by the method
outlined in Ref. 5 of Chapter 1.*

### EXAMPLE 2.20    Using the enthalpy-concentration chart to find heat load to process sulphuric acid-water.

*Problem*: Use *Figure 2.06* to calculate the net change in enthalpy that occurs by (mixing +
cooling) in the liquid phase 1 kg of 80 wt% $H_2SO_4$ at 101.3 kPa(abs), 60°C with
1 kg of water at 101.3 kPa(abs), 80°C to form 2 kg of 40 wt% $H_2SO_4$ at 20°C,
101.3 kPa(abs).

*Solution*: From *Figure 2.06*, reference condition is the pure liquid components at
273.15 K, zero vapour pressure.

Initial enthalpy = $H_{in}$ = enthalpy of 1 kg of 80 wt% $H_2SO_4$ at
60°C + 1 kg of water at 80°C
= (1 kg )(–160 kJ/kg) + (1 kg)(334 kJ/kg)     = + 174 kJ

Final enthalpy = $H_{fin}$ = enthalpy of 2 kg 40 wt% $H_2SO_4$ at
20°C = (2 kg) (–200 kJ/kg)                          = –400 kJ

Change in enthalpy  = $H_{fin} - H_{in}$ = –400 – (174)          = **–574 kJ**

Note that the tie-line for the initial materials shows that under adiabatic conditions the
mixture would reach about 110°C (i.e. nearly boiling).

## PSYCHROMETRIC CHARTS

Psychrometric charts contain thermodynamic data for gas/vapour mixtures at a fixed pressure. The most common pyschrometric chart is that for water vapour in air (i.e. humid air), which is used, for example, in the design of air conditioning systems and water-cooling towers. *Figure 2.07* shows a psychrometric chart for water–air mixtures.

*Figure 2.07* shows the following data for water–vapour–air mixtures at the standard pressure of 101.3 kPa(abs):

- dry-bulb temperature           °C
- wet-bulb temperature           °C
- relative humidity               %
- moisture content of humid air   kg/kg dry air
- specific volume of humid air    m³/kg dry air
- enthalpy of humid air           kJ/kg dry air, ref. liquid water and gaseous air at 101.3 kPa(abs), 273.15 K

*Dry-bulb temperature* ($T_{DB}$) is the ordinary temperature as measured by a dry thermometer.

*Wet-bulb temperature* ($T_{WB}$) is the temperature measured with a thermometer whose bulb is covered by a water-wet wick and moved rapidly through the air (>5 m/s). This procedure establishes a balance between heat transfer from air to bulb and mass transfer of water from bulb to air (see *Refs. 2 and 4*). Evaporation of water into the air lowers the wet-bulb below the dry-bulb temperature to a degree that depends on the relative humidity of the air. If the relative humidity is 100% then the wet-bulb temperature equals the dry-bulb temperature.

*Relative humidity* is defined as:

$$\mathbf{rh = p_w/p_w^*}$$                                             *Equation 2.71*

where:

   rh  = relative humidity                                                               –
   $p_w$  = partial pressure of water vapour in the water–vapour–air mixture    kPa(abs)
   $p_w^*$  = saturated vapour pressure of water at the dry-bulb temperature    kPa(abs)

Note that the enthalpy values in *Figure 2.07* are per kg of dry air, plus the *enthalpy* is defined relative to liquid water and gaseous air at 101.3 kPa(abs), 0°C (273.15 K), and *does not include the heat of formation* of water.

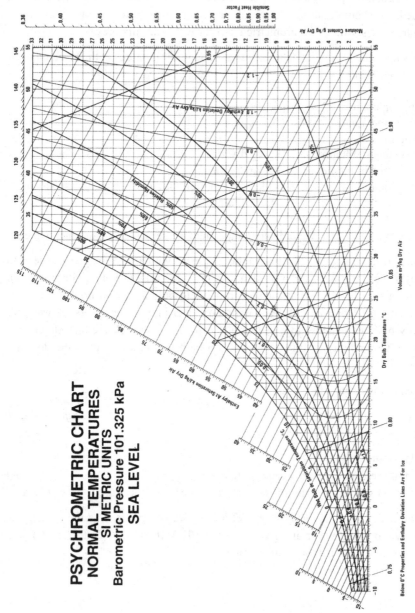

**Figure 2.07. Psychrometric chart for H₂O vapour–Air.**
Reference states: $H_2O$ (l), Dry Air (g) at 101.3 kPa(abs), 273.15 K.
[Reprinted with permission of Carrier Corporation]

**EXAMPLE 2.21   *Using the psychrometric chart to find properties of humid air.***

5 kg of humid air with dry-bulb temperature = 30°C and wet-bulb temperature = 20°C is cooled to 10°C.

**Problem**: Use the psychrometric chart to find:
   A.   The relative humidity of the initial humid air at 101.3 kPa(abs).
   B.   The moisture content of the initial humid air.
   C.   The enthalpy of the initial humid air.
   D.   The dew-point temperature of the air at 101.3 kPa(abs).
   E.   The mass of water condensed from the air when cooled from 30 to 10°C at 101.3 kPa(abs).

**Solution**:
   A.   Locate the intersection of the initial dry-bulb (vertical) and wet-bulb (diagonal) temperature lines. Interpolate the relative humidity (curved) lines to get:
        Initial rh ≈ **40%**

   B.   Locate the intersection of the initial dry-bulb (vertical) and wet-bulb (diagonal) temperature lines. Follow the constant moisture content (horizontal) line across to the moisture content scale to get:
        Initial moisture content = **0.0106 kg water/kg dry air**

   C.   Locate the intersection of the initial dry-bulb (vertical) and wet-bulb (diagonal) temperature lines. Follow the diagonal line (constant wet-bulb temperature) back to the enthalpy scale to get:
        Enthalpy at saturation = 57.6 kJ/kg dry air
        Interpolate the enthalpy deviation lines at the dry-bulb/wet-bulb point to get:
        Enthalpy deviation = –0.35 kJ/kg dry air
        Add the enthalpy deviation to the enthalpy at saturation to get:
        Enthalpy = –0.35 + 57.6 ≈ **57.3 kJ/kg dry air**

   D.   Locate the intersection of the initial dry-bulb (vertical) and wet-bulb (diagonal) temperature lines. Follow the constant moisture content (horizontal) line back to the saturation curve to get: $T_D$ = **14.5°C**

   E.   The final air temperature (10°C) is lower than the initial dew point (14.5°C), so at 10°C the air will be saturated. In the final condition: Dry-bulb temperature = wet-bulb temperature = 10°C.
        Locate the final condition ($T_{DB}$ = $T_{WB}$ = 10°C) and follow the constant moisture content line to get:
        Final moisture content = 0.0076 kg water/kg dry air
        The mass of dry air is fixed, while the mass of water it contains is decreased on cooling from 30 to 10°C.
        Mass of dry air = (5 kg humid air)(1/(1 + 0.0106)) kg dry
                                    air/kg humid air                      = 4.948 kg dry air
        Initial mass of water in air = (4.948 kg dry air)(0.0106 kg
                                                   water/kg dry air)         = 0.0524 kg water

Final mass of water in air = (4.948 kg dry air)(0.0076 kg
water/kg dry air)                     = 0.0376 kg water
Mass of water condensed = Initial mass − final mass     **= 0.0148 kg water**

## SUMMARY

[1] Keep your calculations true by ensuring equations have consistent units, check that the units "balance" at each step of the calculation and report results with their units, rounded to an appropriate number of significant figures.

[2] Pressure (P) and temperature (T) are usually measured as relative values (e.g. kPa(gauge) and °C in the SI system), whereas most thermodynamic calculations (e.g. gas law) require P and T as absolute values (e.g. kPa(abs) and K in the SI system). The relations between these values (assuming the standard atmospheric pressure) are:

|  | **SI System** | **Imperial-American System** |
|---|---|---|
| Pressure: | – | – |
| Super-atmospheric | kPa(abs) = kPa(g) + 101.3 | psia = psig + 14.7 |
| Sub-atmospheric | kPa(abs) = 101.3 – kPa(vac) | psia = 14.7 – psi(vac) |
| Temperature: | K = °C + 273.1 | °R = °F + 459.6 [°R = 1.8K] |

[3] The physical condition of any material can be described by a range of properties, including the phase, density, specific volume, molar volume, specific gravity, concentration (by mass, moles or volume), etc. You should know what these measures mean and have an idea of the ranges of their values for common materials.

[4] The ideal gas equation ($pV = nRT$) is an equation of state that applies with sufficient accuracy for most purposes to any gas when the reduced pressure $P_r = P/P_c \leq 10$ and reduced temperature $T_r = T/T_c \geq 2$.

For gases at conditions near the critical point ($P_c, T_c$) the ideal gas equation is modified to $pV = znRT$, where the gas compressibility factor "z" ranges from about 0.2 to 4 depending on the proximity to the critical state, at which: $P_r = T_r = 1$.

Common gases such $H_2$, $N_2$, $O_2$, $CH_4$, CO and NO have critical temperature below 200 K and can (for most purposes) be treated as ideal gases when the partial pressure is below 1E4 kPa(abs) and temperature above 273 K. Other gases such as $H_2O$, $CO_2$, $C_4H_{10}$, $N_2O$, $NO_2$, $SO_2$ and $SO_3$ have critical temperatures above 300K and may be condensed near ambient conditions but have critical pressure above 4E3 kPa(abs) and can usually be treated as ideal gases at partial pressure below the vapour pressure and temperature above 273 K.

**The ideal gas law does not apply to liquids or solids.**

[5] For liquids and solids the equation of state involves relatively small effects of pressure and temperature that are expressed as the coefficients of compressibility ($\delta = -1/v_o \, (dv/dP)$) and of thermal expansion ($\beta = 1/v_o \, (dv/dT)$), whose respective values are typically below 1E-6 $kPa^{-1}$ and 1E-3 $K^{-1}$.

[6] Heat capacity is specified at constant pressure ($C_p$) or at constant volume ($C_v$).

Liquids and solids:   $C_p \approx C_v$
both values are nearly independent of pressure and increase with temperature.

Ideal gases:           $C_p = C_v + R$
both values are independent of pressure and increase with temperature.

Non-ideal gases:    $C_p$ and $C_v$ both depend on pressure and temperature and become infinite at the critical state.

The heat capacity of a material depends strongly on its phase and is not defined over a phase change.

[7] Specific enthalpy and internal energy are state functions that are used to measure the energy content of materials relative to an arbitrarily specified reference condition. Two common reference conditions are:

A.   The *elements* of the material in their standards states at 101.3 kPa(abs), 298.15 K.
B.   The *compounds* of the material at a specified condition, such as liquid water at its triple point (0.61 kPa(abs), 273.16 K), as used in the steam table.

[8] The enthalpy (h*) defined in 7(A) contains terms for sensible heat, latent heat and heat of formation and is the most useful in energy balance calculations for processes with chemical reactions. The enthalpy (h) defined in 7(B), which does not include the heat of formation, is the value reported in most diagrams and tables of thermodynamic properties and commonly used in energy balance calculations on processes that do not involve chemical reactions.

[9] The phase rule [$F^\# = C^\# - \Pi^\# + 2$] defines a thermodynamic constraint on the number of intensive variables ($F^\#$), needed to fully-specify the state of each phase in a system, the number of components in the system ($C^\#$) and the number of phases in the system ($\Pi^\#$) at equilibrium. For non-reactive systems the number of components is the number of chemical species in the system, but in reactive systems becomes the number of chemical species minus the number of independent reactions.

[10] Mixtures can take the form of multi-phase or single-phase systems. Multiphase mixtures are called dispersions and are exemplified by systems such as emulsions (L/L), foams (G/L) and slurries (S/L) in which the disperse phase consists of drops, bubbles or particles with diameter exceeding about 1 nm. Single-phase mixtures are called solutions and are classified as either ideal or non-ideal mixtures (a.k.a. solutions) depending on the degree of interaction between the constituent molecules.

[11] The formation of non-ideal mixtures (a.k.a. solutions) involves an enthalpy change called the "heat of mixing" that may be positive or negative and must be added to the pure component enthalpies to get the enthalpy of a non-ideal mixture. In many cases the heat of mixing is a relatively small effect that can be ignored in energy balances, but in some cases (such as sulphuric acid/water) the heat of mixing is substantial and should be included in energy balance calculations.

[12] The physical properties of substances depend strongly on phase and in most process calculations it is necessary to know the phases of the substances involved. For single pure substances, the phase can be obtained from thermodynamic diagrams or tables, or by calculations with equilibrium relations such as the Antoine equation for vapour pressure. The phase relations for mixtures of substances (i.e. in multi-species, multi-phase systems) may be simple or complex, depending on the degree of non-ideality of the mixture. Vapour/liquid equilibrium in multi-phase (G/L) systems is given by Raoult's law for ideal mixtures and can be approximated by Henry's law for non-ideal mixtures with low solute concentrations (i.e. < 5 mole%). Henry's law is useful particularly to calculate solubilities on non-condensable gases in liquids.

Equilibrium in other multi-phase systems is characterised by the adsorption isotherm (G/S systems), the distribution coefficient (L/L systems) and the solubility (L/S systems), all of which are strongly temperature dependent and usually a non-linear function of composition.

[13] The calculation of amounts of materials consumed and generated in a chemical reaction begins with a balanced stoichiometric equation:

$$\alpha A + \beta B \rightarrow \chi C + \delta D$$

The numbers $\alpha$, $\beta$, $\chi$ and $\delta$ are the stoichiometric coefficients of the reaction, in which:

$\alpha$ moles A + $\beta$ moles B    react to generate $\chi$ moles C + $\delta$ moles D

The conservation of mass dictates that for any balanced chemical reaction:

$$\Sigma \, [v(j)M(j)] = 0$$

where:

  $M(j)$ = molar mass
  $v(j)$  = stoichiometric coefficient (−ve for reactants, +ve for products)

[14] The progress of chemical reactions is measured by conversion, extent of reaction, selectivity and yield. Conversion and selectivity are dimensionless ratios that measure respectively; the fraction of a reactant consumed and the fraction of a reactant converted to a specified product. The yield of a specific product is then the reactant conversion multiplied by the corresponding selectivity. The extent of reaction is the moles of a species consumed or generated by the reaction, divided by its stoichiometric coefficient. The extent of reaction has dimensions of moles and has the same value for all species in a given reaction.

For a single reaction the conversion and the extent of reaction are essentially measures of the same quantity and either may be used in material balance calculations involving chemical reactions.

[15] Reversible chemical reactions tend to an equilibrium condition defined by the equilibrium conversion ($X_{eq}$). All reactions occur at finite net rates that decrease to zero at equilibrium, so the equilibrium conversion can never be reached in a real reaction. The actual conversion ($X_{act}$) depends on the reaction rate and residence time and is always below the equilibrium conversion. The ratio $X_{act}/X_{eq}$ can exceed 0.99 in many practical processes, but in other equally important processes $X_{act}/X_{eq}$ may be as low as 0.05.

[16] The thermodynamic data needed for process calculations comes from equations or from thermodynamic diagrams and tables. The equations used to estimate specific enthalpy are:

$$h^* \approx \int_{T_{ref}}^{T} C_p.dT + h^{\circ}_{f,T_{ref}} \approx C_{p,m}(T - T_{ref}) + h^{\circ}_{f,T_{ref}} \qquad \text{(see } Equation\ 2.62\text{)}$$

[Element reference state]            [Respect the phase]

or

$$h \approx \int_{T_{ref}}^{T} C_p.dT + h_{p,T_{ref}} \approx C_{p,m}(T - T_{ref}) + h_{p,T_{ref}} \qquad \text{(see } Equation\ 2.65\text{)}$$

[Compound reference state]           [Respect the phase]

where:

| | | |
|---|---|---|
| $h^*$ | = specific enthalpy of the substance at temperature T in phase $\Pi$ w.r.t. *elements* at the standard state | kJ.kmol$^{-1}$ |
| $h$ | = specific enthalpy of the substance at temperature T in phase $\Pi$ w.r.t. *compounds* at the reference state | kJ.kmol$^{-1}$ |
| $C_p$ | = heat capacity at constant pressure of the substance in phase $\Pi$ | kJ.kmol$^{-1}$.K$^{-1}$ |
| $C_{p,m}$ | = mean heat capacity at constant pressure, w.r.t. $T_{ref}$ in phase $\Pi$ | kJ.kmol$^{-1}$.K$^{-1}$ |
| $h^{\circ}_{f,T_{ref}}$ | = heat (enthalpy) of formation of the substance at $T_{ref}$ in phase $\Pi$ | kJ.kmol$^{-1}$ |
| $h_{p,Tref}$ | = latent heat of any phase changes required to convert the substance from its reference state at $T_{ref}$ to the phase $\Pi$ at $T_{ref}$ | kJ.kmol$^{-1}$ |
| $T$ | = temperature | K |
| $T_{ref}$ | = reference temperature | K |

More accurate values of specific enthalpy (h) are obtained from thermodynamic diagrams and tables. Such values are normally based on the *compound* reference state and thus do not include heats of formation. If the *compound* based enthalpy values are used in energy balances with chemical reactions, it is necessary to add the heats of formation to the enthalpies or to otherwise account for the heat of reaction (see Chapter 5).

## FURTHER READING[35]

[1]   T. Perry and D.Green, *Chemical Engineers' Handbook*, McGraw-Hill, 2018.

[2]   R. Rumble, *Handbook of Chemistry and Physics*, CRC Press, 2021.

[3]   C. L. Yaws, *Handbook of Physical Properties of Hydrocarbons and Chemicals*, Gulf, 2015.

[4]   B. Poling, J. Prausnitz and J. O'Connell, *Properties of Gases and Liquids*, McGraw-Hill, 2001.

[5]   O. Hougen, K. Watson and R. Ragatz, *Principles and Calculations in Chemical Engineering*, Prentice Hall, 1954.

[6]   S. Sandler, *Chemical, Biochemical and Engineering Thermodynamics*, Wiley, 2017.

[7]   J. Noggle, *Physical Chemistry*, Prentice Hall, 1997.

[8]   K. Solen and J. Harb, *Introduction to Chemical Engineering*, Wiley, 2011.

[9]   J. Verret, R. Barghout and R. Qiao, *Foundations of Chemical and Biological Engineering*, Victoria, BC Campus, 2020. (Creative Commons).

[10]  R. Felder, R. Rousseau and L. Bullard, *Elementary Principles of Chemical Processes*, Wiley, 2018.

[11]  J. Mihelcic and J. Zimmerman, *Environmental Engineering*, Wiley, 2021.

[12]  M. Doran, *Bioprocess Engineering Principles*, Academic Press, 2013.

[13]  A. Morris, *Handbook of Material and Energy Balance Calculations for Materials Processing*, Wiley, 2011.

[14]  S. Ashrafizadeh and Z. Tan, *Mass and Energy Balances. Basic Principles for Calculation, Design and Optimization of Macro/Nano Systems*, Springer, 2018.

[15]  N. Gjasen and R. Henda, *Principles of Chemical Engineering Processes. Material and Energy Balances*, CRC Press, 2015.

[16]  S. Skogestad, *Chemical and Energy Process Engineering*, CRC Press, 2009.

---

[35] Much of the thermodynamic data mentioned here can be found on-line.

# CHAPTER THREE

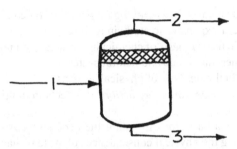

| COMPONENT | M | STREAM (kmol/h) | | |
|---|---|---|---|---|
| | kg/kmol | 1 | 2 | 3 |
| [A] C$_5$H$_{12}$ | 72 | 80 | 66·4 | 13·6 |
| [B] C$_6$H$_5$CH$_3$ | 92 | 40 | 18·4 | 21·6 |
| TOTAL | kmol/h | 120 | 84·8 | 35·2 |
| TOTAL | kg/h | 9440 | 6475 | 2965 |
| PHASE | — | L | G | L |
| PRESSURE | kPa(abs) | 6000 | 800 | 800 |
| TEMPERATURE | K | 510 | 424 | 424 |
| VOLUME | m$^3$/h | 15·0 | 374 | 4·6 |
| ENTHALPY | kJ/h | -870E6 | -7.95E6 | -1.28E6 |

# MATERIAL AND ENERGY BALANCES
# IN PROCESS ENGINEERING

This chapter explains the significance of material and energy balances in process engineering and introduces the concepts of flowsheets, process units and stream tables, which are used throughout subsequent chapters.

## SIGNIFICANCE OF MATERIAL AND ENERGY BALANCES

The Material and Energy (M&E) balance is a primary tool used in the design of industrial chemical processes. Material and energy balance calculations are based on the general balance equation and have several features that make them useful both to process engineers and to others concerned with understanding the behaviour of physical systems, as follows:

- M&E balances provide a basis for *modelling* on paper (or by computer) systems that would be difficult and/or expensive to study in reality.[1]
- M&E balances assist in the *synthesis* of chemical processes and in developing *conceptual paradigms* for other man-made and natural systems.
- M&E balances guide in the *analysis* of physical systems.
- M&E balances result in *quantitative predictions* of effects in multi-variable, interacting, non-linear systems.
- M&E balances provide a basis for estimating the *economic costs and benefits* of a project and for assessing the physical consequences of system change.

As well, material and energy balances can play an important role in assessing the "triple bottom line", of the industrial economy. The "triple bottom line" is a method of accounting for ecological, economic and social factors in the sustainable development of our civilisation.

## FLOWSHEETS

Process engineers usually begin modelling a system by describing it in a flowsheet. Flowsheets may have various levels of complexity, ranging from a simple conceptual flowsheet to a detailed P&ID (process/piping and instrumentation diagram) that is the basis for plant construction. This text is concerned only with conceptual flowsheets.

In a conceptual flowsheet the full system is broken down into a set of singular *process units* interconnected by *streams* that define the paths of material flow between the process

---

[1] The process unit models used in this text are "lumped" models, within which the spatial distribution of matter and energy is not resolved. In other words, each process unit is treated as a "black box". More sophisticated M&E balances, with partial differential equations, can be used to develop "distributed" models that map the distribution of matter and energy inside the process unit.

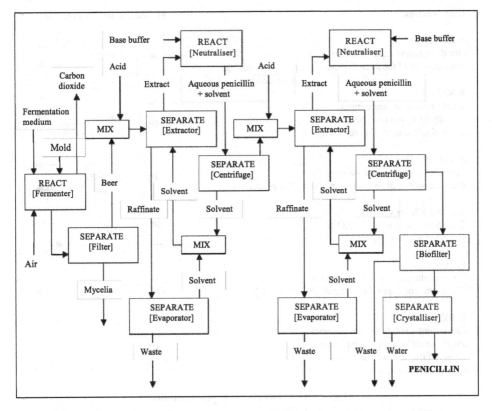

**Figure 3.01.  Conceptual flowsheet for the industrial production of penicillin.**

units. *Figure 3.01* shows a typical conceptual flowsheet for an industrial process used to make penicillin (see *Refs. 2–4*).

## PROCESS UNITS

In process engineering a *process unit* is a piece of hardware that operates on one or more input(s) to produce one or more modified output(s). Actual process units can range from a simple tank to massive and complex pieces of equipment such as fractionation towers, catalytic crackers and multiple effect evaporators (see *Refs. 3–5, 9*).

All process flowsheets can be conceptually simplified to a combination of unit(s), each of which performs one of the six basic functions shown in *Figure 3.02*. The basic unit functions shown in *Figure 3.02* are the building blocks of all conceptual flowsheets and are used throughout this text to model chemical processes. You should study *Figure 3.02*

| Unit function | Flowsheet Representation |
|---|---|

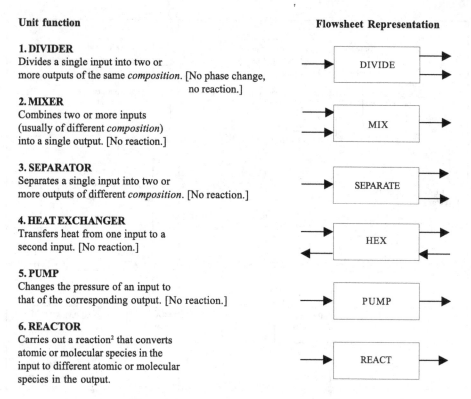

**1. DIVIDER**
Divides a single input into two or
more outputs of the same *composition*. [No phase change,
no reaction.]

**2. MIXER**
Combines two or more inputs
(usually of different *composition*)
into a single output. [No reaction.]

**3. SEPARATOR**
Separates a single input into two or
more outputs of different *composition*. [No reaction.]

**4. HEAT EXCHANGER**
Transfers heat from one input to a
second input. [No reaction.]

**5. PUMP**
Changes the pressure of an input to
that of the corresponding output. [No reaction.]

**6. REACTOR**
Carries out a reaction[2] that converts
atomic or molecular species in the
input to different atomic or molecular
species in the output.

Figure 3.02.    The six basic process units.[2]

carefully, since your understanding of these six basic unit functions will be critical to your ability to formulate and solve M&E balance problems.

In reality a single piece of equipment may perform several basic functions. For example, a "pump" can be used both to pump and mix fluids at the same time. A "separator" such as an evaporator can include a heat exchanger. A "reactor" can be operated to simultaneously mix, heat, react and separate materials. Such complex multi-function units are modelled in commercial process simulators as "user defined" blocks. Nevertheless, for a conceptual flowsheet it is preferable to deconstruct the process into its basic units and to isolate the functions from each other. The advantage of deconstructing a process is that a single algorithm can model each process unit. In this way a complex system can be broken down and analysed as a combination of simple elements.

*Figure 3.02* introduces the terms *composition*, *component*, *species*, and *phase*. These terms have particular meanings in the description of M&E balances that should be adhered to avoid ambiguity.

---

[2] A reaction can be a nuclear reaction that changes the configuration of atomic nuclei or a chemical reaction that changes the configuration of molecules. Only chemical reactions are considered in this text. Chemical reactions may be biochemical, electrochemical, photochemical or thermochemical reactions.

*Composition* = the fractional amount of each *component* in a mixture (e.g. mass fraction or mole fraction)

*Component*[3] = a constituent of a mixture identified as a *phase* or as a *species*

*Species*   = a constituent of a mixture identified by a specific atomic or molecular configuration

*Phase*    = a constituent of a mixture identified as a solid, a liquid or a gas

One of the main difficulties in learning a technical subject such as M&E balances is in the potential for ambiguity incurred by the use of common words with specialised meanings. The specialised meanings are necessary to avoid ambiguity. You should be aware of the specialised meanings of words in your field and take care to "AAA", that is: "Always Avoid Ambiguity".

**Dividers** (and occasionally separators) are sometimes called "splitters". To prevent confusion of their different functions neither dividers nor separators are referred to as splitters in this text.

In a divider all inputs and outputs have the same composition. For a continuous process a divider works like a tee in a pipe and merely sets the flow rates of the output streams in a desired ratio, keeping all their compositions the same as that of the input stream. A divider does not change phases, separate phases or allow reactions to occur.

A **mixer** adds inputs together to produce a single output. A simple mixer may consist of a tank fitted with a stirrer. A mixer does not allow reactions to occur.

A **separator** performs a more complex function than a divider or a mixer. A separator takes a *multi-component* input and manipulates it to produce a set of outputs of different composition. Some examples of separators are: filters, evaporators and fractionating columns. A separator can change phases and/or separate phases but does not allow reactions to occur.

A **heat exchanger** transfers heat between two inputs that are at different temperatures. Heat is transferred from the input at high temperature to the input at low temperature and there must always be a finite difference in temperature between the two streams, such that $T_{hot} > T_{cold}$. A heat exchanger may or may not cause a change in phase in one or both inputs, but does not allow a reaction. There are many types of heat exchangers, in categories such as direct contact, indirect contact, co-current, counter-current, concentric tube, shell and tube, plate, etc. (*Refs. 4, 5, 9*).

A **pump** makes the pressure of an output fluid higher or lower than that of the corresponding input fluid. Pumps that handle liquids or slurries are usually referred to as pumps while those that handle gases are usually called fans, blowers or compressors. Pumps that reduce

---

[3] "Component" has a strict thermodynamic definition, i.e. a species in a mixture whose amount can be independently varied. However the word "component" is used more broadly in some texts to mean either an individual species or a single phase mixture of invariant composition, such as air or sea water.

gas pressure are called "vacuum pumps". Pumps generally convert mechanical energy into fluid flow energy. Working in reverse, pumps, which convert fluid flow energy into mechanical energy, are called turbines or expansion engines. Again there are many types of pumps, in categories such as centrifugal, diaphragm, piston, gear, etc. (see *Refs. 4, 5, 9*).

A **reactor** performs the most complex function of the set of basic process units. Reactors fall into two categories:

- Nuclear reactors, which change the configuration of atomic nuclei (e.g. $U_{235}$ to $U_{236}$ then to $Ba_{144}$ and $Kr_{89}$ by nuclear fission).
- Chemical reactors, which change the configuration of chemical species[4] (e.g. $C + O_2$ to $CO_2$).

The chemical reactors treated in this text and are defined broadly to embrace any type of reaction that changes chemical species — including biochemical, electrochemical, photochemical, plasma and thermochemical reactions. Most reactions with which you are probably familiar are thermochemical reactions, such as:

- Neutralisation (acid/base) reaction    $HCl + NaOH \rightarrow NaCl + H_2O$    *Reaction 3.01*
- Combustion reaction    $CH_4 + 2O_2 \rightarrow CO_2 + 2H_2O$    *Reaction 3.02*
- Organic chlorination reaction    $C_2H_4 + 2Cl_2 \rightarrow C_2H_2Cl_2 + 2HCl$    *Reaction 3.03*

Note that in all chemical reactions the molecules of the reactant species (left hand side) are *transformed* to the products species (right hand side). This transformation of chemical species is the essential feature of a chemical reaction, in which the reactants are *consumed* and the products are *generated*. The properties of the reactants are usually totally different from the properties of the products. For example, in *Reaction 3.01* the reactants are hydrochloric acid (HCl) and sodium hydroxide (NaOH), both corrosive and toxic chemicals, whereas the products are sodium chloride (common salt, NaCl) and water ($H_2O$).

All of the process units described above in the context of process engineering have analogies in natural systems. For example, in the human body:

- The arteries act as flow *dividers*
- The mouth acts as a *mixer*
- The lungs act as a *separator*
- The skin acts as a *heat exchanger*
- The heart acts as a *pump*
- The stomach acts as a *reactor*

---

[4]"Chemical species" includes atoms, molecules, ions, and radicals.

If we consider a system such as the global environment, then:

- A mountain acts as a flow *divider*
- A river confluence acts as a *mixer*
- A lake acts as a *separator*
- Ocean currents act as *heat exchangers*
- The moon's gravity acts as a *pump*
- A forest acts as a *reactor*

## STREAMS AND STREAM TABLES

Process streams are the flows of material that interconnect process units in a flowsheet. In an industrial chemical process the streams usually represent conduits, conveyors, pipes or mobile containers. The stream characteristics show the nature of the process units and reflect the overall process objective. Some specific characteristics that can define a stream are: density, phase, composition, mass, mole or volume flow, pressure, temperature and enthalpy flow.

Stream characteristics are typically recorded in one of two ways:

A.   In a stream table appended to the flowsheet
B.   As sets of data written adjacent to each stream in the flowsheet

This text uses *stream tables* as the primary way to record stream characteristics. In a stream table each stream is identified (by number or name) and the corresponding stream characteristics are listed for quick reference in a single table. Any process can thus be completely defined by a *flowsheet* with its associated *stream table*. *Figure 3.03* shows a flowsheet and stream table for an idealised and simplified continuous process used to produce ammonium nitrate from ammonia and nitric acid by *Reaction 3.04*.

### STREAM TABLE NOTES

1.   The molar masses and flows are rounded to keep the stream table uncluttered.
2.   In reality $NH_3$ would react with $HNO_3$ in the mixer, but in the flowsheet the mixing and reaction function are isolated from each other to clarify and illustrate the calculations. Heats of mixing are assumed negligible.

The stream table in *Figure 3.03* displays the most basic steady-state M&E balance for making ammonium nitrate. Row 1 lists the stream number used to identify each stream on the flowsheet. Column 1 lists the components of the process, which here are the molecular species in each stream, including the solvent water in the feed stream 2. The second column lists the molar mass (molecular weight) of each species, used in the material balance calculations of reaction stoichiometry. Columns 3 to 9 list the composition of each stream in terms of the moles (actually the mole flow rate in kmol/hour) of each species. The

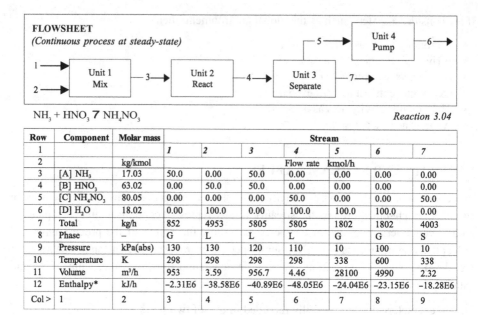

NH$_3$ + HNO$_3$ 7 NH$_4$NO$_3$                                                      *Reaction 3.04*

| Row | Component | Molar mass | Stream | | | | | | |
|-----|-----------|------------|--------|--|--|--|--|--|--|
| | | | 1 | 2 | 3 | 4 | 5 | 6 | 7 |
| 1 | | | *1* | *2* | *3* | *4* | *5* | *6* | *7* |
| 2 | | kg/kmol | | | | Flow rate   kmol/h | | | |
| 3 | [A] NH$_3$ | 17.03 | 50.0 | 0.00 | 50.0 | 0.00 | 0.00 | 0.00 | 0.00 |
| 4 | [B] HNO$_3$ | 63.02 | 0.00 | 50.0 | 50.0 | 0.00 | 0.00 | 0.00 | 0.00 |
| 5 | [C] NH$_4$NO$_3$ | 80.05 | 0.00 | 0.00 | 0.00 | 50.0 | 0.00 | 0.00 | 50.0 |
| 6 | [D] H$_2$O | 18.02 | 0.00 | 100.0 | 0.00 | 100.0 | 100.0 | 100.0 | 0.00 |
| 7 | Total | kg/h | 852 | 4953 | 5805 | 5805 | 1802 | 1802 | 4003 |
| 8 | Phase | – | G | L | L | L | G | G | S |
| 9 | Pressure | kPa(abs) | 130 | 130 | 120 | 110 | 10 | 100 | 10 |
| 10 | Temperature | K | 298 | 298 | 298 | 298 | 338 | 600 | 338 |
| 11 | Volume | m³/h | 953 | 3.59 | 956.7 | 4.46 | 28100 | 4990 | 2.32 |
| 12 | Enthalpy* | kJ/h | -2.31E6 | -38.58E6 | -40.89E6 | -48.05E6 | -24.04E6 | -23.15E6 | -18.28E6 |
| Col > | 1 | 2 | 3 | 4 | 5 | 6 | 7 | 8 | 9 |

**Figure 3.03.   Process flowsheet and stream table for production of ammonium nitrate.**
*Reference condition = elements at standard state, 298K.*

"Total" (row 7) shows the total mass flow in each stream and is used to check the integrity of the material balance.

*Equation 1.06* is observed over each process unit and the whole system i.e. at steady-state:

For:      **Quantity      = Total mass      ACC = 0 = IN – OUT + 0 – 0**
      **Total mass flow IN  =  Total mass flow  OUT = 5805 kg/hour**

Rows 8, 9 and 10 specify the stream conditions of phase, pressure and temperature, which are used to estimate the enthalpy flow of each stream, while row 11 gives the volume flow of each stream. The stream enthalpy is an approximation to the energy content of the stream that is used for chemical process calculations, in which mechanical energy effects are relatively small (see Chapters 2 and 5). The stream enthalpies are used to calculate the thermal loads in the process while the volume flows give the mechanical energy loads and are used to size vessels, pumps and piping.

## PROCESS ECONOMICS

Whether or not the aim is to make a profit, costs and benefits are a major factor in the development of almost all projects. Process engineering always involves a compromise between technical and economic considerations, in which cost and profit are primary measures of the success of a process design. Material and energy balances have a critical role in the estimation of process costs and profitability. For process engineering the *economic viability* of a project is measured by four basic *figures of merit*,[5] the *gross economic potential*, the *net economic potential*, the *capital cost* and the *return on investment*, which are defined as follows:

- Gross economic potential = **GEP = Value of products – Value of feeds** *Equation 3.01*
- Net economic potential  = **NEP = GEP – cost of [*utilities*[6] + *labour* + *maintenance* + *interest*[7]]**                              *Equation 3.02*
- Capital cost          = **Total cost of plant design, construction and installation**
- Return on investment   = **ROI = NEP / Capital cost**                              *Equation 3.03*

The usual units for these quantities are:  GEP[$/year]        Capital cost [$]
                                           NEP[$/year]        ROI [%/year][8]

The calculation of each of these figures of merit hinges on material and energy balances, for example:

- GEP — Quantities of feeds (inputs) and products (outputs) are found by **material balances**
- NEP — Consumption of *utilities* is found by **energy balances**
- Capital cost. The size (and thus the cost) of each process unit is determined by the amounts of material and energy that it handles. The design and costing of process units involves calculations of *process rates*[9] and equipment size that are beyond the scope of this text (see *Refs. 3, 5, 6, 8 and 9* ).

*Example 3.01* demonstrates how M&E balances are used to estimate the economic viability of a chemical process.

## EXAMPLE 3.01   Calculating the economic figures of merit of a chemical process.

*Figure 3.03* shows a conceptual flow-sheet and stream table for a process used to make ammonium nitrate from ammonia and nitric acid, with the stream table values based on an assumption of 100% efficiency for all operations in the process. *Table 3.01* gives some cost data for the process.

---

[5] A *figure of merit* (a.k.a. performance indicator) is a quantity that summarizes value in a specific application.

[6] *Utilities* include items such as electricity, fuel and water needed to operate the plant.

[7] *Interest* is the cost of borrowing money.

[8] This simple ROI is a narrow measure of economic viability. It does not account for the time value of money or the burgeoning environmental costs of "externalities".

[9] *Process rates* include heat transfer, mass transfer and reaction rates (intrinsic kinetics).

**Table 3.01.** Cost data for ammonium nitrate process.

| Cost Data | | | | | |
|---|---|---|---|---|---|
| Component | | $NH_3$ | $HNO_3$ | $NH_4NO_3$ | $H_2O$ |
| Value | $/kg | 0.08 | 0.10 | 0.25 | 0.001 |
| Utilities | | Electricity | Heating | Cooling | |
| Value | $/kWh | 0.06 | 0.05 | 0.001 | |
| Capital cost of plant = 15E6 $US AD 2002. | | | | | |
| Operating period = 8000 hours/year Interest rate = 7%/year. | | | | | |

*Problem:* Estimate: (a) the GEP, (b) the NEP and (c) the ROI of the ammonium nitrate process.

*Solution:*

**A.**  Value of products = (50 kmol/h) (80.05 kg/kmol) (0.25 $/kg)  = 1000$/h
    Value of feeds   = (50 kmol/h) (17.03 kg/kmol) (0.08 $/kg)
                + (50 kmol/h)(63.02 kg/kmol)(0.10 $/kg)  = 383 $/h
    GEP      = Value of products − Value of feeds
            = (1000 $/h − 383 $/h) (8000 h/y)        **GEP = 4.94E6 $/y**

**B.**  Utility loads are calculated from the energy balance:

|       | Load kW | Cost | |
|---|---|---|---|
| Unit 2 | 2000 (cooling) | (2000 kW)(8000 h/y)(0.001 $/kWh) | = 16E3 $/y |
| Unit 3 | 1678 (heating) | (1678 kW)(8000 h/y)(0.05 $/kWh) | = 671E3 $/y |
| Unit 4 | 246 (pumping) | (246 kW)(8000 h/y)(0.06 $/kWh) | = 118E3 $/y |
|       |  | Total utilities | = 805E3 $/y |

Labour + maintenance = 5% of capital cost per year = (0.05 1/y)(15E6 $)   = 750E3 $/y
Interest            = (0.07 1/y)(15E6 $)                              −1050E3 $/y
NEP                 = GEP − cost of [utilities + labour + maintenance + interest]
                    = 4.94E6 − (805E3 + 750E3 + 1050E3)        **NEP = 2.34E6 $/y**

**C.**  ROI     = Annual NEP/Capital cost = 2.34E6 $/y/15E6 $ = 0.16
                                                        **ROI ≈ 16%/y**

Note that the ROI calculated by the method of *Example 3.01* is a very approximate value that would subsequently be refined by more a comprehensive analysis with realistic process efficiencies.[10] The more complex M&E balances required for such work are shown in Chapters 4, 5 and 6 of this text.

---

[10] Better assessments of economic viability account for the time value of money, with measures such as the Net Future Worth (NFW) and Discounted Cash Flow Rate of Return (DCFROR).

## SUMMARY

[1] The M&E balance is the primary tool used in the modelling, design and costing of industrial chemical processes.

[2] M&E balances are calculated from the general balance equation and their results displayed using a combination of:
- A *flowsheet*, which is a diagram showing the process units with their interconnecting material flow paths
- A *stream table*, which tabulates the conditions and the quantities of material and energy in each process stream

[3] All process flowsheets can be conceptually simplified to a combination of one or more process units, each of which performs one of the six basic functions. These functions are: DIVIDE, MIX, SEPARATE, HEAT EXCHANGE, PUMP, REACT. In commercial process simulation, multi-function process units may be modelled by "user" defined blocks that combine two or more of these basic functions.

[4] The information in a process stream table is the basis for calculation of the figures of merit that measure economic viability:
- The *gross economic potential* (GEP) is found from the material balance
- The *net economic potential* (NEP) is found from the GEP plus utility loads calculated by the energy balance
- The *capital cost* is found from design specifications based on both the material and the energy balance
- The *return on investment* (ROI) is estimated as: ROI = NEP/capital cost

## FURTHER READING

[1] S. Smith, *Environmental Economics — a Very Short Introduction*, Oxford Uni Press, 2011.
[2] M. Douglas, *Conceptual Design of Chemical Processes*, McGraw-Hill, 1988.
[3] W. Faith and F. Lowenheim, M.Moran, R.Clark, *Industrial Chemicals*, Wiley, 1975.
[4] Kirh-Othmer, *Kirk-Othmer Encyclopedia of Chemical Technology*, Wiley, 2007.
[5] T. Perry and D.Green, *Chemical Engineers' Handbook*, McGraw-Hill, 2018.
[6] S. Sinnott and G.Towler, *Chemical Engineering Design*, Butterworth-Heinemann, 2019.

[7]   M. Peters, K. Timmerhaus and R.West, *Plant Design and Economics for Chemical Engineers*, McGraw-Hill, 2002.

[8]   G. Ulrich and P. Vasudevan, *Chemical Engineering Process Design and Economics: A Practical Guide*, 2nd Edition, Process Publishing, 2004.

[9]   J. Cooper, W. Penny, J. Fairbanks and S. Walas, *Chemical Process Equipment — Selection and Design*, Butterworth-Heinemann, 2012.

[10]  W. Seider, J. Seader and D. Warren, *Process Design Principles*, Wiley, 1999.

# CHAPTER FOUR

MATERIAL BALANCES

## GENERAL MATERIAL BALANCE

Material balance calculations hinge on the *general balance equation*, introduced in Chapter 1 and repeated below as *Equation 4.01*.

For a defined *system* and a specified *quantity:*

**Accumulation = Input to − Output      + Generation − Consumption**      *Equation 4.01*
**in system            system       from system      in system       in system**

The *system* is a space whose <u>closed boundary</u> is defined to suit the problem at hand. For material balance problems the *quantity* in *Equation 4.01* is an amount of material, usually measured as mass or as moles. *Equation 4.01* then represents the general form of the material balance.

*Equation 4.01*, in abbreviated form, can be an integral balance or differential balance:

| *Integral balance:*            *Equation 4.02* | *Differential balance:*            *Equation 4.03* |
|---|---|
| **ACC = IN − OUT + GEN − CON** | **Rate ACC = Rate IN − Rate OUT + Rate GEN − Rate CON** |
| where: | where: |
| ACC ≡ (final amount of material − initial amount of material) in the system | Rate ACC ≡ rate of accumulation of material in the system, w.r.t. time |
| IN   ≡ amount of material entering the system (input) | Rate IN   ≡ rate of material entering the system, w.r.t. time (input) |
| OUT ≡ amount of material leaving the system (output) | Rate OUT ≡ rate of material leaving the system, w.r.t. time (output) |
| GEN ≡ amount of material generated in the system (source) | Rate GEN ≡ rate of material generated in the system, w.r.t. time (source) |
| CON ≡ amount of material consumed in the system (sink) | Rate CON ≡ rate of material consumed in the system, w.r.t. time (sink) |

In every case the definition of the system boundary is a critical first step in setting up and solving material balance problems. The boundary must be a closed envelope that surrounds the system of interest. All materials crossing the system boundary are considered as input or outputs, and everything occurring inside the system boundary is considered to contribute to the accumulation, generation (source) or consumption (sink) terms in *Equation 4.01*.

Your choice of the system boundary and the quantity balanced can have a big effect on the complexity of the solution of a material balance problem. As a rule you should use the simplest combination of system and quantity that will give the desired result. For some problems the simplest combination is: system = overall process and quantity = mass or moles of specified elements (i.e. an atom balance), whereas more difficult problems may require individual species mole balances on each sub-unit of the process. Sometimes a difficult material balance problem can be easily solved by combining balances on a set of different (carefully chosen) envelopes, each of which contains a part of the system of interest.

In particular cases *Equations 4.02* and *4.03* can be simplified as follows:

|  | Integral balance | Differential balance |
|---|---|---|
| Closed system (batch process): | IN = OUT   = 0 | Rate IN = Rate OUT   = 0 |
| Open system (continuous process) at steady-state: | ACC   = 0 | Rate ACC   = 0 |
| Atoms*: | GEN = CON = 0 | Rate GEN = Rate CON = 0 |
| Mass or moles of a species with no chemical reaction*: | GEN = CON = 0 | Rate GEN = Rate CON = 0 |
| Total mass*: | GEN = CON = 0 | Rate GEN = Rate CON = 0 |
| Total moles with no chemical reaction*: | GEN = CON = 0 | Rate GEN = Rate CON = 0 |

*\*Assumes no nuclear reactions.*

The quantities used in *Equation 4.02* and *4.03* must be clearly specified to avoid ambiguity. For this purpose the nomenclature used in this text is given in *Table 4.01*. With the nomenclature of *Table 4.01* a process *stream table* is described by a two-dimensional matrix, convenient for computer code and spreadsheet calculations.

Table 4.01.   Nomenclature for material balances.

| Symbol | Meaning | Typical Units |
|---|---|---|
| $M(j)$ | molar mass of component $j$ | $kg.kmol^{-1}$ |
| $m(j)$ | mass of component $j$ | kg |
| $n(j)$ | moles of component $j$ | kmol |
| $V(j)$ | volume of component $j$ | $m^3$ |
| $\dot{m}(i, j)$ | mass flow rate of component $j$ in stream $i$ | $kg.h^{-1}$ |
| $\dot{n}(i, j)$ | mole flow rate of component $j$ in stream $i$ | $kmol.h^{-1}$ |
| $\bar{m}(i)$ | total mass flow rate in stream $i$ | $kg.h^{-1}$ |
| $\bar{n}(i)$ | total mole flow rate in stream $i$ | $kmol.h^{-1}$ |
| $\dot{V}(i)$ | total volume flow rate of stream $i$ | $m^3.h^{-1}$ |
| For example: $\dot{n}(5, A)$ = mole flow rate of component A in stream 5. | | |

## CLOSED SYSTEMS (BATCH PROCESSES)

A closed system is defined as one in which, over the time period of interest, there is *zero transfer of material across the system boundary.* Closed systems are also called "controlled mass" systems and correspond to batch processes. Closed systems are common in laboratories and industrial processes but rare in nature. Some examples of closed systems are:

- Batch reactor
- Pressure cooker (before it vents)

- Sealed bottle of beer
- Boiling egg*
- Planet Earth with its atmosphere*

*These natural systems are not strictly closed, since eggs do transpire and some material does enter and leave the perimeter of Earth's atmosphere.*

The integral and differential material balances for a closed system are respectively:

**ACC     = GEN – CON**        (Integral material balance)     *Equation 4.04*
**Rate ACC = Rate GEN – Rate CON**  (Differential material balance)   *Equation 4.05*

*Examples 4.01* and *4.02* illustrate material balance calculations for closed systems.

### EXAMPLE 4.01   Material balance on carbon dioxide in a sealed vessel (*closed system*).

A sealed vessel batch reactor initially contains 3.00 kg carbon + 16.0 kg oxygen + 0.2 $m^3$ 1 M aqueous sodium hydroxide solution. The carbon is ignited and undergoes 100% conversion to carbon dioxide by *Reaction 1*. Carbon dioxide is then absorbed into the solution where it converts 100% of the NaOH to $Na_2CO_3$ by *Reaction 2*.

$C(s) + O_2(g) \rightarrow CO_2(g)$                               *Reaction 1*
$CO_2(g) + 2NaOH(aq) \rightarrow Na_2CO_3(aq) + H_2O(l)$         *Reaction 2*

**Problem:** Calculate the final mass of carbon dioxide in the vessel.

**Solution:** Define the system = sealed vessel

Specify the components:

| Component | C | $O_2$ | $CO_2$ | NaOH | $Na_2CO_3$ |
|---|---|---|---|---|---|
| Number | 1 | 2 | 3 | 4 | 5 |
| Molar mass | 12 | 32 | 44 | 40 | 106 (Rounded to whole numbers) |

Initial amounts:

| | kg | 3 | 16 | 0 | 8 | 0 |
|---|---|---|---|---|---|---|
| | kmol | 0.25 | 0.5 | 0 | 0.2 | 0 |

Mole balance on $CO_2$ [Integral balance], **ACC = IN – OUT + GEN – CON**   *Equation 4.02*

IN   = OUT            = 0                          kmol   (closed system)
GEN = (1) n (1)       = (1) (0.25 kmol)   = 0.25 kmol  (stoichiometry *Reaction 1*)
CON = (0.5) n (4)    = (0.5) (0.2 kmol)   = 0.1  kmol  (stoichiometry *Reaction 2*)
ACC = 0 – 0 + 0.25 – 0.1               = 0.15 kmol  (*Equation 4.02*)
Final mass of $CO_2$ = (0.15 kmol)(44 kg.kmol$^{-1}$) = **6.6 kg**

*EXAMPLE 4.02   Material balance on water in a space station (closed system).*

A space station with a crew of 30 people begins to orbit Earth with a stock of 18.0E3 kg of clean water. During the 500 day period in orbit there is zero import/export of material to/from the space station.

Clean water is generated in the station as follows:

(i)   Recovery (by condensation, distillation and sterilisation) then recycle of water from wastes. This source of clean water is equivalent to 90% of the water used in (iii), plus 90% of the water generated through the metabolic bio-oxidation of 0.5 kg/(person/day) of dry food to $CO_2$ and $H_2O$.
Food is assumed[1] composed only of carbohydrate, whose reaction is:

$$C_6H_{10}O_5 + 6O_2 \rightarrow 6CO_2 + 5H_2O \qquad\qquad Reaction\ 1$$

(ii)   Recombination of metabolic $CO_2$ with all available $H_2$ from water electrolysis:

$$4H_2 + CO_2 \rightarrow CH_4 + 2H_2O \qquad\qquad Reaction\ 2$$

Clean water is consumed in the station as follows:

(iii) Each person uses 8 kg clean water/day for sustenance and hygiene.
(iv) Electrolysis of water to supply 0.6 m³ STP/(person/day) of oxygen gas:
$$2H_2O \rightarrow 2H_2 + O_2 \qquad\qquad Reaction\ 3$$

*Problem:* Calculate the stock of clean water remaining at the end of the 500 day mission. [kg]

*Solution:* Define the system    = space station (a closed system)
       Specify the quantity = amount (kmol) of clean water
       Integral mole balance on clean water, ACC = IN – OUT + GEN – CON
       ACC = ?   (Final amount of clean water – Initial amount of clean water) in stock
       IN   = 0   (closed system)
       OUT = 0   (closed system)

GEN: Recycle   = (0.9)(30 persons)(8 kg/day)(500 days)/(18 kg/kmol)      = 6.00E3 kmol
Bio-oxidation   = (0.9)(30 persons)(500 days)(0.5 kg ($C_6H_{10}O_5$)
           /(person/day)/(162 kg($C_6H_{10}O_5$)/kmol)
           × (5 kmol $H_2O$/kmol ($C_6H_{10}O_5$))      = 208.3 kmol
Recombination = ((30 persons)(0.6 m³$O_2$/day)(500 days)
           /(22.4 m³ STP/kmol)) × (2 kmol $H_2$/kmol $O_2$)
           × (0.5 kmol $H_2O$/kmol $H_2$)      = 401.8 kmol
Total GEN    = Recycle + Bio-oxidation + Recombination
           = 6.00E3 + 208.3 + 401.8      = 6.61E3 kmol
CON: By crew  = (30 persons)(8 kg/day)(500 days)/(18 kg/kmol)      = 6.67E3 kmol

---

[1] For the purpose of this problem. Also assume no change in composition or mass of the crew.

Electrolysis   = $((30 \text{ persons})(0.6 \text{ m}^3/\text{day})(500 \text{ day})/(22.4 \text{ m}^3/\text{kmol}))$
                  $/(0.5 \text{ kmol } O_2/\text{kmol } H_2O)$                                             = 803.6 kmol
Total CON    = Crew + Electrolysis = 6.67E3 + 804                              = 7.47E3 kmol
ACC          = IN – OUT + GEN – CON = 0 – 0 + 6.61E3 – 7.47E3        = –860 kmol
Final amount = Initial amount + ACC = (18.0E3 kg/18 kg/kmol)
                  – 860 kmol                                                                      = 140 kmol
                                                                                              = **2.52E3 kg clean water**

Check $CO_2$ balance:

  GEN  = Metabolism = $(6)(0.5)(30)(500)/(162)$                 = 277.7 kmol (*Reaction 1*)
  CON  = Recombination = $(1/4)(803.6)$                          = 200.9 kmol (*Reaction 2*)
  ACC  = 277.7 – 200.9                                                       = 76.8 kmol

Check $O_2$ balance:

  GEN  = Electrolysis = $(0.5)(803.6)$                                = 401.8 kmol (*Reaction 3*)
  CON  = Metabolism = $(6)(0.5)(30)(500)/(162)$                = 277.7 kmol (*Reaction 1*)
  ACC  = 401.8 – 277.7                                                       = 124.1 kmol

Check $H_2$ balance:

  All $H_2$ generated by electrolysis (*Reaction 3*) is consumed in reaction (*Reaction 2*).
  ACC = 0

These species mole balances predict accumulation of $CO_2$ and of $O_2$ in the system, so further processing (and/or venting) may be needed to control levels of $CO_2$ and $O_2$ in the space station. Similarly, measures should be taken to deal with the $CH_4$ generated in *Reaction 2*.

*NOTE:* Without the water derived from oxidation of food the space station would run out of water in 469 days.


## OPEN SYSTEMS (CONTINUOUS PROCESSES)

An open system is defined as one in which, over the time period of interest, *material is transferred across the system boundary*. Open systems are also called "controlled volume" systems and correspond to continuous processes. Open systems are common to our experience and include, for example:

- Automobile engines
- Industrial processes
- Living cells
- Human body and its organs
- Atmosphere of planet Earth

*Example 4.03* shows a simple material balance on the global carbon cycle to calculate the change in carbon dioxide concentration in the atmosphere of planet Earth. Similar, though more complex, material balances are used by earth scientists to quantify the other geo-chemical cycles (e.g. the oxygen, nitrogen and water cycles) that govern life in our bio-sphere (see *Ref. 1*).

### EXAMPLE 4.03    *Material balance on carbon dioxide in Earth's atmosphere (open system).*

The Earth's atmosphere has a volume of 4.0E18 $m^3$ STP and in 1850 CE contained 0.028 vol% $CO_2$. Over the period 1850 to 1990 fossil fuels containing the equivalent of 220E12 kg C were burned to produce $CO_2$, 440E12 kg $CO_2$ were vented from volcanoes and by decomposition of biomass, 350E12 kg net $CO_2$ were consumed by photosynthesis on land and 370E12 kg $CO_2$ were absorbed by the oceans.

### Problem:
If all $CO_2$ gas is well-mixed into the atmosphere, what is the concentration of $CO_2$ in the atmosphere in 1990? [ppm vol]. Assume total volume of the atmosphere is constant at 4.0 E18 $m^3$ STP.

### Solution:
Define the system          = Atmosphere of Earth (an open system)

Specify the components:     Component   C      $CO_2$

                            Number      1       2

                            Molar mass  12      44      [Rounded to whole numbers]

Initial amount of $CO_2$    = n(2) = PV(2)/RT [Ideal gas law. see *Equation 2.13*]

                            = (101.3 kPa) (4E18 $m^3$)(0.028E-2) / ((8.314 kJ.kmol$^{-1}$.K$^{-1}$) (273K))

                            = 50.0E12 kmol

Mole balance on $CO_2$ (Integral balance): **ACC = IN – OUT + GEN – CON**    *Equation 4.02*

IN          = fossil fuel + (volcano + biomass decomposition)

            = (220E12 kg C)/(12 kg/kmol)

            + 440E12 kg $CO_2$/(44 kg.kmol$^{-1}$)                              = 28.3E12 kmol $CO_2$

OUT         = photosynthesis + absorption

            = 350E12 kg $CO_2$/(44 kg/kmol)

            + 370E12 kg $CO_2$/(44 kg.kmol$^{-1}$)                              = 16.4E12 kmol $CO_2$

GEN         = 0

CON         = 0

ACC         = 28.3E12 – 16.4E12 + 0 – 0                                         = 11.9E12 kmol $CO_2$

Final $CO_2$ = Initial $CO_2$ + ACC $CO_2$ = 50.0E12 + 11.9E12                  = 61.9E12 kmol $CO_2$

Concentration of $CO_2$ in atmosphere in 1990 = 0.028% (61.9E12 kmol $CO_2$/50.0E12 kmol $CO_2$) = 0.0347% (vol)                                                    = **347 ppm (vol)**

As shown in *Example 4.03* and later in *Example 4.16*, the principles used here can be applied to modelling any type of open system. The rest of this chapter focuses on open systems in industrial chemical processes.

Industrial chemical processes normally operate as open systems with continuous or semi-continuous input and output of materials. Large industrial processes (e.g. oil refinery, pulp mill, power plant, integrated circuit plant) operate 24 hours per day for 330+ days per year at conditions close to *steady-state*. Such plants typically run under controls that hold process conditions to *optimise*[2] the production, then shut down for maintenance over a few weeks in an operating cycle that may last from one to five years.

## CONTINUOUS PROCESSES AT STEADY-STATE

Material balances for continuous processes at steady-state use the *differential material balance* with Rate ACC = 0.

**Rate ACC = 0 = Rate IN – Rate OUT**
**+ Rate GEN – Rate CON**
*Equation 4.03*

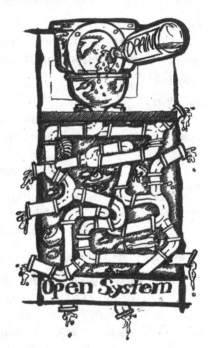

### SINGLE PROCESS UNITS

Chapter 3, *Figure 3.02* describes the six generic process units used to build chemical process flowsheets. Each of these generic units has a material balance specific to its function, as shown in *Figure 4.01* and illustrated in *Example 4.04 A* to *F*.

---

[2] *Optimise* production does not necessarily mean *maximise* production. Process optimisation is beyond the scope of this text but is treated in *Ref. 10*.

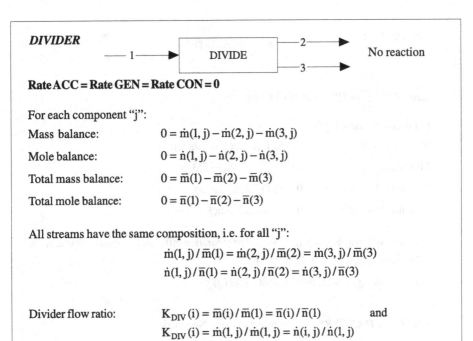

**DIVIDER**

**Rate ACC = Rate GEN = Rate CON = 0**

For each component "j":

| | |
|---|---|
| Mass balance: | $0 = \dot{m}(1,j) - \dot{m}(2,j) - \dot{m}(3,j)$ |
| Mole balance: | $0 = \dot{n}(1,j) - \dot{n}(2,j) - \dot{n}(3,j)$ |
| Total mass balance: | $0 = \overline{m}(1) - \overline{m}(2) - \overline{m}(3)$ |
| Total mole balance: | $0 = \overline{n}(1) - \overline{n}(2) - \overline{n}(3)$ |

All streams have the same composition, i.e. for all "j":

$$\dot{m}(1,j)/\overline{m}(1) = \dot{m}(2,j)/\overline{m}(2) = \dot{m}(3,j)/\overline{m}(3)$$
$$\dot{n}(1,j)/\overline{n}(1) = \dot{n}(2,j)/\overline{n}(2) = \dot{n}(3,j)/\overline{n}(3)$$

Divider flow ratio:    $K_{DIV}(i) = \overline{m}(i)/\overline{m}(1) = \overline{n}(i)/\overline{n}(1)$        and

$$K_{DIV}(i) = \dot{m}(1,j)/\dot{m}(1,j) = \dot{n}(i,j)/\dot{n}(1,j)$$

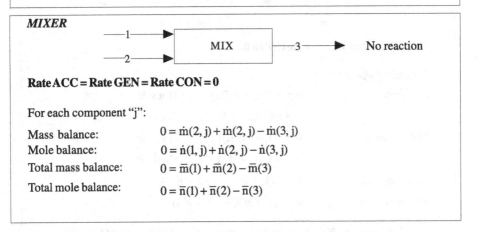

**MIXER**

**Rate ACC = Rate GEN = Rate CON = 0**

For each component "j":

| | |
|---|---|
| Mass balance: | $0 = \dot{m}(2,j) + \dot{m}(2,j) - \dot{m}(3,j)$ |
| Mole balance: | $0 = \dot{n}(1,j) + \dot{n}(2,j) - \dot{n}(3,j)$ |
| Total mass balance: | $0 = \overline{m}(1) + \overline{m}(2) - \overline{m}(3)$ |
| Total mole balance: | $0 = \overline{n}(1) + \overline{n}(2) - \overline{n}(3)$ |

**Figure 4.01.   Material balances on generic process units — continuous processes at steady-state.**

*(Continued)*

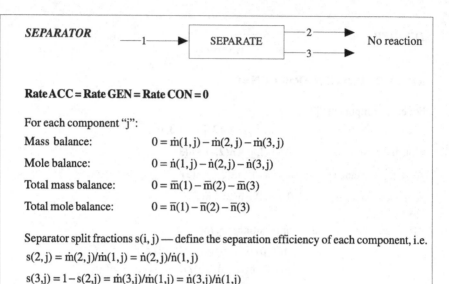

**SEPARATOR**

**Rate ACC = Rate GEN = Rate CON = 0**

For each component "j":

Mass balance: $\qquad 0 = \dot{m}(1,j) - \dot{m}(2,j) - \dot{m}(3,j)$

Mole balance: $\qquad 0 = \dot{n}(1,j) - \dot{n}(2,j) - \dot{n}(3,j)$

Total mass balance: $\qquad 0 = \bar{m}(1) - \bar{m}(2) - \bar{m}(3)$

Total mole balance: $\qquad 0 = \bar{n}(1) - \bar{n}(2) - \bar{n}(3)$

Separator split fractions $s(i,j)$ — define the separation efficiency of each component, i.e.

$s(2,j) = \dot{m}(2,j)/\dot{m}(1,j) = \dot{n}(2,j)/\dot{n}(1,j)$

$s(3,j) = 1 - s(2,j) = \dot{m}(3,j)/\dot{m}(1,j) = \dot{n}(3,j)/\dot{n}(1,j)$

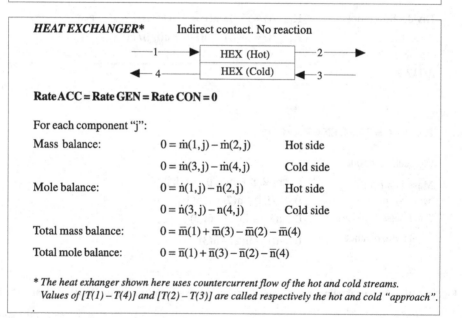

**HEAT EXCHANGER\***      Indirect contact. No reaction

**Rate ACC = Rate GEN = Rate CON = 0**

For each component "j":

Mass balance: $\qquad 0 = \dot{m}(1,j) - \dot{m}(2,j)$      Hot side

$\qquad\qquad\qquad\qquad 0 = \dot{m}(3,j) - \dot{m}(4,j)$      Cold side

Mole balance: $\qquad 0 = \dot{n}(1,j) - \dot{n}(2,j)$      Hot side

$\qquad\qquad\qquad\qquad 0 = \dot{n}(3,j) - n(4,j)$      Cold side

Total mass balance: $\qquad 0 = \bar{m}(1) + \bar{m}(3) - \bar{m}(2) - \bar{m}(4)$

Total mole balance: $\qquad 0 = \bar{n}(1) + \bar{n}(3) - \bar{n}(2) - \bar{n}(4)$

*\* The heat exhanger shown here uses countercurrent flow of the hot and cold streams.*
*Values of $[T(1) - T(4)]$ and $[T(2) - T(3)]$ are called respectively the hot and cold "approach".*

**Figure 4.01.** **Material balances on generic process units — continuous processes at steady-state.**

*(Continued)*

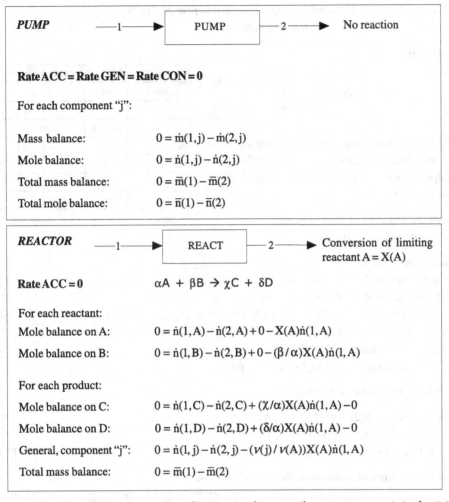

(*Continued*)

*NOTE:* For all process units: $\dot{m}(i, j) = M(j)\dot{n}(i, j)$
$$\bar{m}(i, j) = \Sigma[\dot{m}(i, j)]$$
$$\bar{n}(i) = \Sigma[\dot{n}(i, j)]$$

*In the general reactor balance the value of $\nu(j)$ is −ve for reactants and +ve for products. Thus reactants are <u>consumed</u> and products are <u>generated</u> in a reactor.*

## EXAMPLE 4.04    Material balances on generic process units (open system at steady-state).

### A. Divider*

| Stream Table | | Divider problem | | |
|---|---|---|---|---|
| Species | M | Stream | | |
| | | *1* | *2* | *3* |
| | kg/mol | kmol/h | | |
| A | 20 | 4 | 1 | ? |
| B | 30 | 8 | ? | ? |
| C | 40 | 12 | ? | ? |
| Total | kg/h | 800 | ? | ? |

The steady-state stream tables are presented in the standard spreadsheet format introduced in Chapter 3 and used throughout Chapters 4, 5 and 6 of this text.

**Problem:** Find the unknown flows.

\* *A divider is called a stream "splitter" in some texts.*

**Solution:** [System = divider]

| | | |
|---|---|---|
| Mole balance on A: | $0 = \dot{n}(1, A) - \dot{n}(2, A) - \dot{n}(3, A)$ | [1] |
| Mole balance on B: | $0 = \dot{n}(1, B) - \dot{n}(2, B) - \dot{n}(3, B)$ | [2] |
| Mole balance on C: | $0 = \dot{n}(1, C) - \dot{n}(2, C) - \dot{n}(3, C)$ | [3] |
| Stream mass: | $\bar{m}(2) = M(A)\dot{n}(2, A) + M(B)\dot{n}(2, B) + M(C)\dot{n}(2, C)$ | [4] |
| Stream mass: | $\bar{m}(3) = M(A)\dot{n}(3, A) + M(B)\dot{n}(3, B) + M(C)\dot{n}(3, C)$ | [5] |
| Total mole balance: | $0 = \bar{n}(1) - \bar{n}(2) - \bar{n}(3)$ | [6] |
| Total mass balance: | $0 = \bar{m}(1) - \bar{m}(2) - \bar{m}(3)$ | [7] |

### Divider composition relations:

$$\dot{n}(1, A)/\bar{n}(1) = \dot{n}(2, A)/\bar{n}(2) = \dot{n}(3, A)/\bar{n}(3) \qquad [8]$$
$$\dot{n}(1, B)/\bar{n}(1) = \dot{n}(2, B)/\bar{n}(2) = \dot{n}(3, B)/\bar{n}(3) \qquad [9]$$
$$\dot{n}(1, C)/\bar{n}(1) = \dot{n}(2, C)/\bar{n}(2) = \dot{n}(3, C)/\bar{n}(3) \qquad [10]$$

### Seven independent equations. Seven unknowns.[3]

Equations [1 to 7]  = 5 independent equations
Equations [8 to 10] = 2 independent equations
Substitute: $\dot{n}(1, A)$ = 4 $\dot{n}(1, B)$ = 8 $\dot{n}(1, C)$ =12
    $\dot{n}(2, A) = 1$ $\bar{m}(1) = 800$

Solve for the unknowns and check closure.

| Stream Table | | Divider Solution | | |
|---|---|---|---|---|
| Species | M | Stream | | |
| | | *1* | *2* | *3* |
| | kg/mol | kmol/h | | |
| A | 20 | 4 | 1 | 3 |
| B | 30 | 8 | 2 | 6 |
| C | 40 | 12 | 3 | 9 |
| Total | kg/h | 800 | 200 | 600 |
| Mass IN | = 800 kg/h | Closure | | |
| Mass OUT | = 800 kg/h | 800/800 = 100% | | |

---

[3] Independent equations are those that cannot be obtained by combining other equations in the (independent) set. The remaining equations are redundant. Redundant equations are shown in these examples to illustrate the concepts of *Figure 4.01* and to indicate that the material balance may be solved using different sets.

Solution method:    Solve [9 and 10] for: ṅ(2, B), ṅ(2, C) Solve [1, 2 and 3] for: ṅ(3, A),
                    ṅ(3, B), ṅ(3, C)
                    Solve [4 and 5] for: m̄(2), m̄(3)

## B. Mixer

| Stream Table | | Mixer Problem | | |
|---|---|---|---|---|
| Species | M | Stream | | |
| | | 1 | 2 | 3 |
| | kg/mol | kmol/h | | |
| A | 20 | 4 | 3 | ? |
| B | 30 | 8 | 5 | ? |
| C | 40 | 12 | 7 | ? |
| Total | kg/h | 800 | 490 | ? |

***Problem:*** Complete the stream table material balances.

***Solution:*** [System = mixer]

| | | |
|---|---|---|
| Mole balance on A: | $0 = ṅ(1, A) + ṅ(2, A) - ṅ(3, A)$ | [1] |
| Mole balance on B: | $0 = ṅ(1, B) + ṅ(2, B) - ṅ(3, B)$ | [2] |
| Mole balance on C: | $0 = ṅ(1, C) + ṅ(2, C) - ṅ(3, C)$ | [3] |
| Total stream mass: | $m̄(3) = M(A)ṅ(3, A) + M(B)ṅ(3, B) + M(C)ṅ(3, C)$ | [4] |
| Total mole balance: | $0 = n̄(1) + n̄(2) - n̄(3)$ | [5] |
| Total mass balance: | $0 = m̄(1) + m̄(2) - m̄(3)$ | [6] |

**Four independent equations. Four unknowns.**

Substitute:
$ṅ(1, A) = 4$
$ṅ(1, B) = 8$
$ṅ(1, C) = 12$
$m̄(1) \quad = 800$
$ṅ(2, A) = 3$
$ṅ(2, B) = 5$
$ṅ(2, C) = 7$
$m̄(2) \quad = 490$

| Stream Table | | Mixer Solution | | |
|---|---|---|---|---|
| Species | M | Stream | | |
| | | 1 | 2 | 3 |
| | kg/mol | kmol/h | | |
| A | 20 | 4 | 3 | 7 |
| B | 30 | 8 | 5 | 13 |
| C | 40 | 12 | 7 | 19 |
| Total | kg/h | 800 | 490 | 1290 |
| Mass IN | = | 1290 kg/h | Closure | |
| Mass OUT | = | 1290 kg/h | 1290/1290 = 100% | |

Solve for the unknowns and check closure.

## C. Separator

Split fractions
$s(2, A) = 0.75$
$s(3, B) = 0.50$
$s(3, C) = 0.67$

| Stream Table | | Separator Problem | | |
|---|---|---|---|---|
| Species | M | Stream | | |
| | | 1 | 2 | 3 |
| | kg/kmol | kmol/h | | |
| A | 20 | 4 | ? | ? |
| B | 30 | 8 | ? | ? |
| C | 40 | 12 | ? | ? |
| Total | kg/h | 800 | ? | ? |

**Problem:** Complete the stream table material balance.

**Solution:** [System = separator]

| | | |
|---|---|---|
| Mole balance on A: | $0 = \dot{n}(1, A) - \dot{n}(2, A) - \dot{n}(3, A)$ | [1] |
| Mole balance on B: | $0 = \dot{n}(1, B) - \dot{n}(2, B) - \dot{n}(3, B)$ | [2] |
| Mole balance on C: | $0 = \dot{n}(1, C) - \dot{n}(2, C) - \dot{n}(3, C)$ | [3] |
| Total mole balance: | $0 = \overline{n}(1) - \overline{n}(2) - \overline{n}(3)$ | [4] |
| Total mass balance: | $0 = \overline{m}(1) - \overline{m}(2) - \overline{m}(3)$ | [5] |
| Stream mass: | $\overline{m}(2) = M(A)\dot{n}(2, A) + M(B)\dot{n}(2, B) + M(C)\dot{n}(2, C)$ | [6] |
| | $\overline{m}(3) = M(A)\dot{n}(3, A) + M(B)\dot{n}(3, B) + M(C)\dot{n}(3, C)$ | [7] |
| Split fractions: | $s(2, A) = \dot{n}(2, A)/\dot{n}(1, A)$ | [8] |
| | $s(3, B) = \dot{n}(3, B)/\dot{n}(1, B)$ | [9] |
| | $s(3, C) = \dot{n}(3, C)/\dot{n}(1, C)$ | [10] |

**Eight independent equations. Eight unknowns.**

Equations [1 to 7] = 5 independent equations    Equations [8 to 10] = 3 independent equations

Substitute:

$\dot{n}(1, A) = 4$    $\dot{n}(1, B) = 8$
$\dot{n}(1, C) = 12$    $\overline{m}(1) = 800$
$s(2, A) = 0.75$   $s(2, B) = 0.50$
$s(2, C) = 0.67$

Solve for the unknowns and check closure.

| Stream Table | | Separator Solution | | |
|---|---|---|---|---|
| Species | M | Stream | | |
| | | *1* | *2* | *3* |
| | kg/kmol | kmol/h | | |
| A | 20 | 4 | 3 | 1 |
| B | 30 | 8 | 4 | 4 |
| C | 40 | 12 | 4 | 8 |
| Total | kg/h | 800 | 340 | 460 |
| Mass IN | = 800 kg/h | | | |
| Mass OUT | = 800 kg/h | Closure = 800/800 = 100% | | |

**D. Heat exchanger (indirect contact)**

Indirect contact

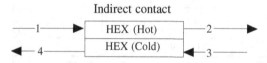

**Problem:** Complete the stream table material balance.

**Solution:** [System = HEX hot side]

| Stream Table | | Heat Exchanger Problem | | | |
|---|---|---|---|---|---|
| Species | M | Stream | | | |
| | | *1* | *2* | *3* | *4* |
| | kg/kmol | kmol/h | | | |
| A | 20 | 4 | ? | 3 | ? |
| B | 30 | 8 | ? | 7 | ? |
| C | 40 | 12 | ? | 9 | ? |
| Total | kg/h | 800 | ? | 630 | ? |

| | | |
|---|---|---|
| Mole balance on A: | $0 = \dot{n}(1, A) - \dot{n}(2, A)$ | [1] |
| Mole balance on B: | $0 = \dot{n}(1, B) - \dot{n}(2, B)$ | [2] |
| Mole balance on C: | $0 = \dot{n}(1, C) - \dot{n}(2, C)$ | [3] |
| Total mole balance: | $0 = \overline{n}(1) - \overline{n}(2)$ | [4] |
| Total mass balance: | $0 = \overline{m}(1) - \overline{m}(2)$ | [5] |
| Stream mass: | $\overline{m}(2) = M(A)\dot{n}(2, A) + M(B)\dot{n}(2, B) + M(C)\dot{n}(2, C)$ | [6] |

[System = HEX cold side]

| | | |
|---|---|---|
| Mole balance on A: | $0 = \dot{n}(3, A) - \dot{n}(4, A)$ | [7] |
| Mole balance on B: | $0 = \dot{n}(3, B) - \dot{n}(4, B)$ | [8] |
| Mole balance on C: | $0 = \dot{n}(3, C) - \dot{n}(4, C)$ | [9] |
| Total mole balance: | $0 = \bar{n}(3) - \bar{n}(4)$ | [10] |
| Total mass balance: | $0 = \bar{m}(3) - \bar{m}(4)$ | [11] |
| Stream mass: | $\bar{m}(4) = M(A)\dot{n}(4, A) + M(B)\dot{n}(4, B) + M(C)\dot{n}(4, C)$ | [12] |

## Eight independent equations. Eight unknowns.

Equations [1 to 6] = 4 independent equations Equations [7 to 12] = 4 independent equations

Substitute:

$\dot{n}(1, A) = 4$    $\dot{n}(1, B) = 8$

$\dot{n}(1, C) = 12$    $\bar{m}(1) = 800$

$\dot{n}(3, A) = 3$    $\dot{n}(3, B) = 7$

$\dot{n}(3, C) = 9$    $\bar{m}(3) = 630$

Solve for the unknowns and check closure.

| Stream Table | | Heat Exchanger Solution | | | |
|---|---|---|---|---|---|
| Species | M | Stream | | | |
| | | 1 | 2 | 3 | 4 |
| | kg/kmol | kmol/h | | | |
| A | 20 | 4 | 4 | 3 | 3 |
| B | 30 | 8 | 8 | 7 | 7 |
| C | 40 | 12 | 12 | 9 | 9 |
| Total | kg/h | 800 | 800 | 630 | 630 |
| | HOT | | COLD | | |
| Mass IN | = 800 kg/h | Closure | 630 kg/h | Closure | |
| Mass OUT | = 800 kg/h | 100% | 630 kg/h | 100% | |

### E. Pump

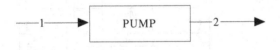

**Problem:** Complete the stream table material balance.

**Solution:** [System = pump]

| Stream Table | | Pump Problem | |
|---|---|---|---|
| Species | M | Stream | |
| | | 1 | 2 |
| | kg/kmol | kmol/h | |
| A | 20 | 4 | ? |
| B | 30 | 8 | ? |
| C | 40 | 12 | ? |
| Total | kg/h | 800 | ? |

| | | |
|---|---|---|
| Mole balance on A: | $0 = \dot{n}(1, A) - \dot{n}(2, A)$ | [1] |
| Mole balance on B: | $0 = \dot{n}(1, B) - \dot{n}(2, B)$ | [2] |
| Mole balance on C: | $0 = \dot{n}(1, C) - \dot{n}(2, C)$ | [3] |
| Total mole balance: | $0 = \bar{n}(1) - \bar{n}(2)$ | [4] |
| Total mass balance: | $0 = \bar{m}(1) - \bar{m}(2)$ | [5] |
| Stream mass: | $\bar{m}(2) = M(A)\dot{n}(2, A) + M(B)\dot{n}(2, B) + M(C)\dot{n}(2, C)$ | [6] |

## Four independent equations. Four unknowns.

Equations [1 to 6] = 4 independent equations

Substitute:

$\dot{n}(1, A) = 4$        $\dot{n}(1, B) = 8$

$\dot{n}(1, C) = 12$        $\bar{m}(1) = 800$

Solve for the unknowns and check closure.

| Stream Table | | Pump Solution | |
|---|---|---|---|
| Species | M | Stream | |
| | | 1 | 2 |
| | kg/kmol | kmol/h | |
| A | 20 | 4 | 4 |
| B | 30 | 8 | 8 |
| C | 40 | 12 | 12 |
| Total | kg/h | 800 | 800 |
| Mass IN | 800 kg/h | Closure | |
| Mass OUT | 800 kg/h | 800/800 = 100% | |

## F. Reactor

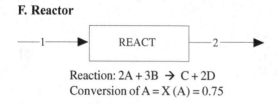

Reaction: $2A + 3B \rightarrow C + 2D$
Conversion of A = X (A) = 0.75

| Stream Table | | *Reactor Problem* | |
|---|---|---|---|
| Species | M | Stream | |
| | | 1 | 2 |
| | kg/kmol | kmol/h | |
| A | 20 | 4 | ? |
| B | 30 | 8 | ? |
| C | 40 | 12 | ? |
| D | 45 | 3 | ? |
| Total | kg/h | 935 | ? |

*Problem:* Complete the stream table material balance.
*Solution:* [System = reactor]

| | | | |
|---|---|---|---|
| Mole balance on A: | $0 = \dot{n}(1, A) - \dot{n}(2, A) + 0 - X(A)\dot{n}(1, A)$ | | [1] |
| Mole balance on B: | $0 = \dot{n}(1, B) - \dot{n}(2, B) + 0 - (3/2) \times (A)\dot{n}(1, A)$ | | [2] |
| Mole balance on C: | $0 = \dot{n}(1, C) - \dot{n}(2, C) + (1/2) \times (A)\dot{n}(1, A) - 0$ | | [3] |
| Mole balance on D: | $0 = \dot{n}(1, D) - \dot{n}(2, D) + \dot{n} (2/2) \times (A)\dot{n}(1, A) - 0$ | | [4] |
| Total mass balance: | $0 = \bar{m}(1) - \bar{m}(2)$ | | [5] |
| Stream total: | $\bar{m}(2) = M(A)\dot{n}(2, A) + M(B)\dot{n}(2, B) + M(C)\dot{n}(2, C)$ | | [6] |

*NOTE:* For reactants: Rate GEN = 0
　　　　　For products: Rate CON = 0

### Five independent equations. Five unknowns.

Equations [1 to 6] = 5 independent equations
Substitute:

$\dot{n}(1, A) = 4$ 　　 $\dot{n}(1, B) = 8$
$\dot{n}(1, C) = 12$ 　 $\dot{n}(1, D) = 3$

Solve for the unknowns and check closure.

| Stream Table | | *Reactor Solution* | |
|---|---|---|---|
| Species | M | Stream | |
| | | 1 | 2 |
| | kg/kmol | kmol/h | |
| A | 20 | 4 | 1 |
| B | 30 | 8 | 3.5 |
| C | 40 | 12 | 13.5 |
| D | 45 | 3 | 6 |
| Total | kg/h | 935 | 935 |
| Mass IN | 935 kg/h | Closure | |
| Mass OUT | 935 kg/h | 935/935 = 100% | |

### SUBSIDIARY RELATIONS

The material balances on mixers, indirect heat exchangers and pumps set out in *Figure 4.01* and *Example 4.04* are straightforward and self-explanatory. The material balances on dividers, separators and reactors use subsidiary relations that need explanation.

### *Dividers*

Dividers convert an input stream to multiple output streams, all with the *same composition* as the inlet stream. The fixed composition of the streams gives a set of equations called the "composition restrictions" that can be manipulated in several ways to solve the material balance. For example, the same composition in each stream requires that:

$\dot{m}(1, j)/m(1) = \dot{m}(2, j)/\bar{m}(2) = \bar{m}(3, j)/\bar{m}(3)$, etc.

These relations can be re-arranged algebraically to give:

$\dot{m}(1, A)/\dot{m}(1, B) = \dot{m}(2, A)/\dot{m}(2, B) = \dot{m}(3, A)/\dot{m}(3, B)$ or
$\dot{m}(1, A)/\dot{m}(1, C) = \dot{m}(2, A)/\dot{m}(2, C) = \dot{m}(3, A)/\dot{m}(3, C)$, etc. where A, B, C are the components in stream 1.

Analogous relations apply for the mole flows. Also, the divider can be specified by the stream flow ratios $K_{DIV}(i)$, where:

$K_{DIV}(i) = \bar{n}(i)/\bar{n}(1) = \bar{m}(i)/\bar{m}(1) = \dot{m}(i, j)/\dot{m}(1, j) = \dot{n}(i, j)/\dot{n}(1, j)$ for each component "j"
and for the sum over all <u>outlet</u> streams: $\Sigma[K_{DIV}(i)] = 1$                          *Equation 4.06*

In some texts the divider is called a stream "splitter". That terminology is not used here because it is ambiguous with respect to the term "split fraction". Split fraction is a term used to specify the performance of separators, which have an entirely different function from that of dividers.


## *Separators*

Separators convert an input stream into multiple outlet streams of *different composition*. They are characterised by the *separation efficiency* for shifting each component into a specified output stream.

$$\frac{\text{Separation efficiency}}{\text{for component "j"}} = \frac{\text{amount of "j" in specified outlet stream}}{\text{amount of "j" in intlet stream}}$$

Separation efficiency is often indicated in process flowsheets as a *split fraction*, which is defined by *Equation 4.07:*

$$\mathbf{s(i, j) = \dot{n}(i, j)/\dot{n}(1, j) = \dot{m}(i, j)/\dot{m}(1, j)}$$                          *Equation 4.07*

where:

$s(i, j)$ = split fraction for component j, from inlet stream 1 to outlet stream i

Separators usually function by concentrating the components into different phases and then separating the phases (e.g. by gravity). Separation efficiency in real separators depends on several factors such as the phase equilibria, mass transfer rates and entrainment rates.[4]

---

[4] Mass transfer is the transport of material across a phase boundary, driven by concentration gradients. Entrainment is the capture of bubbles, drops or particles from one phase into the phase from which it is being separated, for example, micro-drops of liquid from bursting vapour bubbles, that are carried into the gas phase above a boiling liquid.

In simplified material balance calculations the separator split fractions are often assumed to be the values set by the equilibrium distribution of components between the phases leaving the separator (i.e. an *equilibrium separator*). This equilibrium distribution can be calculated from the inter-phase equilibrium relations given in Chapter 2, as shown in *Example 4.05*.

### EXAMPLE 4.05   Calculation of separator split fractions from phase equilibria.

**A. Liquid/liquid separator from the L/L distribution coefficient**

The solute C partitions between two mutually insoluble solvents A and B by the distribution coefficient:

$\quad$ D(C) = 2.57 (see *Equation 2.39*).

This system is used in the liquid/liquid separator of the flowsheet and stream table below.

**Solvent extractor**

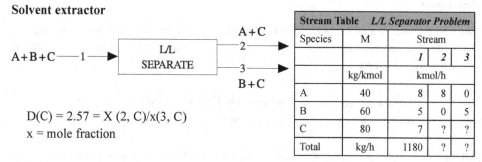

$\quad$ D(C) = 2.57 = X (2, C)/x(3, C)

$\quad$ x = mole fraction

| Stream Table | | L/L Separator Problem | | |
|---|---|---|---|---|
| Species | M | Stream | | |
| | | *1* | *2* | *3* |
| | kg/kmol | kmol/h | | |
| A | 40 | 8 | 8 | 0 |
| B | 60 | 5 | 0 | 5 |
| C | 80 | 7 | ? | ? |
| Total | kg/h | 1180 | ? | ? |

**Problem:** 1. Calculate the split fraction s(2, C) for the separation of C from stream 1 into stream 2.

$\qquad$ 2. Complete the stream table material balance.

**Solution:** [System = separator]

Mole balance on C: $\quad$ 0 = 7– ṅ(2, C) – ṅ(3, C) $\hfill$ [1]

Equilibrium composition: D(C) = 2.57 = ṅ(2, C)/[8+ ṅ(2, C)]/[ṅ(3, C)/[5 + ṅ(3, C)]] $\hfill$ [2]

Stream totals: $\quad$ m̄(3) = (40)(0) + (60)(5) + 80ṅ(3, C) $\hfill$ [3]

$\qquad\qquad\quad$ m̄(2) = (40)(8) + (60)(0) + 80ṅ(2, C) $\hfill$ [4]

**Four independent equations. Four unknowns.**

Simultaneous solution is required for equations [1 and 2].
Combine equations [1 and 2].

Solve the resulting non-linear (quadratic) equation to get:

$\quad$ ṅ(2, C) = 6 kmol/h

Split fraction: $s(2, C) = \dot{n}(2, C)/\dot{n}(1, C) = 6/7 = $ **0.86**

Solve for the unknowns and check closure.

| Stream Table | | *L/L Separator Solution* | | |
|---|---|---|---|---|
| Species | M | Stream | | |
| | | *1* | *2* | *3* |
| | kg/kmol | kmol/h | | |
| A | 40 | 8 | 8 | 0 |
| B | 60 | 5 | 0 | 5 |
| C | 80 | 7 | 6 | 1 |
| Total | kg/h | 1180 | 800 | 380 |
| Mass IN    1180 kg/h | | | | |
| Mass OUT 1180 kg/h  Closure = 1180/1180 = 100% | | | | |

### B. Solid/liquid separator from the S/L solubility

The solute B has a solubility[5] in solvent A of 40 wt% (i.e. 40 kg B/100 kg solution = 66.7 kg B/100 kg A) at 353 K and 20 wt% (i.e. 20 kg B/100 kg solution = 25 kg B/100 kg A) at 293 K.

This system is used in the solid/liquid separator of the flowsheet and stream table below.

*Problem:*

1. Calculate the split fraction $s(3, B)$ for the separation of B from stream 1 into stream 3.

2. Complete the stream table material balance.

| Stream Table | | *S/L Separator Problem* | | |
|---|---|---|---|---|
| Species | M | Stream | | |
| | | *1* | *2* | *3* |
| | kg/kmol | kg/s | | |
| A | 40 | 12 | ? | 0 |
| B | 60 | 8 | ? | ? |
| Total | kg/s | 20 | ? | ? |

*Solution:* [System = separator]

| Mass balance on A: | $0 \quad = 12 - \dot{m}(2, A) - 0$ | [1] |
|---|---|---|
| Mass balance on B: | $0 \quad = 8 - \dot{m}(2, B) - \dot{m}(3, B)$ | [2] |
| Equilibrium composition at 293 K: | $0.2 \quad = \dot{m}(2, B)/[\dot{m}(2, A) + \dot{m}(2, B)]$ | [3] |
| Stream totals: | $\dot{m}(2) = \dot{m}(2, A) + \dot{m}(2, B)$ | [4] |
| | $\dot{m}(3) = 0 + \dot{m}(3, B)$ | [5] |

---

[5] Solubility = concentration of solute in solution (i.e. in the liquid phase) at equilibrium with excess solid solute, at a specified temperature. Note that "solubility" may be expressed as:

   Mass solute/mass solution OR

   Mass solute/mass solvent OR

   Mass solute/volume solution (or solvent)   A "wt% solution" means 100 [mass solute/mass solution].

**Five independent equations. Five unknowns.**

Simultaneous solution is required for equations [1, 2 and 3].

Split fraction:

$s(3, B) = \dot{m}(3, B)/\dot{m}(1, B) = 5/8 = \mathbf{0.63}$

Solve for the unknowns and check closure.

Mass IN = 20 kg/s, Mass OUT = 20 kg/s, Closure = 100%

| Stream Table *S/L Separator Solution* | | | | |
|---|---|---|---|---|
| Species | M | Stream | | |
| | | *1* | *2* | *3* |
| | kg/kmol | kg/s | | |
| A | 40 | 12 | 12 | 0 |
| B | 60 | 8 | 3 | 5 |
| Total | kg/s | 20 | 15 | 5 |

### *Reactors*

Reactors differ from other process units because they transform material through chemical reactions, so the mass (or moles) of each *species* involved in the reaction is not conserved. However, there are no nuclear reactions in chemical process reactors so the *total mass is conserved*, as required for *Equation 4.08*.

Chemical reactor calculations can be based on either the conversion (*Equation 2.43*) or the extent of reaction (*Equation 2.44*), but the dimensionless conversion is used for most of the examples in this text. The reactor material balances shown in *Figure 4.01* and *Example 4.04* hinge on the conversion of a single reactant A. The consumption and generation of every other reactant and product is linked to the conversion of A by the reaction stoichiometry. If reactants are initially in stoichiometric proportions then the conversion will have the same value for all reactants, otherwise each reactant will have a different conversion — so it is important to specify which reactant conversion is considered in the balance equations. To simplify the algebra you should write the individual species material balances for a reactor in terms of mole quantities (not masses).

Conversion in real chemical reactors depends on the reaction rate and residence time[6] of the reaction mixture in the reaction vessel, which drive the reaction to approach equilibrium. In simplified material balance calculations the conversion is often assumed to be the *equilibrium conversion* ($X_{eq}$) calculated from the reaction equilibrium constant *Equation 2.48* (i.e. an *equilibrium reactor*). For more realistic reactor design the actual conversion ($X_{act}$) is measured or is calculated from the reaction rate as illustrated in *Example 4.06*. Actual conversion calculations are also shown in *Examples 2.16* and *7.03*.

---

[6] Residence time is the time spent in the reactor by the reaction mixture. In a continuous flow process: Residence time = (reactor volume)/(volume flow rate of reaction mixture).

**EXAMPLE 4.06    *Conversion in an isothermal continuous stirred tank reactor (CSTR) at steady-state.***

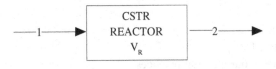

```
        ┌─────────────┐
        │    CSTR     │
──1──▶  │   REACTOR   │ ──2──▶
        │     V_R     │
        └─────────────┘
```

CSTR is "perfectly mixed"
[A] = concentration of A kmol.m$^{-3}$

Reaction: A → Products
*Irreversible reaction with first order kinetics.*

| | | |
|---|---|---|
| Reaction rate | = k[A] | kmol.m$^{-3}$.s$^{-1}$ |
| Rate constant | = k = 0.20 | s$^{-1}$ |
| Volume flow rate | = $\dot V(1) = \dot V(2) = 0.40$ | m$^3$s$^{-1}$ |
| Reactor volume | = $V_R = 5.0$ | m$^3$ |
| Residence time | = $V_R/\dot V(2) = 5/0.4 = 12.5$ | s |

CONVERSION IN A STIRRED TANK

In an ideal stirred tank the reactants are "perfectly mixed" and the composition of the outlet Stream 2 is the same as that of the fluid in the tank.

**Problem:** Calculate the steady-state conversion of A.
**Solution:** [System = Reactor]

## 0 = Rate IN – Rate OUT + Rate GEN – Rate CON                    *Equation 4.03*

Mole balance on A: $0 = \dot n(1, A) - \dot n(2, A) + 0 - k[\dot n(2, A)/\dot V(2)]V_R$          kmol.s$^{-1}$ [1]

Solve for $\dot n(2,A)$:    $\dot n(2, A) = \dot n(1, A)/[1 + kV_R/\dot V(2)]$          kmol.s$^{-1}$ [2]
Conversion[7] of A:    $X(A) = [\dot n(1, A) - \dot n(2, A)]/\dot n(1, A) = 1 - 1/[1 + kV_R/\dot V(2)] -$     [3]
Substitute:    $k = 0.20$ s$^{-1}$     $\dot V(2)$ 0.40 m$^3$s$^{-1}$     $V_R = 5.0$ m$^3$

$$X(A) = 1 - 1[1 + (0.2/0.4)5] = \underline{\mathbf{0.71}}$$

### CLOSURE OF MATERIAL BALANCES

For continuous operation at steady-state the material balance always reduces to *Equation 4.08*.

*Equation 4.08* is the overall MASS BALANCE that applies to all chemical processes and individual process units at steady-state.

Closure

---

[7]Equation [3] in *Example 4.06* is the "design equation" for a CSTR with first order (irreversible) reaction kinetics. This equation is used to find the size of a reactor for a desired conversion and production rate.

$$0 = \Sigma[\bar{m}(i)_{in}] - \Sigma[\bar{m}(i)_{out}]$$     [Steady-state mass balance]     *Equation 4.08*

*assumes no nuclear reactions*

where:

$[\bar{m}(i)_{in}]$ = sum of all input mass flow rates

$[\bar{m}(i)_{out}]$ = sum of all output mass flow rates

*Equation 4.08* gives a simple way to check your material balances.

For *steady-state* material balances you should always check that:

Total mass flow rate in to the system = Total mass flow rate out of the system

[i.e. the *closure* defined below is within acceptable limits of 100%]

The *system* can be an individual process unit, a collection of process units or a complete plant and must be clearly defined for every case. In practical systems where material flows are based on measurements there is nearly always a discrepancy between mass in and mass out. This discrepancy is measured by the *closure* of the mass balance, defined in *Equation 4.09*:

$$e_M = 100\Sigma[\bar{m}(i)_{out}]/\Sigma[\bar{m}(i)_{in}]$$     *Equation 4.09*

where:$e_M$ = mass balance closure %.

The acceptable value of "$e_M$" depends on the type and purpose of the material balance. For exact modelling calculations such as in *Example 4.05* the closure should be 100.0%, but with crude measurements on a real operating system, a closure of 90% to 110% may be acceptable. For chemical process design calculations[8] the closure is typically set above 99.9%. Closure is an important issue in the iterative modelling calculations described later in this chapter.

### SIMULTANEOUS EQUATIONS

Steady-state material balances typically generate a set of algebraic equations that must be solved for one or more unknowns. The cases in *Example 4.04* gave simple sets of uncoupled linear algebraic equations that were solved explicitly for each unknown. More complex cases may involve simultaneous linear or non-linear algebraic equations, as shown in *Example 4.07*.

Simultaneous linear equations can be solved by standard procedures in programmable calculators or spreadsheets (see *Ref. 11*). However, in many cases a formidable looking set of material balance linear equations can be reduced to parts that are be easily solved by "hand". The solution of single non-linear equations usually resorts to machine calculations with numerical methods, such as Newton's method or bisection (see *Refs. 3, 11*). Numerical

---

[8] Detailed process design calculations should have: 99.9% < $e_M$ < 100.1% or better. Improper closure can mean neglecting small amounts of critical materials that have serious negative effects on the process operation. Preliminary (conceptual) process design may be less rigorous, with say: 90% < $e_M$ < 110%, although a value of $e_M$ much above 100% is usually a sign that something is amiss with the material balance equations.

methods are used, for example, in the calculator and spreadsheet routines called "Goal Seek" and "Solver". Simultaneous non-linear equations usually require more sophisticated numerical methods of solution, some of which are described in *Refs. 3, 6, 8, 11.*

*EXAMPLE 4.07   Material balances with simultaneous equations (open system at steady-state).*

**A. Separator with simultaneous linear equations**        (Continuous process at steady-state)

| Stream Table *Separator Problem* | | | | |
|---|---|---|---|---|
| Species | M | Stream | | |
| | | *1* | *2* | *3* |
| | kg/kmol | kmol/h | | |
| A | 40 | 40 | ? | 0 |
| B | 80 | 30 | ? | ? |
| Total | kg/h | 4000 | ? | ? |

*Problem:*

Complete the stream table material balances.

*Solution:* [System = separator]

| | | |
|---|---|---|
| Mole balance on A: | $0 = \dot{n}(1, A) - \dot{n}(2, A) - \dot{n}(3, A) + 0 - 0$ | [1] |
| Mole balance on B: | $0 = \dot{n}(1, B) - \dot{n}(2, B) - \dot{n}(3, B) + 0 - 0$ | [2] |
| Overall mole balance: | $0 = \bar{n}(1) - \bar{n}(2) - \bar{n}(3) + 0 - 0$ | [3] |
| Overall mass balance: | $0 = \bar{m}(1) - \bar{m}(2) - \bar{m}(3) + 0 - 0$ | [4] |
| Stream mass: | $\bar{m}(2) = M(A)\dot{n}(2, A) + M(B)\dot{n}(2, B)$ | [5] |
| | $\bar{m}(3) = M(A)\dot{n}(3, A) + M(B)\dot{n}(3, B)$ | [6] |
| Stream moles: | $\bar{n}(2) = \dot{n}(2, A) + \dot{n}(2, B)$ | [7] |
| | $\bar{n}(3) = \dot{n}(3, A) + \dot{n}(3, B)$ | [8] |
| Stream compositions: | $x(2, A) = \dot{n}(2,A)/\bar{n}(2)$ | [9] |
| $x(i, j)$ = mole fraction j in i | $x(3, A) = \dot{n}(3,A)/\bar{n}(3)$ | [10] |
| | $1 = x(2, A) + x(2, B)$ | [11] |
| | $1 = x(3, A) + x(3, B)$ | [12] |

**Six independent equations. Six unknowns. Simultaneous solution is required.**

[Note that 6 of equations [1 to 12] are redundant]
Combine equations [9–12] with equations [1 and 2]

$$0 = \dot{n}(1, A) - x(2,A)\bar{n}(2) - x(3,A)\bar{n}(3) \qquad [13]$$
$$0 = \dot{n}(1, B) - (1 - x(2, A))\bar{n}(2) - (1 - x(3, A))\bar{n}(3) \qquad [14]$$

Substitute:

$\dot{n}(1, A) = 40 \qquad \dot{n}(1, B) = 30$

$x(2, A) = 0.75 \quad x(3, A) = 0.33$

$0 = 40 - 0.75\bar{n}(2) - 0.33\bar{n}(3) \qquad [15]$

$0 = 30 - 0.25\bar{n}(2) - 0.67\bar{n}(3) \qquad [16]$

Solve simultaneous linear equations [15 and 16].
Check closure.

| Stream Table *Separator   Solution* | | | | |
|---|---|---|---|---|
| Species | M | Stream | | |
| | | *1* | *2* | *3* |
| | kg/kmol | kmol/h | | |
| A | 40 | 40 | 30 | 10 |
| B | 80 | 30 | 10 | 20 |
| Total | kg/h | 4000 | 2000 | 2000 |
| Mass IN   = 4000 kg/h Closure % | | | | |
| Mass OUT = 4000 kg/h 4000/4000 = 100 | | | | |

## B. Separator with simultaneous non-linear equations    (isothermal flash split)

Ideal vapour/liquid system

| Stream Table | *Separator* | *Problem* | | |
|---|---|---|---|---|
| Species | M | Stream | | |
| | | *1* | *2* | *3* |
| | kg/kmol | kmol/h | | |
| A | 40 | 40 | ? | 0 |
| B | 80 | 30 | ? | ? |
| Total | kg/h | 4000 | ? | ? |

Fixed pressure and temperature:

k (A) = p*(2, A)/P = 0.1
k (B) = p*(2, B)/P = 10

*Solution:* [System = Separator]

| | | |
|---|---|---|
| Mole balance on A: | $0 = \dot{n}(1, A) - \dot{n}(2, A) - \dot{n}(3, A) + 0 - 0$ | [1] |
| Mole balance on B: | $0 = \dot{n}(1, B) - \dot{n}(2, B) - \dot{n}(3, B) + 0 - 0$ | [2] |
| Overall mole balance: | $0 = \bar{n}(1) - \bar{n}(2) - \bar{n}(3) + 0 - 0$ | [3] |
| Overall mass balance: | $0 = \bar{m}(1) - \bar{m}(2) - \bar{m}(3) + 0 - 0$ | [4] |
| Stream mass: | $\bar{m}(2) = M(A)\dot{n}(2, A) + M(B)\dot{n}(2, B)$ | [5] |
| | $\bar{m}(3) = M(A)\dot{n}(3, A) + M(B)\dot{n}(3, B)$ | [6] |
| Stream moles: | $\bar{n}(2) = \dot{n}(2, A) + \dot{n}(2, B)$ | [7] |
| | $\bar{n}(3) = \dot{n}(3, A) + \dot{n}(3, B)$ | [8] |
| Stream composition: | $y(2, A) = \dot{n}(2, A)/\bar{n}(2)$ | [9] |
| | $x(3, A) = \dot{n}(3, A)/\bar{n}(3)$ | [10] |
| | $y(2, B) = \dot{n}(2, B)/\bar{n}(2)$ | [11] |
| | $x(3, B) = \dot{n}(3, B)/\bar{n}(3)$ | [12] |
| V/L equilibrium: | $y(2, A) = k(A) \times (3, A)$    Raoult's law (see *Equation 2.33*) | [13] |
| | $y(2, B) = k(B) \times (3, B)$ | [14] |
| | $y(i, j)$ = mole fraction j in gas i    $1 = y(2, A) + y(2, B)$ | [15] |
| | $x(i, j)$ = mole fraction j in liquid i    $1 = x(3, A) + x(3, B)$ | [16] |

**Six independent equations. Six unknowns. Simultaneous solution is required.**

[Note that 10 of equations [1 to 16] are redundant]
Combine equations [1 and 2] with [9 and 10].

$$0 = \dot{n}(1, A) - y(2, A)\bar{n}(2) - x(3, A)\bar{n}(3) \qquad [17]$$
$$0 = \dot{n}(1, B) - y(2, B)\bar{n}(2) - x(3, B)\bar{n}(3) \qquad [18]$$

Combine equations [13 and 14] with [17 and 18]

$$0 = \dot{n}(1, A) \, k(A) \times (3, A)\bar{n}(2) - \times (3, A)\bar{n}(3) \qquad [19]$$
$$0 = \dot{n}(1, B) - k(A) \times (3, B)\bar{n}(2) - \times (3, B)\bar{n}(3) \qquad [20]$$

Solve equations [19 and 20] for ×(3, A) and ×(3, B)

$$x(3, A) = \dot{n}(1, A)/[k(A)\bar{n}(2) + \bar{n}(3)] \qquad [21]$$
$$x(3, B) = \dot{n}(1, B)/[k(B)\bar{n}(2) + \bar{n}(3)] \qquad [22]$$

Combine equations [3 and 16] with [21 and 22]

$$1 = \dot{n}(1, A)/[\bar{n}(2)(k(A) - 1) + \bar{n}(1)] + \dot{n}(1, B)/[\bar{n}(2)(k(B) - 1) + \bar{n}(1)] \qquad [23]$$

Substitute:

$\dot{n}(1, A) = 40$ $\dot{n}(1, B) = 30$ $\bar{n}(1) = 70$ $k(A) = 0.1$ $k(B) = 10$

$1 = 40/[\bar{n}(2)(0.1 - 1) + 70] + 30/[\bar{n}(2)(10 - 1) + 70]$                    [24]

Solve the non-linear equation [24] for $\bar{n}(2)$.

Substitute into prior equations for the full result.

*NOTE:* Equation [23] is the classic "isothermal flash split", often seen as:

$1 = \Sigma\{z(j)/[v(k(j) - 1) + 1]\}$

where:

z(j) = mole fraction of j in feed (stream 1)

k(j) = equilibrium constant of j at the conditions in
   the separator

v = moles vapour/moles feed = $\bar{n}(2)/\bar{n}(1)$

This method can be extended to a multi-component
flash with (J > 2), where conditions (P, T) are
between the bubblepoint and dew-point.

| Stream Table | | *Separator Solution* | | |
|---|---|---|---|---|
| Species | M | Stream | | |
| | | *1* | *2* | *3* |
| | kg/kmol | kmol/h | | |
| A | 40 | 40.0 | 2.6 | 37.4 |
| B | 80 | 30.0 | 26.3 | 3.7 |
| Total | kmol/h | 70 | 28.9 | 41.1 |
| Total | kg/h | 4000 | 2207 | 1793 |
| Mass IN | kg/h | 4000 | Closure % | |
| Mass OUT | kg/h | 4000 | 100 | |

### SPECIFICATION OF MATERIAL BALANCE PROBLEMS

When you are working on a material balance problem with many variables and equations it can be easy to lose track of what is known and what is unknown. Process designers call this issue the "specification" of a process, and it is summarised in the following set of relations:

   Number of unknown quantities > number of independent equations: UNDER-SPECIFIED
   Number of unknown quantities = number of independent equations: FULLY-SPECIFIED
   Number of unknown quantities < number of independent equations: OVER-SPECIFIED

**Degrees of freedom** = **number of unknown** – **number of independent**   *Equation 4.10*
**(D of F)**                      **quantities**                      **equations**

An *under-specified* problem (D of F > 0) is essentially a *design problem*, in which the designer manipulates the values of the process variables to satisfy the design objective. Process design problems often become *optimisation* problems, where the design objective is to maximise a *figure of merit* such as the *return on investment.*

A *fully-specified* problem (D of F = 0) has no scope for manipulation. All values are uniquely fixed by the conditions specified in the problem statement, so the solution is just a matter of writing and solving the equations. This type of problem is usually stated in the form "given the inputs — find the single correct output" and is most familiar to under-graduate students in engineering and science (until they encounter a course in design).

An *over-specified* problem (D of F < 0) is sometimes caused by redundancy of correct information, but usually it is the result of an error. An over-specification error will give you the impossible task of finding results that are mutually incompatible. It is easy to inadvertently and incorrectly over-specify a complex material balance problem, then to go into a hopeless loop trying to find the solution. Try to avoid this type of error.

As you approach a material balance problem you should compare the number of unknown quantities with the number of *independent equations*[9] by the criteria listed above. When the problem is *fully-specified* then you can proceed to find a unique solution. When the problem is *under-specified* you will not be able to find a unique solution without more information. If insufficient data are available then you may treat the problem as a *design problem.* Design problems are beyond the scope of this text but you can read about them in *Refs. 8–10.*

When the problem is *over-specified* you should suspect an error and check the validity of each equation, then remove any invalid equations. If all independent equations of an over-specified problem are valid, then you can choose the set that yields the most efficient or the most accurate solution.

*Example 4.08* illustrates the concept of process specification with a single process unit and *Table 4.02* summarises some rules of thumb for determining the degrees of freedom of multi-unit processes.

Unless indicated otherwise, all problems in this text are *fully-specified.*

***EXAMPLE 4.08 Specification of material balance problems (open system at steady-state).***

**A. Under-specification of a mixer**

| **Stream Table** *Mixer Specification* | | | | |
|---|---|---|---|---|
| Species | M | Stream | | |
| | | *1* | 2 | *3* |
| | kg/kmol | kmol/h | | |
| A | 40 | 8 | ? | 14 |
| B | 60 | 5 | ? | ? |
| Total | kg/h | 620 | ? | ? |

***Problem:*** Complete the material balance stream table.
***Solution:***

Mole balance on A: $0 = 8 + \dot{n}(2, A) - 14$ [1]

Mole balance on B: $0 = 5 + \dot{n}(2, B) - \dot{n}(3, B)$ [2]

$\bar{m}(2) = 40\dot{n}(2, A) + 60\dot{n}(2, B)$ [3]

$\bar{m}(3) = 40\dot{n}(14) + 60\dot{n}(3, B)$ [4]

**Five unknowns. Four independent equations.**

D of F = 5 – 4 = 1 D of F > 0 **UNDER-SPECIFIED**

or (*not counting stream totals*)

---

[9] An *independent equation* is one that cannot be obtained by combining other equations in the set.

Number of stream variables = IJ + L = 6
Number (independent equations + specified values) = 5      D of F = 6 – 5 = 1
There is an infinite number of (potential) solutions to problem A.
A process design aims for the "optimum" solution.

## B. Full-specification of a mixer

| Stream Table | Mixer Specification | | | |
|---|---|---|---|---|
| Species | M | Stream | | |
| | | *1* | *2* | *3* |
| | kg/kmol | kmol/h | | |
| A | 40 | 8 | ? | 14 |
| B | 60 | 5 | 2 | ? |
| Total | kg/h | 620 | ? | ? |

**Problem:** Complete the material balance stream table.
**Solution:**

Mole balance on A:    $0 = 8 + ṅ(2, A) – 14$                                     [1]
Mole balance on B:    $0 = 5 + 2 – ṅ(3, B)$                                       [2]
                      $m̄(3) = (40)ṅ(2, A) + 60ṅ(3, B)$                           [3]
                      $m̄(3) = (40)(14) + 60ṅ(3, B)$                              [4]

**Four unknowns. Four independent equations.**

D of F = 4 – 4 = 0 D of F = 0          **FULLY-SPECIFIED**
or (*not counting stream totals*)

Number of stream variables = IJ + L = 6
Number (independent equations + specified values) = 6      D of F = 6 – 6 = 0

## C. Over-specification of a mixer (redundant)

| Stream Table | Mixer Specification | | | |
|---|---|---|---|---|
| Species | M | Stream | | |
| | | *1* | *2* | *3* |
| | kg/kmol | kmol/h | | |
| A | 40 | 8 | ? | 14 |
| B | 60 | 5 | 2 | ? |
| Total | kg/h | 620 | ? | ? |

Stream 3 density = 870.2 kg/m3

Density A = 900 kg/m³, B = 797 kg/m³
Ideal liquid mixtures.

**Problem:** Complete the material stream table.
**Solution:**

Mole balance on A:  $0 = 8 + ṅ(2, A) – 14$              [1]
Mole balance on B:  $0 = 5 + ṅ(2, B) – ṅ(3, B)$                                 [2]
                    $m̄(2) = 40ṅ(2, A) + 60ṅ(2, B)$                              [3]
                    $m̄(3) = (40)(14) + 60ṅ(3, B)$                               [4]
Stream 3 density:   $ρ(3) = 870.2 \ kg/m^3$                                     [5]

**Four unknowns. Five "independent" equations.**

D of F = 4 – 5 = –1 D of F < 0      **OVER-SPECIFIED and REDUNDANT**
or (*not counting stream totals*)
Number of stream variables = IJ + L = 6
Number (independent equations + specified values) = 7      D of F = 6 – 7 = –1

### D. Over-Specification of a mixer (incorrect)

Stream 3 velocity = 2 m/s

Stream 3 pipe ID = 0.02 m

| Stream Table | *Mixer Specification* | | |
|---|---|---|---|
| Species | M | Stream | | |
| | | *1* | *2* | *3* |
| | kg/kmol | kmol/h | | |
| A | 40 | 8 | ? | 14 |
| B | 60 | 5 | 2 | ? |
| Total | kg/h | 620 | ? | ? |

Density A = 900 kg/m$^3$, B = 797 kg/m$^3$

Ideal liquid mixtures.

***Problem:*** Complete the material stream table.

***Solution:***

Mole balance on A: $0 = 8 - \dot{n}(2, A) - 14$           [1]

Mole balance on B: $0 = 5 + 2 - \dot{n}(2, B)$           [2]

       $\bar{m}(2) = 40\dot{n}(2, A) + 60\dot{n}(2, B)$        [3]

       $\bar{m}(3) = (40)(14) + 60\dot{n}(3, B)$        [4]

Stream 3 velocity in 0.02 m internal diameter ID outlet pipe = $\tilde{u}(3) = 2$ m/s [5]

**Four unknowns. Five independent equations.**

D of F = 4 – 5 = –1 D of F < 0 **OVER-SPECIFIED and INCORRECT** (*a.k.a. incon-*

                           *sistent*)

or (*not counting stream totals*)

 Number of stream variables = IJ + L = 6

 Number (independent equations + specified values) = 7  D of F = 6 – 7 = –1

### MULTIPLE PROCESS UNITS

Most chemical processes are systems with multiple process units joined in sequence through their process streams. Material balances on systems with multiple process units are obtained by simply adding the single unit balances, in the order of the process sequence. In these systems the output of one unit becomes the input of the next unit in the sequence, so individual units are *coupled* through their process streams.

There are two common methods used to solve the material balance equations for multiple process units:

- The method of *sequential modular solution*. Equations for each process unit are solved in sequence.
- The method of *simultaneous solution*. Equations describing the whole system are solved simultaneously.

Once it has been "set up" the method of simultaneous solution is faster and more efficient than the method of sequential solution. However the method of sequential solution is used for most examples in this text because it is easier to apply, more transparent and less error prone than the method of simultaneous solution. The method of sequential modular solution of multi-unit material balance problems is illustrated in *Examples 4.09A* and *4.10A*, while *Examples 4.09B* and *4.10B* illustrate the method of simultaneous solution.

**Table 4.02.   Specification (degrees of freedom) of multi-unit process material balance.**

| Process unit | Divide | Mix | Separate | Hex | Pump | React | OVERALL |
|---|---|---|---|---|---|---|---|
| Stream variables | IJ | IJ | IJ | IJ | IJ | IJ + L | IJ + L |
| Species balance equations | J | J | J | J | J | J | J K[10] |
| Element balance equations | A | A | A | A | A | $< = A^{10}$ | $< = A K^{10}$ |
| Subsidiary relations | Composition restrictions (D – 1)(J – 1) Flow ratios | None | Sep. efficiency Split fractions Flow ratios | None | None | Conversion Extent Selectivity Yield | Sum all units |
| Specified quantities Flows Compositions | BASIS of calculations.[11] Stream or component flows in mass, moles or volume per time Stream compositions in mass, mole or volume fraction (or %) | | | | | | |

A = number of elements  J = number of components (usually chemical species)
D = number of streams out of a divider  K = number of process units
I = number of streams  L = number of independent chemical reactions[12]

For balances in multi-unit processes it is easiest to begin sequential calculations at a unit that is fully specified (i.e. D of F = 0). However, it is not necessary that there will be a unit with D of F = 0

The D of F for a material balance can be determined by either of two methods.

**A.** [Preferred method] Count the total number of stream variables and subtract the number of independent relations.

D of F = number of stream variables − number of independent relations

The number of stream variables for each process unit is tabulated above and includes the known component flows.

Independent relations ≡ Basis + known stream specifications + material balances + subsidiary relations

Stream specifications ≡ Total stream flows, component flows, stream compositions

Subsidiary relations  ≡ Divider composition restrictions and flow ratios, separator split fractions and stream equilibria, reactant conversion, extent of reaction, selectivity, yield, etc.

In method A the independent relations include the basis of calculations and the known stream specifications. When atom balances are used the number of stream variables *excludes* the number of reactions (L).

**B.** Count the unknown flows and subtract the number of independent equations.

D of F = number of unknowns − number of independent equations

Independent equations ≡ Stream compositions + material balances + subsidiary relations

In method B, if total stream flows are used then the independent equations include the stream summations:

$\bar{n}(i) = \sum[\dot{n}(i, j)]$ or $\bar{m}(i) = \sum[M(j)\dot{n}(i, j)]$

When atom balances are used the number of unknowns *excludes* the number of reactions (L).

---

[10] When atoms or molecules occur in fixed ratios, the individual species balances are not independent equations.

[11] Setting a *basis of calculations* means arbitrarily fixing the value of one unknown quantity, without over-specifying the problem. e.g. fixing the total mass, total flow rate, mass or flow of one species, etc.

[12] Independent reactions are those whose stoichiometry cannot be obtained by combining other reactions in the set.

## Example 4.09　Material balances on a multi-unit process (open system at steady-state).

### A. Sequential modular solution

Continuous process at steady-state

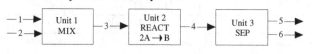

| Specifications | | Stream |
|---|---|---|
| | 1 (kmol/h) | 2 (kmol/h) |
| A (M=40) | 10.0 | 5.0 |
| B (M=80) | 0.0 | 1.0 |
| Conversion of A:X(A) | 0.6 | |
| Split | s(5,A) = | 0.9 |
| fractions | s(5,B) = | 0.2 |

**Problem:** Complete the stream table material balance.

| Specifications | Unit 1 | Unit 2 | Unit 3 | Overall |
|---|---|---|---|---|
| # stream variables | IJ = 6 | IJ + L = 5 | IJ = 6 | IJ + L = 9 |
| # specified values | 4 | 1 | 2 | 4 |
| + subsidiary relations | | | | |
| # species balances | 2 | 2 | 2 | 2 |
| D of F | 0 | 2 | 2 | 3 |

I = No. of streams, J = No. of components, L = No. of reactions

A.  Total stream variables = (6)(2) + 1 = 13
　　　Total independent relations (Units 1, 2, 3) = 13
　　　D of F = 13 – 13 = 0

B.  Total unknowns in the initial problem = (4)(2) + 1 = 9
　　　Total independent equations = 9 (includes split faction and
　　　conversion)
　　　D of F = 9 – 9 = 0

**Problem is FULLY-SPECIFIED**

This procedure does not count the stream total mass or mole flows
as unknowns, they are assumed known as the sum of the component flows.

### Solution:

UNIT 1 (Mixer)

Mole balance on A

$$0 = \dot{n}(1, A) + \dot{n}(2, A) - \dot{n}(3, A) \ [1]$$

Mole balance on B

$$0 = \dot{n}(1, B) + \dot{n}(2, B) - \dot{n}(3, B) \ [2]$$

UNIT 2 (Reactor)

Mole balance on A

$$0 = \dot{n}(3, A) - \dot{n}(4, A) + 0$$
$$- X(A)\dot{n}(3, A) \qquad\qquad [3]$$

Mole balance on B

$$0 = \dot{n}(3, B) - \dot{n}(4, B)$$
$$+ (1/2)X(A)\dot{n}(3, A) \qquad\qquad [4]$$

UNIT 3(Separator)

Mole balance on A

$$0 = \dot{n}(4, A) - \dot{n}(5, A) - \dot{n}(6, A) \qquad\qquad [5]$$

Mole balance on B

$$0 = \dot{n}(4, B) - \dot{n}(5, B) - \dot{n}(6, B) \qquad\qquad [6]$$

Specified values: s(5, A) = 0.9 = $\dot{n}(5, A)/\dot{n}(4, A)$　　　　　　[7]

Split fractions: s(5, B) = 0.2 = $\dot{n}(5, B)/\dot{n}(4, B)$　　　　　　[8]

Reactor conversion: X(A) = 0.6　　　　　　　　　　　　　　[9]

Stream flows:

$\dot{n}(1, A) = 10$

$\dot{n}(1, B) = 0$

$\dot{n}(2, A) = 5$

$\dot{n}(2, B) = 1$ kmol/h

Solve the process units in sequence — beginning at Unit 1.

| Stream Table | | *Mixer-Reactor-Separator* | | | *Solution* | | |
|---|---|---|---|---|---|---|---|
| Species | M | Stream | | | | | |
| | | *1* | *2* | *3* | *4* | *5* | *6* |
| | kg/kmol | | | kmol/h | | | |
| A | 40 | 10.0 | 5.0 | **15.0** | **6.0** | **5.4** | **0.6** |
| B | 80 | 0.0 | 1.0 | **1.0** | **5.5** | **1.1** | **4.4** |
| Total | kg/h | 400 | 280 | 680 | 680 | 304 | 376 |
| Check. | UNIT 1 | | UNIT 2 | | UNIT 3 | | OVERALL |
| Mass IN | 680 kg/h | Closure | 680 kg/h | Closure | 680 kg/h | Closure | 680 kg/h |
| Mass OUT | 680 kg/h | 100% | 680 kg/h | 100% | 680 kg/h | 100% | 680 kg/h |
| Overall yield of B from A = | | 2 (4.4)/15 = | | | 0.6 = 60% | | |

$\dot{n}(3, A) = 10 + 5 = 15$ kmol/h          $\dot{n}(3, B) = 0 + 1 = 1$ kmol/h

$\dot{n}(4, A) = (15)(1 - 0.6) = 6$ kmol/h   $\dot{n}(4, B) = 1 + (1/2)(0.6)(15) = 5.5$ kmol/h

$\dot{n}(5, A) = (0.9)(6) = 5.4$ kmol/h       $\dot{n}(5, B) = (0.2)(5.5) = 1.1$ kmol/h

$\dot{n}(6, A) = 6 - 5.4 = 0.6$ kmol/h        $\dot{n}(6, B) = 5.5 - 1.1 = 4.4$ kmol/h

## B. Simultaneous equation method

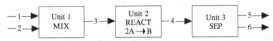

*Problem:* Complete the stream table material balance.

*Solution:*

| *Specifications* | | Stream |
|---|---|---|
| | *1* (kmol/h) | *2* (kmol/h) |
| A (M = 40) | 10.0 | 5.0 |
| B (M = 80) | 0.0 | 1.0 |
| Conversion of A: X(A) | | 0.6 |
| Split | s(5, A) = | 0.9 |
| fractions | s(5, B) = | 0.2 |

UNIT 1 (Mixer)

Mole balance on A:   $0 = \dot{n}(1, A) + \dot{n}(2, A) - \dot{n}(3, A)$                  [1]

Mole balance on B:   $0 = \dot{n}(1, B) + \dot{n}(2, B) - \dot{n}(3, B)$                  [2]

UNIT 2 (Reactor)

Mole balance on A:   $0 = \dot{n}(3, A) - \dot{n}(4, A) + 0 - X(A) \dot{n}(3, A)$        [3]

Mole balance on B:   $0 = \dot{n}(3, B) - \dot{n}(4, B) + (1/2)X(A) \dot{n}(3, A)$      [4]

UNIT 3 (Separator)

Mole balance on A:   $0 = \dot{n}(4, A) - \dot{n}(5, A) - \dot{n}(6, A)$                  [5]

Mole balance on B:   $0 = \dot{n}(4, B) - \dot{n}(5, B) - \dot{n}(6, B)$                  [6]

Specified values:    $s(5, A) = 0.9 = \dot{n}(5, A)/\dot{n}(4, A)$                         [7]

Split fractions:      $s(5, B) = 0.2 = \dot{n}(5, B)/\dot{n}(4, B)$                         [8]

Reactor conversion:  $X(A) = 0.6$                                                            [9]

Stream flows:        $\dot{n}(1, A) = 10$   $\dot{n}(1, B) = 0$   $\dot{n}(2, A) = 5$   $\dot{n}(2, B) = 1$ kmol/h

Specifications same as *Problem 4.09A*. By method B, *Table 4.02*: **Nine unknowns. Nine independent equations.** Write the set of eight simultaneous linear equations (*Equation [9]* is included in the reactor mole balances). Solve the set by matrix algebra.[13]

$$-15 = -(1)\dot{n}(3, A) + (0)\dot{n}(3, B) + (0)\dot{n}(4, A) + (0)\dot{n}(4, B) + (0)\dot{n}(5, A) + (0)\dot{n}(5, B)$$
$$+ (0)\dot{n}(6, A) + (0)\dot{n}(6, B) \tag{1}$$

$$-1 = +(0)\dot{n}(3, A) - (1)\dot{n}(3, B) + (0)\dot{n}(4, A) + (0)\dot{n}(4, B) + (0)\dot{n}(5, A) + (0)\dot{n}(5, B)$$
$$+ (0)\dot{n}(6, A) + (0)\dot{n}(6, B) \tag{2}$$

$$0 = +(0.4)\dot{n}(3, A) + (0)\dot{n}(3, B) - (1)\dot{n}(4, A) + (0)\dot{n}(4, B) + (0)\dot{n}(5, A) + (0)\dot{n}(5, B)$$
$$+ (0)\dot{n}(6, A) + (0)\dot{n}(6, B) \tag{3}$$

$$0 = +(0.3)\dot{n}(3, A) + (1)\dot{n}(3, B) + (0)\dot{n}(4, A) - (1)\dot{n}(4, B) + (0)\dot{n}(5, A) + (0)\dot{n}(5, B)$$
$$+ (0)\dot{n}(6, A) + (0)\dot{n}(6, B) \tag{4}$$

$$0 = +(0)\dot{n}(3, A) + (0)\dot{n}(3, B) + (1)\dot{n}(4, A) + (0)\dot{n}(4, B) - (1)\dot{n}(5, A) + (0)\dot{n}(5, B)$$
$$- (1)\dot{n}(6, A) + (0)\dot{n}(6, B) \tag{5}$$

$$0 = +(0)\dot{n}(3, A) + (0)\dot{n}(3, B) + (0)\dot{n}(4, A) + (1)\dot{n}(4, B) + (0)\dot{n}(5, A) - (1)\dot{n}(5, B)$$
$$+ (0)\dot{n}(6, A) - (1)\dot{n}(6, B) \tag{6}$$

$$0 = +(0)\dot{n}(3, A) + (0)\dot{n}(3, B) + (0.9)\dot{n}(4, A) + (0)\dot{n}(4, B) - (1)\dot{n}(5, A) + (0)\dot{n}(5, B)$$
$$+ (0)\dot{n}(6, A) + (0)\dot{n}(6, B) \tag{7}$$

$$0 = +(0)\dot{n}(3, A) + (0)\dot{n}(3, B) + (0)\dot{n}(4, A) + (0.2)\dot{n}(4, B) + (0)\dot{n}(5, A) - (1)\dot{n}(5, B)$$
$$+ (0)\dot{n}(6, A) + (0)\dot{n}(6, B) \tag{8}$$

| Coefficient Matrix [A] [8 by 8] | | | | | | | | Constant |
|---|---|---|---|---|---|---|---|---|
| $\dot{n}(3, A)$ | $\dot{n}(3, B)$ | $\dot{n}(4, A)$ | $\dot{n}(4, B)$ | $\dot{n}(5, A)$ | $\dot{n}(5, B)$ | $\dot{n}(6, A)$ | $\dot{n}(6, B)$ | Vector[C] |
| −1 | 0 | 0 | 0 | 0 | 0 | 0 | 0 | −15 |
| 0 | −1 | 0 | 0 | 0 | 0 | 0 | 0 | −1 |
| 0.4 | 0 | −1 | 0 | 0 | 0 | 0 | 0 | 0 |
| 0.3 | 1 | 0 | −1 | 0 | 0 | 0 | 0 | 0 |
| 0 | 0 | 1 | 0 | −1 | 0 | −1 | 0 | 0 |
| 0 | 0 | 0 | 1 | 0 | −1 | 0 | −1 | 0 |
| 0 | 0 | 0.9 | 0 | −1 | 0 | 0 | 0 | 0 |
| 0 | 0 | 0 | 0.2 | 0 | −1 | 0 | 0 | 0 |
| **Inverse Matrix [A⁻¹] [8 by 8]** | | | | | | | | **Solution** |
| | | | | | | | | **Vector[X]** |
| −1 | 0 | 0 | 0 | 0 | 0 | 0 | 0 | 15 $\dot{n}(3, A)$ |
| 0 | −1 | 0 | 0 | 0 | 0 | 0 | 0 | 1 $\dot{n}(3, B)$ |
| −0.4 | 0 | −1 | 0 | 0 | 0 | 0 | 0 | 6 $\dot{n}(4, A)$ |
| −0.3 | −1 | 0 | −1 | 0 | 0 | 0 | 0 | 5.5 $\dot{n}(4, B)$ |
| −0.36 | 0 | −0.9 | 0 | 0 | 0 | −1 | 0 | 5.4 $\dot{n}(5, A)$ |
| −0.06 | −0.2 | 0 | −0.2 | 0 | 0 | 0 | −1 | 1.1 $\dot{n}(5, B)$ |
| −0.04 | 0 | −0.1 | 0 | −1 | 0 | 1 | 0 | 0.6 $\dot{n}(6, A)$ |
| −0.24 | −0.8 | 0 | −0.8 | 0 | −1 | 0 | 1 | 4.4 $\dot{n}(6, B)$ |

[13] Solution vector = X = A⁻¹ C where A⁻¹ = inverse of coefficient matrix; C = constant vector.

| Stream Table | | Mixer-Reactor-Separator | | Solution | | | |
|---|---|---|---|---|---|---|---|
| Species | M | Stream | | | | | |
| | | *1* | *2* | *3* | *4* | *5* | *6* |
| | kg/kmol | | | kmol/h | | | |
| A | 40 | 10.0 | 5.0 | **15.0** | **6.0** | **5.4** | **0.6** |
| B | 80 | 0.0 | 1.0 | **1.0** | **5.5** | **1.1** | **4.4** |
| Total | kg/h | 400 | 280 | 680 | 680 | 304 | 376 |
| Check. | UNIT 1 | | UNIT 2 | | UNIT 3 | | OVERALL |
| Mass IN | 680 kg/h | Closure | 680 kg/h | Closure | 680 kg/h | Closure | 680 kg/h |
| Mass OUT | 680 kg/h | 100% | 680 kg/h | 100% | 680 kg/h | 100% | 680 kg/h |
| Overall yield of B from A = | | | 2 (4.4) /15 = | | 0.6 = 60% | | |

## RECYCLING, ACCUMULATION AND PURGING

Recycling is commonly used in chemical plants, both to conserve materials and for process control. The presence of a recycle stream in a process flowsheet complicates the material balance by coupling the output of one unit to the input of a previous unit in the sequence. Recycle balance problems can be solved indirectly by iterating the sequential modular method or directly by the method of simultaneous solution. Both methods are illustrated with a simple recycle flowsheet in *Example 4.10*.

The iterative sequential modular method is the basis of most commercial software for process modelling by computer. By this method one stream in the recycle loop is designated as the "tear stream" and given an arbitrary initial condition (e.g. zero flow). The material balance is then calculated in sequence and iterated through the tear point with each new condition of the tear stream. The iteration is terminated when the mass balance closure (*Equation 4.09*) converges to an acceptable value and/or shows little change with the number of iterations. Depending on the complexity of the problem the number of iterations required to reach acceptable closure (e.g. 99.9% or higher) can range from about 5 to 10,000. In bad cases the iteration may be unstable and never converge to a satisfactory point, though such cases can usually be resolved by numerical techniques such as *relaxation*.[14]

Recycling is a good way to increase the profitability and/or reduce the environmental impact of a chemical process. However the tendency of undesired materials to *accumulate* in recycle loops is a major problem in recycling. If they are not controlled, small amounts of undesired substances, such as inert "non-process elements",[15] catalyst poisons, etc. can

---

[14] *Relaxation* (specifically under-relaxation) means applying a damping factor to the progressive incremental change in an iterated value.

[15] *Non-process elements* are trace impurities that are not required for operation of the process, for example, *silicon* that is introduced with wood in the chemical pulping process.

build up in recycle loops to high levels that make the system inoperable. One way to control this accumulation is to *purge* a fraction of some stream(s) from the system to balance the rate of input of the undesired substance(s).

*Example 4.11A* shows how a small amount of inert substance (C) in a feed stream can accumulate in a recycle loop to virtually take over the process, then *Example 4.11B* shows that the accumulation is controlled by purging part of the recycle stream.

### EXAMPLE 4.10    *Material balances on a multi-unit recycle process (open system at steady-state).*

**A. Iterative sequential modular method**

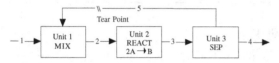

| Specifications | | Stream |
|---|---|---|
| | M | 1 |
| | kg/kmol | kmol/h |
| A | 40 | 10.0 |
| B | 80 | 0.0 |
| Conversion of A: X(A) | | 0.6/pass |
| Split | s(4, A) = | 0.1 |
| fractions | s(4, B) = | 0.8 |

**Problem:** Complete the stream table material balance.
**Solution:**

| UNIT 1 (Mixer) | Mole balance on A: | $0 = 10 + \dot{n}(5, A) - \dot{n}(2, A)$ | [1] |
|---|---|---|---|
| | Mole balance on B: | $0 = 0 + \dot{n}(5, B) - \dot{n}(2, B)$ | [2] |
| UNIT 2 (Reactor) | Mole balance on A: | $0 = \dot{n}(2, A) - \dot{n}(3, A) - 0.6\dot{n}(2, A)$ | [3] |
| | Mole balance on B: | $0 = \dot{n}(2, B) - \dot{n}(3, B) + (1/2)(0.6\dot{n}(2, A))$ | [4] |
| UNIT 3 (Separator) | Mole balance on A: | $0 = \dot{n}(3, A) - \dot{n}(4, A) - \dot{n}(5, A)$ | [5] |
| | Mole balance on B: | $0 = \dot{n}(3, B) - \dot{n}(4, B) - \dot{n}(5, B)$ | [6] |
| Split fractions: | | $0.1 = \dot{n}(4, A) / \dot{n}(3, A)$ | [7] |
| | | $0.8 = \dot{n}(4, B) / \dot{n}(3, B)$ | [8] |
| Conversion: | | Specified in the reactor balances | |

**FULLY-SPECIFIED. By method B, *Table 4.02.* Eight unknowns. Eight independent**
**equations.**

Equations are coupled through the recycle stream. Note also that the flows of A and B are coupled through the reactor. This procedure does not count the stream total mass or mole flows as unknowns, since they are assumed known as the sum of the component flows. Solve by the iterative sequential modular method shown below. Begin iterations with tear stream 5 "empty", i.e. $\dot{n}(5, A) = 0$.

**Sequenced explicit solution**  |  **Iteration number** kmol/h

| | 0 | 1 | 2 | 3 | 4 | 5 |
|---|---|---|---|---|---|---|
| ṅ(1, A) = 10 | 10 | 10 | 10 | 10 | 10 | 10 |
| ṅ(1, B) = 0 | 0 | 0 | 0 | 0 | 0 | 0 |
| ṅ(2, A) = 10 + ṅ(5, A) | 10.00 | 13.60 | 14.90 | 15.36 | 15.53 | 15.59 |
| ṅ(2, B) = 0 + ṅ(5, B) | 0.00 | 0.60 | 0.94 | 1.08 | 1.14 | 1.16 |
| ṅ(3, A) = ṅ (2, A) − 0.6 ṅ(2, A) | 4.00 | 5.44 | 5.96 | 6.15 | 6.21 | 6.24 |
| ṅ(3, B) = (2, B) + (1/2)0.6ṅ(2, A) | 3.00 | 4.68 | 5.40 | 5.69 | 5.80 | 5.84 |
| ṅ(4, A) = 0.1ṅ(3, A) | 0.40 | 0.54 | 0.60 | 0.61 | 0.62 | 0.62 |
| ṅ(4, B) = 0.8ṅ(3, B) | 2.40 | 3.74 | 4.32 | 4.55 | 4.64 | 4.67 |
| ṅ(5, A) = ṅ(3, A) − ṅ(4, A) | 3.60 | 4.90 | 5.36 | 5.53 | 5.59 | 5.61 |
| ṅ(5, B) = ṅ(3, B) − ṅ(4, B) | 0.60 | 0.94 | 1.08 | 1.14 | 1.16 | 1.17 |
| **Overall closure %** | **52.0** | **80.3** | **92.4** | **97.1** | **99.0** | **99.6** |

*NOTE:* The iterations in *Examples 4.10A*, *4.11A* and *4.11B* are recorded here to only illustrate the sequential iterative method. You will normally do such iterative calculations by computer (e.g. spreadsheet) as shown later in this text. It is sometimes useful, but generally not necessary to record the iteration value. With increasing number of iterations the closure approaches 100%, and the solution converges on the simultaneous solution of *Example 4.10 B*.

**Overall yield of B from A = 2 (4.67)/10 = 0.94 = 94%**

| Stream Table | | | *Mixer-Reactor-Separator+Recycle Solution at Iteration no. 5* | | | | |
|---|---|---|---|---|---|---|---|
| Species | M | | | | Stream | | |
| | | *1* | *2* | *3* | *4* | *5* | |
| | kg/kmol | | | kmol/h | | | |
| A | 40 | 10.00 | 15.59 | **6.24** | **0.62** | **5.61** | Overall |
| B | 80 | 0.00 | 1.16 | **5.84** | **4.67** | **1.17** | Closure % |
| Total | kg/h | 400.0 | 716.4 | 716.4 | 398.5 | 317.9 | 99.6 |
| Check | UNIT 1 | | UNIT 2 | | UNIT 3 | | OVERALL |
| Mass IN | 717.9 | Closure | 716.4 | Closure | 716.4 | Closure | 400.0 |
| Mass OUT | 716.4 | 99.8 | 716.4 | 100.0 | 716.4 | 100.0 | 398.5 |

Overall yield of B from A = 2(4.67)/10 = 0.94 = 94%

**B. Simultaneous equation method**

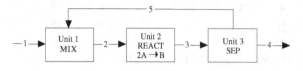

| Specifications | | Stream |
|---|---|---|
| | M | 1 |
| | kg/kmol | kmol/h |
| A | 40 | 10.0 |
| B | 80 | 0.0 |
| Conversion of A: X(A) | | 0.6/pass |
| Split | s(4, A) = | 0.1 |
| fractions | s(4, B) = | 0.8 |

*Problem:* Complete the stream table material balance.

*Solution:*

| | | |
|---|---|---|
| UNIT 1 (Mixer) | Mole balance on A: $0 = 10 + \dot{n}(5, A) - \dot{n}(2, A)$ | [1] |
| | Mole balance on B: $0 = 0 + \dot{n}(5, B) - \dot{n}(2, B)$ | [2] |
| UNIT 2 (Reactor) | Mole balance on A: $0 = \dot{n}(2, A) - \dot{n}(3, A) - 0.6\dot{n}(2, A)$ | [3] |
| | Mole balance on B: $0 = \dot{n}(2, B) - \dot{n}(3, B) + (1/2)(0.6\dot{n}(2, A))$ | [4] |
| UNIT 3 (Separator) | Mole balance on A: $0 = \dot{n}(3, A) - \dot{n}(4, A) - \dot{n}(5, A)$ | [5] |
| | Mole balance on B: $0 = \dot{n}(3, B) - \dot{n}(4, B) - \dot{n}(5, B)$ | [6] |
| Split fractions: | $0.1 = \dot{n}(4, A)/\dot{n}(3, A)$ | [7] |
| | $0.8 = \dot{n}(4, B)/\dot{n}(3, B)$ | [8] |
| Conversion: | Specified in the reactor balances | |

**FULLY-SPECIFIED. By method B, *Table 4.02*. Eight unknowns. Eight independent equations.**

Equations are coupled through the recycle stream. Note also the the flows of A and B are coupled through the reactor. This procedure does not count the stream total mass or mole flows as unknowns, since they are easily found as the sum of the component flows.

Simultaneous solution:

Combine equations [7 and 8] with [5 and 6]:  $0 = \dot{n}(3, A) - (0.1)\dot{n}(3, A) - \dot{n}(5, A)$   [9]

Combine equations [9 and 10] with [3 and 4]:  $0 = \dot{n}(3, B) - (0.8)\dot{n}(3, B) - \dot{n}(5, B)$   [10]

$0 = \dot{n}(2, A) - \dot{n}(5, A) / 0.9 - 0.6\dot{n}(2, A)$   [11]

$0 = \dot{n}(2, B) - \dot{n}(5, B) / 0.2 + (1/2)0.6\dot{n}(2, A)$   [12]

Combine equations [11 and 12] with [1 and 2]:  $0 = 10 + \dot{n}(5, A) - \dot{n}(5, A)/(0.36)$   [13]

$0 = 0 - 4\dot{n}(5, B) + 833\dot{n}(5, A)$   [14]

Yields the solution: $\dot{n}(5, A) = 5.625$ $\dot{n}(2, A) = 15.625$ $\dot{n}(3, A) = 6.250$ $\dot{n}(4, A) = 0.625$
$\dot{n}(5, B) = 1.172$ $\dot{n}(2, B) = 1.172$ $\dot{n}(3, B) = 5.860$ $\dot{n}(4, B) = 4.688$ kmol/h

| Stream Table | | | | Mixer-Reactor-Separator + Recycle Simultaneous Solution | | | |
|---|---|---|---|---|---|---|---|
| Species | M | Stream | | | | | |
| | | *1* | *2* | *3* | *4* | *5* | *6* |
| | kg/kmol | kmol/h | | | | | |
| A | 40 | 10.000 | 15.625 | **6.250** | **0.625** | **5.625** | Overall* |
| B | 80 | 0.000 | 1.172 | **5.860** | **4.688** | **1.172** | Closure % |
| Total | kg/h | 400.0 | 718.8 | 718.8 | 400.0 | 318.8 | 100.01 |
| Check | UNIT 1 | | UNIT 2 | | UNIT 3 | | OVERALL |
| Mass IN | 718.8 | Closure % | 718.8 | Closure % | 718.8 | Closure % | 400.0 |
| Mass OUT | 718.8 | 100.00 | 718.8 | 100.01 | 718.8 | 100.00 | 400.0 |

*Due to rounding of the stream flow values the overall closure is not exactly 100%.

**Overall yield of B from A = 2(4.688/10) = 0.94 = 94%**

*EXAMPLE 4.11   Material balance on a multi-unit recycle process (open system at steady-state).*

**A. Accumulation of non-process species**

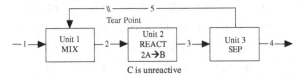

C is unreactive

| Specifications | Stream |
|---|---|
| | M | 1 |
| | kg/kmol | kmol/h |
| A | 40 | 10.0 |
| B | 80 | 0.0 |
| C | 50 | 0.1 |
| Conversion of A: X(A) | 0.6/pass |
| Split fractions | s(4,A) = | 0.1 |
| | s(4,B) = | 0.8 |
| | s(4,C) = | 0.05 |

*Problem:* Complete the stream table material balance.
*Solution:*

| | | | |
|---|---|---|---|
| UNIT 1 (Mixer) | Mole balance on A: | $0 = 10 + \dot{n}(5, A) - \dot{n}(2, A)$ | [1] |
| | Mole balance on B: | $0 = 0 + \dot{n}(5, B) - \dot{n}(2, B)$ | [2] |
| | Mole balance on C: | $0 = 0.1 + \dot{n}(5, C) - \dot{n}(2, C)$ | [3] |
| UNIT 2 (Reactor) | Mole balance on A: | $0 = \dot{n}(2, A) - \dot{n}(3, A) - 0.6\dot{n}(2, A)$ | [4] |
| | Mole balance on B: | $0 = \dot{n}(2, B) - \dot{n}(3, B) + (1/2)(0.6\dot{n}(2, A))$ | [5] |
| | Mole balance on C: | $0 = \dot{n}(2, C) - \dot{n}(3, C)$ [Unreactive] | [6] |
| UNIT 3 (Separator) | Mole balance on A: | $0 = \dot{n}(3, A) - \dot{n}(4, A) - \dot{n}(5, A)$ | [7] |
| | Mole balance on B: | $0 = \dot{n}(3, B) - \dot{n}(4, B) - \dot{n}(5, B)$ | [8] |
| | Mole balance on C: | $0 = \dot{n}(3, C) - \dot{n}(4, C) - \dot{n}(5, C)$ | [9] |
| Split fractions: | $0.1 = \dot{n}(4, A)/\dot{n}(3, A)$ | | [10] |
| | $0.8 = \dot{n}(4, B)/\dot{n}(3, B)$ | | [11] |
| | $0.05 = \dot{n}(4, C)/\dot{n}(3, C)$ | | [12] |
| | Conversion: | Specified in reactor balances | |

**FULLY-SPECIFIED. By method B, *Table 4.02*. 12 independent equations. 12 unknowns.**

Solve by the iterative sequential modular method. Begin iterations with tear stream 5 "empty".

| Sequenced explicit solutions | | | Iteration number kmol/h | | | |
|---|---|---|---|---|---|---|
| | 0 | 1 | 2 | 3 | 4 | 100 |
| $\dot{n}(1, A) = 10$ | 10 | 10 | 10 | 10 | 10 | 10 |
| $\dot{n}(1, B) = 0$ | 0 | 0 | 0 | 0 | 0 | 0 |
| $\dot{n}(1, C) = 0.1$ | 0.1 | 0.1 | 0.1 | 0.1 | 0.1 | 0.1 |
| $\dot{n}(2, A) = 10 + \dot{n}(5, A)$ | 10.00 | 13.60 | 14.90 | 15.36 | 15.53 | 15.63 |
| $\dot{n}(2, B) = 0 + \dot{n}(5, B)$ | 0.00 | 0.60 | 0.94 | 1.08 | 1.14 | 1.17 |
| $\dot{n}(2, C) = 0.1 + \dot{n}(5, C)$ | 0.10 | 0.20 | 0.29 | 0.37 | 0.45 | 2.00 |
| $\dot{n}(3, A) = \dot{n}(2, A) - 0.6\dot{n}(2, A)$ | 4.00 | 5.44 | 5.96 | 6.15 | 6.21 | 6.25 |
| $\dot{n}(3, B) = \dot{n}(2, B) + (1/2)0.6\dot{n}(2, A)$ | 3.00 | 4.68 | 5.40 | 5.69 | 5.80 | 5.85 |
| $\dot{n}(3, C) = \dot{n}(2, C)$ | 0.10 | 0.20 | 0.29 | 0.37 | 0.45 | 2.00 |
| $\dot{n}(4, A) = 0.1\dot{n}(3, A)$ | 0.40 | 0.54 | 0.60 | 0.61 | 0.62 | 0.63 |
| $\dot{n}(4, B) = 0.8\dot{n}(3, B)$ | 2.40 | 3.74 | 4.32 | 4.55 | 4.64 | 4.68 |
| $\dot{n}(4, C) = 0.05\dot{n}(3, C)$ | 0.005 | 0.010 | 0.014 | 0.019 | 0.023 | 0.100 |
| $\dot{n}(5, A) = \dot{n}(3, A) - \dot{n}(4, A)$ | 3.60 | 4.90 | 5.36 | 5.53 | 5.59 | 5.63 |
| $\dot{n}(5, B) = \dot{n}(3, B) - \dot{n}(4, B)$ | 0.60 | 0.94 | 1.08 | 1.14 | 1.16 | 1.17 |
| $\dot{n}(5, C) = \dot{n}(3, C) - \dot{n}(4, C)$ | 0.095 | 0.185 | 0.271 | 0.352 | 0.430 | 1.900 |
| **Overall closure %** | **51.42** | **79.45** | **91.47** | **96.21** | **98.02** | **99.93** |

| Stream Table | | *Mixer-Reactor-Separator+Recycle+Accumulation Solution at itn100* | | | | | |
|---|---|---|---|---|---|---|---|
| Species | M | | Stream | | | | |
| | | 1 | 2 | 3 | 4 | 5 | |
| | kg/kmol | | | kmol/h | | | |
| A | 40 | 10.00 | 15.63 | 6.25 | 0.63 | 5.63 | |
| B | 80 | 0.00 | 1.17 | 5.85 | 4.68 | 1.17 | Overall |
| C | 50 | 0.10 | 2.00 | 2.00 | 0.10 | 1.90 | Closure % |
| Total | kg/h | 405.0 | 818.4 | 818.4 | 404.7 | 413.4 | 99.93 |
| | UNIT 1 | | UNIT 2 | | UNIT 3 | | OVERALL |
| Mass IN | 818.4 | Closure % | 818.4 | Closure % | 818.4 | Closure % | 405.00 |
| Mass OUT | 818.4 | 100.0 | 818.4 | 100.0 | 818.1 | 100.0 | 404.71 |
| Overall yield of B from A = | | 2(4.68)/10 = 0.94 = 94% | | | | | |
| Note the accumulation of C in the recycle stream 5. | | | | | | | |

## B. Accumulation and purge of non-process species

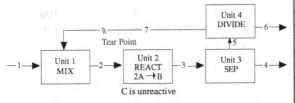

| Specifications | | Stream |
|---|---|---|
| | M | 1 |
| | kg/kmol | kmol/h |
| A | 40 | 10.0 |
| B | 80 | 0.0 |
| C | 50 | 0.1 |
| Conversion of A: X(A) | | 0.6/pass |
| Split fractions | s(4, A) = 0.1 | Divider |
| | s(4, B) = 0.8 | n(6)/n(5) = |
| | s(4, C) = 0.05 | $K_{DIV}$ (6) = 0.1 |

*Problem:* Complete the stream table material balance.

*Solution:*

| | | | |
|---|---|---|---|
| UNIT 1 (Mixer) | Mole balance on A: | $0 = 10 + \dot{n}(7, A) - \dot{n}(2, A)$ | [1] |
| | Mole balance on B: | $0 = 0 + \dot{n}(7, B) - \dot{n}(2, B)$ | [2] |
| | Mole balance on C: | $0 = 0.1 + \dot{n}(7, C) - \dot{n}(2, C)$ | [3] |
| UNIT 2 (Reactor) | Mole balance on A: | $0 = \dot{n}(2, A) - \dot{n}(3, A) - 0.6\dot{n}(2, A)$ | [4] |
| | Mole balance on B: | $0 = \dot{n}(2, B) - \dot{n}(3, B) + (1/2)(0.6\dot{n}(2, A))$ | [5] |
| | Mole balance on C: | $0 = \dot{n}(2, C) - \dot{n}(3, C)$ [Unreactive] | [6] |
| | Conversion: | Specified in reactor balances. | |
| UNIT 3 (Separator) | Mole balance on A: | $0 = \dot{n}(3, A) - \dot{n}(4, A) - \dot{n}(5, A)$ | [7] |
| | Mole balance on B: | $0 = \dot{n}(3, B) - \dot{n}(4, B) - \dot{n}(5, B)$ | [8] |
| | Mole balance on C: | $0 = \dot{n}(3, C) - \dot{n}(4, C) - \dot{n}(5, C)$ | [9] |
| | Split fractions: | $0.1 = \dot{n}(4, A)/\dot{n}(3, A)$ | [10] |
| | | $0.8 = \dot{n}(4, B)/\dot{n}(3, B)$ | [11] |
| | | $0.05 = \dot{n}(4, C)/\dot{n}(3, C)$ | [12] |
| UNIT 4 (Divider) | Mole balance on A: | $0 = \dot{n}(5, A) - \dot{n}(6, A) - \dot{n}(7, A)$ | [13] |
| | Mole balance on B: | $0 = \dot{n}(5, B) - \dot{n}(6, B) - \dot{n}(7, B)$ | [14] |
| | Mole balance on C: | $0 = \dot{n}(5, C) - \dot{n}(6, C) - \dot{n}(7, C)$ | [15] |
| | Divider ratios: | $0.1 = \dot{n}(6, A)/\dot{n}(5, A) = \dot{n}(6, B)/\dot{n}(5, B)$ | |
| | | $= \dot{n}(6, C)/\dot{n}(5, C)$ | [16][17][18] |

### FULLY-SPECIFIED. By method B, *Table 4.02.* 18 unknowns.
### 18 independent equations.

Solve by the *iterative sequential modular method* shown on next page. Begin iterations with tear stream "empty".

| Sequenced explicit solutions | | Iteration number kmol/h | | | | |
|---|---|---|---|---|---|---|
| | 0 | 1 | 2 | 3 | 4 | 100 |
| $\dot{n}(1, A)$ | 10 | 10 | 10 | 10 | 10 | 10 |
| $\dot{n}(1, B)$ | 0 | 0 | 0 | 0 | 0 | 0 |
| $\dot{n}(1, C) = 0.1$ | 0.1 | 0.1 | 0.1 | 0.1 | 0.1 | 0 |
| $\dot{n}(2, A) = 10 + \dot{n}(7, A)$ | 10.00 | 13.24 | 14.29 | 14.63 | 14.74 | 14.79 |
| $\dot{n}(2, B) = 0 + \dot{n}(7, B)$ | 0.00 | 0.54 | 0.81 | 0.92 | 0.96 | 0.97 |
| $\dot{n}(2, C) = 0.1 + \dot{n}(7, C)$ | 0.10 | 0.19 | 0.26 | 0.32 | 0.37 | 0.69 |

**Sequenced explicit solutions**

**Iteration number** kmol/h

| | 0 | 1 | 2 | 3 | 4 | 100 |
|---|---|---|---|---|---|---|
| $\dot{n}(3, A) = \dot{n}(2, A) - 0.6\dot{n}(2, A)$ | 4.00 | 5.30 | 5.72 | 5.85 | 5.90 | 5.92 |
| $\dot{n}(3, B) = \dot{n}(2, B) + (1/2)0.6\dot{n}(2, A)$ | 3.00 | 4.51 | 5.10 | 5.31 | 5.38 | 5.41 |
| $\dot{n}(3, C) = \dot{n}(2, C)$ | 0.10 | 0.19 | 0.26 | 0.32 | 0.37 | 0.69 |
| $\dot{n}(4, A) = 0.1\dot{n}(3, A)$ | 0.40 | 0.53 | 0.57 | 0.59 | 0.59 | 0.59 |
| $\dot{n}(4, B) = 0.8\dot{n}(3, B)$ | 2.40 | 3.61 | 4.08 | 4.25 | 4.30 | 4.33 |
| $\dot{n}(4, C) = 0.05\dot{n}(3, C)$ | 0.005 | 0.009 | 0.013 | 0.016 | 0.019 | 0.034 |
| $\dot{n}(5, A) = \dot{n}(3, A) - \dot{n}(4, A)$ | 3.60 | 4.77 | 5.14 | 5.27 | 5.31 | 5.33 |
| $\dot{n}(5, B) = \dot{n}(3, B) - \dot{n}(4, B)$ | 0.60 | 0.90 | 1.02 | 1.06 | 1.08 | 1.08 |
| $\dot{n}(5, C) = \dot{n}(3, C) - \dot{n}(4, C)$ | 0.095 | 0.176 | 0.246 | 0.305 | 0.356 | 0.655 |
| $\dot{n}(6, A) = 0.1\dot{n}(5, A)$ | 0.36 | 0.48 | 0.51 | 0.53 | 0.53 | 0.53 |
| $\dot{n}(6, B) = 0.1\dot{n}(5, B)$ | 0.060 | 0.090 | 0.102 | 0.106 | 0.108 | 0.108 |
| $\dot{n}(6, C) = 0.1\dot{n}(5, C)$ | 0.010 | 0.018 | 0.025 | 0.031 | 0.036 | 0.066 |
| $\dot{n}(7, A) = \dot{n}(5, A) - \dot{n}(6, A)$ | 3.24 | 4.29 | 4.63 | 4.74 | 4.78 | 4.79 |
| $\dot{n}(7, B) = \dot{n}(5, B) - \dot{n}(6, B)$ | 0.54 | 0.81 | 0.92 | 0.96 | 0.97 | 0.97 |
| $\dot{n}(7, C) = \dot{n}(5, C) - \dot{n}(6, C)$ | 0.086 | 0.159 | 0.221 | 0.275 | 0.320 | 0.59 |
| **Overall closure %** | **56.28** | **83.35** | **93.78** | **97.51** | **98.83** | **100.00** |

| Stream Table | *Mixer-Reactor-Separator+Recycle+Accumulation+Purge.* | | | | | | | *Solution at itn100* |
|---|---|---|---|---|---|---|---|---|
| Species | M | | | | Stream | | | |
| | | 1 | 2 | 3 | 4 | 5 | 6 | 7 | |
| | kg/kmol | | | | kmol/h | | | | |
| A | 40 | 10.00 | 14.79 | 5.92 | 0.59 | 5.33 | 0.53 | 4.79 | |
| B | 80 | 0.00 | 0.97 | 5.41 | 4.33 | 1.08 | 0.11 | 0.97 | **Overall** |
| C | 50 | 0.10 | 0.69 | 0.69 | 0.03 | 0.66 | 0.07 | 0.59 | **Closure %** |
| Total | kg/h | 405.0 | 704.1 | 704.1 | 371.8 | 332.4 | 33.2 | 299.1 | **100.0** |
| Check | UNIT 1 | | UNIT 2 | | UNIT 3 | | UNIT 4 | | OVERALL |
| Mass IN | 704.1 | Closure % | 704.1 | Closure % | 704.1 | Closure % | 332.4 | Closure % | 405.0 |
| Mass OUT | 704.1 | 100.0 | 704.1 | 100.0 | 704.1 | 100.0 | 332.4 | 100.0 | 405.0 |

**Overall yield of B from A = (2)(4.33)/10 = 0.87 = 87%**

**MATERIAL BALANCE CALCULATIONS BY COMPUTER (SPREADSHEET)**

Material balance problems often require a lot of calculations. The most efficient way to do these calculations is by computer. There are three ways that you can use a computer to help solve material balance problems:

1. By spreadsheet (e.g. Excel™, Quattro Pro™)
2. By writing your own "in house" code in a high level language (e.g. Basic, C, Fortran, MATLAB™)
3. By commercial "process simulation" software (e.g. ASPENPLUS™/HYSIS™, PRO/ II™, CHEMCAD™)

Process engineers use all of the above three methods, but this text focuses only on method 1, i.e. solving material (and energy) balances by *spreadsheet.*

As outlined in Chapter 3 there are two ways to set up a spreadsheet for material balances:

A.  As a separate *stream table* attached to the labelled flowsheet
B.  As sets of data adjacent to each stream in the flowsheet

In option A the *spreadsheet* is used as a two-dimensional matrix of cells in which the column and row values correspond to the stream and quantity entries in a process *stream table*. In option B the stream quantities are calculated and displayed in cells adjacent to each process stream.

For both options A and B the values in each cell are assigned or calculated from the appropriate balance equations. In simple cases each equation is solved explicitly for the corresponding cell value. In more complex cases (e.g. non-linear equations) the cell value may be found by a "GoalSeek" or "Solver" routine,[16] or calculated through a *macro* via Visual Basic code (see *Ref. 11*).

Iterative calculations in the spreadsheet first appear as "circular references". These are handled in Excel for example, by ticking the "iterations" box in the "tools-options-calculation" window. On the "calculate" command the computer will then run through the set number of iterations, or it can be programmed to terminate the iterations if the balance reaches acceptable *closure,* where the closure may be defined as in *Equation 4.09.* The acceptable closure value can be set as the *convergence criterion* in a conditional command (e.g. IF $e_M$ > 99.9% THEN stop iterating).

*Example 4.12* shows the spreadsheet solution of the recycle material balance problem in *Example 4.10* by both the method of option A and the method of option B. In this text option A is preferred for its superior transparency, and will be used in all subsequent spreadsheet calculation of steady-state material and energy balances.

*Examples 4.13* to *4.16* illustrate the set-up and solution of some more complex material balance problems. Each of these problems is solved by spreadsheet, using the iterative sequential modular method to produce a stream table attached to the process flowsheet.

*Examples 4.13, 4.14* and *4.15* show examples in process engineering, being respectively a biochemical process, an electrochemical process and a thermochemical process, each in continuous operation at steady-state. *Example 4.16* presents a case related to the Earth's environment and sustainable development.

*Example 4.17* shows how the concepts of conversion, extent of reaction and selectivity, together with atom and mole balances, can be used in different methods to calculate the material balance on a continuous combustion process at steady-state.

---

[16] "GoalSeek" can solve single linear and non-linear equations. The "Solver" is a more powerful tool that can solve both single and simultaneous linear and non-linear equations (see *Ref. 11*), although non-linear sets often give difficulties.

The best way to become competent with material balance calculations is to practice solving problems. The illustrative material balance problems on the next pages will get you started. You will find many more practice problems in the appendix and in *Refs. 2–9*.

**EXAMPLE 4.12** *Spreadsheet calculation of a material balance (open system at steady-state).*

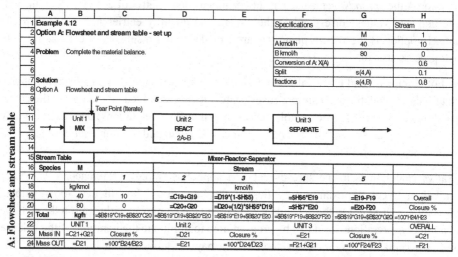

A: Flowsheet and stream table

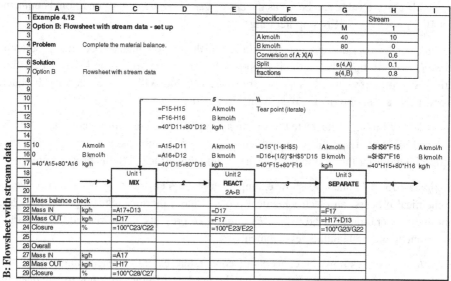

B: Flowsheet with stream data

**A: Flowsheet and stream table — solution**

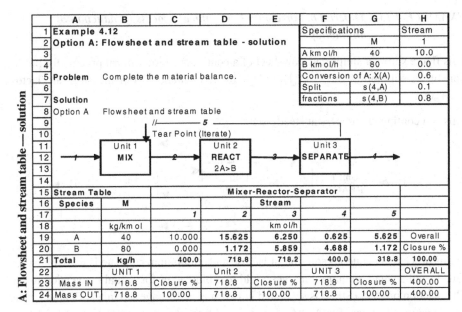

| | A | B | C | D | E | F | G | H |
|---|---|---|---|---|---|---|---|---|
| 1 | Example 4.12 | | | | | Specifications | | Stream |
| 2 | Option A: Flowsheet and stream table - solution | | | | | | M | 1 |
| 3 | | | | | | A km ol/h | 40 | 10.0 |
| 4 | | | | | | B km ol/h | 80 | 0.0 |
| 5 | Problem | Complete the material balance. | | | | Conversion of A: X(A) | | 0.6 |
| 6 | | | | | | Split | s (4,A) | 0.1 |
| 7 | Solution | | | | | fractions | s (4,B) | 0.8 |
| 8 | Option A | Flowsheet and stream table | | | | | | |
| 9 | | | | | | | | |
| 10 | | | | Tear Point (Iterate) | | | | |
| 11 | | | Unit 1 | | Unit 2 | | Unit 3 | |
| 12 | | | MIX | | REACT | | SEPARATE | |
| 13 | | | | | 2A>B | | | |
| 14 | | | | | | | | |
| 15 | Stream Table | | | | Mixer-Reactor-Separator | | | |
| 16 | Species | M | | | Stream | | | |
| 17 | | | 1 | 2 | 3 | 4 | 5 | |
| 18 | | kg/km ol | | | km ol/h | | | |
| 19 | A | 40 | 10.000 | 15.625 | 6.250 | 0.625 | 5.625 | Overall |
| 20 | B | 80 | 0.000 | 1.172 | 5.859 | 4.688 | 1.172 | Closure % |
| 21 | Total | kg/h | 400.0 | 718.8 | 718.2 | 400.0 | 318.8 | 100.00 |
| 22 | | UNIT 1 | | Unit 2 | | UNIT 3 | | OVERALL |
| 23 | Mass IN | 718.8 | Closure % | 718.8 | Closure % | 718.8 | Closure % | 400.00 |
| 24 | Mass OUT | 718.8 | 100.00 | 718.8 | 100.00 | 718.8 | 100.00 | 400.00 |

**B: Flowsheet with stream data — solution**

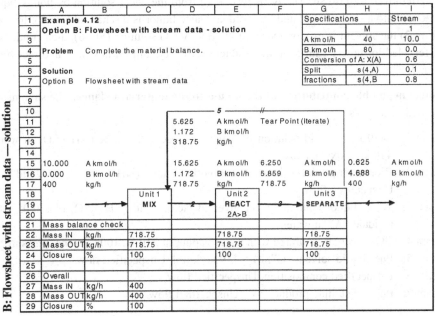

| | A | B | C | D | E | F | G | H | I |
|---|---|---|---|---|---|---|---|---|---|
| 1 | Example 4.12 | | | | | | Specifications | | Stream |
| 2 | Option B: Flowsheet with stream data - solution | | | | | | | M | 1 |
| 3 | | | | | | | A km ol/h | 40 | 10.0 |
| 4 | Problem | Complete the material balance. | | | | | B km ol/h | 80 | 0.0 |
| 5 | | | | | | | Conversion of A: X(A) | | 0.6 |
| 6 | Solution | | | | | | Split | s (4,A) | 0.1 |
| 7 | Option B | Flowsheet with stream data | | | | | fractions | s (4,B | 0.8 |
| 8 | | | | | | | | | |
| 9 | | | | | | | | | |
| 10 | | | | | 5 | // | | | |
| 11 | | | | 5.625 | A km ol/h | Tear Point (Iterate) | | | |
| 12 | | | | 1.172 | B km ol/h | | | | |
| 13 | | | | 318.75 | kg/h | | | | |
| 14 | | | | | | | | | |
| 15 | | 10.000 | A km ol/h | | 15.625 | A km ol/h | 6.250 | A km ol/h | 0.625 | A km ol/h |
| 16 | | 0.000 | B km ol/h | | 1.172 | B km ol/h | 5.859 | B km ol/h | 4.688 | B km ol/h |
| 17 | | 400 | kg/h | | 718.75 | kg/h | 718.75 | kg/h | 400 | kg/h |
| 18 | | | | Unit 1 | | Unit 2 | | Unit 3 | |
| 19 | | | | MIX | | REACT | | SEPARATE | |
| 20 | | | | | | 2A>B | | | |
| 21 | Mass balance check | | | | | | | | |
| 22 | Mass IN | kg/h | 718.75 | | 718.75 | | 718.75 | | |
| 23 | Mass OUT | kg/h | 718.75 | | 718.75 | | 718.75 | | |
| 24 | Closure | % | 100 | | 100 | | 100 | | |
| 25 | | | | | | | | | |
| 26 | Overall | | | | | | | | |
| 27 | Mass IN | kg/h | 400 | | | | | | |
| 28 | Mass OUT | kg/h | 400 | | | | | | |
| 29 | Closure | % | 100 | | | | | | |

Note: In row 15-17 of Option B the first group shows stream 1 (A km ol/h 10.000, B km ol/h 0.000, kg/h 400), the second group stream 2 (15.625, 1.172, 718.75), the third group stream 3 (6.250, 5.859, 718.75), and the fourth group stream 4 (0.625, 4.688, 400).

### EXAMPLE 4.13    Material balance on a biochemical process (production of antibiotic).

This figure shows a simplified flowsheet of a continuous biochemical process for production of an antibiotic. "A" = $C_{16}H_{18}O_5N_2S$ by fermentation from the substrate lactose, $C_{12}H_{22}O_{11}$.

This is a continuous process at steady-state.

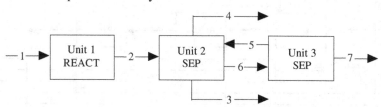

In this process, an aqueous solution of lactose and ammonium sulphate is fed with oxygen (Str. 1) and a small amount of active mould to a bioreactor for partial conversion to "A". The reaction product (Str. 2) is separated in two stages. In stage 1 (Unit 2) $CO_2$ and excess $O_2$ are removed as gas (Str. 4), "A" is recovered from the water solution by liquid–liquid extraction into a solvent "G", and the residual waste liquor is discharged (Str. 3). In stage 2 (Unit 3) the solvent is evaporated from the rich A + G solution (Str. 6), then condensed and recycled to stage 1 (Str. 5). Pure product "A" is crystallised in Unit 3 and recovered in Str. 7.

**Assume negligible contribution of the mould to the material balance.** The streams are specified as follows:

Stream 1:  3420 kg/h liquid solution of 5 wt% lactose + 0.4 wt% $(NH_4)_2SO_4$ in water, 448 kg/h oxygen gas.

Stream 2:  Bioreactor product mixture of antibiotic, lactose, $(NH_4)_2SO_4$, $O_2$, $CO_2$ and $H_2O$. Unspecified flow.

Stream 3:  Waste liquor containing unconverted lactose and $(NH4)_2SO_4$ plus 0.05 wt% residual "A" in water.

Stream 4:  $CO_2 + O_2$ gas saturated with water vapour at 310 K, 100 kPa(abs) pressure.

Stream 5:  Pure liquid solvent "G". M = 119 kg/kmol. Unspecified flow.

Stream 6:  Unspecified composition. Unspecified flow.

Stream 7:  Pure crystalline product "A". Unspecified flow.

| | |
|---|---|
| Conversion Lactose | 90% |
| Selectivity for "A" | 10% |
| L/L distribution coefficient | D(A) = x(6, A) / x(3, A) = 30 |

*Problem:*
A.  Make a degrees of freedom analysis of the problem.
B.  Use a spreadsheet to complete the stream table for the steady-state material balance.

**Solution:**

A. Examine the specification of each unit and of the overall process:

|  | Unit 1 | Unit 2 | Unit 3 | OVERALL |
|---|---|---|---|---|
| Number of unknowns (stream variables) | IJ + L = 14 | IJ = 35 | IJ = 21 | IJ + L = 30 |
| Number of independent equations (see below) | 14 | 24 | 19 | 30 |
| D of F | 0 | 11 | 2 | 0 |

B. Write the equations. Define the species as shown in the stream table, column 1:

UNIT 1 (React) **[Six species, excluding the solvent, two streams, two reactions]**

Atom balance on:

C: $0 = 16\dot{n}(1, A) + 12\dot{n}(1, B) + 1\dot{n}(1, E) - 16\dot{n}(2, A) - 12\dot{n}(2, B) - 1\dot{n}(2, E)$           [1]

H: $0 = 18\dot{n}(1, A) + 22\dot{n}(1, B) + 8\dot{n}(1, C) + 2\dot{n}(1,F) - 18\dot{n}(2, A) - 22\dot{n}(2, B) - 1\dot{n}(2, E)$   [2]

O: $0 = 5\dot{n}(1, A) + 11\dot{n}(1, B) + 4\dot{n}(1, C) - 2\dot{n}(1, D) + 2\dot{n}(1, E) + 1\dot{n}(1, F) - 5\dot{n}(2, A)$
   $- 1\dot{n}(2, B) - 4\dot{n}(2, C) - 2\dot{n}(2, E) - 1\dot{n}(2, F)$           [3]

N: $0 = 2\dot{n}(1, A) + 2\dot{n}(2, C) - 2\dot{n}(2, A) - 2\dot{n}(2, C)$           [4]

S: $0 = 1\dot{n}(1, A) + 1\dot{n}(1, C) - 1\dot{n}(2, A) - 1\dot{n}(2, C)$           [5]

*NOTE:* Equations [4 and 5] make only 1 independent equation.

Stream compositions:

Stream 1:   $\dot{n}(1, A) = 0$           [6]

$\dot{n}(1, B) = (0.05)(3420/342) = 0.5$           [7]

$\dot{n}(1, C) = (0.004)(3420/132) = 0.104$           [8]

$\dot{n}(1, D) = 448/32 = 14$           [9][10]

$\dot{n}(1, F) = (1 - 0.05 - 0.004)(4320/18) = 179.7$           [11]

$\dot{n}(1, G) = 0$           [12]

Stream 2:   $\dot{n}(2, G) = 0$   $\dot{n}(1, E) = 0$           [13]

Conversion: $X(B) = 0.90 = (\dot{n}(1, B) - \dot{n}(2, B))/(\dot{n}(1, B)$           [14]

Selectivity   $S(A) = 0.10 = (16/12)(\dot{n}(2,A) - \dot{n}(1,A))/(\dot{n}(1,B) - \dot{n}(2,B))$           [15]
[based on carbon]

UNIT 2 (Separate) **[Seven species, including the solvent, five streams, no reactions]**

Mole balance on:

| | | |
|---|---|---|
| Antibiotic: | $0 = \dot{n}(2, A) + \dot{n}(5, A) - \dot{n}(3, A) - \dot{n}(4, A) - \dot{n}(6, A)$ | [16] |
| Lactose: | $0 = \dot{n}(2, B) + \dot{n}(5, B) - \dot{n}(3, B) - \dot{n}(4, B) - \dot{n}(6, B)$ | [17] |
| $(NH_4)_2SO_4$: | $0 = \dot{n}(2, C) + \dot{n}(5, C) - \dot{n}(3, C) - \dot{n}(4, C) - \dot{n}(6, C)$ | [18] |
| $O_2$: | $0 = \dot{n}(2, D) + \dot{n}(5, D) - \dot{n}(3, D) - \dot{n}(4, D) - \dot{n}(6, D)$ | [19] |
| $CO_2$: | $0 = \dot{n}(2, E) + \dot{n}(5, E) - \dot{n}(3, E) - \dot{n}(4, E) - \dot{n}(6, E)$ | [20] |
| $H_2O$: | $0 = \dot{n}(2, F) + \dot{n}(5, F) - \dot{n}(3, F) - \dot{n}(4, F) - \dot{n}(6, F)$ | [21] |
| Solvent: | $0 = \dot{n}(2, G) + \dot{n}(5, G) - \dot{n}(3, G) - \dot{n}(6, G)$ | [22] |

Stream compositions:

Stream 3:    $0.0005 = 350\dot{n}(3, A)/350\dot{n}(3, A) + 342\dot{n}(3, B) + 132(3, C)$

$+ 18\dot{n}(3, F))$    [23]

$\dot{n}(3, D) = 0 \; \dot{n}(3, E) = 0 \; \dot{n}(3, G) = 0$    [24][25][26]

Stream 4:    $\dot{n}(4, A) = 0 \; \dot{n}(4, B) = 0$    [27][28]

$\dot{n}(4, C) = 0 \; \dot{n}(4, G) = 0$    [29][30]

Vapour pressure: $H_2O$ at 310 K = 6.2 kPa(abs)    see *Equation 2.32* or steam table

$\dot{n}(4, F)/(\dot{n}(4, D) + \dot{n}(4, E) + \dot{n}(4, F)) = 6.23/100 = 0.062$    [31]

Stream 5:    $\dot{n}(5, A) = 0 \; \dot{n}(5, B) = 0 \; \dot{n}(5, C) = 0$    [32][33][34]

$\dot{n}(5, D) = 0 \; \dot{n}(5, E) = 0 \; \dot{n}(5, F) = 0$    [35][36][37]

L/L distribution coefficient: $[\dot{n}(6, A)/(\dot{n}(6, A) + \dot{n}(6, G))]/[\dot{n}(3, A)/(\dot{n}(3, A)$

$+ \dot{n}(3, B) + (3, C) + \dot{n}(3, F))] = 30$    [38]

## UNIT 3 (Separate) [Seven species, including the solvent, three streams, no reactions]

Mole balance on:

Antibiotic:  $0 = \dot{n}(6, A) - \dot{n}(5, A) - \dot{n}(7, A)$    [39]

Lactose:    $0 = \dot{n}(6, B) - \dot{n}(5, B) - \dot{n}(7, B)$    [40]

$(NH_4)_2SO_4$:  $0 = \dot{n}(6, C) - \dot{n}(5, C) - \dot{n}(7, C)$    [41]

$O_2$:    $0 = \dot{n}(6, D) - \dot{n}(5, D) - \dot{n}(7, D)$    [42]

$CO_2$:    $0 = \dot{n}(6, E) - \dot{n}(5, E) - \dot{n}(7, E)$    [43]

$H_2O$:    $0 = \dot{n}(6, F) - \dot{n}(5, F) - \dot{n}(7, F)$    [44]

Solvent:   $0 = \dot{n}(6, G) - \dot{n}(5, G) - \dot{n}(7, G)$    [45]

**FULLY-SPECIFIED. By method A, *Table 4.02*. 45 independent equations.**
**45 stream variables.**

*The problem is fully-specified and can be solved in a spreadsheet by the sequential modular method.*

| Stream Table | *Example 4.13* | | *Biosynthesis of an Antibiotic from Lactose* | | | | *Solution* |
|---|---|---|---|---|---|---|---|
| Species | M | | | Stream | | | |
| | | *1* | *2* | *3* | *4* | *5* | *6* | *7* |
| | kg/kmol | | | kmol/h | | | | |
| [A] $C_{16}H_{18}O_5N_2S$ | 350 | 0.000 | 0.034 | 0.005 | 0.000 | 0.000 | 0.029 | 0.029 |
| [B] $C_{12}H_{22}O_{11}$ | 342 | 0.500 | 0.050 | 0.050 | 0.000 | 0.000 | 0.000 | 0.000 |
| [C] $(NH_4)_2SO_4$ | 132 | 0.104 | 0.070 | 0.070 | 0.000 | 0.000 | 0.000 | 0.000 |
| [D] $O_2$ | 32 | 14.000 | 9.208 | 0.000 | 9.208 | 0.000 | 0.000 | 0.000 |
| [E] $CO_2$ | 44 | 0.000 | 4.860 | 0.000 | 4.860 | 0.000 | 0.000 | 0.000 |
| [F] $H_2O$ | 18 | 179.700 | 184.481 | 183.551 | 0.930 | 0.000 | 0.000 | 0.000 |
| [G] Solvent | 119 | 0.000 | 0.000 | 0.000 | 0.000 | 37.064 | 37.064 | 0.000 |
| Total | kg/h | 3867 | 3867 | 3332 | 525 | 4411 | 4421 | 10 |
| Mass balance checks | Unit 1 | | Unit 2 | | Unit 3 | | Overall | |
| Mass IN | kg/h | 3867 | Closure % | 8278 | Closure % | 4421 | Closure % | 3867 | Closure % |
| Mass OUT | kg/h | 3867 | 100.0 | 8278 | 100.0 | 4421 | 100.0 | 3867 | 100.0 |

*EXAMPLE 4.14   Material balance on an electrochemical process (production of sodium chlorate).*

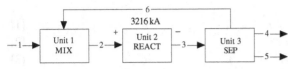

This figure, shows a simplified flowsheet of a continuous electrochemical process for the production of sodium chlorate from salt by the overall cell *Reaction 1*, assumed to occur at 100% current efficiency. This is a continuous process at steady-state.

$$NaCl + 3H_2O \xrightarrow{6F} NaClO_3 + 3H_2 \qquad [3216\ kA] \qquad\qquad Reaction\ 1$$

In this process a fresh feed of NaCl plus water (Str. 1) is mixed with a recycle mother liquor (Str. 6) and fed to electrochemical reactors where NaCl undergoes 20% conversion to $NaClO_3$ by Reaction 1, using 3216 kA. The reaction product mix (Str. 3) is separated to give $H_2$ gas (Str. 4), $NaClO_3$ crystals (Str. 5) and a mother liquor (Str. 6) which is recycled to the process. The streams are specified as follows:

Stream 1:  NaCl + $H_2O$                                                  Unspecified flow
Stream 2:  Unspecified composition                              Unspecified flow
Stream 3:  Unspecified composition                              Unspecified flow
Stream 4:  $H_2$ gas saturated with water vapour at 330 K, total pressure
　　　　　100 kPa(abs)                                                   Unspecified flow
Stream 5:  Pure $NaClO_3$ crystals                                   Unspecified flow
Stream 6:  Mother liquor = liquid solution of 20 wt% NaCl + 30 wt%
　　　　　$NaClO_3$ in $H_2O$                                           Unspecified flow

*Problem:*
**A.**  Make a degrees of freedom analysis of the problem.
**B.**  Use a spreadsheet to complete the stream table for the steady-state material balance.

*Solution:*
**A.** Examine the specification of each unit and of the overall process.

|                                                       | Unit 1     | Unit 2       | Unit 3      | OVERALL        |
|-------------------------------------------------------|------------|--------------|-------------|----------------|
| Number of unknowns (stream variables)                 | IJ = 12    | IJ + L = 9   | IJ = 16     | IJ + L = 13    |
| Number of independent equations (see below)           | 9          | 6            | 10          | 13             |
| D of F                                                | 3          | 3            | 6           | 0              |

B.　Write the equations. Define the species as shown in the stream table, column 1.

UNIT 1 (Mix)　　[**Four species, three streams, no reactions**]
Mole balance on:

| | | |
|---|---|---|
| NaCl: | $0 = \dot{n}(1, A) + \dot{n}(6, A) - \dot{n}(2, A)$ | [1] |
| NaClO$_3$: | $0 = \dot{n}(1, B) + \dot{n}(6, B) - \dot{n}(2, B)$ | [2] |
| H$_2$O: | $0 = \dot{n}(1, C) + \dot{n}(6, C) - \dot{n}(2, C)$ | [3] |
| H$_2$: | $0 = \dot{n}(1, D) + \dot{n}(6, D) - \dot{n}(2, D)$ | [4] |

Stream compositions:

| | | | |
|---|---|---|---|
| Stream 1: | $\dot{n}(1, B) = 0$ | $\dot{n}(1, D) = 0$ | [5][6] |
| Stream 6: | $\dot{n}(6, A)/\dot{n}(6, B) = (20/50)(106.5/58.5) = 0.728$ | | [7] |
| | $\dot{n}(6, C)/\dot{n}(6, B) = (30/50)(106.5/18) = 3.550$ | | [8] |
| | $\dot{n}(6, D) = 0$ | | [9] |

Unit 2 (REACT) [**Four species, two streams, one reaction**]

Let　　　　　　　$I'$ = current = 3216 kA = kC/s
Mole balance on:

| | | |
|---|---|---|
| NaCl: | $0 = \dot{n}(2, A) - \dot{n}(3, A) + 0 - X (A)\dot{n}(2, A)$ | [10] |
| NaClO$_3$: | $0 = \dot{n}(2, B) - \dot{n}(3, B) + X (A)\dot{n}(2, A) - 0$ | [11] |
| H$_2$O: | $0 = \dot{n}(2, C) - \dot{n}(3, C) + 0 - 3X(A)\dot{n}(2, A)$ | [12] |
| H$_2$: | $0 = \dot{n}(2, D) - \dot{n}(3, D) + 3X(A)\dot{n}(2, A) - 0$ | [13] |
| Conversion: | $X(A) = 0.2$ | [14] |
| Faraday's law: | $X(A) \dot{n}(2, A) = (3600)(I')/(6F)$ | [15] |
| | $F$ = Faraday's number = 96485 kC/kmol | |

Unit 3 (SEPARATE) [**Four species, four streams, no reactions**]

Mole balance on:

| | | |
|---|---|---|
| NaCl: | $0 = \dot{n}(3, A) - \dot{n}(4, A) - \dot{n}(5, A) - \dot{n}(6, A) + 0 - 0$ | [16] |
| NaClO$_3$: | $0 = \dot{n}(3, B) - \dot{n}(4, B) - \dot{n}(5, B) - \dot{n}(6, B) + 0 - 0$ | [17] |
| H$_2$O: | $0 = \dot{n}(3, C) - \dot{n}(4, C) - \dot{n}(5, C) - \dot{n}(6, C) + 0 - 0$ | [18] |
| H$_2$: | $0 = \dot{n}(3, D) - \dot{n}(4, D) - \dot{n}(5, D) - \dot{n}(6, D) + 0 - 0$ | [19] |

Stream compositions:

| | | | |
|---|---|---|---|
| Stream 4: | $\dot{n}(4, A) = 0$ | $\dot{n}(4, B) = 0$ | [20][21] |

Vapour pressure: H$_2$O at 330K = 17.2 kPa(abs)　　　　see *Equation 2.32* or steam table
　　　　　　　$\dot{n}(4, C)/\dot{n}(4, D) = 0.172 /(1- 0.172) = 0.208$　　see *Equation 2.07*　　[22]

| | | |
|---|---|---|
| Stream 5: | $\dot{n}(5, A) = 0 \; \dot{n}(5, C) = 0 \; \dot{n}(5, D) = 0$ | [23][24][25] |

**FULLY-SPECIFIED. By method A,** *Table 4.02.* **25 independent equations.**
**25 stream variables.**

*The problem is fully-specified and can be solved by the sequential modular method.*

| Stream Table | *Example 4.14* | *Electro-synthesis of Sodium Chlorate* | | | | *Solution* | |
|---|---|---|---|---|---|---|---|
| Species | M | Stream | | | | | |
| | | *1* | *2* | *3* | *4* | *5* | *6* |
| | kg/kmol | kmol/h | | | | | |
| [A] NaCl | 58.5 | 20.0 | 100.0 | 80.0 | 0.0 | 0.0 | 80.0 |
| [B] NaClO₃ | 106.5 | 0.0 | 109.9 | 129.9 | 0.0 | 20.0 | 109.9 |
| [C] H₂O | 18 | 72.5 | 462.6 | 402.6 | 12.5 | 0.0 | 390.1 |
| [D] H₂ | 2 | 0.0 | 0.0 | 60.0 | 60.0 | 0.0 | 0.0 |
| Total | kg/h | 2475 | 25880 | 25880 | 345 | 2130 | 23405 |
| Checks | Unit 1 | | Unit 2 | | Unit 3 | | Overall |
| Mass IN | 25880 | Closure % | 25880 | Closure % | 25880 | Closure % | 2475 |
| Mass OUT | 25880 | 100 | 25880 | 100 | 25880 | 100 | 2475 |

*EXAMPLE 4.15   Material balance on a thermochemical process (gas sweetening).*

This figure shows a simplified flowsheet of a continuous thermochemical process for the removal of hydrogen sulphide gas from natu-

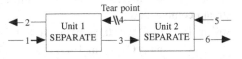

ral gas by a cyclic absorption-desorption pro-
cedure. In this process the contaminated natural gas (Str. 1) is treated in an absorption column (Unit 1) where the $H_2S$ is absorbed into a countercurrent stream of a liquid amine absorbent (M = 111 kg/kmol) (Str. 4). Sweetened natural gas leaves the column as Str. 2 and the $H_2S$ rich absorbent (Str. 3) is passed to a stripping column (Unit 2) where the $H_2S$ is desorbed into steam (Str. 5) and leaves the process (Str. 6). The lean absorbent (Str. 4) then recycles to the absorber. The streams are specified as follows *[this is a continuous process at steady-state]*:

Stream 1:  10,000 kg/h of natural gas containing 95 vol% $CH_4$ + 5 vol% $H_2S$
Stream 2:  Sweetened natural gas with $H_2S$ in equilibrium[17] with
           stream 4 at 120 kPa(abs)                                    Unspecified flow
Stream 3:  Rich amine containing dissolved $H_2S$                      Unspecified flow
Stream 4:  Lean amine containing dissolved $H_2S$                      Unspecified flow
Stream 5:  Steam                                                       Unspecified flow
Stream 6:  Steam containing 6 wt% $H_2S$                               Unspecified flow
Henry's constant $H_2S$ in absorbent $K_H(A)$ = 20 kPa. Assume gas streams behave as ideal gases.

---

[17] That is, an equilibrium separator.

**Problem:**
**A.** Make a degrees of freedom analysis of the problem.
**B.** Use a spreadsheet to complete the stream table for the steady-state material balance.

**Solution:**
**A.** Examine the specification of each unit and of the overall process.

|  | Unit 1 | Unit 2 | OVERALL |
|---|---|---|---|
| Number of unknowns (stream variables) | IJ = 16 | IJ = 16 | IJ = 16 |
| Number of independent equations (see below) | 15 | 14 | 16 |
| D of F | 1 | 2 | 0 |

**B.** Write the equations. Define the species as shown in the stream table, column 1.

UNIT 1 (Separate)
**[Four species including $H_2O$, four streams, no reactions]**

Mass balance on:

| | | |
|---|---|---|
| $H_2S$: | $0 = \dot{m}(1, A) + \dot{m}(4, A) - \dot{m}(2, A) - \dot{m}(3, A)$ | [1] |
| $CH_4$: | $0 = \dot{m}(1, B) + \dot{m}(4, B) - \dot{m}(2, B) - \dot{m}(3, B)$ | [2] |
| $H_2O$: | $0 = \dot{m}(1, C) + \dot{m}(4, C) - \dot{m}(2, C) - \dot{m}(3, C)$ | [3] |
| Absorbent: | $0 = \dot{m}(1, D) + \dot{m}(4, D) - \dot{m}(2, D) - \dot{m}(3, D)$ | [4] |

Stream compositions:

Stream 1:  $y(1, A) = 0.05$  $\bar{m}(1) = 10000$                    [5][6]
          $\dot{m}(1, C) = 0$  $\dot{m}(1, D) = 0$                     [7][8]
Stream 2:  $\dot{m}(2, C) = 0$  $\dot{m}(2, D) = 0$                    [9][10]
Stream 3:  $\dot{m}(3, B) = 0$  $\dot{m}(3, C) = 0$                    [11][12]
Stream 4:  $\dot{m}(4, B) = 0$  $\dot{m}(4, C) = 0$                    [13][14]

Equilibrium separator: (Henry's law)
          $y(2, A) = K_H (A)x(4, A)/P(2) = 20x(4, A)/120$ see *Equation 2.34*     [15]
          $y$ = mole fraction in gas
          $x$ = mole fraction in liquid

UNIT 2 (Separate)
**[Four species, four streams, no reactions]**
Mass balance on:

| | | |
|---|---|---|
| $H_2S$: | $0 = \dot{m}(3, A) + \dot{m}(5, A) - \dot{m}(4, A) - \dot{m}(6, A)$ | [16] |
| $CH_4$: | $0 = \dot{m}(3, B) + \dot{m}(5, B) - \dot{m}(4, B) - \dot{m}(6, B)$ | [17] |
| $H_2O$: | $0 = \dot{m}(3, C) + \dot{m}(5, C) - \dot{m}(4, C) - \dot{m}(6, C)$ | [18] |
| Absorbent: | $0 = \dot{m}(3, D) + \dot{m}(5, D) - \dot{m}(4, D) - \dot{m}(6, D)$ | [19] |

Stream compositions:

Stream 5:  $\dot{m}(5, A) = 0$          $\dot{m}(5, B) = 0$      $\dot{m}(5, D) = 0$     [20][21][22]
Stream 6:  $\dot{m}(6, A)/\bar{m}(6, A) = 0.06$   $\dot{m}(6, C) = 0$                      [23][24]

**FULLY-SPECIFIED. By method A, *Table 4.02*. 24 independent equations.
24 stream variables.**

*The problem is fully-specified and can be solved by the sequential modular method.*
Convert vol% $H_2S$ (i.e. mol%) in Str.1 to mass fraction:

w(1, A) = (0.05)(34)/[(0.05)(34) + (0.95)(16)]
= 0.10            see *Equation 2.28*

Convert wt% $H_2S$ in Str. 4 to mole fraction:
x(4, A) = 0.0007 see *Equation 2.27*

Henry's law:
y(2, A) = 0.0001 see *Equation 2.34*

| Stream Table | | Example 4.15 | | | Gas Sweetening Process Solution | | |
|---|---|---|---|---|---|---|---|
| Species | M | Stream | | | | | |
| | | 1 | 2 | 3 | 4 | 5 | 6 |
| | kg/kmol | kg/h | | | | | |
| [A] $H_2S$ | 34 | 1000 | 2 | 999 | 1 | 0 | 998 |
| [B] $CH_4$ | 16 | 9000 | 9000 | 0 | 0 | 0 | 0 |
| [C] $H_2O$ | 18 | 0 | 0 | 0 | 0 | 15634 | 15634 |
| [D] Absorbent | 111 | 0 | 0 | 5000 | 5000 | 0 | 0 |
| Total | kg/h | 10000 | 9002 | 5999 | 5001 | 15634 | 16632 |
| Mass balance check | | Unit 1 | | Unit 2 | | Overall | |
| Mass IN | kg/h | 15001 | Closure % | 21633 | Closure % | 25634 | Closure % |
| Mass OUT | kg/h | 15001 | 100.0 | 21633 | 100.0 | 25634 | 100.0 |

*EXAMPLE 4.16   Material balance on an environmental system
(ethanol from biomass).*

This figure shows a conceptual flowsheet for a "greenhouse gas neutral" closed carbon cycle fuel production system in which gasoline is replaced by ethanol from biomass as the fuel for the world's motor vehicles. In this proposed system energy from the Sun is used in the photosynthesis of selected plants (Unit 1) to produce cel-

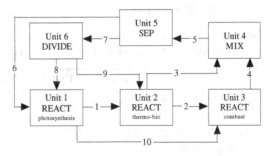

lulose biomass (Str. 1) which is converted to ethanol (Str. 2) and $CO_2$ (Str. 3) by thermal hydrolysis and fermentation (Unit 2). The ethanol is then used in automobiles (Unit 3) where it burns with oxygen from the air (Str. 10) to produce $CO_2$ and $H_2O$ (Str. 4). The auto-exhaust gas (Str. 4) mixes with $CO_2$ from Unit 2 in the atmosphere (Unit 4, Str. 5).

Water separates as rain (Unit 5) and is divided (Unit 6) to recycle to Unit 1 (Str. 8) and Unit 2 (Str. 9), while the $CO_2$ recycles (Str. 6) to support photosynthesis in Unit 1. Oxygen from the photosynthesis recycles (Str. 10) to support the combustion in Unit 3. The reactions and process specifications are as follows: *[This is assumed to be a continuous process at steady-state with 100% conversion of stoichiometric reactants in all reactors and 100% separator efficiency]*

Unit 1:   Reaction 1 $6CO_2 + 5H_2O \rightarrow C_6H_{10}O_5 + 6O_2$                photosynthesis

Unit 2:   Reaction 2 $C_6H_{10}O_5 + H_2O \rightarrow (C_6H_{12}O_6) \rightarrow 2C_2H_5OH + 2CO_2$   hydrolysis +

                                                         fermentation

Unit 3:   Reaction 3 $C_2H_5OH + 3O_2 \rightarrow 2CO_2 + 3H_2O$                    combustion

| | | |
|---|---|---|
| Stream 1: | Cellulose (biomass) $C_6H_{10}O_5$ | Flow unspecified |
| Stream 2: | $C_2H_5OH$ | 3.50E+12 kg/year |
| Stream 3: | $CO_2$ | Flow unspecified |
| Stream 4: | $CO_2 + H_2O$ | Flow unspecified |
| Stream 5: | Composition unspecified | Flow unspecified |
| Stream 6: | $CO_2$ | Flow unspecified |
| Stream 7: | $H_2O$ | Flow unspecified |
| Stream 8: | Composition unspecified | Flow unspecified |
| Stream 9: | Composition unspecified | Flow unspecified |
| Stream 10: | $O_2$ | Flow unspecified |

| | |
|---|---|
| Unit 1: | Converts 600 tonne $CO_2/(km^2.year)$ to cellulose by photosynthesis. |
| Unit 3: | Contains 7.00E+08 motor vehicles, each consuming 5000 kg/year ethanol. |

*NOTE:* Nitrogen (from air) is assumed unreactive and is not considered in this balance.

**Problem:**
A.  Make a degrees of freedom analysis of the problem.
B.  Use a spreadsheet to complete the stream table for the steady-state material balance.
C.  Calculate the area of agricultural land required to supply the ethanol for 700 million (= 7.00E8) automobiles.

**Solution:**
A.  Examine the specification of each unit and of the overall process.

| | Unit 1 | Unit 2 | Unit 3 | Unit 4 | Unit 5 | Unit 6 | OVERALL |
|---|---|---|---|---|---|---|---|
| Number of unknowns (stream variables) | IJ + L = 21 | IJ + L = 21 | IJ + L = 16 | IJ = 15 | IJ = 15 | IJ = 15 | IJ + L = 3 |
| Number of independent equations (below) | 15 | 16 | 15 | 10 | 11 | 11 | 3 |
| D of F | 6 | 5 | 1 | 5 | 4 | 4 | 0 |

B.  Write the equations. Define species as in the stream table, column 1.

## Unit 1 (REACT) [Five species, four streams, one reaction]

Atom balance on:

| | | |
|---|---|---|
| C: | $0 = \dot{n}(6, A) - 6\dot{n}(1, D)$ | [1] |
| H: | $0 = 2\dot{n}(8, B) - 10\dot{n}(1, D)$ | [2] |
| O: | $0 = 2\dot{n}(6, A) + \dot{n}(8, B) - 5\dot{n}(1, D) - 2\dot{n}(10, C)$ | [3] |

Stream compositions:

| | | |
|---|---|---|
| Stream 1: | $\dot{n}(1, A) = 0 \ \dot{n}(1, B) = 0 \ \dot{n}(1, C) = 0 \ \dot{n}(1, E) = 0$ | [4][5][6][7] |
| Stream 6: | $\dot{n}(6, B) = 0 \ \dot{n}(6, C) = 0 \ \dot{n}(6, D) = 0 \ \dot{n}(6, E) = 0$ | [8][9][10][11] |
| Stream 10: | $\dot{n}(10, A) = 0 \ \dot{n}(10, B) = 0 \ \dot{n}(10, D) = 0 \ \dot{n}(10, E) = 0$ | [12][13][14][15] |

## UNIT 2 (React) [Five species, four streams, one reaction]

Atom balance on:

| | | |
|---|---|---|
| C: | $0 = 6\dot{n}(1, D) - 2\dot{n}(2, E) - \dot{n}(3, A)$ | [16] |
| H: | $0 = 10\dot{n}(1, D) + 2\dot{n}(9, B) - 6\dot{n}(2, E)$ | [17] |
| O: | $0 = 5\dot{n}(1, D) + \dot{n}(9, B) - \dot{n}(2, E)$ | [18] |

Stream compositions:

| | | |
|---|---|---|
| Stream 2: | $\dot{n}(2, A) = 0 \ \dot{n}(2, B) = 0 \ \dot{n}(2, C) = 0 \ \dot{n}(2, D) = 0$ | [19][20][21][22] |
| | $\dot{n}(2, E) = 9.11E{+}06 \ kmol/h$ | [23] |
| Stream 3: | $\dot{n}(3, B) = 0 \ \dot{n}(3, C) = 0 \ \dot{n}(3, D) = 0 \ \dot{n}(3, E) = 0$ | [24][25][26][27] |

## UNIT 3 (React) [Five species, three streams, one reaction]

Atom balance on:

| | | |
|---|---|---|
| C: | $0 = 2\dot{n}(2, E) - \dot{n}(4, A)$ | [28] |
| H: | $0 = 6\dot{n}(2, E) - 2\dot{n}(4, B)$ | [29] |
| O: | $0 = \dot{n}(2, E) + 2\dot{n}(10, C) - 2\dot{n}(4, A) - \dot{n}(4, B)$ | [30] |

Stream compositions

| | | |
|---|---|---|
| Stream 4: | $\dot{n}(4, C) = 0 \ \dot{n}(4, D) = 0 \ \dot{n}(4, E) = 0$ | [31][32][33] |

## UNIT 4 (Mix) [Five species, three streams, zero reactions]

Atom balance on:

| | | |
|---|---|---|
| C: | $0 = \dot{n}(3, A) + \dot{n}(4, A) - \dot{n}(5, A)$ | [34] |
| H: | $0 = 2\dot{n}(2, B) - 2\dot{n}(4, B)$ | [35] |
| O: | $0 = 2\dot{n}(3, A) + 2\dot{n}(4, A) + \dot{n}(4, B) - 2\dot{n}(5, A) - \dot{n}(5, B)$ | [36] |

## UNIT 5 (Separate) [Five species, three streams, zero reactions]

Atom balance on:

| | | |
|---|---|---|
| C: | $0 = \dot{n}(5, A) + 6\dot{n}(5, D) + 2\dot{n}(5, E) - \dot{n}(6, A) - \dot{n}(7, A)$ | [37] |
| H: | $0 = 2\dot{n}(5, B) + 10\dot{n}(6, D) + 6\dot{n}(5, E) - 2\dot{n}(6, B) - 2\dot{n}(7, B)$ | [38] |
| O: | $0 = 2\dot{n}(5, A) + \dot{n}(5, B) + 2\dot{n}(5, C) + 5\dot{n}(5, D) + \dot{n}(5, E) - 2\dot{n}(6, A)$ | |
| | $\quad - \dot{n}(7, A) - \dot{n}(7, B)$ | [39] |

Stream compositions:

| | | |
|---|---|---|
| Stream 7: | $\dot{n}(7, A) = 0 \ \dot{n}(7, C) = 0 \ \dot{n}(7, D) = 0 \ \dot{n}(7, E) = 0$ | [40][41][42][43] |

UNIT 6 (Divide) **[Five species, three streams, zero reactions]**

Atom balance on:

C:          $0 = \dot{n}(7, A) - \dot{n}(8, A) - 6\dot{n}(8, D) - 2\dot{n}(8, E) - \dot{n}(9, A) - 6\dot{n}(9, D) - 2\dot{n}(9, E)$ [44]

H:          $0 = 2\dot{n}(7, B) - 2\dot{n}(8, B) - 10\dot{n}(8, D) - 6\dot{n}(8, E) - 2\dot{n}(9, B)$
                 $- 10\dot{n}(9, D) - 6\dot{n}(9, E)$ [45]

O:          $0 = 2\dot{n}(7, A) + \dot{n}(7, B) - 2\dot{n}(8, A) - \dot{n}(8, B) - 2\dot{n}(2, C) - 5\dot{n}(8, D)$ [46]
                 $- \dot{n}(8, E) - 2\dot{n}(9, A) - \dot{n}(9, B) - 2\dot{n}(9, C) - 5\dot{n}(9, D) - \dot{n}(9, E)$

Divider relations: $\dot{n}(8, A)/\overline{n}(8) = \dot{n}(7, A)/\overline{n}(7)$     $\dot{n}(8, C)/\overline{n}(8) = \dot{n}(7, C)/\overline{n}(7)$ [47][48]
                     $\dot{n}(8, D)/\overline{n}(8) = \dot{n}(7, D)/\overline{n}(7)$     $\dot{n}(8, E)/\overline{n}(8) = \dot{n}(7, E)/\overline{n}(7)$ [49][50]

**FULLY-SPECIFIED. By method A, *Table 4.02*. 50 independent equations.**
                 **50 stream variables.**

| Stream Table | | *Example 4.16 Material Balance on Carbon Cycle for Ethanol Fuel from Biomass* | | | | | | | | | *Solution* |
|---|---|---|---|---|---|---|---|---|---|---|---|
| Species | M | Stream | | | | | | | | | |
| | | 1 | 2 | 3 | 4 | 5 | 6 | 7 | 8 | 9 | 10 |
| | kg/ kmol | kmol/hr | | | | | | | | | |
| [A]CO$_2$ | 44 | 0 | 0 | 9.1E+06 | 1.8E+07 | 2.7E+07 | 2.7E+07 | 0 | 0 | 0 | 0 |
| [B]H$_2$O | 18 | 0 | 0 | 0 | 2.73E+07 | 2.73E+07 | 0 | 2.7E+07 | 2.3E+07 | 4.6E+06 | 0 |
| [C]O$_2$ | 32 | 0 | 0 | 0 | 0 | 0 | 0 | 0 | 0 | 0 | 2.7E+07 |
| [D] C$_6$H$_{10}$O$_5$ | 162 | 4.6E+06 | 0 | 0 | 0 | 0 | 0 | 0 | 0 | 0 | 0 |
| [E] C$_2$H$_5$OH | 46 | 0 | 9.1E+06 | 0 | 0 | 0 | 0 | 0 | 0 | 0 | 0 |
| Total | kg/h | 7.38E+08 | 4.19E+08 | 4.01E+08 | 1.29E+09 | 1.69E+09 | 1.20E+09 | 4.92E+08 | 4.10E+08 | 8.19E+07 | 8.74E+08 |
| Mass balance check | | Unit 1 | Unit 2 | Unit 3 | Unit 4 | Unit 5 | Unit 6 | Overall | | | |
| Mass IN | | 1.61E+09 | 8.19E+08 | 1.29E+09 | 1.69E+09 | 1.69E+09 | 4.92E+08 | 0 | | | |
| Mass OUT | | 1.61E+09 | 8.19E+08 | 1.29E+09 | 1.69E+09 | 1.69E+09 | 4.92E+08 | 0 | | | |
| Closure % | | 100.0 | 100.0 | 100.0 | 100.0 | 100.0 | 100.0 | 100.0 | | | |

C.   Land area to supply ethanol for 7.00E+08 vehicles = **1.75E+07 km²**
     Fraction of land area of Earth              = **11.7%**

*EXAMPLE 4.17*    *Material balance on a combustion process, using various methods of solution.*

*Problem:* 100 kmol/h of the fuel ethane ($C_2H_6$) undergoes partial combustion in excess dry air to give a product gas with 0.501 vol% $C_2H_6$, 1.803 vol% CO and a dew-point of 325 K at 100 kPa(abs). Calculate the complete material balance stream table for this continuous process at steady-state. Assume negligible production of nitrogen oxides.

Reactions:    $C_2H_6 + 3.5O_2 \rightarrow 2CO_2 + 3H_2O$    [1]    $C_2H_6 + 2.5O_2 \rightarrow 2CO + 3H_2O$    [2]

| Component | $C_2H_6$ | $O_2$ | $N_2$ | $CO_2$ | CO | $H_2O$ |
|-----------|----------|-------|-------|--------|-----|--------|
| Symbol | A | B | C | D | E | F |

Fuel ——1——► REACT ——3——► Combustion product gas
Air ——2——►

## Solution:

Rate ACC = Rate IN − Rate OUT + Rate GEN − Rate CON

$\dot{n}(1, A) = 100$ kmol/h. Vapour pressure of water at 325 K = 13.53 kPa(abs) (see *Equation 2.32*):

Define: X = overall conversion of $C_2H_6$    S = selectivity for $CO_2$ from $C_2H_6$

Define: $\varepsilon(1)$ = extent of reaction 1    $\varepsilon(2)$ = extent of reaction 2 (kmol/h)

### Method A[18]

Atom balances.

$H_1$: $0 = 6\dot{n}(1, A) - [6\dot{n}(3, A) + 2\dot{n}(3, F)]$

$O_1$: $0 = 2\dot{n}(2, B) - [2\dot{n}(3, B) + 2\dot{n}(3, D) + 1\dot{n}(3, E) + 1\dot{n}(3, F)]$

$C_1$: $0 = 2\dot{n}(1, A) - [2\dot{n}(3, A) + 1\dot{n}(3, D) + 1\dot{n}(3, E)]$

$N_1$: $0 = 2\dot{n}(2, C) - 2\dot{n}(3, C)$

Stream composition:

Stream 2: $\dot{n}(2, B)/\dot{n}(2, C) = 21/79$

Stream 3: $y(3, A) = \dot{n}(3, A)/\bar{n}(3) = 0.00501$

$y(3, E) = \dot{n}(3, E)/\bar{n}(3) = 0.01803$

$y(3, F) = \dot{n}(3, F)/\bar{n}(3)$

$= p(3, F)/P(3)$

$= 0.1353$

### Method B

Atom balances using extents of reaction.

$0 = 6\dot{n}(1, A) - [6\dot{n}(3, B) + 6\varepsilon(1) + 6\varepsilon(2)]$  [1]

$0 = 2\dot{n}(2, B) - [2\dot{n}(3, B) + 7\varepsilon(1) + 5\varepsilon(2)]$  [2]

$0 = 2\dot{n}(1, A) - [2\dot{n}(3, A) + 2\varepsilon(1) + 2\varepsilon(2)]$  [3]

$0 = 2\dot{n}(2, C) - 2\dot{n}(3, C)$  [4]

Stream composition:

$\dot{n}(2, B)/\dot{n}(2, C) = 21/79$  [5]

$y(3, A) = \dot{n}(3, A)/\bar{n}(3) = 0.00501$  [6]

$y(3, E) = \dot{n}(3, E)/\bar{n}(3) = 0.01803$  [7]

$y(3, F) = \dot{n}(3, F)/\bar{n}(3) = p(3, F)/P(3)$

$= 0.1353$  [8]

### Method C

Mole balances using conversion and selectivity.

$C_2H_6$  $0 = \dot{n}(1, A) - \dot{n}(3, A) - X\dot{n}(1, A)$

$O_2$:  $0 = \dot{n}(2, B) - \dot{n}(3, B) - X[3.5S + 2.5(1-S)]\dot{n}(1, A)$

$N_2$:  $0 = \dot{n}(2, C) - \dot{n}(3, C)$

$CO_2$:  $0 = 0 - \dot{n}(3, D) + 2XS\dot{n}(1, A)$

CO:  $0 = 0 - \dot{n}(3, E) + 2X(1-S)\dot{n}(1, A)$

$H_2O$:  $0 = 0 - \dot{n}(3, F) + 3X\dot{n}(1, A)$

Stream composition:

Stream 2: $\dot{n}(2, B)/\dot{n}(2, C) = 21/79$

Stream 3: $y(3, A) = \dot{n}(3, A)/\bar{n}(3) = 0.00501$

$y(3, E) = \dot{n}(3, E)/\bar{n}(3) = 0.01803$

$y(3, F) = \dot{n}(3, F)/\bar{n}(3)$

$= p(3, F)/P(3) = 0.1353$

### Method D

Mole balances using extents of reaction.

$0 = \dot{n}(1, A) - \dot{n}(3, A) - \varepsilon(1) - \varepsilon(2)$  [1]

$0 = \dot{n}(2, B) - \dot{n}(3, B) - 3.5\varepsilon(1) - 2.5\varepsilon(2)$  [2]

$0 = \dot{n}(2, C) - \dot{n}(3, C)$  [3]

$0 = 0 - \dot{n}(3, D) + 2\varepsilon(1)$  [4]

$0 = 0 - \dot{n}(3, E) + 2\varepsilon(2)$  [5]

$0 = 0 - \dot{n}(3, F) + 3\varepsilon(1) + 3\varepsilon(2)$  [6]

Stream composition:

$\dot{n}(2, B)/\dot{n}(2, C) = 21/79$  [7]

$y(3, A) = \dot{n}(3, A)/\bar{n}(3) = 0.00501$  [8]

$y(3, E) = \dot{n}(3, E)/\bar{n}(3) = 0.01803$  [9]

$y(3, F) = \dot{n}(3, F)/\bar{n}(3) = p(3, F)/P(3)$

$= 0.135$  [10]

---

[18] Equation numbers appear on the right of the page.

In all methods:

$$\bar{n}(3) = \dot{n}(3, A) + \dot{n}(3, B) + \dot{n}(3, C) + \dot{n}(3, D) + \dot{n}(3, E) + \dot{n}(3, F) \qquad [11]$$

Method A:   From [1, 6 and 8], $\dot{n}(3) = 1997$ kmol/h. Equations [6, 7, 8] then [2, 3, 4, 5 and 11] give the rest.

Method B:   Equations [1 and 3] are identical. Seven equations with eight unknowns. No solution.

Method C:   From [1 and 6], $X = 0.9$, $\dot{n}(3) = 1997$ kmol/h.
From [5], $S = 0.8$.
Equations [2, 3, 4, 7 and 11] give the rest.

Method D:   From [1 and 6], $[\varepsilon(1) + \varepsilon(2)] = 90$ kmol/h,
$\dot{n}(3) = 1997$ kmol/h
From [5], $\varepsilon(2) = 18$, $\varepsilon(1) = 72$ kmol/h.
Equations [2, 3, 4, 7 and 11] give the rest.

| Stream Table | | Combustion Solution | | | |
|---|---|---|---|---|---|
| Component | M | Stream [kmol/h] | | | |
| | kg/kmol | *1* | *2* | *3* | ymol % |
| [A] $C_2H_6$ | 30 | 100 | 0 | 10 | 0.50 |
| [B] $O_2$ | 32 | 0 | 385 | 88 | 4.41 |
| [C] $N_2$ | 28 | 0 | 1448 | 1448 | 72.55 |
| [D] $CO_2$ | 44 | 0 | 0 | 144 | 7.21 |
| [E] CO | 28 | 0 | 0 | 36 | 1.80 |
| [F] $H_2O$ | 18 | 0 | 0 | 270 | 13.52 |
| Total | kg/h | 3000 | 52873 | 55873 | |
| Mass IN  kg/h   55873 Mass OUT kg/h   55873 | | | | | |
| Closure   %    100 | | | | | |

## SUMMARY

[1]   Material balance calculations hinge on the general balance equation (GBE), which can be used in either the integral or differential form, depending on the type of problem:

**Integral form of GBE:**   [For a defined system and a specified quantity]
Material ACC = Material IN – Material OUT + Material GEN – Material CON

**Differential form of GBE:** [For a defined system and a specified quantity]
Rate of material ACC = Rate of material IN – Rate of material OUT
                            + Rate of material GEN – Rate of material CON

[2]   Material balance calculations with the GBE are mostly a matter of bookkeeping, backed by knowledge of mathematics, physical properties, phase equilibria, and stoichiometry.
The main steps involved in doing material balances are as follows:

**a.** Interpreting the jargon and resolving ambiguities that sometimes obscure the problem.

**b.** Locating the process units and the known conditions and/or [in practical "on the job" situations] obtaining reliable data.

**c.** Translating a word problem into a flowsheet, with a corresponding stream table showing the known quantities of the material balance.

**d.** Defining the system envelope(s) that give the simplest solution.

**e.** Specifying the quantities of interest, along with a set of symbols that represent the quantities without ambiguity.

**f.** Writing a set of equations that relate the known quantities to the unknown quantities needed to complete the stream table.

**g.** Determining the degrees of freedom of the system, and avoiding the trap of inconsistent over-specification.

**h.** Arranging and sequencing the equations for a computer (e.g. spreadsheet) calculation.

**i.** Translating the output of the calculations into a practical result with the correct units and the appropriate number of significant figures.

**j.** Presenting the output in a transparent and unambiguous format, such as stream table together with a labelled flowsheet.

[3] Choosing the system envelope(s) and the quantity(s) are critical in starting a material balance, and can have a big effect on the complexity of subsequent calculations. The simplest cases use atom balances on overall systems, while more difficult problems may need atom and/or species balances on individual process units and/or combinations of units.

[4] Material balance calculations (in non-nuclear processes) may use either conserved or non-conserved quantities.
*Conserved quantities* (e.g. total mass, atoms, moles or mass of individual species without chemical reaction): GEN = CON = 0
*Non-conserved quantities* (e.g. moles or mass of individual species in a chemical reaction): GEN and/or CON ≠ 0

[5] Each generic process unit (DIVIDE, MIX, SEPARATE, HEAT EXCHANGE, PUMP, REACT) is represented by a set of material balance equations plus subsidiary relations that uniquely define its function. A complete process flowsheet can be constructed by sequencing these generic process units with interconnecting process streams.

[6] Always check your material balance solutions for *closure* of the mass balance. Closing the mass balance means accounting for all mass entering, leaving and accumulating in the system, to ensure agreement with the principle of conservation of mass.

*Closure of the mass balance is a necessary (but not sufficient) check on the correctness of material balance calculations.*

For a continuous process at steady state the closure (%) is defined as:

Closure $= e_M = 100$ [Total mass flow rate OUT/Total mass flow rate IN].

Exact balance calculations should give closure $= 100\%$, although practical process design calculations may be satisfied by a closure of 99.9% or lower, depending on the application. A tight material balance (i.e. closure $\cong 100\%$) is usually needed when the process involves trace materials such as catalyst poisons and/or environmental contaminants that are problematic in low (e.g. ppm or ppb) concentrations.

[7] The *specification* of a material balance problem involves a comparison between the number of *stream variables* and the number of independent relations that connect the variables to known values and to each other.

The *degrees of freedom (D of F)* of the material balance is then defined as:

D of F = number of stream variables – number of (known values of stream variables + independent relations between stream variables)

= number of unknown quantities – number of independent equations

In material balance problems the *stream variables* are the flows (or amounts) of each species in each process stream. The number of independent chemical reactions is included with the stream variables for species (i.e. mass or mole) balances, but excluded for element (i.e. atom) balances.

The independent relations include the individual species or elemental balances plus subsidiary relations defining the function of each process unit.

If, D of F > 0

The material balance is *under-specified* and has no unique solution.

Typical of *design problems*, whose solution involves optimisation.

If, D of F = 0

The material balance is *fully-specified* and has one unique solution.

Typical of problems in beginners' engineering and science courses.

If, D of F < 0

The material balance is *over-specified* and is either redundant or incorrect.

When a balance is over-specified look at it carefully for errors.

[8] The solution of integral material balances and of differential balances at steady-state requires only algebra, whereas unsteady-state differential balances are solved using calculus. Algebraic balances may involve linear, non-linear and simultaneous equations and sometimes require numerical methods for their solution.

[9] Full material balances for multi-unit processes are set up by combining the individual unit material balances in sequence. The resulting multi-unit balances can be treated by either:

a. The method of *sequential modular solution,* where equations for each process unit are solved in sequence, OR

b. The method of *simultaneous solution,* where equations for the whole system are solved simultaneously.

The sequential modular method, which is used in most commercial process simulation software, is employed throughout this text.

**[10]** The inclusion of recycle streams in a process flowsheet complicates the material balance by coupling the equations for all process units in each recycle loop. Recycle material balances are done with the sequential modular method by adopting a *tear stream* and iterating the calculations until the balance is converged. *Convergence* of the material balance means that the mass balance is closed to an acceptable tolerance.

**[11]** In practical recycle systems there is usually an accumulation of undesirable material(s) in the recycle loops that leads to difficulties in long-term operation of the system. An undesired accumulating substance may be removed from the recycle system by purging a stream containing the substance (preferably in high concentration).

**[12]** A computer should be used for all except the most simple material balance calculations. The computer solution can use a spreadsheet, in-house code or commercial process simulation software.

The spreadsheet solution of material balances takes two common forms:
**a.** A *stream table* attached to a process flowsheet, with designated process units, streams and species, OR
**b.** Sets of stream variable values adjacent to each stream in the process flowsheet.
This text uses spreadsheets with form (a) to set-up, solve and present material balances (and also energy balances — see Chapter 5).

## FURTHER READING

[1] A. Ford, *Modeling the Environment*, Island Press, 2009.
[2] K. Solen and J. Harb, *Introduction to Chemical Engineering*, Wiley, 2011.
[3] R. Felder, R. Rousseau and L. Bullard, *Elementary Principles of Chemical Processes*, Wiley, 2018.
[4] J. Mihelcic and J. Zimmerman, *Environmental Engineering*, Wiley, 2021.
[5] P. Doran, *Bioprocess Engineering Principles*, Elsevier, 2013.
[6] D. Himmelblau and J. Riggs, *Basic Principles and Calculations in Chemical Engineering*, Pearson, 2022.
[7] G. V. Reklaitis, *Introduction to Material and Energy Balances*, Wiley, 1983.
[8] R. Murphy, *Introduction to Chemical Processes*, McGraw-Hill, 2007.
[9] A. Morris, *Handbook of Material and Energy Calculations for Materials Processing*, Wiley, 2012.
[10] W. Seider, D. Lewin, J. Seider, S. Widagdo, R. Gani and K. Ng, *Product and Process Design Principles*, Wiley, 2016.
[11] B. Liegme and K. Hekman, *Excel 2016® for Scientists and Engineers*, Academic Press, 2019.

# CHAPTER FIVE

ENERGY BALANCES

## GENERAL ENERGY BALANCE

In this chapter energy balances are divorced from material balances and calculated with the assumption that the corresponding material balance is already known. Energy balance calculations then follow a similar pattern to the material balance calculations of Chapter 4.

Energy balances are based on the general balance equation of Chapter 1, reintroduced here as *Equation 5.01*. In a defined system and for a specified *quantity*:

**Accumulation = Input    – Output     + Generation – Consumption** *Equation 5.01*
**in system       to system   from system   in system      in system**

For energy balance problems the *system* is a space whose boundaries are defined to suit the problem at hand and the *quantity* in *Equation 5.01* is the amount of energy. The energy balance differs from the general material balance in that total energy (in non-nuclear processes) is a *conserved quantity*, so the generation and consumption terms in the total energy balance are both zero.

*Equation 5.01*, written in abbreviated form for the total energy, can be used as an *integral balance*:

**ACC = IN – OUT**                                          *Equation 5.02*

where:
     ACC ≡ (final amount of energy – initial amount of energy) in the system      kJ
     IN    ≡ energy into the system      kJ
     OUT ≡ energy out of the system      kJ
or as a *differential balance*:
     **Rate ACC = Rate IN – Rate OUT**                    *Equation 5.03*

where:
     Rate ACC ≡ rate of accumulation of energy in the system    w.r.t. time (t)    kW
     Rate IN    ≡ rate of energy input to the system            w.r.t. time (t)    kW
     Rate OUT ≡ rate of energy output from the system       w.r.t. time (t)    kW

The rate terms in *Equation 5.03* have units of energy/time, which is equal to power (e.g. kJ/s = kW). *Equations 5.02* and *5.03* are both equivalents of the *first law of thermodynamics*, which states that energy is conserved in any (non-nuclear) process.

Note that the "energy" in *Equations 5.02 and 5.03* embraces all forms of energy related to the system, including the heat and work transferred between the system and its surroundings. In the subsequent *Equations 5.04, 5.05, 5.06 and 5.07*, energy is considered to consist of two parts, the energy associated with the material of the system (E) and the energy *transferred* across the system boundary as heat and work (Q and W). Although E, Q and

W are all measured in the same units (e.g. kJ) these terms differ in the important respect that E is a state function, while Q and W are both path functions.[1]

Energy transferred to or from the system as radiation is usually lumped into the heat term, for example, energy transfer by infra-red radiation is the major source of "heat" in industrial furnaces. Photochemical processes are driven by other types of radiation.

To avoid ambiguity the quantities used in energy balance calculations must be clearly specified. For this purpose the nomenclature used in this text is given in *Table 5.01*.

**Table 5.01.**   Nomenclature for energy balances (see also Chapter 2 and Table 4.01).

| Symbol | Meaning | Typical Units |
|---|---|---|
| E | energy content of the material of the system, with respect to reference condition(s)[2] | kJ |
| $h(j)$ | specific enthalpy of component "j", with respect to *compounds* at a reference condition | $kJ.kg^{-1}$, $kJ.kmol^{-1}$ |
| $h^*(j)$ | specific enthalpy of component "j", with respect to *elements* at standard state | $kJ.kg^{-1}$, $kJ.kmol^{-1}$ |
| H | enthalpy, with respect to a reference condition = U + PV | kJ |
| U | internal energy, with respect to a reference condition | kJ |
| $E_k$ | kinetic energy = $(1E\text{-}3)\ 0.5\ m\tilde{u}^2$ | kJ |
| $E_p$ | potential energy, with respect to a reference level = $(1E\text{-}3)mgL'$ | kJ |
| PV | pressure–volume energy | kJ |
| Q | net heat <u>input</u> to the system | kJ |
| W | net work <u>output</u> from (i.e. work done by) the system | kJ |
| $\dot{H}(i)$ | enthalpy flow of stream "i", with respect to *compounds* at a reference condition | kW |
| $\dot{H}^*(i)$ | enthalpy flow of stream "i", with respect to *elements* at standard state | kW |
| $\dot{H}(i)$ | enthalpy flow of stream "i" w.r.t. elements or compounds at reference conditions | |
| $\dot{E}_k(i)$ | kinetic energy flow of stream "i" = $(1E\text{-}3)\ 0.5\ \bar{m}(i)\tilde{u}^2$ | kW |
| $\dot{E}_p(i)$ | potential energy flow of stream "i" = $(1E\text{-}3)\ \bar{m}(i)gL'$ | kW |
| $\dot{Q}$ | net heat rate of heat <u>input</u> to the system | kW |
| $\dot{W}$ | net rate of work (i.e. power) <u>output</u> from the system | kW |
| $P(i)$ | pressure of stream "i" | kPa(abs) |
| $T(i)$ | temperature of stream "i" | K |
| $T_{ref}$ | reference temperature for calculation of H and U | K |
| $\dot{V}(i)$ | volumetric flow rate of stream "i" | $m^3.s^{-1}$ |
| $\tilde{u}$ | velocity | $m.s^{-1}$ |
| g | gravitation constant | $m.s^{-2}$ |
| $L'$ | height above a reference level | m |

---

[1] A state function depends only on the state and is independent of the path to that state. A path function is one whose value depends on the path between two states.

[2] Some texts refer to E as the "total energy" of the system, but you should be aware that this term is ambiguous w.r.t. its meaning in the "total energy balance".

Calculations with differential energy balances in which [Rate ACC $\neq$ 0] require the use of calculus and are treated in Chapter 7. All other energy balance problems can be solved by arithmetic and algebra, as shown in the examples following in Chapter 5.

## CLOSED SYSTEMS (BATCH PROCESSES)

A closed system is defined as one in which, over the time period of interest, there is zero transfer of *material* across the system boundary. The movement of *energy* across the boundaries of a closed system is permitted and occurs by the transfer of heat (Q) or work (W). The integral and differential energy balances for a closed system are respectively *Equations 5.02* and *5.03*. These equations are simplified to *Equations 5.04* and *5.05*.

Closed system – integral energy balance:

$$\mathbf{E_{final} - E_{initial} = Q - W} \hspace{5cm} \textit{Equation 5.04}$$

Closed system – differential energy balance:

$$\mathbf{dE/dt = \dot{Q} - \dot{W}.} \hspace{5cm} \textit{Equation 5.05}$$

where:

| | | |
|---|---|---|
| E | = $[U + E_k + E_p]$ content of the system | kJ |
| $E_{final}$ | = final value of "E" for the system | kJ |
| $E_{initial}$ | = initial value of "E" for the system | kJ |
| Q | = net heat <u>input</u> to system | kJ |
| $\dot{Q}$ | = net rate of heat <u>input</u> to system | kW |
| t | = time | s |
| W | = net work <u>output</u> from the system | kJ |
| $\dot{W}$ | = net rate of work (i.e. power) <u>output</u> from system | kW |

The definition of energy "E" used here excludes exotic energy terms such as surface energy, potential energy in a magnetic field, etc. that are insignificant in most chemical process calculations.

*Equations 5.04* and *5.05* may be familiar to you as forms of the first law of *thermodynamics* for a closed system.

Note that the work[3] (W) is defined here as positive for net work done <u>BY</u> the system (the opposite sign convention is used in some thermodynamics texts). This work usually takes the form of:

---

[3] The term "work" is used loosely in many texts, to mean either work (i.e. (force)(distance)) or power (i.e. work/time).

- mechanical "work", such as expansion, compression, lifting, pushing or rotation of a shaft.
- electrical "work", for which: Power (kW) = (Voltage (V)) (current (kA)).

*Examples 5.01, 5.02* and *5.03* illustrate some energy balances in closed systems.

## EXAMPLE 5.01   Energy balance on a falling mass (closed system).

A 5.0 kg block of of iron at 298 K ($C_{v,m}$ ref 298 K = 0.50 kJ/(kg.K)) falls to earth from a height of 100 m under a constant gravity of 9.81 m/s$^2$, with air resistance assumed to be zero.

*Problem:* Calculate:

A.  The velocity of the iron just before impact with the earth, assuming no change in temperature during the fall.
B.  The temperature rise in the iron on impact with the earth, assuming zero heat and work transfer to air or earth.

## Solution:

Define the system    = block of iron (a closed system)
Specify the quantity = energy
Reference condition = element iron at T = 298 K, L′ = 0, $\tilde{u}$ = 0
L′                              = height above earth m
T                               = temperature K
$\tilde{u}$                     = velocity m/s

## A. Integral energy balance on the closed system:

$E_{final} - E_{initial}$        = Q – W                     Q = W = 0
Initial condition  = block at L′ = 100 m
Final condition    = block at L′ = 0, just before impact
$E_{initial}$      = m ($C_{v,m}$(T– 298) + $h_{f,298K}$) + mgL′ + $0.5m\tilde{u}^2$
                   = 5 kg(0.5 kJ/(kg.K)(298 – 298) K + 0 kJ/kg) + (5 kg)(9.81 m/s$^2$)
                   (100 m) + (0.5)(5 kg)(0 m/s)$^2$
                   = 4.91E3 J
$E_{final}$        = 5 kg(0.5 kJ/(kg.K)(298 – 298) K + 0 kJ/kg) + (5 kg)(9.81 m/s$^2$)
                   (0 m) + (0.5)(5 kg)(m/s)$^2$
                   = $2.5\tilde{u}^2$ J
i.e.          $2.5\tilde{u}^2 - 4.91E3 = 0 - 0$
              $\tilde{u} = \underline{\textbf{44 m/s}}$

**B. Integral energy balance on the closed system:**

$E_{final} - E_{initial}$        $= Q - W$         $Q = W = 0$

Initial condition        $=$ block at $L' = 100$ m

Final condition        $=$ block at $L' = 0$, just after impact

$E_{initial}$        $= m\ (C_{v,m}(T- 298) + h_{f,298K}) + mgL' + 0.5m\tilde{u}^2$

        $= 5$ kg$(0.5$ kJ/(kg.K)$(298 - 298)$ K $+ 0$ kJ/kg)$(1E3$ J/kJ) $+ (5$ kg)

        $\times (9.81$ m/s$^2)$ $(100$ m) $+ (0.5)(5$ kg)$(0$ m/s$^2) = 4.91E3$ J

$E_{final}$        $= 5$ kg$(0.5$ kJ/(kg.K)$(T - 298)$ K $+ 0$ kJ/kg)$(1E3$ J/kJ) $+ (5$ kg)

        $\times (9.81$ m/s$^2)$ $(0$ m) $+ (0.5)(5$ kg)$(0$ m/s$^2) = 2.5E3$ (T–298) J

$2.5E3(T–298) - 4.91E3 = 0 - 0$   Solve for $T = 300$ K

Temperature rise on impact $=$ **2.0 K**

*NOTE:* Mass cancels from the balances.

The three assumptions used in this problem would not apply in the real situation.

***EXAMPLE 5.02    Energy balance for expansion of gas against a piston (closed system).***

A vertical cylinder of 2.00 m internal diameter contains 1.00 kmol $N_2$ gas at 298 K, enclosed under a 5000 kg frictionless piston. The cylinder sits on the Earth with atmospheric pressure $= 101.3$ kPa(abs) and $g = 9.81$ m/s$^2$. $N_2$ gas $C_{v,m}$ ref 298 K $= 21.2$ kJ/(kmol.K).

***Problem:***
Calculate the amount of heat that must be transferred to the gas to raise the piston by 4.00 m.

***Solution:***
Define the system   $=$ gas inside the cylinder (a closed system)
Specify the quantity $=$ energy
Reference condition $=$ elements at standard state 298 K, $L' = 0$, $\tilde{u} = 0$
Integral energy balance on the closed system: $E_{final} - E_{initial} = Q - W$
Initial condition      $=$ gas at 298 K, piston $L' = 0$ m
Final condition      $=$ gas at $T_f$ K, piston $L' = 4$ m
Initial pressure in
cylinder      $= P_i = 101.3$ kPa(abs) $+ (1E-3$ Pa/kPa)$(5000$ kg)$(9.81$ m/s$^2)/(3.14)(1$ m)$^2$
        $= 101.3 + 15.6$
        $= 117$ kPa(abs)

Initial volume of gas
(Ideal gas, see      $= V_i = nRT_i /P_i = (1$ kmol)$(8.314$ kJ/(kmol.K)$(298$ K)/$(117$ kPa(abs))
        $= 21.2$ m$_3$                                    *Equation 2.13)*
Final pressure      $= P_f =$ initial pressure                  $= 117$ kPa(abs)
Final volume      $= V_f = 21.3$ m$^3 + (3.14)(1$ m)$^2$ $(4$ m) $= 21.3 + 12.6 = 33.9$ m$^3$

Final temperature

(ideal gas)  $= T_f = P_f V_f / nR = (117 \text{ kPa(abs)})(33.9 \text{ m}^3)/(1 \text{ kmol})(8.314 \text{ kJ/(kmol.K)})$

$= 477 \text{ K}$

$E_{initial}$  $= U + E_k + E_p = (1 \text{ kmol})(21.2 \text{ kJ/(kmol.K)})(298-298)\text{K} + 0 \text{ kJ/kmol}) + 0 + 0)$

$= 0 \text{ kJ}$

$E_{final}$  $= U + E_k + E_p = (1 \text{ kmol})(21.2 \text{ kJ/(kmol.K)})(477-298)\text{K} + 0 \text{ kJ/kmol}) + 0 + 0)$

$= 3795 \text{ kJ}$

Q  $= 5278 \text{ kJ (Heat IN)}$

W  $=$ work done in expansion against the atmospheric pressure + weight of piston

$= P_{atm} (V_f - V_i) + mg(L_f' - L_i')$

$= (101.3 \text{ kPa})(33.9 - 21.2) \text{ m}^3 + (1\text{E-3 J/kJ})(5000 \text{ kg})(9.81 \text{ m/s}^2)$

$(4 - 0) \text{ m}$                                                                            $= 1287 + 196$

$= 1483 \text{ kJ}$

Or W  $= P(V_2 - V_1) = 117 \text{ kPa}(33.9 - 21.2) \text{ m}^3$                    $= 1483 \text{ kJ}$

(i.e. work done by expansion of gas against constant pressure)

Energy balance:  $3795 - 0 = Q - 1483$

Solve for                                                                     $\underline{Q = 5278 \text{ kJ}}$ (heat IN)

---

*EXAMPLE 5.03   Energy balance for an electric battery (closed system).*

A fully charged silver-zinc electric battery contains 0.01 kmol $Ag_2O$ with 0.01 kmol Zn. The battery is initially at 298 K then discharged under adiabatic conditions at a constant voltage and current, respectively = 1.2 V and 50 A. By Faraday's law, in 1 hour of operation at 50 A, 9.33% of both reactants are consumed in the net cell discharge reaction:

$$\begin{array}{c} 2F \\ \downarrow \end{array}$$

$Ag_2O + Zn \rightarrow 2Ag + ZnO$                                             *Reaction 1*

Data:

| Component | $Ag_2O$ | Zn | Ag | ZnO |
|---|---|---|---|---|
| $C_{v,m}$ ref 298 K  kJ/(kmol.K) | 80 | 26 | 26 | 42 |
| $h_{f,298K}^o$  kJ/kmol | −29E3 | 0 | 0 | −348E3 |

*Problem:*

Calculate the battery (uniform) internal temperature 1 hour after the current is started.

*Solution:*

Define the system  = battery (a closed system)

Specify the quantity = energy

Reference condition = elements at standard state, 298 K, $L' = 0$, $\tilde{u} = 0$

Integral energy balance on the closed system: $E_{final} - E_{initial} = Q - W$

Initial condition   = fully charged battery, t = 0

Final condition = 9.33% discharged battery, t = 3600 s

$E_{initial}$ = $\Sigma$ [n(j)($C_{v,m}$(j)(298 − 298) + $h^°_{298K}$(j))] + 0 + 0     ($E_k = E_p = 0$)

= (0.01 kmol)((80 kJ/((kmol.K))(298 − 298) K + (−29E3 kJ/kmol))
+ (0.01 kmol)((26 kJ/(kmol.K))(298 − 298) K + (0 kJ/kmol))
= −290 kJ

$E_{final}$ = $\Sigma$ [n(j)($C_{v,m}$(j)(T − 298) + $h^°_{298K}$(j))] + 0 + 0  ($E_k = E_p = 0$)

= (1 − 0.0933)(0.01)((80)(T − 298) + (−29E3)) + (1 − 0.0933)(0.01)((26)
(T − 298) + (0)) + (2)(0.0933)(0.01) ((26))(T − 298) + (0))
+ (0.0933)(0.01) ((42))(T − 298) + (−348E3))
= 1.049(T − 298) − 588 kJ

Q = 0 (adiabatic operation)

W = $E_v$I′t = (1.2 V ) (50 A) (3600 s) = 216E3 J = 216 kJ

Energy balance: [1.049(T − 298) − 588] − (−290) = 0 − 216 Solve for:

T = **376 K** at 1 hour

$E_v$     = voltage     V
I′     = current     A
t     = time     s

*NOTE:* When a battery is discharged, part of the initial chemical energy of the reactants is converted to electricity (work) and the remainder is converted to heat. A discharging battery gets very hot if it is insulated to prevent heat loss, particularly at high discharge rates or short circuit conditions.

## OPEN SYSTEMS (CONTINUOUS PROCESSES)

An open system is defined as one in which, over the time period of interest, *material is transferred across the system boundary*. In this case energy can enter and/or leave the system in the form of the energy content of material, as well as by the transfer of heat and work. The integral and differential energy balances for an open system are now adapted to account for the energy content (E) of material that crosses the system boundary and the net pressure–volume energy ($PV_{output} − PV_{input}$), called the "flow work", needed to move material in and out of the system.

Open system – integral energy balance:

$E_{final} − E_{initial} = E_{input} − E_{output} + Q − W − ( PV_{output} − PV_{input})$     *Equation 5.06*

Open system – differential energy balance:

$dE/dt = \dot{E}_{input} − \dot{E}_{output} + \dot{Q} − \dot{W} − (P\dot{V}_{output} − P\dot{V}_{input})$     *Equation 5.07*

where:     Typical Units

E     = [U + $E_k$ + $E_p$] content of the system (i.e. inside the system boundary)     kJ
$E_{input}$ = [U + $E_k$ + $E_p$] sum for all material inputs to the system     kJ
$E_{output}$ = [U + $E_k$ + $E_p$] sum for all material outputs from the system     kJ
$\dot{E}_{input}$ = [$\dot{U}$+ $\dot{E}_k$ + $\dot{E}_p$] sum for all material inputs to the system     kW
$\dot{E}_{output}$ = [$\dot{U}$+ $\dot{E}_k$+ $\dot{E}_p$] sum for all material outputs from the system     kW

| | | |
|---|---|---|
| $E_{initial}$ | = initial value of $[U + E_k + E_p]$ content of the system (i.e. inside the system boundary) | kJ |
| $E_{final}$ | = final value of $[U + E_k + E_p]$ content of the system (i.e. inside the system boundary) | kJ |
| $PV_{input}$ | = pressure–volume energy for material inputs | kJ |
| $PV_{output}$ | = pressure–volume energy for material outputs | kJ |
| $P\dot{V}_{input}$ | = pressure–volume power for material inputs | kW |
| $P\dot{V}_{output}$ | = pressure–volume power for material outputs | kW |
| $Q$ | = net heat <u>input</u> to system | kJ |
| $\dot{Q}$ | = net rate of heat <u>input</u> to system | kW |
| $t$ | = time | s |
| $W$ | = net work <u>output</u> from the system (a.k.a. "shaft work"), excluding the flow work | kJ |
| $\dot{W}$ | = net rate of work (a.k.a. "shaft power") <u>output</u> from system, excluding the flow power | kW |

You should understand that the Q and W of the above equations are respectively energy inputs and energy outputs and do not correspond to generation and consumption terms, which are both zero in the total energy balance. Also, the word "energy" is often used in thermodynamics texts instead of the more accurate term "power" (i.e. energy/time) for the description of differential energy balances.

*Example 5.04* illustrates an integral energy balance for an open system.
*Examples 5.05* to *5.13*, as well as *Examples 6.01* to *6.05* and *7.04* to *7.07* illustrate a range of problems based on differential energy balances in open systems.

### EXAMPLE 5.04  *Energy balance for a personal rocket pack (open system).*

A personal rocket pack powered by the thermo-chemical decomposition of hydrogen peroxide generates an exhaust of water vapour and oxygen with a nozzle velocity of 1050 m/s at 100 kPa(abs), 700 K.

$$2H_2O_2(l) \rightarrow 2H_2O(g) + O_2(g) \qquad\qquad \text{Reaction 1}$$

A 75 kg RocketPerson is strapped into a 30 kg rocket pack (empty mass) initially filled with 25 kg of 90 wt% $H_2O_2$ in liquid water at 298 K.

Data:

| Component | | $H_2O_2(l)$ | $H_2O(l)$ | $H_2O(g)$ | $O_2(g)$ |
|---|---|---|---|---|---|
| $h^{\circ}_{f,298\ K}$ | kJ/kmol | −188E3 | −286E3 | −242E3 | 0 |
| $C_{p,m}$ref 298 K | kJ/(kmol.K) | 89 | 75 | 36 | 30 |

### Problem:
Calculate the maximum height above ground that RocketPerson could reach in a vertical upward flight that uses all the peroxide (i.e. no peroxide remains for the descent). Assume: temperature of RocketPerson and rocket pack stay at 298 K. Zero air resistance and constant gravity.

### Solution:

First do the material balance for the peroxide decomposition process from the reaction stoichiometry.

| Component | | $H_2O_2(l)$ | $H_2O(l)$ | $H_2O(g)$ | $O_2(g)$ | TOTAL MASS |
|---|---|---|---|---|---|---|
| Molar mass | kg/kmol | 34.0 | 18.0 | 18.0 | 32.0 | |
| Initial | kmol | 0.662 | 0.139 | 0.000 | 0.000 | 25.0 kg |
| Final | kmol | 0.000 | 0.000 | 0.801 | 0.331 | 25.0 kg |

Define the system     = RocketPerson with rocket pack (an open system)
Specify the quantity  = Energy
Reference conditions: RocketPerson and rocket pack = compounds at 298 K
                               Fuel and exhaust = elements at 298 K

Let: $L'$ = maximum height reached by RocketPerson with the rocket pack.
Integral energy balance on the open system: ACC = (Final – Initial) = IN – OUT     (GEN = CON = 0)

Initial energy[4] = Energy in (RocketPerson + rocket pack) + energy in fuel
          $= (U + KE + PE)_{person + pack} + (U + KE + PE)_{fuel}$
          = (0 + 0 + 0) + ((0.662 kmol)(–188E3 kJ/kmol) + (0.139 kmol)(–286E3 kJ/kmol))

          = –164.0E3 kJ

Final energy = Energy in (RocketPerson + rocket pack)
          $= (U + KE + PE)_{person + pack}$
          $= (0 + 0 + (1E–3 \ kJ/J)(105 \ kg)(9.81 \ m/s^2)$
          (L′ m)) kJ

Energy IN     = 0 kJ
Energy OUT[5] $= (H + KE + PE)_{exhaust}$
          = (0.801 kmol)(36 kJ/(kmol.K)(700 – 298) K
          – 242E3 kJ/kmol) + (0.331 kmol)(30 kJ/
          (kmol.K) (700 – 298) K + 0 kJ/kmol)
          $+ (1E–3 \ kJ/J)((0.5)(25 \ kg) (1050 \ m/s))^2$
          $+ (1E–3 \ kJ/J)(0.5)(25 \ kg)(9.81 \ m/s^2)$
          (L′ m) kJ

Energy balance: 1.030L′ + 164.00E3 = 164.40E3
                                        – 0.122L′

Solve the energy balance to get: L′ = **345 m**

---

[4] Internal energy is assumed equal to enthalpy for the liquid fuel.
[5] Potential energy of the exhaust is effectively the PE of the total exhaust mass at elevation = 0.5L′.

## CONTINUOUS PROCESSES AT STEADY-STATE

Energy balances for continuous processes at steady-state use the differential energy balance *Equation 5.07*, with [Rate ACC = dE/dt = 0]. In this case the calculations are simplified by replacing the "U + PV" term by the enthalpy "$\dot{H}$" to give:

$$dE/dt = 0 = [\dot{H} + \dot{E}_k + \dot{E}_p]_{input} - [\dot{H} + \dot{E}_k + \dot{E}_p]_{output} + \dot{Q} - \dot{W} \qquad \textit{Equation 5.08}^6$$

where:                                                                                    Typical Units

| | |
|---|---|
| $[\dot{H} + \dot{E}_k + \dot{E}_p]_{input}$ = sum for all streams entering the system | kW |
| $[\dot{H} + \dot{E}_k + \dot{E}_p]_{output}$ = sum for all streams leaving the system | kW |
| $\dot{Q}$ = net rate of heat transfer <u>into</u> the system | kW |
| $\dot{W}$ = net rate of work (a.k.a. shaft power) <u>output</u> from the system | kW |

*Example 5.05* shows how *Equation 5.08* can be applied to calculate the power requirements of a hypothetical continuous process at steady-state, in which liquid water is pumped, elevated, heated, vaporised, expanded, heated and finally decomposed (i.e. reacted) to produce hydrogen and oxygen. By examination of *Example 5.05* you will see that the kinetic and potential energy terms are relatively small and that the energy balance is dominated by enthalpy and heat terms. This situation is common in energy balances on chemical processes, where the relative magnitude of the energy effects is usually:

*Reaction > latent heat > sensible heat > gas compression or expansion > liquid pumping > kinetic and potential energy > exotics*

### MECHANICAL ENERGY BALANCE

For "mechanical" systems, in which thermal effects such as chemical reaction, phase transformation and heat transfer are not present (or negligible), the energy balance *Equations 5.04* to *5.08* reduce to a form called the "mechanical energy balance", the steady-state version of which is *Equation 5.08A*.

$$0 = [\dot{E}_k + \dot{E}_p]_{in} - [\dot{E}_k + \dot{E}_p]_{out} - \int_{Pin}^{Pout} \dot{V}_{dp} - \dot{F}_e - \dot{W} \qquad \textit{Equation 5.08A}$$

where:

| | | |
|---|---|---|
| $\dot{E}_k, \dot{E}_p$ | = kinetic, potential energy flow | kW |
| $\dot{F}_e$ | = rate of energy degradation by friction in flowing fluid | kW |
| $\dot{W}$ | = rate of work output from the system | kW |
| $\dot{V}$ | = volumetric fluid flow | m$^3$.s$^{-1}$ |
| P | = pressure | kPa |

---

[6] *The $\dot{H}$ in equations 5.08, 5.09 and 5.10 denotes both the $\dot{H}$\* introduced with equation 5.11 and the $\dot{H}$ of equation 5.15.*

For the special case of incompressible fluid flow with zero friction loss and zero work done, *Equation 5.08A* reduces to *Equation 5.08B*, commonly known as the Bernoulli equation.

$$0 = 0.5[(\tilde{u}_{out})^2 - (\tilde{u}_{in})^2] + g[L'_{out} - L'_{in}] + [P_{out} - P_{in}]/\rho \qquad \textit{Equation 5.08B}$$

| | | |
|---|---|---|
| $L'$ = height above a reference level | m |
| g = acceleration of gravity | $ms^{-2}$ |
| P = pressure | Pa |
| $\tilde{u}$ = fluid velocity | $ms^{-1}$ |
| $\rho$ = fluid density | $kg.m^{-3}$ |

The mechanical energy balance is used for fluid flow and pressure changes in pipes and process units.

### EXAMPLE 5.05    *Energy balance for processing water (open system at steady-state).*

Flowsheet [This is a hypothetical process]

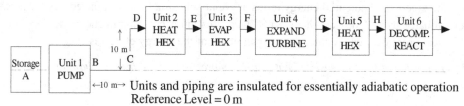

←10 m→   Units and piping are insulated for essentially adiabatic operation
Reference Level = 0 m

| Basis | 1 kmol/s H₂O | Location | A | B | C | D | E | F | G | H | I# |
|---|---|---|---|---|---|---|---|---|---|---|---|
| | [18.02 kg/s H₂O] | Species | $H_2O$ | $H_2O$ | $H_2O$ | $H_2O$ | $H_2O$ | $H_2O$ | $H_2O$ | $H_2O$ | $H_2+1/2O_2$ |
| **Quantity** | **Symbol** | **Units** | | | | | | | | | |
| Temperature | T | K | 273.15 | 273.25 | 273.25 | 273.25 | 453.00 | 453.10 | 423 | 4000 | 4000 |
| Pressure | P | kPa(abs) | 101.33 | 1170 | 1160 | 1050 | 1020 | 1000 | 140 | 120 | 100 |
| Phase | — | | L | L | L | L | L | G | G | G | G |
| Density | $\rho$ | kg/m³ | 1000 | 1001 | 1001 | 1001 | 887 | 5.15 | 0.725 | 0.065 | 0.036 |
| Velocity | $\tilde{u}$ | m/s | 0.00 | 2.00 | 2.00 | 2.00 | 2.00 | 50.0 | 50.0 | 50.0 | 50.0 |
| Elevation | $L'$ | m | 0.00 | 0.00 | 0.00 | 10.0 | 10.0 | 10.0 | 10.0 | 10.0 | 10.0 |
| Kinetic (flow) energy | $\dot{E}_k$ | kW | 0.00 | 0.04 | 0.04 | 0.04 | 0.04 | 22.53 | 22.53 | 22.53 | 22.53 |
| Potential (flow) energy | $\dot{E}_p$ | kW | 0.00 | 0.00 | 0.00 | 1.77 | 1.77 | 1.77 | 1.77 | 1.77 | 1.77 |
| Internal (flow) energy | U | kW | 0.00 | 7.50 | 7.50 | 7.50 | 13720 | 46530 | 46497 | 212332 | 155100 |

*(Continued)*

(*Continued*)

| Basis | 1 kmol/s H₂O [18.02 kg/s H₂O] | Location | A | B | C | D | E | F | G | H | I# |
|---|---|---|---|---|---|---|---|---|---|---|---|
| | | Species | $H_2O$ | $H_2O$ | $H_2O$ | $H_2O$ | $H_2O$ | $H_2O$ | $H_2O$ | $H_2O$ | $H_2+1/2O_2$ |
| Press–Vol (flow) energy | PV | kW | 1.83 | 21.06 | 20.88 | 18.90 | 22.00 | 3497 | 3477 | 33202 | 63215 |
| Enthalpy (flow) | $\dot{H}$ | kW | 1.83 | 28.56 | 28.38 | 26.40 | 13742 | 50027 | 49974 | 245534 | 218315 |
| Enthalpy (flow) | $\dot{H}^*$ | kW | −284998 | −284971 | −284972 | −284974 | −271258 | −234973 | −235026 | −39466 | 218315 |
| Total (flow) energy | $\dot{E}^*$ | kW | −284998 | −284971 | −284972 | −284972 | −271256 | −234949 | −235002 | −39442 | 218339 |

| Process unit loads | | | Unit 1 | Unit 2 | Unit 3 | Unit 4 | Unit 5 | Unit 6 |
|---|---|---|---|---|---|---|---|---|
| Heat in (rate) | $\dot{Q}$ | kW | 0.00 | 13716 | 36285 | 0.00 | 195560 | 257781 |
| Work out (rate) | $\dot{W}$ | kW | −27 | 0.00 | 0.00 | 53 | 0.00 | 0.00 |

The asterisk (*) indicates that the reference condition is the elements at their standard states ($H_2$ and $O_2$). No asterisk means the reference condition is the compounds. The reference pressure and temperature for both the element and compound reference conditions are 101.3 kPa(abs), 273.15 K. The effect of pressure from 0.61 kPa(abs) [steam table data] to 101.3 kPa(abs) [heat of formation data] on the reference condition is assumed negligible. Heat of formation of $H_2O$ (liq) at 273 K = −285000 kJ/kmol. #assumes complete decomposition of $H_2O$ at 4000 K. Values for points H and I are approximate, since thermodynamic data for 4000 K are not available.

## ENTHALPY BALANCE

For chemical process energy balances, *Equation 5.08* is simplified by dropping the kinetic and potential energy terms, to give *Equation 5.09* and its equivalent *Equation 5.10*.

$$0 = \ddot{H}_{input} - \ddot{H}_{output} + \dot{Q} - \dot{W}$$ 
*Equation 5.09*

$$0 = \Sigma[\ddot{H}(i)]in - \Sigma[\ddot{H}(i)]out + \dot{Q} - \dot{W}$$ 
*Equation 5.10*

where:
$\ddot{H}_{input}$ = sum of enthalpy flow carried by all input streams to the system        kW
$\ddot{H}_{output}$ = sum of enthalpy flow carried by all output streams from the system     kW
$\ddot{H}(i)$ = enthalpy flow of stream "i" w.r.t. a specified reference condition           kW
$\dot{Q}$ = thermal *utility load* on the system (+) ≡ heating, (−) ≡ cooling               kW
$\dot{W}$ = mechanical or electrical *utility load* on the system (+) ≡ power out; (−) ≡ power in kW

The utility loads are the energy (a.k.a. power) requirements to operate the system, which are usually supplied by fuel, steam or cooling water and/or by electricity that is used to run machines such as compressors, pumps, heaters and refrigerators.

*Equation 5.10* (called the "enthalpy balance" or when $\dot{W}$ is zero, "heat balance") is the main equation for calculating energy balances on continuous chemical processes at steady-state, where kinetic and potential energy effects are assumed insignificant. This equation is easy to use as long as you obey the rules for energy accounting regarding the reference state. This means that the enthalpy of reference state(s) must cancel out of the equation when the enthalpy OUT is subtracted from the enthalpy IN. The recommended procedure is thus to calculate all enthalpies in *Equation 5.10* with respect to the same reference state. This procedure is inconvenient in cases where enthalpy values are taken from different sources (e.g. steam tables and thermochemical tables), so it is permissible to use different reference states for terms in the enthalpy IN, provided the reference enthalpies cancel from corresponding terms in the enthalpy OUT (or if you know that the error introduced by un-cancelled reference states is acceptable for the objective of the energy balance).

The best reference state to use for energy balances on general chemical processes is the *elements* in their standard state at 298 K (see *Figure 2.05*). The procedure based on this reference state is sometimes called the *heat of formation method* of energy balance calcu-lation. On this basis the enthalpy of each process stream is estimated by *Equation 5.11*, which assumes the stream is an ideal mixture and sums the components enthalpies calcu-lated by *Equation 5.12*, i.e.:

$$\dot{H}^*(i) = \Sigma[\dot{n}(i, j)\, h^*(j)] \qquad \text{[Ideal mixture]} \qquad\qquad \textit{Equation 5.11}$$

and:

$$h^*(j) \approx \int_{T_{ref}}^{T(i)} Cp(j)dT + h^\circ_{f,T_{ref}}(j) \qquad \text{[Respect the phase]} \qquad \textit{Equation 5.12}$$

where:

$C_p(j)$ = heat capacity at constant pressure of the substance "j"
        in phase $\Pi$        kJ.kmol$^{-1}$.K$^{-1}$

$h^*(j)$ = specific enthalpy of the substance "j" at temperature $T(i)$
        in phase $\Pi$ w.r.t. *elements* at standard state (see *Figure 2.05*)    kJ.kmol$^{-1}$

$h^\circ_{f,T_{ref}}(j)$ = standard heat of formation of the substance "j" at $T_{ref}$ in phase $\Pi$   kJ.kmol$^{-1}$

$T(i)$ = temperature of stream "i"        K

$T_{ref}$ = reference temperature = 298 K        K

$\dot{n}(i, j)$ = flow of substance "j" in stream "i"        kmol.s$^{-1}$

$\dot{H}^*(i)$ = total enthalpy flow of stream "i" w.r.t. *elements* at standard state   kW

In *Equation 5.12* the phase $\Pi$ is the actual phase of the substance "j" in stream "i". If substance "j" exists in more than one phase in stream "i", then its flow is assigned to each phase in the appropriate split fractions.

For many practical calculations, *Equation 5.12* can be simplified by replacing the integral of heat capacity by the mean heat capacity defined in *Equation 2.60*, i.e.

$$h^*(j) \approx C_{p,m}(j)[T(i) - T_{ref}] + h^\circ_{f,Tref} \qquad \text{[Respect the phase]} \qquad \textit{Equation 5.13}$$

where:

     $C_{p,m}(j)$ = mean heat capacity of substance "j" in phase $\Pi$ over the temperature range
             $T_{ref}$ to the stream temperature $T(i)$        kJ.kmol$^{-1}$.K$^{-1}$

For non-ideal mixtures a term is added to *Equation 5.11* to account for the heat of mixing.

$$\dot{H}^*(i) = \Sigma[\dot{n}(i,j)h^*(j)] + \dot{H}_{mix} \qquad \text{[Non-ideal mixture]} \qquad \textit{Equation 5.14}$$

where:

$\dot{H}_{mix}$ = enthalpy of mixing (i.e. heat of mixing) for the flowing component   kW

With the heat of formation method, using the enthalpy of *Equations 5.11* (or *5.14*) and *5.12*, the energy balance automatically takes care of any *heat of reaction* in a chemical process, so it is not necessary to add a special term for the heat of reaction. This energy balance still applies to process units where no reactions occur because the standard heats of formation in the enthalpy IN and enthalpy OUT are equal, and so cancel from the balance.

For energy balances on systems that involve only physical processes such as sensible heating, evaporation, compression, etc. (i.e. no chemical reactions) it is convenient and acceptable to calculate enthalpy relative to *compounds* at a reference condition. This enthalpy, defined in *Equations 5.15* and *5.16*, does not include heats of formation and therefore does not compensate the energy balance for heats of reaction.

$$\dot{H}(i) = \Sigma[\dot{n}(i, j)h(j)] \qquad \text{[Ideal mixture]} \qquad \textit{Equation 5.15}$$

$$\text{and: } h(j) \approx \int_{T_{ref}}^{T(i)} C_p(j) \, dT + h_{p,Tref}(j) \qquad \text{[Respect the phase]}$$

$$\text{plus } h(j) \approx C_{p,m}(j)[T(i) - T_{ref}] + h_{p,Tref}(j) \qquad \textit{Equation 5.16}$$

where:

| | |
|---|---|
| $h(j)$ | = specific enthalpy of the substance "j" at temperature T in phase Π w.r.t. the *compound* at the reference state   kJ.kmol$^{-1}$ |
| $C_p(j)$ | = heat capacity at constant pressure of the substance "j" in phase Π  kJ.kmol$^{-1}$.K$^{-1}$ |
| $h_{p,Tref}(j)$ | = latent heat of any phase change(s), measured at $T_{ref}$ of substance "j" from its reference state to phase Π   kJ.kmol$^{-1}$ |
| $T(i)$ | = temperature of stream "i"   K |
| $T_{ref}$ | = reference temperature   K |
| $\dot{n}(i, j)$ | = flow of substance "j" in stream "i"   kmol.s$^{-1}$ |
| $\dot{H}(i)$ | = enthalpy flow of stream "i" w.r.t. *compounds* at the reference state  kW |

In *Equation 5.16* the phase Π is the actual phase of the substance "j" in stream "i". If substance "j" exists in more than one phase in stream "i", then its flow is assigned to each phase in the appropriate split fractions.

The enthalpy estimated by *Equations 5.15* and *5.16* is similar to that obtained from thermodynamic tables and charts such as the steam table and enthalpy-concentration diagrams, which also do not include heats of formation. This enthalpy is normally used for energy balances on "mechanical" systems such as compressors, refrigeration cycles and steam engines, as well as on process operations like heat transfer, mixing and separation (e.g. distillation, extraction, etc.) where reactions do not occur.

If the enthalpy from *Equations 5.15* and *5.16* is used for an energy balance on a chemical reactor then an *extra term* must be added to the balance equation to account for the *heat(s) of reaction*. This *heat of reaction method* is used in many texts (see *Refs. 4–8*) but it is not emphasised in this text due to its relative complexity when dealing with reactive systems with incomplete conversion and multiple reactions. *Table 5.02* compares the "heat of formation" and "heat of reaction" methods of setting up energy balances.

**Table 5.02. Comparison of energy balance reference conditions.**

| Reference condition | Non-reactive (physical) processes | Reactive (chemical) processes | Steady-state energy balance |
|---|---|---|---|
| Elements at 298 K [Respect the phases] | CORRECT. Heats of formation cancel | CORRECT. Heats of formation capture heat(s) of reaction | $0 = \Sigma[\dot{H}*(i)in] - \Sigma[\dot{H}*(i)out] + \dot{Q} - \dot{W}$ *Heat of formation* method |
| Compounds at $T_{ref.}$ [Respect the phases] | CORRECT. Heats of formation not needed | NOT SUFFICIENT. Must add heat(s) of reaction term to balance | $0 = \Sigma[\dot{H}(i)in] - \Sigma[\dot{H}(i)out] + \dot{Q} - \dot{W}$ $- \Sigma[\varepsilon(\ell)h_{mx}(\ell)]$ *Heat of reaction* method |

*Examples 5.06* and *5.09* to *5.13* show the application of *Equations 5.11* and *5.12* to energy balances on systems both with and without chemical reactions.

*Example 5.07 A and B* illustrates the application of *Equations 5.15* and *5.16* to energy balances on systems without chemical reaction. *Example 5.08 A and B* shows how steam tables and enthalpy-concentration diagrams can be used to solve energy balances in systems without chemical reaction.

## HEAT AND WORK

In a continuous chemical process the *heat* ($\dot{Q}$) and *work* ($\dot{W}$) terms[7] of the energy balance are called the *utility loads* on the process. The *utility loads* are the rates of energy transfer (a.k.a. power) across the system envelope required to operate the process. Utilities are usually supplied by fuel, steam and cooling water, as well as by electricity to drive machines such as compressors, pumps, heaters and refrigerators.

Heat is transferred to the system across *heat transfer surfaces*. A *heat transfer surface* may be the shell of a process unit, through which heat transfers from/to the surroundings, or may be contained in a dedicated heat exchanger, where heat is transferred from/to a utility or another process stream.

Heat transfer through the shells of insulated units from/to the surroundings is usually a minor part of the energy balance for industrial chemical processes, accounting for up to about 5% of the total energy flow. However in small-scale systems without insulation (e.g. laboratory apparatus) heat transfer from/to the surroundings can be important, due to the relatively high ratio of surface area to volume for small process units and associated piping.

Heat transfer via dedicated indirect heat exchangers is typically a major feature of the energy balance for industrial chemical processes. The rate of heat transfer can be calculated by *Equation 5.17* (see *Ref. 9*).

---

[7] Remember that heat and work are path functions, whose values depend on the path taken between the inlet and outlet process conditions.

$$\dot{Q} = U_{hex}A_{hex}\Delta T \qquad\qquad\qquad\qquad\qquad\qquad\qquad \textit{Equation 5.17}$$

where:

| | | |
|---|---|---|
| $\dot{Q}$ | = rate of heat transfer into the system (i.e. the *thermal load* or *thermal* | kW |
| | *duty* of the exchanger) | |
| $U_{hex}$ | = heat transfer coefficient, which depends in part on the thermal | |
| | conductivity of the process streams and the heat transfer wall | $kW.m^{-2}.K^{-1}$ |
| $A_{hex}$ | = area of the heat transfer surface | $m^2$ |
| $\Delta T$ | = temperature difference between the hot and cold streams | |
| | (logarithmic mean value[8]) | K |

In this convention $\Delta T$ is (+) when the "system" temperature is below that of the heat transfer medium, and negative (–) in the reverse condition.

Work (a.k.a. power) is usually in the form of mechanical work or electrical work. This work is sometimes called "shaft work" ($W_s$) to separate it from the "flow work" (PV) done on the material entering/exiting the system and/or the work done by changing the system volume. Mechanical work is associated with gas compressors and liquid pumps, as well as with turbines that transfer energy by gas expansion or by a pressure drop in liquid flow. The shaft power in compression and expansion of ideal gases can be calcu-

HEAT IN          WORK OUT

lated by *Equations 5.18* or *5.19*. Power in pumping liquids is given by *Equation 5.20*, which is simpler than the gas equations because liquids are nearly incompressible.

$$\dot{W} = -(P_1\dot{V}_1)(r/(r-1))[P_2/P_1]^{((r-1)/r)} -1] \quad \text{[Adiabatic power, ideal gas, and} \quad \textit{Equation 5.18}$$

$$T_2 = T_1[P_2/P_1]^{((r-1)/r)} \qquad\qquad \text{Adiabatic temperature]}$$

$$\dot{W} = -(P_1\dot{V}_1)\ln(P_2/P_1) \qquad\qquad \text{[Isothermal power, ideal gas]} \qquad \textit{Equation 5.19}$$

$$\dot{W} = -\dot{V}_1(P_2/P_1) \qquad\qquad\qquad \text{[Liquid (any conditions)]} \qquad \textit{Equation 5.20}$$

where:

$\dot{W}$ = rate of mechanical energy transfer (i.e. power) out of the system          kW

$P_1$ = inlet stream pressure          kPa(abs)

$P_2$ = outlet stream pressure          kPa(abs)

$\dot{V}_1$ = inlet stream volumetric flow rate          $m^3.s^{-1}$

---

[8]Logarithmic mean $\Delta T = [\Delta T_1 - \Delta T_2]/\ln[\Delta T_1/\Delta T_2]$ where subscripts 1 and 2 refer to the respective ends of the heat exchanger.

r $= C_p/C_v =$ ratio of gas heat capacities (e.g. r $\cong$ 1.4 for air)            –
$T_1$ = inlet stream temperature                                          K
$T_2$ = outlet stream temperature                                         K

Note that the work (power) from *Equations 5.18, 5.19* and *5.20* is the rate transferred *across the system envelope*. This is the quantity used in the energy balance on the system, but the work (power) in the surroundings must be adjusted by the efficiency of the transfer machine. For example, a gas compressor that transfers 30 kW to the system with an efficiency of 60% requires a power input of (30/0.6) = 50 kW, while a turbine that transfers 50 kW from the system with an efficiency of 60% will give a useful power output of (50 * 0.6) = 30 kW.

Electrical work can take several forms, such as power to a resistance heater or from a photovoltaic cell. Electrical work is also a major part of the energy balance for electro-chemical systems such as batteries, fuel cells and electro-synthesis reactors. In each case the electric work (power) can be calculated from *Equation 5.21*.

$$\dot{W} = E_v I' \qquad \text{[Electric power]} \qquad\qquad Equation\ 5.21$$

where:
$\dot{W}$ = rate of electrical energy transfer (i.e. power) out of the system          kW
$E_v$ = voltage drop across the system                                  Volt
$I'$ = current                                                          kA

For a photovoltaic cell, battery or fuel cell $\dot{W}$ is (+), whereas for an electric heater or electro-synthesis reactor $\dot{W}$ is (−).

## SPECIFICATION OF ENERGY BALANCE PROBLEMS

Material flows in a chemical process are specified by the set of equations and variables summarised in *Table 4.02*. The energy balance adds one independent equation to this set and introduces four more variables, the pressure (P) and temperature (T) needed to find the phase(s) and calculate the enthalpy of each stream, the heat ($\dot{Q}$) and the work ($\dot{W}$). Altogether, the material flow, pressure, temperature and phase(s) of each stream, with the heat transfer and work done, specifies every process unit and fully defines the performance of multiunit processes. The *stream table* for a full M&E balance is thus expanded to include the phase(s), pressure, temperature, volume and enthalpy of each stream, plus the values of $\dot{Q}$ and $\dot{W}$ assigned to each unit in the flowsheet.

*Table 5.03* summarises the number of variables and equations associated with the energy balance on individual process units and on a combined multi-unit process, where all material flows are already known. The degrees of freedom (D of F) of the energy balance is that of *Equation 4.10*.

The comprehensive treatment of energy balance problems is difficult because it requires simultaneous solution of *Equation 5.07* with the equations of state, the entropy balance, momentum balance, heat transfer rate, etc. which is beyond the scope of this text. Fortunately, for most chemical process calculations you can make simplifying assumptions to fully specify the problem and solve the energy balance. The assumptions depend on the situation and can be one or more of the following:

A. Specific enthalpy is independent of pressure and depends only on temperature and phase, as shown in *Equations 5.12* and *5.16*. This approximation is good for solids and liquids, and for gases away from their critical state. The specific enthalpy from thermodynamic tables (e.g. the steam table) incorporates the effect of pressure but still depends mainly on temperature and phase, except near the critical state.

B. Heat transfer to a process unit is zero, i.e. an *adiabatic process* ($\dot{Q} = 0$). This approximation is good for most industrial process units, which are usually insulated to suppress heat transfer with the surroundings, unless they are explicitly designed for that heat transfer.

C. Zero work is done by a process unit ($\dot{W} = 0$). This approximation is usually good for all process units except compressors, expansion turbines, pumps, electric heaters and electrochemical reactors. In some cases it is necessary to include work in energy balances on mixers and bio or thermo-chemical reactors when they use a lot of mechanical energy (a.k.a. power) for mixing.

D. Temperature is unchanged through the unit, i.e. an *isothermal process* (T = constant). This approximation is good for process units such as dividers, some mixers and separators ($\dot{Q} = 0$, $\dot{W} = 0$) and is usually adequate for pumps with a net liquid flow. Other process units such as compressors and reactors can be forced to operate isothermally by the appropriate heat transfer (i.e. $\dot{Q} \neq 0$). Heat exchangers are not isothermal, except in their condensing and vaporising sections.

E. The pressures and the temperatures of all streams leaving a process unit are equal. This approximation is good for dividers and for some separators such as evaporators, centrifuges and clarifiers, which operate with a single stage, but not for multi-stage units such as gas absorbers and fractional distillation columns.

Table 5.03.  **Specification of multi-unit process energy balance.**

| Process unit | Divide | Mix | Separate | Hex | Pump | React | Overall |
|---|---|---|---|---|---|---|---|
| Stream variables (P, T) | 2I | 2I | 2I | 2I | 2I | 2I | 2I |
| Unit variables ($\dot{Q}, \dot{W}$) | 2 | 2 | 2 | 2 | 2 | 2 | 2 |
| Energy balance equations | 1 | 1 | 1 | 1 | 1 | 1 | 1 |
| Entropy balance equations | 1 | 1 | 1 | 1 | 1 | 1 | 1 |
| Momentum balance equations | 1 | 1 | 1 | 1 | 1 | 1 | 1 |
| State equations | I | I | I | I | I | I | I |
| Other relations | Heat transfer rate: $Q = U_{hex}A_{hex}\Delta T$ <br> Mechanical energy balance. Pressure drop. | | | | | | |
| Phase equilibria | Phase rule, Raoult's law, Henry's law, distribution coefficient, etc. | | | | | | |
| Simplifying assumptions [Depend on situation, see notes above] <br> Enthalpy is independent of pressure.  $\dot{Q}=0$  *(adiabatic process)* <br> $\dot{W}=0$ Zero work done  $T$ = constant  *(isothermal process)* <br> P and T the same for all streams leaving a process unit. Pressure drop zero, or estimate by rule of thumb. | | | | | | | |
| Specified quantities | Stream pressures (P)  Stream temperatures (T) <br> Heat transfer ($\dot{Q}$) to process units  Work done ($\dot{W}$) by process units | | | | | | |

I = number of streams.
All material flows are assumed known from an independent material balance.
Phase splits in each stream are not assumed known.
Degrees of freedom of energy balance = number of unknown quantities – number of independent equations.

F.  Pressure drop through a process unit is zero, or is estimated by a rule of thumb.[9] Since most energy balances are insensitive to pressure these assumptions are usually good for all process units except compressors, expansion turbines, pumps, and some separators and reactors that function with a high pressure drop. For the examples in this chapter all pressure drops are estimated by the rules of thumb used in process design (see *Refs. 9, 10*).

The specified stream pressures and temperatures are mostly fixed by design considerations for efficient reaction and separation. Heat transfer and work are then usually determined to meet these process requirements — ultimately through an optimisation procedure with an objective such as maximum return on investment.

### ENERGY BALANCE CALCULATIONS BY COMPUTER (SPREADSHEET)

Energy balance calculations in this chapter start from the assumption that the material balance has been fixed and summarised in a stream table, as shown in Chapter 4. The energy balance is treated here as a fully-specified problem (i.e. D of F = 0) that hinges on

[9] A rule of thumb is an approximation that holds in most circumstances. In process design, rules of thumb are based on trade-offs between capital and operating costs that aim to minimise the total cost and/or maximise return on investment.

| Component | M | Stream | | | |
|---|---|---|---|---|---|
| | kg/kmol | *1* | *2* | *3* | *4* |
| | | Flow (kmol/h) | | | |
| A | | | | | |
| B | | | | | |
| C | | | | | |
| Total | kg/h | | | | |
| Phase | – | | | | |
| Pressure | kPa(abs) | | | | |
| Temperature | K | | | | |
| Volume | m³/h | | | | |
| Enthalpy | kJ/h | | | | |
| | | Unit 1 | Unit 2 | Unit 3 | Overall |
| Q | kJ/h | | | | |
| W | kJ/h | | | | |
| *Mass balance check* | | | | | |
| Mass IN | kg/h | | | | |
| Mass OUT | kg/h | | | | |
| Closure | % | | | | |
| *Energy balance check* | | | | | |
| Energy IN | kJ/h | | | | |
| Energy OUT | kJ/h | | | | |
| Closure | % | | | | |

**Figure 5.01.   Generic material and energy balance stream table.**

calculation of the stream enthalpies. The solved energy balance is added to the material balance to form a complete M&E balance. The M&E balance is then presented in an expanded stream table that shows the material balance plus the phase(s), pressure, temperature, volume and enthalpy of each process stream, together with the heat and work transfers to each process unit. The basic format of the stream table is shown in *Figure 5.01*.[10] The completed stream table should be presented within view of the process flowsheet (e.g. on the same page) so the process can be quickly understood and examined for consistency on both a conceptual and quantitative level.

As a check on the calculations it is recommended to include in the stream table values for the closure of both the mass and the energy balances on each process unit and on the overall process. For a more detailed analysis it is sometimes useful to track down errors with individual element balances (a.k.a. atom balances) across each process unit and on the overall process.

The mass balance closure ($e_M$ = mass out/mass in) is defined in *Equation 4.09*. The energy balance closure is defined in *Equation 5.22*.

$$e_E = 100[\Sigma \dot{H}^*(i)out + \dot{W}]/[\Sigma \dot{H}^*(i)in + \dot{Q}] \qquad\qquad Equation\ 5.22$$

---

[10] Values for other properties, e.g. heat capacity, heat of formation and density are listed in or near the stream table for convenient calculation by routines such as SUMPRODUCT.

where:

| | | |
|---|---|---|
| $e_E$ | = energy balance closure (steady-state) | % |
| $\Sigma\dot{H}^*(i)out$ | = sum of enthalpy rates carried by all output streams to the system | kW |
| $\Sigma\dot{H}^*(i)in$ | = sum of enthalpy rates carried by all input streams from the system | kW |
| $\dot{Q}$ | = net rate of heat transfer <u>into</u> the system | kW |
| $\dot{W}$ | = net rate of work <u>output</u> by the system | kW |

Spreadsheet calculations for the energy balance follow the pattern of those for the material balance outlined in Chapter 4. The major issue is finding the stream enthalpies. To estimate enthalpies by *Equation 5.11* or *5.15* the data needed to calculate the *specific enthalpy* of each component (i.e. $C_p$, $h_p$, $h_f$ ) must be tabulated on the spreadsheet, either as part of the stream table or in a separate "properties" section. If the specific enthalpies are taken from thermodynamic tables or charts you need a "look-up" function that refers to thermodynamic data in the spreadsheet, or better, a mathematical correlation that calculates specific enthalpy from the stream conditions.

Energy balance calculations in the spreadsheet usually fall into one of two categories:

1. Given the conditions (flow, phase, P, T) for each stream, find $\dot{Q}$ and/or $\dot{W}$.
   In this case you first calculate the stream enthalpies to complete the "enthalpy" row of the stream table.
   Next, calculate any accessible and significant mechanical or electrical work (power) transfer by one of the *Equations 5.18* to *5.21* given above. Finally, solve *Equation 5.10* for the unknown $\dot{Q}$ or $\dot{W}$.
2. Given $\dot{Q}$ and $\dot{W}$ plus some stream conditions, find the other stream condition(s).
   In this case you first calculate all the known stream enthalpies to partially complete the "enthalpies" row of the stream table. Next, calculate significant but unspecified mechanical or electrical work (power) transfer. Finally, solve *Equation 5.10* for the unknown condition(s). For example, if the unknown condition is the outlet temperature from a utility HEX heating a single component fluid *without changing its phase,*[11] then *Equation 5.10*, with *Equations 5.11* and *5.12*, can be solved explicitly for that temperature to give:

$$\mathbf{T(out) = T_{ref} + [(\dot{H}^*(in) + \dot{Q} - \dot{W})/\bar{n}(in) - h^o_{f,ref}]/C_{p,m}}$$         *Equation 5.23*

where:

| | | |
|---|---|---|
| T(out) | = outlet stream temperature | K |
| $T_{ref}$ | = reference temperature for enthalpy calculations | K |
| $\dot{H}^*(in)$ | = enthalpy flow of inlet stream | kW |

---

[11] If part or all of the process fluid does change phase in the HEX then this solution must be modified because the value of $C_{p,m}$ is not defined across a phase change, plus the enthalpy change will include a latent heat term.

$\dot{Q}$   = rate of heat transfer (i.e. the utility load)                                                kW
$\dot{W}$   = rate of work done by process unit                                                             kW
$n(in)$ = inlet flow rate of process stream                                                                 $kmol.s^{-1}$
$C_{pm}$  = mean heat capacity of process stream across the HEX                                             $kJ.kmol^{-1}.K^{-1}$
$h^{\circ}_{f,Tref}$ = heat of formation of the process fluid component at
$\quad$ $T_{ref}$ in phase $\Pi$                                                                            $kJ.kmol^{-1}$

Note that phase $\Pi$ here is the phase of the inlet stream, and that the heat of formation includes latent heat for any phase changes from the standard state to the inlet stream condition.

When using a spreadsheet it is not necessary for you to write an explicit solution for the unknown quantity in the energy balance. If the balance is fully-specified the unknown value can be found by setting up the stream table with known conditions and using the "Goal Seek" or "Solver" to find the value of the unknown condition that closes the energy balance at steady-state (i.e. gives $e_E = 100\%$).

## ENERGY BALANCES WITH THE HEAT OF COMBUSTION

An important class of energy balance problems involves reactions with materials such as fuels and biological substrates (i.e. biomass) that are characterised only by an empirical formula and a "heat of combustion" or a "heating value". Examples of these materials are wood, coal, fuel oil, waste mixtures, bio-sludge, bio-oil, enzymes, yeast, etc. There are two ways to handle such materials in the energy balance, the first (A) has general application; the second (B) applies only to combustion. These calculations correspond respectively to the "heat of formation" and the "heat of reaction" methods of *Table 5.02*.

A. Calculate the effective heat of formation of the material from its heat of combustion and use this heat of formation in the general energy balance.

To find the effective heat of formation, first write the stoichiometry of the combustion reaction from the empirical formula of the reactant, e.g. for a material with empirical formula $C_aH_bS_cN_d$:

$$C_aH_bS_cN_d + (a + b/4 + c)O_2(g) \rightarrow (a)CO_2(g) + (b/2)H_2O(l) + (c)SO_2(g) + (d/2)N_2(g)$$

$$\textit{Reaction 5.01}^{12}$$

*(The material balance is most easily calculated by atom balances on C, H, S, N and O.)*

Next, solve *Equation 2.53* for the heat of formation of the reactants:

$$\Sigma(v(j)h^{\circ}_f (j)) \text{ reactants} = -h_{rxn} + \Sigma(v(j)h^{\circ}_f (j)) \text{ products}$$

---

[12] This stoichiometry ignores the formation of nitrogen oxides ($NO$, $NO_2$, $N_2O$), that normally accounts for a small fraction of the nitrogen fed to combustion processes.

Substitute:

$v(j) = 1$ for the reactant (i.e. the "fuel" of the combustion reaction). Then:

$\Sigma(v(j)h_f^o(j))$ reactants $= h_f^o$ (fuel)

$h_{f,O_2(g)}^o = h_{f,N_2(g)}^o = 0 \qquad$ and $\qquad h_{rxn} = h_c^o$

then

$h_{f,C_aH_bS_cN_d}^o = -h_c^o + [(a)\, h_{f,O_2(g)}^o + (b/2)\, h_{f,H_2O(l)}^o + (c)\, h_{f,SO_2(g)}^o + (d/2)\, h_{f,N_2(g)}^o]$

where:

$h_{f,C_aH_bS_cN_d}^o$ = standard heat of formation of the reactant (a.k.a. the fuel)  kJ.kmol$^{-1}$

$h_c^o$       = standard *gross* heat of combustion of the reactant (a.k.a. the fuel)  kJ.kmol$^{-1}$

$h_{f,CO_2(g)}^o, h_{f,H_2O(l)}^o, h_{f,SO_2(g)}^o, h_{f,N_2(g)}^o$ = standard heats of formation of the

                                 reaction products [note phases]  kJ.kmol$^{-1}$

a, b, c, d = stoichiometric coefficients in the combustion *Reaction 5.01*  –

Note that $h_c^o$ above is the *gross* heat of combustion.

If you use the *net* heat of combustion here the $h_{f,H_2O(l)}^o$ must be replaced by $h_{f,H_2O(g)}^o$.

B. In typical "combustion" processes the fuel undergoes complete combustion in air or oxygen to give $CO_2$, $H_2O$, $SO_2$, $N_2$, etc. by a reaction such as *5.01*. Method A can be used for energy balances with complete or incomplete combustion, but for complete combustion it is sometimes easier to use the "heat of reaction" method to solve combustion problems. For the "heat of reaction" method the heat of combustion is added as an extra term to the energy balance, in which the stream enthalpies *do not include heats of formation.* *Equation 5.10* then becomes:

$$0 = \Sigma[\dot{H}(i)]\text{in} - \Sigma[\dot{H}(i)]\text{out} + \dot{Q} - \dot{W} - \dot{n}_{fuel}\, h_c^o \qquad\qquad \text{Equation 5.24}$$

where:

$\Sigma[\dot{H}(i)]$in   = sum of enthalpy rates for all inlet streams, w.r.t.

               compounds at reference state  kW

$\Sigma[\dot{H}(i)]$in   = sum of enthalpy rates of all outlet streams, w.r.t.

               compounds at reference state  kW

$\dot{Q}$            = net rate of heat transfer <u>into</u> the system  kW

$\dot{W}$            = net rate of work transfer <u>out</u> of the system  kW

$\dot{n}_{fuel}$        = rate of fuel flow into the system  kmol.s$^{-1}$

$h_c^o$          = standard heat of combustion of fuel  kJ.kmol$^{-1}$

*Equation 5.24* can cause difficulty because the reference state for the enthalpy terms [$\dot{H}(i)$] may not be same as the standard state for the heat of combustion (usually 101.3 kPa(abs), 298 K). Strictly the references states should be corrected to cancel from the balance. However the heat of combustion is typically so large that the relatively small difference in the usual reference states (e.g. 273 K vs. 298 K) has little effect on the energy balance.

In almost all practical combustion problems, the product water leaves the system as a gas, in which case the heat of combustion in *Equation 5.24* must be the *net* heat of combustion. The net heat of combustion can be found from the gross heat of combustion by *Equation 5.25*:

$$h^\circ_{c,net} = h^\circ_{c,gross} + (b/2)h_{v,H_2O} \qquad\qquad\qquad Equation\ 5.25$$

where:

$h^\circ_{c,net}$   = standard net heat of combustion of the fuel (= −lower heating value)   kJ.kmol$^{-1}$

$h^\circ_{c,gross}$ = standard gross heat of combustion of the fuel

(= −higher heating value)                                                      kJ.kmol$^{-1}$

b/2= moles of $H_2O$ produced per mole of fuel (cf. *Reaction 5.01*)            −

$h_{v,H_2O}$ = heat of vaporisation of water (e.g. 44E3 kJ/kmol at 298K)        kJ.kmol$^{-1}$

Beware that combustion is an *exothermic* reaction, so $h^\circ_c$ should always be negative (−), whereas the "heating value" is usually given as positive (+).

SINGLE PROCESS UNITS

Chapter 3 describes six generic process units used to build chemical process flowsheets and Chapter 4 *(Figure 4.01 and Example 4.04)* shows the steady-state material balance for each generic unit. The corresponding steady-state energy balances on each of these generic process units are illustrated in *Example 5.06 A, B, C, D, E and F*.

*Example 5.06* uses only the *heat of formation method* for energy balances, in which stream enthalpies include the heats of formation of the components, as in *Equations 5.11* and *5.12*. For the five generic process units without chemical reaction (i.e. DIVIDE, MIX, SEPARATE, HEAT EXCHANGE and PUMP) the heats of formation are redundant and cancel from the energy balances. However the heats of formation are essential in the energy balance on the REACTOR, to account for the heat liberated or absorbed by chemical reaction(s).

The heat of formation method is preferred for general chemical process energy balance calculations because it automatically deals with both non-reactive and reactive processes. Nonetheless, for purely non-reactive processes it is sometimes convenient to calculate stream enthalpies without the heats of formation, as in *Equations 5.15* and *5.16*. This situation arises particularly with non-reactive processes when enthalpies are obtained from thermodynamic diagrams or tables (e.g. the steam table) whose reference conditions are not the elements at standard state. *Example 5.07 A* and *B* shows such energy balance calculations on two non-reactive systems (SEPARATOR and PUMP) using *Equations 5.15* and *5.16* to estimate the stream enthalpies. *Example 5.08 A and B* illustrates non-reactive energy balance calculations on a MIXER and a HEAT EXCHANGER using thermodynamic diagrams and tables as the source of enthalpy values.

*Example 5.09* shows a more "advanced" problem involving a chemical reactor whose temperature is controlled by indirect heat transfer from condensing steam. In this case, since the steam does not mix with the reactants, it is convenient and acceptable to take the *element* reference state for the reactor side and the *compound* reference state for the steam side of the REACT/HEX unit. *Example 5.09* (as well as *Example 5.08B*) demonstrates that enthalpy values can be taken from sources with different reference conditions, provided the individual reference conditions cancel from the energy balance.

## MULTIPLE PROCESS UNITS

Energy balance calculations on multiple process units (i.e. complete process flowsheets) follow the same procedure as material balances on multiple process units described in Chapter 4. As for material balances the system should first be checked for its specification (degrees of freedom), guided by the rules of *Table 5.03*. The coupled energy balances of a fully-specified system may then be solved by either the sequential modular or the simultaneous solution method. In the sequential modular method the solution is iterated to convergence through tear streams to deal with recycle loops.

*Example 5.10* demonstrates an energy balance for a simple process flowsheet with a mixer-reactor-separator + recycle pump, based on the material balance of *Example 4.10*.

Note that in *Example 5.10* the unknown values of T(2), T(6), $\dot{Q}(2)$ and $\dot{Q}(3)$ are obtained by solving the energy balances on the respective process units. These results can come from either:

A. Explicit algebraic solution.

or

B. Using the spreadsheet "Goal Seek" or "Solver" tool with the unknown value as a variable to set closure: $e_E = 100\%$ (see *Equation 5.22*) in the appropriate energy balance.

If necessary, more difficult cases may be dealt with by programming a spreadsheet macro with a numerical method for solving non-linear equations (e.g. bisection, Newton's method, etc.).

*Examples 5.11, 5.12* and *5.13* show the spreadsheet solutions to a set of material and energy balances on respectively, a biochemical, an electrochemical and a thermo-chemical process, each with three process units coupled through a recycle stream. In each of *Examples 5.11, 5.12* and *5.13* the material balance is independent of the energy balance, so the separate material and energy balances are solved in sequence. As in *Example 5.10* the unknown values in the energy balance are found using the "Goal Seek" or "Solver" tool to set closure = 100% in *Equation 5.22*. This procedure is exemplified in *Example 5.13* by the solution for T(2) = 2148 K, which is effectively the *adiabatic flame temperature* in the fuel oil combustion process.

### EXAMPLE 5.06   Energy balances for generic process units (open system at steady-state).

Note that these problems are simplified by using mean heat capacities instead of integrating the $C_p$ polynomials (see *Equations 2.59, 2.60* and *2.62*).

### A. Divider

No phase change
No reaction
For material balance, see *Example 4.04*

| Stream Table | | Divider | | Problem | |
|---|---|---|---|---|---|
| Species | M | Stream | | | |
| | | | *1* | *2* | *3* |
| | kg/kmol | | kmol/h | | |
| A (liq) | 20 | | 4 | 1 | 3 |
| B (liq) | 30 | | 8 | 2 | 6 |
| C (liq) | 40 | | 12 | 3 | 9 |
| Total | kg/h | | 800 | 200 | 600 |
| Phase | | | L | L | L |
| Temp. | K | | 400 | ? | ? |
| Press. | kPa(abs) | | 150 | 140 | 140 |
| Enthalpy | kJ/h | | −1.05E+06 | ? | ? |

| Properties | Species | A (liq) | A (gas) | B (liq) | B (gas) | C (liq) | C (gas) |
|---|---|---|---|---|---|---|---|
| $C_{p,m}$(ref.298K) | kJ/kmol.K | 50 | 35 | 70 | 45 | 90 | 55 |
| $h_{f,298K}$ | kJ/kmol | −2.00E+05 | −1.80E+05 | −1.00E+05 | −7.00E+04 | 3.00E+04 | 4.00E+04 |
| $h_{v,298K}$ | kJ/kmol | – | 2.00E+04 | – | 3.00E+04 | – | 1.00E+04 |

*Solution:* Energy balance.     Reference condition = *elements* at standard state, 298K.

$$0 = \Sigma[\dot{H}*(in)] - \Sigma[\dot{H}*(out)] + \dot{Q} - \dot{W}$$

For a divider $\dot{Q} = \dot{W} = 0$ and T(2) = T(3)   i.e.

$$0 = \dot{H}*(1) - \dot{H}*(2) - \dot{H}*(3) + 0 - 0 \qquad\qquad [1]$$

where:
$\dot{H}*(i) = \Sigma[\dot{n}(i,j)h*(j)]$ and $h*(j) = C_{p,m}(j)(T(i) - T_{ref}) + h^{o}_{f,Tref}(j)$     [Respect the phase]

$\dot{H}*(1)$ = (4 kmol/h)((50 kJ/kmol.K)(400 K − 298 K) + (−2E5 kJ/kmol))
     + (8 kmol/h)((70 kJ/kmol.K)(400 K − 298 K) + (−1E5 kJ/kmol))
     + (12 kmol/h)((90 kJ/kmol.K)(400 K − 298 K) + (3E4 kJ/kmol)) = −1.05E+06 kJ/h

$\dot{H}*(2)$ = (1 kmol/h)((50 kJ/kmol.K)(T(2) K − 298 K) + (−2E5 kJ/kmol))
     + (2 kmol/h)((70 kJ/kmol.K)(T(2) K − 298K) + (−1E5 kJ/kmol))
     + (3 kmol/h)((90 kJ/kmol.K)(T(2) K − 298 K) + (3E4 kJ/kmol)) = f (T(2))   kJ/h

$\dot{H}*(3)$ = (3 kmol/h)((50 kJ/kmol.K)(T(3) K − 298 K) + (−2E5 kJ/kmol))
     + (6 kmol/h)((70 kJ/kmol.K)(T(3) K − 298K) + (−1E5 kJ/kmol))
     + (9 kmol/h)((90 kJ/kmol.K)(T(3) K − 298 K) + (3E4 kJ/kmol)) = f (T(3))   kJ/h

Solve equation [1] for:T(2) = T(3) = 400 K

| Stream Table | | Divider | Solution | |
|---|---|---|---|---|
| Species | M | | Stream | |
| | | 1 | 2 | 3 |
| | kg/kmol | | kmol/h | |
| A (liq) | 20 | 4 | 1 | 3 |
| B (liq) | 30 | 8 | 2 | 6 |
| C (liq) | 40 | 12 | 3 | 9 |
| Total | kg/h | 800 | 200 | 600 |
| Phase | | L | L | L |
| Temp. | K | 400 | 400 | 400 |
| Press. | kPa(abs) | 150 | 140 | 140 |
| Enthalpy* | kJ/h | −1.05E+06 | −2.63E+05 | −7.89E+05 |
| Energy balance check | | | | |
| Energy IN | kJ/h | −1.05E+06 | Closure % | |
| Energy OUT | kJ/h | −1.05E+06 | 100.0 | |

*Enthalpy values include heats of formation.*

## B. Mixer

No phase change
No reaction
For material balance, see *Example 4.04*

| Stream | | Mixer | | |
|---|---|---|---|---|
| Species | M | | Stream | |
| | | 1 | 2 | 3 |
| | kg/kmol | | kmol/h | |
| A (liq) | 20 | 4 | 3 | 7 |
| B (liq) | 30 | 8 | 5 | 13 |
| C (liq) | 40 | 12 | 7 | 19 |
| Total | kg/h | 800 | 490 | 1290 |
| Phase | | L | L | L |
| Temp. | K | 400 | 300 | ? |
| Press. | kPa(abs) | 150 | 150 | 140 |
| Enthalpy | kJ/h | −1.05E+06 | −8.88E+05 | ? |

| Properties | Species | A (liq) | A (gas) | B (liq) | B (gas) | C (liq) | C (gas) |
|---|---|---|---|---|---|---|---|
| $C_{p,m}$(ref.298K) | kJ/kmol.K | 50 | 35 | 70 | 45 | 90 | 55 |
| $h_{f,298K}$ | kJ/kmol | −2.00E+05 | −1.80E+05 | −1.00E+05 | −7.00E+04 | 3.00E+04 | 4.00E+04 |
| $h_{v,298K}$ | kJ/kmol | – | 2.00E+04 | – | 3.00E+04 | – | 1.00E+04 |

## Solution:

Energy balance.

Reference condition = *elements* at standard state, 298K.

$0 = [\dot{H}^*(in)] - [\dot{H}^*(out)] + \dot{Q} - \dot{W}$ For a mixer $\dot{Q} = \dot{W} = 0$  i.e.

$0 = \dot{H}^*(1) - \dot{H}^*(2) - \dot{H}^*(3) + 0 - 0$ [1]

where:

$\dot{H}*(i) = \Sigma[\dot{n}(i, j)h*(j)]$ and $h*(j) = C_{p,m}(j)(T(i) - T_{ref}) + h^\circ_{f,Tref}(j)$        [Respect the phase]

$\dot{H}*(1) = (4 \text{ kmol/h})((50 \text{ kJ/kmol.K})(400 \text{ K} - 298 \text{ K}) + (-2E5 \text{ kJ/kmol}))$
        $+ (8 \text{ kmol/h})((70 \text{ kJ/kmol.K})(400 \text{ K} - 298 \text{ K}) + (-1E5 \text{ kJ/kmol}))$
        $+ (12 \text{ kmol/h})((90 \text{ kJ/kmol.K})(400 \text{ K} - 298 \text{ K}) + (3E4 \text{ kJ/kmol})) = -1.05E+06$   kJ/h

$\dot{H}*(2) = (3 \text{ kmol/h})((50 \text{ kJ/kmol.K})(300K - 298 \text{ K}) + (-2E5 \text{ kJ/kmol}))$
        $+ (5 \text{ kmol/h})((70 \text{ kJ/kmol.K})(300K - 298K) + (-1E5 \text{ kJ/kmol}))$
        $+ (7 \text{ kmol/h})((90 \text{ kJ/kmol.K})(300K - 298 \text{ K}) + (3E4 \text{ kJ/kmol})) = -8.88E+05$   kJ/h

$\dot{H}*(3) = (7 \text{ kmol/h})((50 \text{ kJ/kmol.K})(T(3) \text{ K} - 298 \text{ K}) + (-2E5 \text{ kJ/kmol}))$
        $+ (13 \text{ kmol/h})((70 \text{ kJ/kmol.K})(T(3) \text{ K} - 298 \text{ K}) + (-1E5 \text{ kJ/kmol}))$
        $+ (19 \text{ kmol/h}) \times ((90 \text{ kJ/kmol.K})(T(3) \text{ K} - 298 \text{ K}) + (3E4 \text{ kJ/kmol})) = f(T(3))$   kJ/h

Solve equation [1] for: T(3) = **362 K**

| Stream | | Mixer | | |
|---|---|---|---|---|
| Species | M | | Stream | |
| | | 1 | 2 | 3 |
| | kg/kmol | kmol/h | | |
| A (liq) | 20 | 4 | 3 | 7 |
| B (liq) | 30 | 8 | 5 | 13 |
| C (liq) | 40 | 12 | 7 | 19 |
| Total | kg/h | 800 | 490 | 1290 |
| Phase | | L | L | L |
| Temp. | K | 400 | 300 | 362 |
| Press. | kPa(abs) | 150 | 150 | 140 |
| Enthalpy* | kJ/h | −1.05E+06 | −8.88E+05 | −1.94E+06 |
| Energy balance check | | | | |
| Energy IN | kJ/h | −1.94E+06 | Closure % | |
| Energy OUT | kJ/h | −1.94E+06 | 100.0 | |

*Enthalpy values <u>include</u> heats of formation.*

## C. Separator

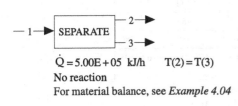

$\dot{Q} = 5.00E+05$ kJ/h       T(2) = T(3)

No reaction

For material balance, see *Example 4.04*

| Stream Table | | Separator | Problem | |
|---|---|---|---|---|
| Species | M | | Stream | |
| | | 1 | 2 | 3 |
| | kg/kmol | kmol/h | | |
| A | 20 | 4 | 3 | 1 |
| B | 30 | 8 | 4 | 4 |
| C | 40 | 12 | 4 | 8 |
| Total | kg/h | 800 | 340 | 460 |
| Phase | | L | G | L |
| Temp. | K | 300 | ? | ? |
| Press. | kPa(abs) | 600 | 580 | 580 |
| Enthalpy | kJ/h | −1.24E+06 | ? | ? |

| Properties | Species | A (liq) | A (gas) | B (liq) | B (gas) | C (liq) | C (gas) |
|---|---|---|---|---|---|---|---|
| $C_{p,m}$(ref.298K) | kJ/kmol.K | 50 | 35 | 70 | 45 | 90 | 55 |
| $h_{f,298K}$ | kJ/kmol | –2.0E+05 | –1.80E+05 | –1.0E+05 | –7.0E+04 | 3.00E+04 | 4.00E+04 |
| $h_{v,298K}$ | kJ/kmol | – | 2.00E+04 | – | 3.00E+04 | – | 1.00E+04 |

**Solution:** Energy balance.   Reference condition = *elements* at standard state, 298K.

$$0 = [\dot{H}^*(in)] - [\dot{H}^*(out)] + \dot{Q} - \dot{W} \qquad \text{For a Separator} \qquad \dot{W} = 0 \qquad \text{i.e.}$$

$$0 = \dot{H}^*(1) - \dot{H}^*(2) - \dot{H}^*(3) + \dot{Q} - 0 \qquad\qquad\qquad\qquad\qquad [1]$$

where:

$$\dot{H}^*(i) = \Sigma[\dot{n}(i,j)h^*(j)] \text{ and } h^*(j) = C_{p,m}(j)(T(i) - T_{ref}) + h^\circ_{f,Tref}(j) \qquad \text{[Respect the phase]}$$

$\dot{H}^*(1) = (4 \text{ kmol/h})((50 \text{ kJ/kmol.K})(300 \text{ K} - 298 \text{ K}) + (-2E5 \text{ kJ/kmol}))$
$\qquad\quad + (8 \text{ kmol/h})((70 \text{ kJ/kmol.K})(300 \text{ K} - 298 \text{ K}) + (-1E5 \text{ kJ/kmol}))$
$\qquad\quad + (12 \text{ kmol/h})((90 \text{ kJ/kmol.K})(300 \text{ K} - 298 \text{ K}) + (3E4 \text{ kJ/kmol})) \quad = -1.24E+06 \text{ kJ/h}$

$\dot{H}^*(2) = (3 \text{ kmol/h})((35 \text{ kJ/kmol.K})(T(2) \text{ K} - 298 \text{ K}) + (-1.8E5 \text{ kJ/kmol}))$
$\qquad\quad + (4 \text{ kmol/h})((45 \text{ kJ/kmol.K})(T(2) \text{ K} - 298\text{K}) + (-7E4 \text{ kJ/kmol}))$
$\qquad\quad + (4 \text{ kmol/h})((55 \text{ kJ/kmol.K})(T(2) \text{ K} - 298 \text{ K}) + (4E4 \text{ kJ/kmol})) = f(T(2)) \quad \text{kJ/h}$

$H^*(3) = (1 \text{ kmol/h})((50 \text{ kJ/kmol.K})(T(3) \text{ K} - 298 \text{ K}) + (-2E5 \text{ kJ/kmol}))$
$\qquad\quad + (4 \text{ kmol/h})((70 \text{ kJ/kmol.K})(T(3) \text{ K} - 298\text{K}) + (-1E5 \text{ kJ/kmol}))$
$\qquad\quad + (8 \text{ kmol/h})((90 \text{ kJ/kmol.K})(T(3) \text{ K} - 298 \text{ K}) + (3E4 \text{ kJ/kmol})) = f(T(3)) \quad \text{kJ/h}$

Solve equation [1] for: $T(2) = T(3) = \textbf{481K}$

| Stream | *Separator* | | *Solution* | |
|---|---|---|---|---|
| Species | M | | Stream | |
| | | 1 | 2 | 3 |
| | kg/kmol | | kmol/h | |
| A | 20 | 4 | 3 | 1 |
| B | 30 | 8 | 4 | 4 |
| C | 40 | 12 | 4 | 8 |
| Total | kg/h | 800 | 340 | 460 |
| Phase | | L | G | L |
| Temp. | K | 300 | 481 | 481 |
| Press. | kPa(abs) | 600 | 580 | 580 |
| Enthalpy | kJ/h | –1.24E+06 | –5.68E+05 | –1.68E+05 |
| Energy balance check | | | | |
| Energy IN | kJ/h | –7.36E+05 | Closure % | |
| Energy OUT | kJ/h | –7.36E+05 | 100.0 | |

*Enthalpy values <u>include</u> heats of formation.*

## D. Heat exchanger (indirect contact)

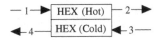

Indirect contact
No reaction
For material balance, see *Example 4.04*

| Stream Table | | *Heat Exchanger* | | | *Problem* |
|---|---|---|---|---|---|
| Species | M | Stream | | | |
| | | *1* | *2* | *3* | *4* |
| | kg/kmol | kmol/h | | | |
| A | 20 | 4 | 4 | 3 | 3 |
| B | 30 | 8 | 8 | 7 | 7 |
| C | 40 | 12 | 12 | 9 | 9 |
| Total | kg/h | 800 | 800 | 630 | 630 |
| Phase | | G | L | L | G |
| Temp. | K | 500 | 420 | 300 | ? |
| Press. | kPa(abs) | 600 | 570 | 200 | 170 |
| Enthalpy | kJ/h | | | | |

| Properties | Species | A (liq) | A (gas) | B (liq) | B (gas) | C (liq) | C (gas) |
|---|---|---|---|---|---|---|---|
| $C_{p,m}$(ref.298K) | kJ/kmol.K | 50 | 35 | 70 | 45 | 90 | 55 |
| $h_{f,298K}$ | kJ/kmol | −2.0E+05 | −1.8E+05 | −1.0E+05 | −7.0E+04 | 3.00E+04 | 4.00E+04 |
| $h_{v,298K}$ | kJ/kmol | − | 2.00E+04 | − | 3.00E+04 | − | 1.00E+04 |

*Solution:*

Energy balance.
Reference condition = *elements* at standard state, 298K.

$$0 = \Sigma[\dot{H}^*(in)] - \Sigma[\dot{H}^*(out)] + \dot{Q} - \dot{W} \qquad \text{For a heat exchanger} \quad \dot{Q} = \dot{W} = 0 \quad \text{i.e.}$$

$$0 = \dot{H}^*(1) - \dot{H}^*(2) + \dot{H}^*(3) - \dot{H}^*(4) + 0 - 0 \quad \text{[overall HEX balance]} \qquad [1]$$

where:

$$\dot{H}^*(i) = \Sigma[\dot{n}(i, j)h^*(j)] \text{ and } h^*(j) = C_{p,m}(j)(T(i) - T_{ref}) + h^o_{f,Tref(j)} \qquad \text{[Respect the phase]}$$

$\dot{H}^*(1) = $ (4 kmol/h)((35 kJ/kmol.K)(500 K − 298 K)
     $+ $ (−1.8E5 kJ/kmol))$+ $ (8 kmol/h)((45 kJ/kmol.K)
     $\times$ (500 K − 298 K) + (−7E4 kJ/kmol))$+ $ (12 kmol/h)
     $\times$ ((55 kJ/kmol.K)(500 K − 298 K) + (4E4 kJ/kmol))   $= -5.66E+05$   kJ/h

$\dot{H}^*(2) = $ (4 kmol/h)((50 kJ/kmol.K)(400 K − 298 K)
     $+ $ (−2E5 kJ/kmol))$+ $ (8 kmol/h)((70 kJ/kmol.K)
     $\times$ (400 K − 298 K) + (−1E5 kJ/kmol))$+ $ (12 kmol/h)
     $\times$ ((90 kJ/kmol.K)(400 K − 298 K) + (3E4 kJ/kmol))   $= -1.08E+06$   kJ/h

$\dot{H}^*(3) = $ (3 kmol/h)((50 kJ/kmol.K)(350 K − 298 K)
     $+ $ (−2E5 kJ/kmol))$+ $ (7 kmol/h)((70 kJ/kmol.K)
     $\times$ (350 K − 298 K) + (−1E5 kJ/kmol))$+ $ (9 kmol/h)
     $\times$ ((90 kJ/kmol.K)(350 K − 298 K) + (3E4 kJ/kmol))   $= -1.03E+06$   kJ/h

$\dot{H}^*(4) = $ (3 kmol/h)((35 kJ/kmol.K)(T(4) K − 298K)
     $+ $ (−1.8E5 kJ/kmol))$+ $ (7 kmol/h)((45 kJ/kmol.K)
     $\times$ (T(4) K − 298 K) + (−7E4 kJ/kmol))$+ $ (9 kmol/h)
     $\times$ ((90 kJ/kmol.K)(T(4) K − 298K)+ (4E4 kJ/kmol))   $= f(T(4))$   kJ/h

Solve equation [1] for T(4) = 473K

Thermal duty of HEX = $\dot{H}*(1) - \dot{H}*(2) = \dot{H}*(4) - \dot{H}*(3)$
                   = 5.18E+05 kJ/h = **144kW**

| Stream Table | | *Heat Exchanger* | | *Problem* | |
|---|---|---|---|---|---|
| Species | M | | Stream | | |
| | | *1* | *2* | *3* | *4* |
| | kg/kmol | | kmol/h | | |
| A | 20 | 4 | 4 | 3 | 3 |
| B | 30 | 8 | 8 | 7 | 7 |
| C | 40 | 12 | 12 | 9 | 9 |
| Total | kg/h | 800 | 800 | 630 | 630 |
| Phase | | G | L | L | G |
| Temp. | K | 500 | 420 | 300 | 473 |
| Press. | kPa(abs) | 600 | 570 | 200 | 170 |
| Enthalpy* | kJ/h | −5.66E+05 | −1.08E+06 | −1.03E+06 | −5.10E+05 |
| Energy balance check | | | | | |
| Energy IN | kJ/h | −1.59E+06 | Closure % | | |
| Energy OUT | kJ/h | −1.59E+06 | 100.0 | | |

*Enthalpy values <u>include</u> heats of formation.*

## E. Pump

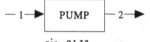

$\dot{W}$ =? kJ/h

For material balance, see *Example 4.04*

| Stream Table | | *Pump* | *Problem* |
|---|---|---|---|
| Species | M | Stream | |
| | | *1* | *2* |
| | kg/kmol | kmol/h | |
| A | 20 | 4 | 4 |
| B | 30 | 8 | 8 |
| C | 40 | 12 | 12 |
| Total | kg/h | 800 | 800 |
| Phase | | L | L |
| Temp. | K | 300 | ? |
| Press. | kPa(abs) | 150 | 2150 |
| Volume | m³/h | 0.89 | 0.89 |
| Enthalpy | kJ/h | −1.24E+06 | ? |

| Properties | Species | A (liq) | A (gas) | B (liq) | B (gas) | C (liq) | C (gas) |
|---|---|---|---|---|---|---|---|
| $C_{p,m}$(ref.298K) | kJ/kmol.K | 50 | 35 | 70 | 45 | 90 | 55 |
| $h_{f,298K}$ | kJ/kmol | −2.0E+05 | −1.8E+05 | −1.0E+05 | −7.0E+04 | 3.00E+04 | 4.00E+04 |
| $h_{v,298K}$ | kJ/kmol | − | 2.00E+04 | − | 3.00E+04 | − | 1.00E+04 |
| Density | kg/m³ | 900 | − | 900 | − | 900 | − |

***Solution:***

Energy balance.

Reference condition = *elements* at standard state, 298K.

$0 = \Sigma[\dot{H}*(in)] - \Sigma[\dot{H}*(out)] + \dot{Q} - \dot{W}$     For a pump     $\dot{Q} = 0$

$0 = \dot{H}*(1) - \dot{H}*(2) + 0 - \dot{W}$                                                          [1]

where:

$$\dot{H}^*(i) = \Sigma[\dot{n}(i,j)h^*(j)] \text{ and } h^*(j) \; C_{p,m}(j)(T(i) - T_{ref}) + h^\circ_{f,Tref}(j) \qquad \text{[Respect the phase]}$$

$\dot{H}^*(1) = (4 \text{ kmol/h})((50 \text{ kJ/kmol.K})(300 \text{ K} - 298 \text{ K}) + (-2E5 \text{ kJ/kmol}))$

$\qquad + (8 \text{ kmol/h})((70 \text{ kJ/kmol.K})(300 \text{ K} - 298 \text{ K}) + (-1E5 \text{ kJ/kmol}))$

$\qquad + (12 \text{ kmol/h})((90 \text{ kJ/kmol.K})(300 \text{ K} - 298 \text{ K}) + (3E4 \text{ kJ/kmol})) = -1.24E{+}06$  kJ/h

$\dot{H}^*(2) = (4 \text{ kmol/h})((50 \text{ kJ/kmol.K})(T(2) \text{ K} - 298 \text{ K}) + (-2E5 \text{ kJ/kmol}))$

$\qquad + (8 \text{ kmol/h})((70 \text{ kJ/kmol.K})(T(2) \text{ K} - 298 \text{ K}) + (-1E5 \text{ kJ/kmol}))$

$\qquad + (12 \text{ kmol/h})((90 \text{ kJ/kmol.K})(T(2) \text{ K} - 298 \text{ K}) + (3E4 \text{ kJ/kmol})) = f(T(2))$     kJ/h

$\dot{W} = \dot{V}(1)[P(1) - P(2)] = (0.89 \text{ m}^3/\text{h})(150 \text{ kPa} - 2150 \text{ kPa})$          $= -1780$     kJ/h

Solve equation [1] for T(2) = **301K**

| Stream Table | | *Pump* | *Solution* |
|---|---|---|---|
| Species | M | Stream | |
| | | *1* | *2* |
| | kg/kmol | kmol/h | |
| A | 20 | 4 | 4 |
| B | 30 | 8 | 8 |
| C | 40 | 12 | 12 |
| Total | kg/h | 800 | 800 |
| Phase | | L | L |
| Temp. | K | 300 | 301 |
| Press. | kPa(abs) | 150 | 2150 |
| Volume | m³/h | 0.89 | 0.89 |
| Enthalpy* | kJ/h | −1.24E+06 | −1.23E+06 |
| Energy balance check | | | |
| Energy IN | kJ/h | −1.24E+06 | Closure % |
| Energy OUT | kJ/h | −1.24E+06 | 100.0 |

*Enthalpy values <u>include</u> heats of formation.*

## F. Reactor

$\dot{Q} = ? \text{ kJ/h}$

*Reaction:*          $2A + 3B \rightarrow C + 2D$

Conversion of A = 75%

For material balance, see *Example 4.04*

| Stream Table | | *Reactor* | *Problem* |
|---|---|---|---|
| Species | M | Stream | |
| | | *1* | *2* |
| | kg/kmol | kmol/h | |
| A | 20 | 4 | 1 |
| B | 30 | 8 | 3.5 |
| C | 40 | 12 | 13.5 |
| D | 45 | 3 | 6 |
| Total | kg/h | 935 | 935 |
| Phase | | L | G |
| Temp. | K | 400 | 500 |
| Press. | kPa(abs) | 500 | 450 |
| Enthalpy | kJ/h | ? | ? |

**Solution:**

| Properties | Species | A (liq) | A (gas) | B (liq) | B (gas) | C (liq) | C (gas) | D (liq) | D (gas) |
|---|---|---|---|---|---|---|---|---|---|
| $C_{p,m}$(ref.298K) | kJ/kmol.K | 50 | 35 | 70 | 45 | 90 | 55 | 105 | 65 |
| $h_{f,298K}$ | kJ/kmol | –2.0E+05 | –1.8E+05 | –1.0E+05 | –7.0E+04 | 3.00E+04 | 4.00E+04 | –1.00E+05 | –6.00E+04 |
| $h_{v,298K}$ | kJ/kmol | – | 2.00E+04 | – | 3.00E+04 | – | 1.00E+04 | – | 4.00E+04 |

Energy balance.

Reference condition = *elements* at standard state, 298K.

$$0 = \Sigma[\dot{H}*(in)] - \Sigma[\dot{H}*(out)] + \dot{Q} - \dot{W} \quad \text{For a reactor} \quad \dot{W} = 0 \quad \text{i.e.}$$

$$0 = \dot{H}*(1) - \dot{H}*(2) + \dot{Q} - 0 \tag{1}$$

where:

$$\dot{H}*(i) = \Sigma[\dot{n}(i, j)h*(j)] \text{ and } h*(j) = C_{p,m}(j)(T(i) - T_{ref}) + h^{\circ}_{f,Tref(j)} \quad \text{[Respect the phase]}$$

$\dot{H}*(1)$ = (4 kmol/h)((50 kJ/kmol.K)(400 K – 298 K)
  + (–2E5 kJ/kmol))+ (8 kmol/h)((70 kJ/kmol.K)
  × (400 K – 298 K) + (–1E5 kJ/kmol))+ (12 kmol/h)
  × ((90 kJ/kmol.K)(400 K – 298 K) + (3E4 kJ/kmol))
  + (3 kmol/h)((105 kJ/kmol.K)(400 K – 298 K)
  + (–1E5 kJ/kmol))                           = –1.32E+06   kJ/h

$\dot{H}*(2)$ = (1 kmol/h)((35 kJ/kmol.K)(500 K – 298 K)
  + (–1.8E5 kJ/kmol))+ (3.5 kmol/h)((45 kJ/kmol.K)
  × (500 K – 298 K) + (–7E4 kJ/kmol)) + (13.5 kmol/h)
  × ((55 kJ/kmol.K)(500 K – 298 K) + (4E4 kJ/kmol))
  + (6 kmol/h)((65 kJ/kmol.K)(500 K – 298 K)
  + (–6E4 kJ/kmol))                           = 2.27E+04   kJ/h

| Stream Table | | *Reactor* | *Solution* |
|---|---|---|---|
| Species | M | Stream | |
| | | *1* | *2* |
| | kg/kmol | kmol/h | |
| A | 20 | 4 | 1 |
| B | 30 | 8 | 3.5 |
| C | 40 | 12 | 13.5 |
| D | 45 | 3 | 6 |
| Total | kg/h | 935 | 935 |
| Phase | | L | G |
| Temp. | K | 400 | 500 |
| Press. | kPa(abs) | 500 | 450 |
| Enthalpy* | kJ/h | *–1.32E+06* | *2.27E+04* |
| Energy balance check | | | |
| Energy IN | kJ/h | 2.27E+04 | Closure % |
| Energy OUT | kJ/h | 2.27E+04 | 100.0 |

*Enthalpy values <u>include</u> heats of formation.*

Solve equation [1] for

Q = $\dot{H}*(2)$ – $\dot{H}*(1)$ = **1.34E+06 kJ/h = 373 kW**

## EXAMPLE 5.07   *Energy balances without chemical reaction (open system at steady-state) using equations to estimate enthalpy.*

### A. Compressor

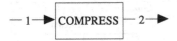

$\dot{W}$ = ? kJ/h

Adiabatic compression. Ideal gas.
No reaction.

*Solution:*

Energy Balance. Reference condition = *compounds* at standard state, 298K.

$0 = \Sigma[\dot{H}]in - \Sigma[\dot{H}]out + \bar{Q} - \dot{W}$   For
an adiabatic compressor $\dot{Q} = 0$ i.e.
$0 = \dot{H}(1) - \dot{H}(2) + \dot{Q} - \dot{W}$          [1]

| Stream Table | | Compressor | Problem |
|---|---|---|---|
| Species | M | Stream | |
| | | 1 | 2 |
| | kg/kmol | kg/kmol | |
| [A] $N_2$ | 28 | 4 | 4 |
| [B] $O_2$ | 32 | 8 | 8 |
| [C] $CO_2$ | 44 | 12 | 12 |
| Total | kg/h | 896 | 896 |
| Phase | | G | G |
| Temp. | K | 300 | ? |
| Press. | kPa(abs) | 100 | 1100 |
| Volume | m³/h | 599 | ? |
| Enthalpy | kJ/h | ? | ? |

| Properties | Species | A (gas) | A (gas) | B (gas) |
|---|---|---|---|---|
| $C_{p,m}$(ref.298K) | kJ/kmol.K | 29.1 | 29.4 | 37.2 |

where:

$\dot{H}(i) = \Sigma[\dot{n}(i, j)h(j)]$

$h(j) = C_{p,m}(j)(T(i) - T_{ref}) + h_{p,Tref}(j)$     [Respect the phase]

$\dot{H}(1) = (4 \text{ kmol/h})((29.1 \text{ kJ/kmol.K})(300 \text{ K} - 298 \text{ K}) + 0)$
$\qquad + (8 \text{ kmol/h})((29.4 \text{ kJ/kmol.K})(300 \text{ K} - 298 \text{ K}) + 0)$
$\qquad + (12 \text{ kmol/h})((37.2 \text{ kJ/kmol.K})(300 \text{ K} - 298 \text{ K}) + 0) = 1596$          kJ/h

$\dot{H}(2) = (4 \text{ kmol/h})((29.1 \text{ kJ/kmol.K})(T(2) \text{ K} - 298 \text{ K}) + 0)$
$\qquad + (8 \text{ kmol/h})((29.4 \text{ kJ/kmol.K})(T(2) \text{ K} - 298 \text{ K}) + 0)$
$\qquad + (12 \text{ kmol/h})((37.2 \text{ kJ/kmol.K})(T(2) \text{ K} - 298 \text{ K}) + 0) = f (T(2))$          kJ/h

$C_{p,m}$ gas mixture $= \Sigma[y(j)C_{p,m}(j)] = (4/24)(29.1) + (8/24)(29.4) + (12/24)(37.2)$
$\qquad\qquad\qquad = 33.3$ kJ/kmol.K

$r = C_p/C_v = C_p/(C_p - R) = 33.3/(33.3 - 8.31)$
$\quad = 1.333 \dot{W}$

$\dot{W} = [(P_1\dot{V}_1 (r/(r-1))[(P_2/P_1)^{((r-1)/r)} - 1]$   (see *Equation* 5.18)

$= -(100 \text{ kPa})(599 \text{ m}^3/\text{h})(1.333/(1.333 -1))$
$\quad [(1100 \text{ kPa}/100 \text{ kPa})^{((1.333 -1)/1.333)} - 1]$

$= -1.97E+05$ kJ/h          $= -54.7$ kW

Solve equation [1] for T(2) = **546 K**

or alternatively, by *Equation 5.18*:

| Stream Table | | Compressor | Solution |
|---|---|---|---|
| Species | M | Stream | |
| | | 1 | 2 |
| | kg/kmol | kg/kmol | |
| [A] $N_2$ | 28 | 4 | 4 |
| [B] $O_2$ | 32 | 8 | 8 |
| [C] $CO_2$ | 44 | 12 | 12 |
| Total | kg/h | 896 | 896 |
| Phase | | G | G |
| Temp. | K | 300 | *546* |
| Press. | kPa(abs) | 100 | 1100 |
| Volume | m³/h | 599 | *99* |
| Enthalpy* | kJ/h | *1.60E+03* | *1.98E+05* |
| Energy balance check | | | |
| Energy IN | kJ/h | 1.98E+05 | Closure % |
| Energy OUT | kJ/h | 1.98E+05 | 100 |

[#] *Enthalpy values <u>exclude</u> heats of formation.*

$T(2) = T(1) [P(2)/P(1)]^{((r-1)/r)} = (300) (1100/100)^{((1.333-1)/1.333)}$

= **546 K**

## B. Separator

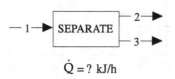

$\dot{Q}$ = ? kJ/h

### Solution:

Energy Balance.

Reference condition = *compounds* at standard state, 298K.

| Stream Table | *Separator* | *Problem* | | |
|---|---|---|---|---|
| Species | M | Stream | | |
| | | 1 | 2 | 3 |
| | kg/kmol | kmol/h | | |
| A | 20 | 4 | 3 | 1 |
| B | 30 | 8 | 4 | 4 |
| C | 40 | 12 | 4 | 8 |
| Total | kg/h | 800 | 340 | 460 |
| Phase | | L | G | L |
| Temp. | K | 300 | 500 | 500 |
| Press. | kPa(abs) | 400 | 200 | 200 |
| Enthalpy | kJ/h | ? | ? | ? |

| Properties | Species | A (liq) | A (gas) | B (liq) | B (gas) | C (liq) | C (gas) |
|---|---|---|---|---|---|---|---|
| $C_{p,m}$ wrt.298K | kJ/kmol.K | 50 | 35 | 70 | 45 | 90 | 55 |
| $h_{v,298K}$ | kJ/kmol | – | 2.00E+04 | – | 3.00E+04 | – | 1.00E+04 |

$0 = \Sigma[\dot{H}(in)] - \Sigma[\dot{H}(out)] + \dot{Q} - \dot{W}$      For a separator:      $\dot{W} = 0$

$0 = \dot{H}(1) - \dot{H}(2) - \dot{H}(3) + \dot{Q} - \dot{W}$ [1]

where:

$\dot{H}(i) = \Sigma[\dot{n}(i, j)h(j)]$

$h(j) = C_{p,m}(j)(T(i) - T_{ref}) + h_{p,Tref}(j)$      [Respect the phase]

$\dot{H}(i)$ = (4 kmol/h)((50 kJ/kmol.K)(300 K – 298 K) + 0) + (8 kmol/h)((70 kJ/kmol.K)
      (300 K – 298 K) + 0) + (12 kmol/h)((90 kJ/kmol.K)(300 K – 298 K) + 0)
      = 3.68E+03   kJ/h

$\dot{H}(2)$ = (3 kmol/h)((35 kJ/kmol.K)(500 K – 298 K) + 2E4) + (4 kmol/h)
      ((45 kJ/kmol.K)(500 K – 298 K) + 3E4) + (4 kmol/h)((55 kJ/kmol.K)
      (500 K – 298 K) + 1E4)
      = 1.02E+05   kJ/h

$\dot{H}(3)$ = (1 kmol/h)((50 kJ/kmol.K)(500 K – 298 K) + 0) + (4 kmol/h)((70 kJ/kmol.K)
      (500 K – 298 K) + 0) + (8 kmol/h)((90 kJ/kmol.K)(500 K – 298 K) + 0)
      = 2.12E+05   kJ/h

| Stream Table | *Separator* | *Solution* | | |
|---|---|---|---|---|
| Species | M | Stream | | |
| | | 1 | 2 | 3 |
| | kg/kmol | kmol/h | | |
| A | 20 | 4 | 3 | 1 |
| B | 30 | 8 | 4 | 4 |
| C | 40 | 12 | 4 | 8 |
| Total | kg/h | 800 | 340 | 460 |
| Phase | | L | G | L |
| Temp. | K | 300 | 500 | 500 |
| Press. | kPa(abs) | 400 | 200 | 200 |
| Enthalpy[#] | kJ/h | *3.68E+03* | *1.02E+05* | *2.12E+05* |

| Energy balance check | | | |
|---|---|---|---|
| Energy IN | kJ/h | 3.14E+05 | Closure % |
| Energy OUT | kJ/h | 3.14E+05 | 100.0 |

[#]*Enthalpy values <u>exclude</u> heats of formation.*

Solve equation [1] for

$$\dot{Q} = \dot{H}(2) + \dot{H}(3) - \dot{H}(1) = \underline{\mathbf{3.10E+05 \ kJ/h}}$$

$$= \underline{\mathbf{86.2 \ kW}}$$

### EXAMPLE 5.08   *Energy balances without chemical reaction (open system at steady-state) using thermodynamic diagrams and tables for enthalpy values.*

*Using enthalpy values from thermodynamic diagrams and tables*

**A. Mixer**

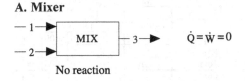

$$\dot{Q} = \dot{W} = 0$$

No reaction

**Solution:**

Energy Balance.
Reference condition = $H_2SO_4$ and $H_2O(l)$ at 273 K, 101.3 kPa(abs).

| Stream Table | *Mixer* | *Problem* | | |
|---|---|---|---|---|
| Species | M | Stream | | |
| | | 1 | 2 | 3 |
| | | kg/h | | |
| [A] $H_2O$ | 18 | 50 | 10 | 60 |
| [B] $H_2SO_4$ | 98 | 0 | 90 | 90 |
| Total | kg/h | 50 | 100 | 150 |
| Phase | | L | L | L |
| Temp. | K | 311 | 298 | ? |
| Press. | kPa(abs) | 150 | 150 | 140 |
| Enthalpy | kJ/h | ? | ? | ? |

$$0 = \Sigma[\dot{H}(in)] - \Sigma[\dot{H}(out)] + \dot{Q} - \dot{W} \qquad\qquad [1]$$

$$0 = \dot{H}(1) + \dot{H}(2) - \dot{H}(3) + 0 - 0$$

where:

$\dot{H}(i) = \bar{m}(i)h(i)$

Values of h(i) are taken from the enthalpy-concentration chart for $H_2SO_4 - H_2O$  *Figure 2.06*

Stream 1: $H_2O$ liquid at 311K, 150 kPa(abs)

$\dot{H}(1) = (50 \text{ kg/h})(149 \text{ kJ/kg}) = 7450$    kJ/h

Stream 2: 90 wt% $H_2SO_4 - H_2O$ liquid at 289 K, 150 kPa(abs)

$\dot{H}(2) = (100 \text{ kg/h})(-186 \text{ kJ/kg}) = -18600$ kJ/h

Stream 3: 60 wt% $H_2SO_4 - H_2O$ at T(3), 140 kPa(abs)

$\dot{H}(3) = (150 \text{ kg/h})(h(3) \text{ kJ/kg}) = f(T(3))$ kJ/h

Solve equation [1] for

h(3) = –74    kJ/kg liquid

By examination of the enthalpy-concentration chart, *Figure 2.06:*

T(3) = **344 K**

| Stream Table | | Mixer | | Solution |
|---|---|---|---|---|
| Species | M | Stream | | |
| | | 1 | 2 | 3 |
| | | kg/h | | |
| [A]$H_2O$ | 18 | 50 | 10 | 60 |
| [B]$H_2SO_4$ | 98 | 0 | 90 | 90 |
| Total | kg/h | 50 | 100 | 150 |
| Phase | | L | L | L |
| Temp | K | 310 | 298 | *344* |
| Press | kPa(abs) | 150 | 150 | 140 |
| Enthalpy[#] | kJ/h | *7450* | *–18600* | *–11150* |
| Energy balance check | | | | |
| Energy IN | kJ/h | –11150 | Closure % | |
| Energy OUT | kJ/h | –11150 | 100.0 | |

[#]*Enthalpy values exclude heats of formation.*

## B. Heat exchanger (indirect contact)

— 1 ▶ | HEX (Hot) | ─ 2 ▶
◀ 4 ─ | HEX (Cold) | ◀ 3 ─

No reaction

**Solution:**

Energy balance. Reference conditions: Streams 1 and 2 *compounds* at triple point of water = 273.16 K, 0.61 kPa(abs) Streams 3 and 4 *compounds* at 273 K, 101.3 kPa(abs)

| Stream Table | Heat exchanger | | Problem | |
|---|---|---|---|---|
| Species | M | Stream | | |
| | | *1* | *2* | *3* | *4* |
| | kg/kmol | kmol/h | | |
| [A] $H_2O$ | 18 | 50 | 50 | 100 | 100 |
| [B] $H_2SO_4$ | 98 | 0 | 0 | 900 | 900 |
| Total | kg/h | 50 | 50 | 1000 | 1000 |
| Phase | | ? | ? | L | L |
| Temp. | K | 450 | 350 | 289 | ? |
| Press. | kPa(abs) | 931 | 900 | 200 | 170 |
| Enthalpy | kJ/h | ? | ? | ? | ? |

$$0 = \Sigma[\dot{H}(in)] - \Sigma[\dot{H}(out)] + \dot{Q} - \dot{W} \qquad \text{For a heat exchanger } \dot{Q} = \dot{W} = 0 \quad \text{i.e.}$$

$$0 = \dot{H}(1) - \dot{H}(2) + \dot{H}(3) - \dot{H}(4) + 0 - 0 \quad \text{[overall HEX balance]} \qquad\qquad [1]$$

where:

$\dot{H}(i) = \bar{m}(i)h(i)$                    h(i) = specific enthalpy

For streams 1 and 2 values of h(i) are taken from the steam table (*Table 2.20*). For streams 3 and 4 values of h(i) are taken from the $H_2SO_4 - H_2O$ enthalpy-concentration chart (*Figure 2.06*). Note that different reference conditions are acceptable here because streams 1–2 and 3–4 are kept separate and the individual reference conditions cancel from equation [1].

Stream 1: From the steam table: at 450 K the vapour pressure of $H_2O$ = 931.5 kPa(abs) Since P(1) < p*$H_2O$ at 450 K, 931 kPa(abs), stream 1 is a *GAS*.

$\dot{H}(1)$ = (50 kg/h)(2775 kJ/kg) = 138750 kJ/h

| Stream Table | | Heat Exchanger | | Solution | |
|---|---|---|---|---|---|
| Species | M | | Stream | | |
| | | 1 | 2 | 3 | 4 |
| | kg/kmol | | kg/h | | |
| [A]$H_2O$ | 18 | 50 | 50 | 100 | 100 |
| [B]$H_2SO_4$ | 98 | 0 | 0 | 900 | 900 |
| Total | kg/h | 50 | 50 | 1000 | 1000 |
| Phase | | G | L | L | L |
| Temp. | K | 450 | 350 | 289 | 355 |
| Press. | kPa(abs) | 931 | 900 | 200 | 170 |
| Enthalpy* | kJ/h | 1.39E+05 | 1.62E+04 | -1.86E+05 | -6.34E+04 |
| Energy balance check | | | | | |
| Energy IN    kJ/h   -47250 | | | Closure % | | |
| Energy OUT   kJ/h   -47250 | | | 100.0 | | |

*Enthalpy values exclude heats of formation.

Stream 2: From the steam table:

At 350 K the vapour pressure of $H_2O$ = 41.7 kPa(abs) Since P(2) > p*$H_2O$ at 350 K, 900 kPa(abs), stream 2 is a *LIQUID*.

$\dot{H}(2)$ = (50 kg/h)(323 kJ/kg)        = 16150 kJ/h

Stream 3: 90 wt% $H_2SO_4$ at 289 K, 200 kPa(abs)

$\dot{H}(3)$ = (1000 kg/h)(-186 kJ/kg)       = -186000 kJ/h

Stream 4: 90 wt% $H_2SO_4$ at T(4) K, 170 kPa(abs)

$\dot{H}(4)$ = (1000 kg/h)(h(4))        = f (T(4)) kJ/h

Solve equation [1] for h(4) = -63 kJ/kg

By examination of the $H_2SO_4 - H_2O$ enthalpy concentration chart (*Figure 2.06*), **T(4) = 355 K**

## EXAMPLE 5.09　Energy balance with chemical reaction (open system at steady-state) using both heat of reaction and heat of formation methods.

Using enthalpy values based on both the *element* and the *compound* reference states.

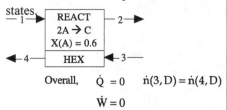

| Stream Table | Reactor with Steam Heating | | Problem | | | |
|---|---|---|---|---|---|---|
| Species | M | | Stream | | | |
| | | | 1 | 2 | 3 | 4 |
| | kg/kmol | | kmol/h | | | |
| A | 20 | | 10 | 4 | 0 | 0 |
| B | 40 | | 0 | 3 | 0 | 0 |
| C (H₂O) | 18 | | 0 | 0 | ? | ? |
| Total | kg/h | | 200 | 200 | ? | ? |
| Phase | | | L | G | G | L |
| Temp. | kPa(abs) | | 200 | 150 | 400 | 400 |
| Press. | K | | 300 | 380 | 417 | 390 |
| Enthalpy | kJ/h | | −2.00E+06 | −5.75E+05 | ? | ? |

*Solution:* Energy Balance. Reference conditions:

Streams 1 and 2 (h*) *Elements* at standard state, 298 K

Streams 3 and 4 (h) *Compound* liquid water at triple-point, 273 K

$$0 = \Sigma[\dot{H}(in)] - \Sigma[\dot{H}(out)] + \dot{Q} - \dot{W} \qquad \text{Overall balance.}$$

$$0 = \dot{H}*(1) + \dot{H}(3) - \dot{H}*(2) - \dot{H}(4) + 0 - 0$$

Note that the two separate reference states both cancel out of this energy balance.

Reference state cancels out　　　Reference state cancels out

where:　　　　　　　　　　　　　　　　　　　　　　　　　Reference State

$\dot{H}*(i) = \Sigma[\dot{n}(i, j)h*(j)]$ and $h*(j) = C_{p,m}(j)(T(i) - T_{ref}) + h^{\circ}_{f,Tref}(j)$　*Elements* [Respect the phase]

$\dot{H}(i) \quad = \Sigma[\dot{n}(i, j)h(j)]$ and $h(j) = C_{p,m}(j)(T(i) - T_{ref}) + h_{p,Tref}(j)$　*Compound* [Respect the phase]

$\dot{H}*(1) = (10 \text{ kmol/h})((50 \text{ kJ/kmol.K})(300 \text{ K} - 298 \text{ K}) + (-2.0\text{E}5))$
　　　　　$+ (0 \text{ kmol/h})((90 \text{ kJ/kmol.K})(300 \text{ K} - 298 \text{ K}) + (3.0\text{E}4)) = -2.00\text{E}+06$　kJ/h

$\dot{H}*(2) = (4 \text{ kmol/h})((35 \text{ kJ/kmol.K})(300 \text{ K} - 298 \text{ K}) + (-1.8\text{E}5))$
　　　　　$+ (3 \text{ kmol/h})((55 \text{ kJ/kmol.K})(300 \text{ K} - 298 \text{ K}) + (4.0\text{E}4)) = -5.75\text{E}+05$　kJ/h

$\dot{H}(3) \quad = (18 \text{ kg/kmol})(\dot{n}(3,E) \text{ kmol/h})(2739^{\#} \text{ kJ/kg}) \qquad = 4.93\text{E}+04 \; \dot{n}(3,D)$　kJ/h

$\dot{H}(4) \quad = (18 \text{ kg/kmol})(\dot{n}(4,E) \text{ kmol/h})(490.4^{\#} \text{ kJ/kg}) \qquad = 8.83\text{E}+03 \; \dot{n}(4,D)$　kJ/h

　　　$\#$ = *specific enthalpy from the steam table (<u>excludes</u> heat of formation)*

Solve equation [1] for

$\dot{n}(3,D) = \dot{n}(4,D) = \textbf{35.2 kmol/h}$

## COMMENT:

The energy balance finds the flow of condensing steam required to supply the heat of the endo-thermic reaction, with incomplete conversion, plus to vaporise the reaction product mixture.

Steam heating utility load:

$$\dot{H}^*(2) - \dot{H}^*(1) = \dot{H}(3) - \dot{H}(4) = \underline{\textbf{396 kW}}$$

| Stream Table | | Reactor with Steam Heating Solution | | | |
|---|---|---|---|---|---|
| Species | M | Stream | | | |
| | | 1 | 2 | 3 | 4 |
| | kg/kmol | kmol/h | | | |
| A | 20 | 10 | 4 | 0 | 0 |
| C | 40 | 0 | 3 | 0 | 0 |
| D (H₂O) | 18 | 0 | 0 | 35.2 | 35.2 |
| Total | kg/h | 200 | 200 | 633.3 | 633.3 |
| Phase | | L | G | G | L |
| Pressure | kPa(abs) | 200 | 150 | 400 | 400 |
| Temp. | K | 300 | 380 | 417 | 390 |
| Enthalpy | kJ/h | −2.00E+06 | −5.75E+05 | 1.73E+06 | 3.11E+05 |
| Energy balance check | | | | | |
| Energy IN | kJ/h | −2.64E+05 | | | |
| Energy OUT | kJ/h | −2.64E+05 | Closure = 100% | | |

## EXAMPLE 5.10   Energy balance on a multi-unit recycle process (open system at steady-state).

**Problem:** Complete the stream table energy balance.

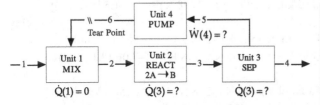

| Specifications | Stream |
|---|---|
| M | 1 |
| kg/kmol | kmol/h |
| A        40 | 10.0 |
| B        80 | 0.0 |
| Conversion of A: X(A)    0.6/pass | |
| Split fractions:  s(4, A) =    0.1 | |
| s(4, B) =    0.8 | |

| Properties | Species | A (liq) | A (gas) | B (liq) | B (gas) |
|---|---|---|---|---|---|
| C_{p,m} (ref.298K) | kJ/kmol.K | 50 | 35 | 70 | 45 |
| h_{f,298K} | kJ/kmol | −2.0E+05 | −1.8E+05 | −1.0E+05 | −7.0E+04 |
| h_{v,298K} | kJ/kmol | – | 2.00E+04 | – | 3.00E+04 |
| Density | kg/m³ | 900 | – | 900 | – |

*Material and Energy Balances for Engineers and Environmentalists*

| Stream Table | | Mixer-reactor-separator + recycle | | | | Problem | |
|---|---|---|---|---|---|---|---|
| Species | M | Stream | | | | | |
| | | 1 | 2 | 3 | 4 | 5 | 6 |
| | kg/kmol | kmol/h | | | | | |
| A | 40 | 10.00 | 15.63 | 6.25 | 0.63 | 5.63 | 5.63 |
| B | 80 | 0.00 | 1.17 | 5.86 | 4.69 | 1.17 | 1.17 |
| *Total* | kg/h | *400* | *719* | *719* | *400* | *319* | *319* |
| Phase | | L | L | G | L | L | L |
| Temp. | K | 300 | ? | 500 | 400 | 400 | ? |
| Press. | kPa(abs) | 200 | 180 | 150 | 130 | 130 | 200 |
| Volume | m³/h | ? | ? | ? | ? | ? | ? |
| Enthalpy | kJ/h | ? | ? | ? | ? | ? | ? |

### Solution:

Energy balance. Reference condition = *elements* at standard state, 298K.

$$0 = \Sigma[\dot{H}*(in)] - \Sigma[\dot{H}*(out)] + \dot{Q} - \dot{W}$$

where:

$$\dot{H}*(i) = \Sigma[n(i,j) \, h*(j)]$$

$$h*(j) = C_{p,m}(j)(T(i) - T_{ref}) + h^{\circ}_{f,Tref}(j) \qquad \text{[Respect the phase]}$$

$\dot{H}*(1)$ = (10.00 kmol/h)[(50 kJ/kmol.K)(300 K − 298 K) + (−2.0E5 kJ/kmol)]
    + (0.00 kmol/h)[(70 kJ/kmol.K)(300 K − 298 K) + (−1.0E5 kJ/kmol)]
    = −2.00E + 06                                                   kJ/h

$\dot{H}*(2)$ = (15.59 kmol/h)[(50 kJ/kmol.K)(T(2) K − 298 K) + (−2.0E5 kJ/kmol)]
    + (1.16 kmol/h)[(70 kJ/kmol.K)(T(2) K − 298 K) + (−1.0E5 kJ/kmol)]
    = f (T(2))                                                         kJ/h

$\dot{H}*(3)$ = (6.24 kmol/h)[(35 kJ/kmol.K)(500 K − 298 K) + (−1.8E5 kJ/kmol)]
    + (5.84 kmol/h)[(45 kJ/kmol.K)(500 K − 298 K) + (−7.0E4 kJ/kmol)]
    = −1.44E+06                                                 kJ/h

$\dot{H}*(4)$ = (0.62 kmol/h)[(50 kJ/kmol.K)(400 K − 298 K) + (−2.0E5 kJ/kmol)]
    + (4.67 kmol/h)[(70 kJ/kmol.K)(400 K − 298 K) + (−1.0E5 kJ/kmol)]
    = −5.57E+05                                                 kJ/h

$\dot{H}*(5)$ = (5.61 kmol/h)[(50 kJ/kmol.K)(400 K − 298 K) + (−2.0E5 kJ/kmol)]
    + (1.17 kmol/h)[(70 kJ/kmol.K)(400 K − 298 K) + (−1.0E5 kJ/kmol)
    = −1.21E+06                                                 kJ/h

$\dot{H}*(6)$ = (6.24 kmol/h)[(50 kJ/kmol.K)(T(6) K − 298 K) + (−2.0E5 kJ/kmol)]
    + (5.84 kmol/h)[(70 kJ/kmol.K)(T(6) K − 298 K) + (−1.0E5 kJ/kmol)]
    = f (T(6))                                                         kJ/h

Iterative sequential modular solution of recycle material and energy balance. For unit "k":

$$0 = \Sigma[\dot{H}^*(i)]in - \Sigma[\dot{H}^*(i)]out + \dot{Q}(k) - \dot{W}(k)$$

| | | | |
|---|---|---|---|
| UNIT 1 (Mixer): | $\dot{Q}(1) = 0 \ \dot{W}(1) = 0$ | $0 = \dot{H}^*(1) + \dot{H}^*(6) - \dot{H}^*(2) + 0 - 0$ | [1] |
| UNIT 2 (Reactor): | $\dot{Q}(2) = ? \ \dot{W}(2) = 0$ | $0 = \dot{H}^*(2) + \dot{H}^*(3) + \dot{Q}(2) - 0$ | [2] |
| UNIT 3 (Separator): | $\dot{Q}(3) = ? \ \dot{W}(3) = 0$ | $0 = \dot{H}^*(3) - \dot{H}^*(4) - \dot{H}^*(5) + \dot{Q}(3) - 0$ | [3] |
| UNIT 4 (Pump): | $\dot{Q}(4) = 0 \ \dot{W}(4) = ?$ | $0 = \dot{H}^*(5) - \dot{H}^*(6) + 0 - \dot{W}(4)$ | [4] |

For each liquid stream: $\dot{V}(i) = \bar{m}(i)/\rho(i)$ e.g. $\dot{V}(1) = (400 \text{ kg/h}) /(900 \text{ kg/m3}) = 0.44 \ \text{m}^3/\text{h}$

For each gas stream(ideal gas): $\dot{V}(i) = \bar{n}(i)RT(i)/P(i)$

e.g. $\dot{V}(3) = (12.08 \text{ kmol/h})(8.31 \text{ kJ/kmol.K})(500 \text{ K})/(150 \text{ kPa}) = 336 \text{ m}^3/\text{h}$

$\dot{W}(4) = -(0.44 \text{ m}^3/\text{h})(200 \text{ kPa} - 130 \text{ kPa})$
    $= -31 \text{ kJ/h}$

Solve equation [4] for T(6) = **400.01** K
Solve equation [1] for T(2) = **342** K
Solve equation [2] for $\dot{Q}(2)$
       = **1.77E+06**    kJ/h
       = 4.91E+02     kW heating

Solve equation [3] for $\dot{Q}$ (3)
       = **-3.25E+05**   kJ/h
       = -9.02E+01    kW cooling

| Stream Table | | Mixer-reactor-separator + recycle | | | | | Solution |
|---|---|---|---|---|---|---|---|
| Species | M | Stream | | | | | |
| | | 1 | 2 | 3 | 4 | 5 | 6 |
| | kg/kmol | | | kmol/h | | | |
| A | 40 | 10.00 | 15.63 | 6.25 | 0.63 | 5.63 | 5.63 |
| B | 80 | 0.00 | 1.17 | 5.86 | 4.69 | 1.17 | 1.17 |
| Total | kg/h | 400 | 719 | 719 | 400 | 319 | 319 |
| Phase | | L | L | G | L | L | L |
| Temp. | K | 300 | 342 | 500 | 400 | 400 | 400.01 |
| Press. | kPa(abs) | 200 | 180 | 150 | 130 | 130 | 200 |
| Volume | m³/h | 0.44 | 0.80 | 336 | 0.44 | 0.35 | 0.35 |
| Enthalpy* | kJ/h | -2.00E+06 | -3.20E+06 | -1.44E+06 | -5.57E+05 | -1.21E+06 | -1.21E+06 |
| Energy balance check | | Unit 1 | Unit 2 | Unit 3 | Unit 4 | Overall | |
| Energy IN   kJ/h | | -3.20E+06 | -1.44E+06 | -1.76E+06 | -1.21E+06 | -5.57E+05 | |
| Energy OUT  kJ/h | | -3.20E+06 | -1.44E+06 | -1.76E+06 | -1.21E+06 | -5.57E+05 | |
| Closure % | | 100.0 | 100.0 | 100.0 | 100.0 | 100.0 | |

* Enthalpy values include heats of formation.

**EXAMPLE 5.11   Material and energy balance on a biochemical process (ethanol from glucose).**

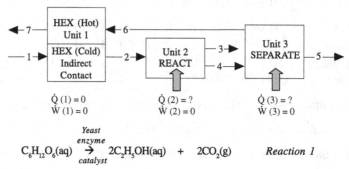

The figure shows a simplified flowsheet of a biochemical process for the production of ethanol by fermentation of glucose. In this process a solution of glucose in water *(Stream 1)* is preheated prior to delivery to a reactor *(Unit 2)* where glucose is converted to ethanol by *Reaction 1*. Carbon dioxide gas is released from the reactor *(Stream 3)* and the reaction liquid product mixture *(Stream 4)* is separated by evaporation *(Unit 3)* to a waste liquor *(Stream 5)* and an ethanol rich vapour *(Stream 6)*, which is used to preheat the process feed *(Stream 1)* in a heat exchanger *(Unit 1)*.

The process specifications are as follows:

Stream 1: 5 wt% liquid solution of      40E3 kg/h                  140 kPa(abs),
          glucose in water                                         288 K
Stream 2: 5 wt% liquid solution of      40E3 kg/h                  110 kPa(abs),
          glucose in water                                         T(2) K
Stream 3: $CO_2$ gas with ethanol and              Flow           110 kPa(abs),
          water vapour                             unspecified    313 K
Stream 5: Liquid solution of glucose    Zero $CO_2$   Flow         110 kPa(abs),
          + ethanol in water                         unspecified   363 K
Stream 6: Vapour mixture of ethanol     96   wt% Flow              140   kPa(abs),
          + water                        ethanol   unspecified     363 K
Stream 7: Liquid mixture of ethanol     96   wt% Flow              110   kPa(abs),
          + water                        ethanol   unspecified     298 K
Unit 2:   Conversion of glucose = 90%
Unit 3: Separation efficiency of ethanol from Stream 4 into Stream 6 = 95%

| Properties | Species | $C_6H_{12}O_6$ | $C_2H_5OH(l)$ | $C_2H_5OH(g)$ | $CO_2(g)$ | $H_2O(l)$ | $H_2O(g)$ |
|---|---|---|---|---|---|---|---|
| $C_{p,m}$ (ref.298K) | kJ/kmol.K | 100 | 66 | *40* | 39 | 75 | 34 |
| $h_{f,298K}$ | kJ/kmol | −7.0E+05 | −2.78E+05 | −2.35E+05 | −3.94E+05 | −2.86E+05 | −2.42E+05 |
| Density | kg/m³ | 1500 | 789 | gas law | gas law | 1000 | gas law |

Vapour pressure of water:    $p^* = \exp[16.5362 - 3985.44/(T - 38.9974)]$    kPa,   T = K
Vapour pressure of ethanol:  $p^* = \exp[16.1952 - 3423.53/(T - 55.7152)]$    kPa,   T = K

Assume that ethanol-water forms an ideal liquid mixture that follows Raoult's law and has negligible heat of mixing.

## Problem:

A.   Make a degrees of freedom analysis of the problem.
B.   Use a spreadsheet to calculate the stream table for the steady-state material and energy balance.

## Solution:

### Material balance

In this example the material balance is independent of the energy balance.The material balance problem is fully defined and is solved by the methods of Chapter 4, using mole balances on each species, stream compositions and reaction stoichiometry to give the following stream table. The compositions of streams 3 and 4 are obtained by iterating on the vapour–liquid equilibria.

Vapour pressure at 313 K; Ethanol = 18.0   kPa,Water = 7.3   kPa

| Stream Table | *Bio-synthesis of Ethanol* | | | *Material Balance* | | *Solution* | |
|---|---|---|---|---|---|---|---|
| Species | M | | | Stream | | | |
| | | *1* | *2* | *3* | *4* | *5* | *6* | *7* |
| | kg/kmol | | | kmol/h | | | |
| [A] $C_6H_{12}O_6$ | 180 | 11.11 | 11.11 | 0.00 | 1.11 | 1.11 | 0.00 | 0.00 |
| [B] $C_2H_5OH$ | 46 | 0.00 | 0.00 | 0.03 | 19.97 | 1.00 | 18.97 | 18.97 |
| [C] $CO_2$ | 44 | 0.00 | 0.00 | 20.00 | 0.00 | 0.00 | 0.00 | 0.00 |
| [D] $H_2O$ | 18 | 2111.11 | 2111.11 | 1.41 | 2109.70 | 2107.70 | 2.00 | 2.00 |
| Total | kg/h | *40000* | *40000* | *907* | *39093* | *38185* | *909* | *909* |

| Mass balance check | Unit 1 | Unit 2 | Unit 3 | Overall |
|---|---|---|---|---|
| Mass IN | 40909 | 40000 | 39093 | 40000 |
| Mass OUT | 40909 | 40000 | 39093 | 40000 |
| Closure % | 100.0 | 100.0 | 100.0 | 100.0 |

### Energy balance

A. Examine the specification of each unit and of the overall process (see *Table 5.03*).

| | Unit 1 | Unit 2 | Unit 3 | OVERALL |
|---|---|---|---|---|
| Number of unknowns (stream + unit variables) | 2I + 2 = 10 | 2I + 2 = 8 | 2I + 2 = 8 | 2I + 2 = 10 |
| Number of independent equations (see below) | 10 | 8 | 8 | 10 |
| D of F | 0 | 0 | 0 | 0 |

Overall unknowns       $= 2I + 2 = (2)(4) + 2 = 10$

Independent equations  $= 1$ energy balance $+ 1$ work $+ 8$ specified T and P $=10$

                                                    FULLY-SPECIFIED

**B.** Write the equations.

Reference condition = *elements* at standard state, 298K.

$$\dot{H}^*(i) = \Sigma[\dot{n}(i,j)(C_{p,m}(j)(T(i) - T_{ref}) + h^\circ_{f,Tref}(j))] \qquad \text{[Respect the phase]}$$

UNIT 1 (Heat exchanger)

$$0 = \dot{H}^*(1) + \dot{H}^*(6) - \dot{H}^*(2) - \dot{H}^*(7) + \dot{Q}(1) - \dot{W}(1) \qquad [1]$$

Stream conditions:

| | | | |
|---|---|---|---|
| Stream 1: | P(1) = 140 kPa, | T(1) = 288 K | [2][3] |
| Stream 2: | P(2) = 110 kPa, | | [4] |
| Stream 6: | P(6) = 140 kPa, | T(6) = 363 K | [5][6] |
| Stream 7: | P(7) = 110 kPa, | T(7) = 298 K | [7][8] |

Unit loads: $\dot{Q}(1) = 0$ $\qquad \dot{W}(1) = 0$ [9][10]

UNIT 2 (Reactor)

$$0 = \dot{H}^*(2) - \dot{H}^*(3) - \dot{H}^*(4) + \dot{Q}(2) - \dot{W}(2) \qquad [11]$$

Stream conditions:

| | | | |
|---|---|---|---|
| Stream 3: | P(3) = 110 kPa, | T(3) = 313 K | [12][13] |
| Stream 4: | P(4) = 110 kPa, | T(4) = 313 K | [14][15] |

Unit loads: $\dot{W}(2) = 0$ [16]

UNIT 3 (Separator)

$$0 = \dot{H}^*(4) - \dot{H}^*(5) - \dot{H}^*(6) + \dot{Q}(3) - \dot{W}(3) \qquad [17]$$

Stream conditions:

Stream 5:   P(5) = 110 kPa,      T(5) = 363 K [18][19]
Unit loads: $\dot{W}(3) = 0$ [20]

| Stream Table | | *Bio-synthesis of Ethanol* | | | *Material and Energy Balance* | | | *Solution* |
|---|---|---|---|---|---|---|---|---|
| Species | M | | | | Stream | | | |
| | | *1* | *2* | *3* | *4* | *5* | *6* | *7* |
| | kg/kmol | | | | kmol/h | | | |
| [A] $C_6H_{12}O_6$ | 180 | 11.11 | 11.11 | 0.00 | 1.11 | 1.11 | 0.00 | 0.00 |
| [B] $C_2H_5OH$ | 46 | 0.00 | 0.00 | 0.03 | 19.97 | 1.00 | 18.97 | 18.97 |
| [C] $CO_2$ | 44 | 0.00 | 0.00 | 20.00 | 0.00 | 0.00 | 0.00 | 0.00 |
| [D] $H_2O$ | 18 | 2111.11 | 2111.11 | 1.41 | 2109.70 | 2107.70 | 2.00 | 2.00 |
| Total | kg/h | *40000* | *40000* | *907* | *39093* | *38185* | *909* | *909* |
| Phase | | L | L | G | L | L | G | L |
| Temp. | K | 288 | *294* | 313 | 313 | 363 | 363 | 298 |
| Press. | kPa(abs) | 140 | 110 | 110 | 110 | 110 | 140 | 110 |
| Volume | m³/h | 39.3 | 39.3 | 507 | 39.3 | 38.1 | 452 | 1.1 |
| Enthalpy* | kJ/h | −6.13E+08 | −6.12E+08 | −8.22E+06 | −6.07E+08 | −5.94E+08 | −4.89E+06 | −5.84E+06 |
| Mass balance check | | *Unit 1* | *Unit 2* | *Unit 3* | *Overall* | | | |
| Mass IN | kg/h | 40909 | 40000 | 39093 | 40000 | | | |
| Mass OUT | kg/h | 40909 | 40000 | 39093 | 40000 | | | |
| Closure % | | 100.0 | 100.0 | 100.0 | 100.0 | | | |
| Energy balance check | | | | | | | | |
| Energy IN | kJ/h | −6.18E+08 | −6.16E+08 | −5.98E+08 | −6.08E+08 | | | |
| Energy OUT | kJ/h | −6.18E+08 | −6.16E+08 | −5.98E+08 | −6.08E+08 | | | |
| Closure % | | 100.0 | 100.0 | 100.0 | 100.0 | | | |

*Enthalpies <u>include</u> heats of formation.

**EXAMPLE 5.12    *Material and energy balance on an electrochemical process (hydrogen/air fuel cell).***

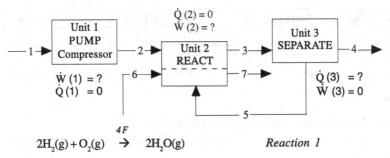

$$2H_2(g) + O_2(g) \quad \rightarrow \quad 2H_2O(g) \qquad\qquad \textit{Reaction 1}$$

This figure shows a simplified flowsheet for a hydrogen/air fuel cell used to power a bus. In this process atmospheric air *(Stream 1 – Stream 2)* is taken by an on-board compressor *(Unit 1)* and delivered to the fuel cell *(Unit 2)*, where part of the $O_2$ is consumed by the *4F* fuel cell *Reaction 1*. Exhaust air from the fuel cell *(Stream 3)* containing the reaction product water is passed through a separator *(Unit 3)* where part of the water is recovered then recycled *(Stream 5)* to cool the fuel cell and to keep the fuel cell membrane properly humidified. Hydrogen from on-board compressed gas tanks *(Stream 6)* is delivered to the fuel cell where it undergoes partial conversion by *Reaction 1*, to generate electric power [$\dot{w}(2)$]. In this example both the exhaust air *(Stream 4)* and the unconverted hydrogen *(Stream 7)* are rejected to atmosphere, but in practice the energy in these streams would be recovered to help run the fuel cell.

*NOTE:* The fuel cell is an electrochemical reactor in which the air and the hydrogen reactant streams are kept apart by a water permeable proton exchange membrane.

| Stream 1: | Air at 30% relative humidity | Flow unspecified | 100 kPa(abs), 290 K |
|---|---|---|---|
| Stream 2: | Air | Flow and T(2) unspecified | 350 kPa(abs), T(2) |
| Stream 3: | Exhaust "air" + $H_2O$ | Flow and phase unspecified | 320 kPa(abs), 370 K |
| Stream 4: | Exhaust "air" + $H_2O$ | Flow and phase unspecified | 310 kPa(abs), 350 K |
| Stream 5: | Recycle liquid $H_2O$ | Flow to match the electro-osmotic flux | 310 kPa(abs), 350 K |
| Stream 6: | 20 kg/h $H_2$ | | 310 kPa(abs), 290 K |
| Stream 7: | Exhaust $H_2$ saturated with $H_2O$ vapour | Flow unspecified | 280 kPa(abs), 370 K |

Air stoichiometry $O_2$ feed  = 3 times the stoichiometric rate for *Reaction 1*
$H_2$ conversion            = 90%

Fuel cell operating voltage = 0.6 Volt/cell

Electro-osmotic flux = 2 moles $H_2O$ per mole $H^+$ crossing the membrane

| Properties | Species | $O_2(g)$ | $N_2(g)$ | $H_2(g)$ | $H_2O(l)$ | $H_2O(g)$ |
|---|---|---|---|---|---|---|
| $C_{p,m}$ (ref.298K) | kJ/kmol.K | 30 | 29 | 29 | 75 | 34 |
| $h_{f,298K}$ | kJ/kmol | 0.00E+00 | 0.00E+00 | 0.00E+00 | −2.86E+05 | −2.42E+05 |
| Density | kg/m³ | Ideal gas | Ideal gas | Ideal gas | 1000 | Ideal gas |

*Problem:*

**A.** Make a degrees of freedom analysis of the problem.

**B.** Use a spreadsheet to calculate the stream table for the steady-state material and energy balance.

*Solution:*

**Material balance**

In this example the material balance is independent of the energy balance. The material balance problem is fully defined and is solved by the methods of Chapter 4, using mole balances on each species, stream compositions and reaction stoichiometry to give the following stream table.

| Stream Table | | *Fuel-cell Material Balance* | | | | | *Solution* | |
|---|---|---|---|---|---|---|---|---|
| Species | M | Stream | | | | | | |
| | | *1* | *2* | *3* | *4* | *5* | *6* | *7* |
| | kg/kmol | kmol/h | | | | | | |
| [A] $O_2$ | 32 | 13.50 | 13.50 | 9.00 | 9.00 | 0.00 | 0.00 | 0.00 |
| [B] $N_2$ | 28 | 50.79 | 50.79 | 50.79 | 50.79 | 0.00 | 0.00 | 0.00 |
| [C] $H_2$ | 2 | 0.00 | 0.00 | 0.00 | 0.00 | 0.00 | 10.00 | 1.00 |
| [D] $H_2O$ | 18 | 1.25 | 1.25 | 45.78 | 9.78 | 36.00 | 0.00 | 0.47 |
| Total | kg/h | *1876* | *1876* | *2534* | *1886* | *648* | *20* | *10* |

**Energy balance**

**A.** Examine the specification of each unit and of the overall process.

| | Unit 1 | Unit 2 | Unit 3 | OVERALL |
|---|---|---|---|---|
| Number of unknowns (stream + unit variables) | 2I + 2 = 6 | 2I + 2 = 12 | 2I + 2 = 8 | 2I + 2 = 10 |
| Number of independent equations (see below) | 6 | 11 | 8 | 10 |
| D of F | 0 | 1 | 0 | 0 |

Overall Energy Balance

Unknowns = 2I + 2 = (2)(4) + 2 = 10

Independent equations = 1 energy balance + 1 work + 8 specified T and P = 10

FULLY-SPECIFIED

**B.** Write the equations. *Reference condition = elements* at standard state, 298K.

$$\dot{H}^*(i) = \Sigma[\dot{n}(i, j)(C_{p,m} (j)(T(i) - T_{ref}) + h^\circ_{f,Tref}(j))] \qquad \text{[Respect the phase]}$$

UNIT 1 (Pump)

$$0 = \dot{H}^*(1) - \dot{H}^*(2) + \dot{Q}(1) - \dot{W}(1) \tag{1}$$

Stream conditions:

| | | | |
|---|---|---|---|
| Stream 1: | P(1) = 100 kPa, | T(1) = 290 K | [2][3] |
| Stream 2: | P(2) = 350 kPa, | | [4] |

Compressor: $\dot{Q}(1) = 0 \ \dot{W}(1) = -P(1) \ \dot{V}(1)(r/(r-1)) [((P(2)/P(1))^{((r-1)/r)} - 1]$

$$= \underline{-2.36E+05 \text{ kJ/h}} \tag{5][6}$$

UNIT 2 (Reactor)

$$0 = \dot{H}^*(2) + \dot{H}^*(5) + \dot{H}^*(6) - \dot{H}^*(3) - \dot{H}^*(7) + \dot{Q}(2) - \dot{W}(2) \tag{7}$$

Stream conditions:

| | | | |
|---|---|---|---|
| Stream 3: | P(3) = 320 kPa, | T(3) = 370 K | [8][9] |
| Stream 5: | P(5) = 310 kPa, | T(5) = 350 K | [10][11] |
| Stream 6: | P(6) = 310 kPa, | T(6) = 290 K | [12][13] |
| Stream 7: | P(7) = 280 kPa, | T(7) = 370 K | [14][15] |

Faraday's law:   $I' = 2F(\dot{n}(6,C) - \dot{n}(7,C))/3600 = (2)(96485)(10 - 1)/3600 = 482\text{kA}$

Electric power:   $\dot{W}(2) = E_V I' = (0.6 \text{ Volt})(482 \text{ kA}) = 289 \text{ kW} = \underline{1.04E+06 \text{ kJ/h}}$   [16]

UNIT 3 (Separator)

$$0 = \dot{H}^*(3) - \dot{H}^*(4) - \dot{H}^*(5) + \dot{Q}(3) - \dot{W}(3) \tag{17}$$

Stream conditions:

| | | | |
|---|---|---|---|
| Stream 4: | P(4) = 310 kPa, | T(4) = 350 K | [18][19] |
| Separator: | $\dot{W}(3) = 0$ | | [20] |

*NOTE:* Calculation of $\dot{H}^*(3)$ and $\dot{H}^*(4)$ requires the distribution of $H_2O$ between the liquid and gas phase. This is obtained from the Antoine equation assuming that when $(y(D))P > p^*$, the gas phase of *Stream 3* and *Stream 4* is saturated with $H_2O$ vapour.

Vapour pressure of $H_2O$: $p(D)^* = \exp[16.5362 - 3985.44/(T - 38.9974)]$  kPa      [21]

| Stream | p*(D)(kPa) |
|---|---|
| 3 | 89.6 |
| 4 | 41.3 |
| 7 | 89.6 |

Solve for:

Q(2) = $\underline{-5.94E+05}$ kJ/h (cooling)

Q(3) = $\underline{-6.76E+05}$ kJ/h (cooling)

| Stream Table | | *Fuel-cell Material and Energy Balance* | | | | | | *Solution* |
|---|---|---|---|---|---|---|---|---|
| Species | M | Stream | | | | | | |
| | | 1 | 2 | 3 | 4 | 5 | 6 | 7 |
| | kg/kmol | kmol/h | | | | | | |
| [A] $O_2$(g) | 32 | 13.50 | 13.50 | 9.00 | 9.00 | 0.00 | 0.00 | 0.00 |
| [B] $N_2$(g) | 28 | 50.79 | 50.79 | 50.79 | 50.79 | 0.00 | 0.00 | 0.00 |
| [C] $H_2$(g) | 2 | 0.00 | 0.00 | 0.00 | 0.00 | 0.00 | 10.00 | 1.00 |
| [D] $H_2O$(g) | 18 | 1.25 | 1.25 | 23.26 | 9.19 | 0.00 | 0.00 | 0.47 |
| $H_2O$(l) | 18 | 0.00 | 0.00 | 22.51 | 0.58 | 36.00 | 0.00 | 0.00 |
| Total | kg/h | 1876 | 1876 | 2534 | 1886 | 648 | 20 | 10 |
| Phase | | G | G | G+L | G+L | L | G | G |
| Temp. | K | 290 | 413 | 370 | 350 | 350 | 290 | 370 |
| Press. | kPa(abs) | 100 | 350 | 320 | 310 | 310 | 310 | 280 |
| Volume | m³/h | 1579 | 643 | 1014 | 508 | 0.648 | 78 | 16 |
| Enthalpy* | kJ/h | -3.17E+05 | -8.04E+04 | -1.18E+07 | -2.28E+06 | -1.02E+07 | -2.32E+03 | -1.11E+05 |

| Mass balance check | | Unit 1 | Unit 2 | Unit 3 | Overall |
|---|---|---|---|---|---|
| Mass IN | kg/h | 1876 | 2544 | 2534 | 1896 |
| Mass OUT | kg/h | 1876 | 2544 | 2534 | 1896 |
| Closure % | | 100.0 | 100.0 | 100.0 | 100.0 |
| Energy balance check | | | | | |
| Energy IN | kJ/h | -8.04E+04 | -1.08E+07 | -1.24E+07 | -1.59E+06 |
| Energy OUT | kJ/h | -8.04E+04 | -1.08E+07 | -1.24E+07 | -1.59E+06 |
| Closure % | | 100.0 | 100.0 | 100.0 | 100.0 |

*Enthalpies <u>include</u> heats of formation.

### EXAMPLE 5.13    Material and energy balance on a thermochemical process (oil fired boiler).

This figure shows a simplified flowsheet for part of a power generation cycle in an oil burning power station. In this process a fuel oil *(Stream 1)* is burned with excess preheated air *(Stream 6)* in a furnace *(Unit 1)*, where the oil undergoes complete combustion to $CO_2$, $H_2O$ and $SO_2$ *(Stream 2)*. The hot combustion product gas then passes through a heat exchanger called a "boiler" *(Unit 2)* where heat is transferred to water inside steel tubes. Water enters the tubes as a liquid *(Stream 7)* and leaves as steam *(Stream 8)* which is subsequently used to drive turbines that generate electricity. The cooled combustion gas *(Stream 3)* passes through a second heat exchanger *(Unit 3)* where part of the remaining heat is transferred to the incoming air *(Stream 5)* to raise the thermal efficiency of the system. The exhaust gas *(Stream 4)* is then rejected through a stack into the atmosphere. The process specifications are as follows:

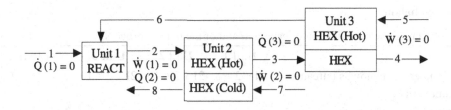

Stream 1:  Fuel oil with empirical formula  50,000 kg/h                    130 kPa(abs), 298 K
           $CH_2S_{0.1}$

Stream 2:  Combustion product gas mixture   Flow unspecified               110 kPa(abs), T(2) K

Stream 3:  Combustion product gas mixture   Flow unspecified               100 kPa(abs), 700 K

Stream 4:  Combustion product gas mixture   Flow unspecified               100 kPa(abs), 500 K

Stream 5:  Air at 60% relative humidity     20% excess for oil             140 kPa(abs), 300 K
                                            combustion

Stream 6:  Preheated air                    Flow unspecified               130 kPa(abs), T(6) K

Stream 7:  Water                            489,381 kg/h                   2650 kPa(abs), 300 K

Stream 8:  Water                            489,381 kg/h                   2637 kPa(abs), T(8) K

Fuel oil gross heat of combustion           $h^\circ_{c,298K} = -3.00E+04$   kJ/kg

| Properties | Species | Fuel oil | $O_2(g)$ | $N_2(g)$ | $H_2O(l)$ | $H_2O(g)$ | $CO_2(g)$ | $SO_2(g)$ |
|---|---|---|---|---|---|---|---|---|
| $C_{p,m}$ (ref.298K) | kJ/kmol.K | – | 31 | 30 | 75 | 35 | 42 | 43 |
| $h_{f,298K}$ | kJ/kmol | – | 0.0E+00 | 0.0E+00 | –2.86E+05 | –2.42E+05 | –3.94E+05 | –2.97E+05 |
| Density | kg/m$^3$ | 900 | Ideal gas | Ideal gas | 1000 | Ideal gas | Ideal gas | Ideal gas |

*Problem:*

A.  Make a degrees of freedom analysis of the problem.
B.  Use a spreadsheet to calculate the stream table for the steady-state material and energy
    balance.

*Solution:*

**Material balance**

In this example the material balance is independent of the energy balance. The material
balance problem is fully defined and is solved by the methods of Chapter 4, using atom
balances on each element and the known stream compositions. The material balance solution is given in the top seven rows of the M&E balance stream table presented below.

**Energy balance**

**A.** Examine the specification of each unit and of the overall process.

|  | Unit 1 | Unit 2 | Unit 3 | OVERALL |
|---|---|---|---|---|
| Number of unknowns (stream + unit variables) | $2I + 2 = 8$ | $2I + 2 = 10$ | $2I + 2 = 10$ | $2I + 2 = 12$ |
| Number of independent equations (see below) | 7 | 9 | 10 | 12 |
| D of F | 1 | 1 | 0 | 0 |

Total number of stream + unit variables $= 2I + 2K = (2)(8) + 2(3) = 22$
Requires 22 independent equations
Overall Energy Balance: Unknowns $= 2I + 2 = (2)(5) + 2 = 12$
Independent equations $= 1$ energy balance $+ 1$ heat $+ 1$ work $+ 9$ specified T and P $= 12$,
FULLY-SPECIFIED.

**B.** Reference conditions:
Stream 1 to 6 ref. = *elements* at standard state 298 K

$$\dot{H}*(i) = \Sigma[\dot{n}(i, j)(C_{p,m}(j)(T(i) - T_{ref}) + h^{\circ}_{f,Tref}(j))] \qquad \text{[Respect the phase]}$$

Stream 7 and 8 Ref. = *compound* water (liquid) at its triple-point, i.e. 273.16 K, 0.61 kPa(abs)

$$\dot{H}*(i) = \Sigma[\dot{n}(i, j)h(j)] \quad \text{Specific enthalpy } h(j) \text{ obtained from steam table}$$

Note that the different reference conditions are no problem here since they cancel from the energy balance. UNIT 1 (Adiabatic reactor)

UNIT 1 (Adiabatic reactor)

$$0 = \dot{H}*(1) + \dot{H}*(6) - \dot{H}*(2) + \dot{Q}(1) - \dot{W}(1) \tag{1}$$

Stream conditions.

| | | | |
|---|---|---|---|
| Stream 1: | $P(1) = 130$ kPa(abs) | $T(1) = 300$ K | [2][3] |
| Stream 2: | $P(2) = 110$ kPa(abs) | | [4] |
| Stream 6: | $P(6) = 130$ kPa(abs) | | [5] |
| | $\dot{Q}(1) = 0$ | $\dot{W}(1) = 0$ | [6][7] |

UNIT 2 (Heat exchanger)

$$0 = \dot{H}*(2) + \dot{H}*(7) - \dot{H}*(3) - \dot{H}*(8) + \dot{Q}(2) - \dot{W}(2) \tag{8}$$

Stream conditions:

| | | | |
|---|---|---|---|
| Stream 3: | $P(3) = 100$ kPa(abs) | $T(3) = 700$ K | [9][10] |
| Stream 7: | $P(7) = 2650$ kPa(abs) | $T(7) = 300$ K | [11][12] |

Stream 8:         P(8) = 2637 kPa(abs)                                              [13]
                  $\dot{Q}(2) = 0$              $\dot{W}(2) = 0$                     [14][15]

UNIT 3 (Heat exchanger)

$$0 = \dot{H}*(3) + \dot{H}*(5) - \dot{H}*(4) - \dot{H}*(6) + \dot{Q}(3) - \dot{W}(3)$$   [16]

Stream conditions:

Stream 4:         P(4) = 100 kPa(abs)    T(4) = 500 K                               [17][18]
Stream 5:         P(5) = 140 kPa(abs)    T(5) = 300 K                               [19][20]

                  $\dot{Q}(3) = 0$             $\dot{W}(3) = 0$                      [21][22]

Note that T(2) is effectively the *adiabatic flame temperature*, calculated on the simplifying assumptions of constant heat capacities and stable reaction products. These assumptions inflate T(2) by about 200 K.

Fuel oil (see *Equation 2.53* and *Reaction 5.01*)

$$h^{\circ}_{f,298K} = -(-30E3)(17.2) + [(1)(-3.94E5) + (2/2)(-2.86E5) + (0.1)(-2.97E5)] = -1.50E+05$$
$$\text{kJ/kmol}$$

T(8) is found by solving the energy balance for the specific enthalpy of steam in Stream 8 = 2803 kJ/kg, and matching it in the steam table at P(8).

| Stream Table | | Furnace (thermo-reactor) and Boiler | | | Material and Energy Balance | | | Solution |
|---|---|---|---|---|---|---|---|---|
| Species | M | | | | Stream | | | |
| | | *1* | *2* | *3* | *4* | *5* | *6* | *7* | *8* |
| | kg/kmol | | | | kmol/h | | | | |
| [A] CH₂S₀.₁ | 17.2 | 2907 | 0 | 0 | 0 | 0 | 0 | 0 | 0 |
| [B] O₂ | 32 | 0 | 930 | 930 | 930 | 5581 | 5581 | 0 | 0 |
| [C] N₂ | 28 | 0 | 20997 | 20997 | 20997 | 20997 | 20997 | 0 | 0 |
| [D] H₂O | 18 | 0 | 3315 | 3315 | 3315 | 408 | 408 | 27188 | 27188 |
| [E] CO₂ | 44 | 0 | 2907 | 2907 | 2907 | 0 | 0 | 0 | 0 |
| [F] SO₂ | 64 | 0 | 291 | 291 | 291 | 0 | 0 | 0 | 0 |
| Total | kg/h | 50000 | 823860 | 823860 | 823860 | 773860 | 773860 | 489381 | 489381 |
| Phase | | L | G | G | G | G | G | L | G |
| Temp. | K | 298 | *2148* | 700 | 500 | 300 | *523* | 300 | *500* |
| Press | kPa(abs) | 130 | 110 | 100 | 100 | 140 | 130 | 2650 | 2637 |
| Volume | m³/h | 56 | 4615714 | 1654345 | 1181675 | 480550 | 901429 | 489 | 42839 |
| Enthalpy* | kJ/h | −4.36E+08 | −3.51E+08 | −1.67E+09 | −1.85E+09 | −9.72E+07 | 8.47E+07 | 5.47E+07 | 1.37E+09 |

| *Mass Balance Check* | | Unit 1 | Unit 2 | Unit 3 | Overall |
|---|---|---|---|---|---|
| Mass IN | kg/h | 823860 | 1313241 | 1597721 | 1313241 |
| Mass OUT | kg/h | 823860 | 1313241 | 1597721 | 1313241 |
| Closure % | | 100.0 | 100.0 | 100.0 | 100.0 |
| *Energy Balance Check* | | | | | |
| Energy IN | kJ/h | −3.51E+08 | −2.97E+08 | −1.77E+09 | −4.79E+08 |
| Energy OUT | kJ/h | −3.51E+08 | −2.97E+08 | −1.77E+09 | −4.79E+08 |
| Closure % | | 100.0 | 100.0 | 100.0 | 100.0 |

*Enthalpies <u>include</u> heats of formation, except Streams 7 and 8.*

## SUMMARY

[1] In the general M&E balance problem the energy balance may be coupled to or uncoupled from the material balance. This chapter treats the energy balance as uncoupled from the material balance and assumes that the material balance has been solved before the energy balance is considered.

[2] Energy balances are derived from the general balance equation (GBE) in which the specified quantity is "energy" in all of its forms. Since (in non-nuclear processes) energy[13] is a conserved quantity, the integral and differential forms of the GBE are simplified by dropping the generation and consumption terms, as follows:

Integral form of GBE:      Energy ACC = Energy IN − Energy OUT
Differential form of GBE:  Rate of energy ACC = Rate of energy IN − Rate of energy OUT

[3] As a rule of thumb, in chemical process calculations the relative order of magnitude of terms in the energy balance is:

Reaction (heat of reaction) > phase change (latent heat) > temperature change (sensible heat) > gas compression/expansion > liquid pumping > kinetic and potential energy > exotics

Consequently energy balances on chemical processes can often be simplified by neglecting the exotic energy terms and some or all mechanical energy effects. The latter simplification is not valid in thermal power generation cycles, where about 40% of the chemical energy (heat of combustion) is converted to mechanical energy via an expansion turbine. Also in electric batteries, fuel cells and electro-synthesis processes the chemical energy terms are balanced by large electrical energy effects.

[4] The energy balance on a closed system (batch process) translates to a commonly identified case of *the first law of thermodynamics*, for which the integral form is:

$$E_{final} - E_{initial} = Q - W \qquad \text{(see Equation 5.04)}$$

where:                                                          Typical units
$E$       = [$U + E_k + E_p$] content of the system              kJ
$E_{final}$   = final value of "E" for the system                    kJ
$E_{initial}$ = initial value of "E" for the system                  kJ
$Q$       = net heat <u>input</u> to system                      kJ
$W$       = net work <u>output</u> from the system               kJ

[5] The energy balance on an open system (continuous process) includes the energy content of *material that crosses the system boundary*, so the integral form of this energy balance is:

$$E_{final} - E_{initial} = E_{input} - E_{output} + Q - W - (PV_{output} - PV_{input}) \qquad \text{(see Equation 5.06)}$$

---

[13] a.k.a. the total energy, considered with *Equations 5.02* and *5.03*.

where: Typical units

$E_{final}$ = final value of $[U + E_k + E_p]$ content of the system     kJ

$E_{initial}$ = initial value of $[U + E_k + E_p]$ content of the system     kJ

$E_{input}$     = $[U+E_k+E_p]$ sum for all material inputs to the system     kJ

$E_{output}$     = $[U+E_k+E_p]$ sum for all material outputs from the system     kJ

$(PV_{output} - PV_{input})$ = net (flow) work to move material in/out of system     kJ

Q     = net heat input to system     kJ

W     = net work output from the system (a.k.a. shaft work)     kJ

[6] The energy balance on an open system for a chemical process at steady-state is usually simplified by replacing U + PV with H and dropping the kinetic and potential energy terms, to give the differential form:

$$dE/dt = 0 = \ddot{H}_{input} - \ddot{H}_{output} + \dot{Q} - \dot{W} \qquad \text{(see } Equation\ 5.09\text{)}$$

where: Typical units

$E \cong U$ = internal energy of the system     kJ

$\ddot{H}_{input}$ = sum of enthalpy flow carried by all material input streams to the system     kW

$\ddot{H}_{output}$ = sum of enthalpy flow carried by all material output streams from the system     kW

$\dot{Q}$ = thermal *utility load* on the system (+) heating, (–) cooling     kW

$\dot{W}$ = mechanical or electrical *utility load* on the system (+) power out, (–) power in     kW

t = time     s

[7] The best reference state to use for energy balances on general chemical processes is the *elements* in their standard state at 298K. The procedure based on this reference state is called the *heat of formation method* of energy balance calculation because the energy of each stream, as defined in *Equations 5.11* and *5.12*, includes the heat of formation of the stream components. The heat of formation method automatically accounts for the energy effects of chemical reactions, so it does not require addition of an extra term to the energy balance to deal with the heat of reaction.

[8] An alternative reference state for energy balances is the relevant *compounds* at some reference condition. The compound reference state is common in thermodynamic diagrams and tables such as enthalpy-concentration charts, psychrometric charts and the steam table.

Energy balances with non-reactive systems (i.e. physical processes) can use either the element or the compound references state without adjustment. However when the compound reference state is used with reactive systems it is necessary to add an extra term to the energy balance to deal with the heat of reaction. The procedure based on the compound reference state is called the *heat of reaction method* of energy balance calculation because the energy of each stream, as defined in *Equations 5.15* and *5.16*, does not include the heat of formation of the stream components, so this method requires that the heat of reaction be added as an explicit term in the energy balance.

[9] The heat (Q) and work (W) terms in the energy balance correspond to transfer of energy without the transfer of material across the system envelope. The Q and W terms usually represent respectively heat transfer across a heat exchange surface and mechanical work done by a rotating shaft or piston or electrical work done by an electrochemical, photoelectric or resistive system.

[10] The energy balance gives a single equation for each process unit that requires the specification of five variables: the phase (Π), pressure (P) and temperature (T) of each process stream, plus the heat (Q) and work (W) crossing the system envelope. Fully specifying the energy balance may require simplifying assumptions such as: adiabatic or isothermal conditions, zero work done, enthalpy independent of pressure, etc. as in *Table 5.03*. When such simplifications are not available, a more comprehensive analysis is called for, using relations such as: the entropy balance, momentum balance, heat transfer rate and pressure drop factors (not covered in this text), possibly combined with rules of thumb common to process design (see *Refs. 9–10*). The example energy balance problems in this text are made fully-specified by a combination of simplifying assumptions and rules of thumb.

[11] Energy balance problems that involve a *heat of combustion* can be treated by either:
- The *heat of formation method*, first calculating the heat of formation from the heat of combustion.
- The *heat of reaction method*, where the heat of combustion is added as the heat of reaction term in combustion processes.

[12] As for material balances, energy balances can be written for each generic process unit (DIVIDE, MIX, SEPARATE, HEAT EXCHANGE, PUMP and REACT) and sequenced to model a complete process flowsheet. For a fully-specified continuous multi-unit process operating at steady-state the set of energy balance equations is conveniently solved in a spreadsheet by the iterative sequential modular method. The "Goal Seek" and "Solver" are handy spreadsheet tools that can be used to solve energy balances by making: *closure* = $e_E$ = 100% at steady-state.

[13] The solution to the steady-state energy balance is presented in an expanded M&E balance stream table (see *Figure 5.01*) that shows the material balance with the phase(s), pressure, temperature, plus the volume and enthalpy flow of each process stream, the rates of heat and work transfer to each process unit and the *closure* of both the mass and the energy balances on each process unit and on the overall process.

## FURTHER READING

[1]  S. Skogestad, *Chemical and Energy Process Engineering*, CRC Press, 2009.
[2]  International Energy Agency, *Global Energy Review 2021*, IEA, 2021.
[3]  F. Kreith and D. Goswami, *Handbook of Energy Efficiency and Renewable Energy*, CRC Press, 2007.

[4]  R. Felder, R. Rousseau and L. Bullard, *Elementary Principles of Chemical Processes*, Wiley, 2018.

[5]  D. Himmelblau and J. Riggs, *Basic Principles and Calculations in Chemical Engineering*, Pearson, 2022.

[6]  P. Doran, *Bioprocess Engineering Principles*, Elsevier, 2013.

[7]  A. Morris, *Handbook of Material and Energy Calculations for Materials Processing*, Wiley, 2012.

[8]  H. Fogler, *Elements of Chemical Reaction Engineering*, Pearson, 2020.

[9]  J. Cooper, W. Penny, J. Fairbanks and S. Walas, *Chemical Process Equipment — Selection and Design*, Butterworth-Heinemann, 2012.

[10] S. Sinnott and G. Towler, *Chemical Engineering Design*, Butterworth-Heinemann, 2019.

# CHAPTER SIX

# SIMULTANEOUS MATERIAL AND ENERGY BALANCES

## SIMULTANEOUS BALANCES

In Chapters 4 and 5, the material balance and the energy balance on a system were introduced as separate problems that are solved in sequence:

(1)   Material balance $\rightarrow$ (2) Energy balance

However, for some systems the material balance and the energy balance are coupled in such a way that it is not possible to solve them separately. In these cases the material balance and the energy balance must be solved simultaneously to obtain the distribution of material and energy in the process.

You can encounter simultaneous material and energy balance problems in many practical situations. In chemical process flowsheets, for example, such problems may arise in mixers, separators, reactors and direct contact heat exchangers. A characteristic of these process units is that the stream flows interact with the pressure and/or temperature in the unit. For example, the outlet component flows from a reactor depend on the reactor temperature, since temperature affects the reaction rate and determines the reactant conversion. Similarly, the outlet stream flows from a gas/liquid separator are affected by both the pressure and the temperature in the separator, which set the gas phase partial pressures for the phase "split".[1]

## SPECIFICATION OF SIMULTANEOUS MATERIAL AND ENERGY BALANCES

When you approach a problem involving simultaneous material and energy balances the first thing you usually notice is that the material balance is *under-specified* (i.e. D of F > 0). The under-specification of a material balance can have two causes:

A.   The material balance is part of a fully-specified M&E balance but must be solved simultaneously with the energy balance.
B.   The material balance is part of a *design problem,* in which the M&E balance is truly under-specified.

The specification (i.e. D of F) of a M&E balance problem can be checked by combining the criteria of *Tables 4.02* and *5.03*. With this check you will often find that a material balance with one D of F can be resolved when an energy balance is added to the set of equations for the system, so the complete M&E balance becomes fully-specified. If the M&E balance is truly under-specified then its solution requires design and optimisation procedures outside the scope of this text (see *Refs. 5, 6*).

---

[1] The *phase split* is the distribution of components between the phases in a process stream or from a process unit.

*Examples 6.01* to *6.10* and associated comments illustrate some typical problems involving simultaneous material and energy balances. These examples show both closed (batch) and open (continuous) systems and are set up as fully-specified M&E balances.

## CLOSED SYSTEMS (BATCH PROCESSES)

### *EXAMPLE 6.01   Simultaneous M&E balance for vapour/liquid equilibrium (closed system).*

A rigid, closed and evacuated vessel initially holds a sealed container filled with 5 kg of liquid water at 100 kPa(abs), 300 K. The container is subsequently opened and 4380 kJ of heat is transferred into the vessel's contents, whose total volume is fixed at 1.00 m³. No work is transferred. Assume mass and volume of the container material are negligible.

*Problem:* Find the final (equilibrium) mass of *liquid* water in the vessel (i.e. find the phase split).

*Solution:* The final mass of *liquid* water will depend on the (unknown) final pressure and temperature in the vessel.

Define the *system* = closed vessel and contents. Let
$m_l$ = final mass of water liquid                        kg
$m_v$ = final mass of water vapour                      kg

**Material balance** on water *liquid* in the closed system:
ACC = IN – OUT + GEN – CON         (IN = OUT = 0 for material in a closed system)
$m_l - 5 = 0 - 0 + 0 - m_v$         (mass balance)                        [1]

Note that the material balance is under-specified, i.e. 1 equation, 2 unknowns ($m_l$ and $m_v$).

**Energy balance** on the closed system. Reference condition = *liquid water* at its triple-point.

Using the steam table, specific internal energy of liquid water at 100 kPa(abs), 300 K = 112.5 kJ/kg
ACC = IN – OUT + GEN – CON   Q = 4830 kJ,   W = 0   (GEN = CON = 0 for energy)
$(m_l\, u_l + m_v\, u_g) - (5\text{ kg})(112.5\text{ kJ/kg}) = 4830\text{ kJ} - 0 + 0 - 0$ (energy balance)         [2]

Also, since the total volume of the mixture is fixed at 1 m³:        $m_l\, v_l + m_v v_g = 1$   [3]

where:

$u_l$ = specific internal energy of water liquid at final conditions    kJ.kg$^{-1}$

$u_g$ = specific internal energy of water vapour at final conditions    kJ.kg$^{-1}$

$v_l$ = specific volume of water liquid at final conditions    m$^3$.kg$^{-1}$

$v_g$ = specific volume of water vapour at final conditions    m$^3$.kg$^{-1}$

The final pressure and temperature are related by the state equations for water, which are embodied in the steam table (*Table 2.20*). Equations [1 to 3] are solved simultaneously by trial and error with the help of the steam table, to give:

At P = 246 kPa(abs),   T = 400 K          $u_l$ = 532.6 kJ/kg      $u_g$ = 2536.2 kJ/kg

$\qquad\qquad\qquad$ $v_l$ = 0.001067 m$^3$/kg   $v_g$ = 0.7308 m$^3$/kg   $m_l$ = **3.64 kg**

## EXAMPLE 6.02 Simultaneous M&E balance for a batch chemical reactor (closed system).

The reversible liquid phase thermo-chemical reaction and its temperature dependent equilibrium constant are defined by:

$\quad$ A(l) $\leftrightarrow$ B(l)   $K_{eq}$ = [B]/[A] = exp(200/T)

where:

| | | |
|---|---|---|
| [A] | = concentration of A | kmol.m$^{-3}$ |
| [B] | = concentration of B | kmol.m$^{-3}$ |
| $K_{eq}$ | = equilibrium constant | – |
| T | = reaction equilibrium temperature | K |

| Data Component | | A | B |
|---|---|---|---|
| $h^o_{f,298k}$ | kJ/kmol | –10E3 | –20E3 |
| $C_{vm}$, ref. 298 K | kJ/kmol.K | 80 | 80 |

A batch reactor of fixed volume contains initially only pure liquid A at 298 K. The reaction is then initiated and allowed to go essentially to equilibrium (i.e. X ≈ $X_{eq}$) under adiabatic conditions, without a change of phase or transfer of work.

**Problem**: Calculate the final conversion of A. Assume the reaction vessel has zero heat capacity.

**Solution**: The final conversion of A will depend on the (unknown) final reaction temperature.

Define the *system* = closed reactor plus contents. Let:

$\quad$ X = final (equilibrium) conversion of A = (initial moles A – final moles A)/initial moles A

$\quad$ $T_f$ = final (equilibrium) temperature $\qquad\qquad\qquad\qquad\qquad$ K

**Material balance on A** in the closed system

ACC          = IN − OUT + GEN − CON [IN = OUT = 0 for material in a closed system]

$- [A]_i \, V_R \, X = 0 - 0 + 0 - K_{eq} \, [A]_f V_R = - K_{eq} \, [A]_i \, V_R \, (1 - X)$

$X \qquad = K_{eq} \, (1 - X) = (exp \, (200/T_f)) \, (1 - X)$                                    [1]

where:

   $[A]_i$ = initial concentration of A                                    $kmol.m^{-3}$

   $[A]_f$ = final (equilibrium) concentration of A                        $kmol.m^{-3}$

   $V_R$ = reactor volume (= fixed volume of reaction mixture)   $m^3$

Note that the material balance is under-specified, i.e. 1 equation, 2 unknowns (X and $T_f$).

**Energy balance** on the closed system. *Reference condition = elements at standard state, 298 K*

ACC   = IN − OUT + GEN − CON

$U_f - U_i = 0 - 0 + 0 - 0 = 0$      [Energy is conserved. Q = 0, W = 0]                     [2]

$U_i$      = initial internal energy of system          kJ

$U_f$      = final internal energy of system          kJ

$U_i$      = $[A]_i \, V_R \, (C_{vm} \, (A) \, (T_i - 298) + h^{\circ}_{f,298K} \, (A)) = 1$ kmol (80 kJ/(kmol.K)(298 − 298)

          K + (−10E3 kJ/kmol))

          = −10E3 kJ

$U_f$      = $[A]_f \, V_R \, (C_{vm} \, (A) \, (T_f - 298) + h^{\circ}_{f,298K} \, (A) + [B]_f \, V_R \, (C_{vm} \, (B) \, (T_f - 298) + h^{\circ}_{f,298K} \, (B))$

          = (1−X) (80 kJ/(kmol.K)($T_f$ − 298) K + (−10E3 kJ/kmol)) + X(80 kJ/(kmol.K)($T_f$

          − 298) K + (−20E3 kJ/kmol))

Note that the reactor volume does not affect the equilibrium conversion of A and drops from the material and energy balance equations.

Equations [1 and 2] make the system fully-specified and are solved simultaneously (by spreadsheet "Solver") to give: **$T_f$ = 334 K    $K_{eq}$ = 1.82    $X_{eq}$ = 0.65**

## OPEN SYSTEMS (CONTINUOUS PROCESSES)

### SINGLE PROCESS UNITS

*Example 6.03 A, B, C, D, E and F* illustrates simultaneous M&E balance problems for the generic process units: MIX, SEPARATE, HEAT EXCHANGE (indirect and direct), PUMP and REACT in open systems operating at steady-state. In single unit problems of this sort a common feature is an interdependent combination of an unknown flow (or flows) with an unknown stream temperature, as typified in *Example 6.03B*. In less common cases where enthalpy is a function of pressure an unknown stream pressure may also be combined with an unknown flow. Such features are usually easy to see in a "problem statement" stream table that lays out all known and unknown values of the case at hand. Simultaneous M&E balances do not arise in the generic process unit DIVIDE because by definition the composition and the temperature are equal for all streams in and out of this unit.

## MULTIPLE PROCESS UNITS

Simultaneous material and energy balances are more difficult to identify and to solve when they occur in multi-unit processes, especially in processes with recycle streams. In some cases such problems can be solved by the *overall balance* approach in which the overall material balance is combined with the overall energy balance to give the desired result. A simple case of the overall balance approach is shown in *Example 6.04*.

When a complete stream table is needed (e.g. for process design) simultaneous M&E balances may have to be solved for some or all of the process units in the flowsheet. *Example 6.05* shows a recycle M&E balance similar to that of *Example 5.10*, with the complication that the conversion in the reactor depends on the reaction temperature, which comes in turn from energy balances on the pump, mixer and reactor in the recycle loop. Problems such as this can be solved by an extension of the *iterative sequential modular method* shown in *Chapters 4* and *5*, with provision to deal with the coupled material and energy equations that arise in some process units. The spreadsheet solution of this type of problem may use manual iteration and/or macros to close the simultaneous balances on individual process units.

*Example 6.05* demonstrates effects that are characteristic of chemical processes. It also provides an insight to the relevance of complex systems with respect to the "ingenuity gap". The ingenuity gap is the divide between the *need for* and the *supply of* ideas to solve the problems facing society in the 21st century. This gap relates to ecological, economic and technological systems whose complexity exceeds the ability of humans to predict and/ or control their behaviour. Such complexity is due to the connectedness between many sub-systems, with non-linearity and interaction of multiple variables, driven by feedback loops among the system parts.

*Example 6.05* is a simple illustration of the concepts of coupling, non-linearity, interaction[2] and recycling in a multi-unit, multi-variable system. You can see that even at this low level of complexity it is difficult to have an intuitive understanding of the system's behaviour because *everything depends on everything else*. The simultaneous M&E balance provides a predictive model of the system that helps to narrow the ingenuity gap. Simultaneous M&E balances are used in engineering design, with sophisticated chemical process simulation software and complex mathematical algorithms linked to large data banks and associated correlations. On a larger scale, simultaneous differential M&E balances are at

---

[2] An interaction exists between two variables A and B when the effect of variable A on the objective function depends on the value of variable B.

the heart of modelling the global climate change on planet Earth (see *Example 7.07*). In this case the high level of complexity leads to the possibility of system instabilities with "threshold effects"[3] that threaten to devastate our civilisation!

Finally, *Example 6.06* shows a simultaneous M&E balance on a hypothetical "environmentally balanced" process in which solar radiation and natural rainfall are integrated with hydrogen fuel cells to power the world's automobiles. This example shows how simultaneous M&E balances may be used to predict some of the consequences of a "sustainable" energy policy on the global environment.

## EXAMPLE 6.03   Simultaneous M&E balances for generic process units (open system at steady-state).

This example shows how the known output temperature is used to find an unknown input flow.

### A.   Mixer

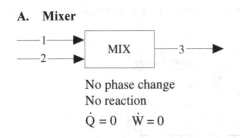

No phase change
No reaction
$\dot{Q} = 0 \quad \dot{W} = 0$

| Stream Table | | Mixer Problem | | |
|---|---|---|---|---|
| Species | M | Stream | | |
| | | *1* | *2* | *3* |
| | kg/kmol | kmol/h | | |
| A (liq) | 20 | 4 | 3 | ? |
| B (liq) | 30 | 8 | 5 | ? |
| C (liq) | 40 | 12 | ? | ? |
| Total | kg/h | 800 | ? | ? |
| Phase | | L | L | L |
| Temp. | K | 400 | 300 | 350 |
| Press. | kPa(abs) | 150 | 150 | 140 |
| Enthalpy | kJ/h | −1.05E+06 | ? | ? |

| Properties | Species | A (liq) | A(gas) | B(liq) | B(gas) | C(liq) | C(gas) |
|---|---|---|---|---|---|---|---|
| $C_{p,m}$(ref.298K) | kJ/kmol.K | 50 | 35 | 70 | 45 | 90 | 55 |
| $h_{f,298K}$ | kJ/kmol | −2.0E+05 | −1.8E+05 | −1.0E+05 | −7.0E+04 | 3.00E+04 | 4.00E+04 |
| $h_{v,298K}$ | kJ/kmol | − | 2.00E+04 | − | 3.00E+04 | − | 1.00E+04 |

*Solution*: **Material balance**

| Mole balance on A | $0 = 4 + 3 - \dot{n}(3, A)$ | kmol/h | [1] |
|---|---|---|---|
| Mole balance on B | $0 = 8 + 5 - \dot{n}(3, B)$ | kmol/h | [2] |
| Mole balance on C | $0 = 12 + \dot{n}(2, C) - \dot{n}(3, C)$ | kmol/h | [3] |

Three independent equations. Four unknowns. The material balance is UNDER-SPECIFIED.

**Energy balance**  Reference condition = *elements* at standard state, 298K

$$0 = \Sigma[\dot{H}^*(in)] - \Sigma[\dot{H}^*(out)] + \dot{Q} - \dot{W}$$

---

[3] A threshold effect is a sudden change in behaviour of a system that occurs when process variable(s) cross a critical value.

For a mixer $\dot{Q} = \dot{W} = 0$   i.e.

$$0 = \dot{H}*(1) + \dot{H}*(2) - \dot{H}*(3) + 0 - 0 \qquad\qquad [4]$$

where:

$$\dot{H}*(i) = \Sigma \left[ \dot{n}(i,j)(C_{p,m}(j)(T(i) - T_{ref}) + h^\circ_{f,Tref}(j)) \right] \qquad \text{[Respect the phase]}$$

$\dot{H}*(1) = (4 \text{ kmol/h})((50 \text{ kJ/kmol.K})(400 \text{ K} - 298 \text{ K}) + (-2E5 \text{ kJ/kmol}))$
$\qquad + (8 \text{ kmol/h})((70 \text{ kJ/kmol.K})(400 \text{ K} - 298 \text{ K}) + (-1E5 \text{ kJ/kmol}))$
$\qquad + (12 \text{ kmol/h})((90 \text{ kJ/kmol.K})(400 \text{ K} - 298 \text{ K}) + (3E4 \text{ kJ/kmol}))$
$\qquad = -1.05E+06 \text{ kJ/h}$

$\dot{H}*(2) = (3 \text{ kmol/h})((50 \text{ kJ/kmol.K})(300 \text{ K} - 298 \text{ K}) + (-2E5 \text{ kJ/kmol}))$
$\qquad + (5 \text{ kmol/h})((70 \text{ kJ/kmol.K})(300 \text{ K} - 298 \text{ K}) + (-1E5 \text{ kJ/kmol}))$
$\qquad + (\dot{n}(2, C) \text{ kmol/h})((90 \text{ kJ/kmol.K})(300 \text{ K} - 298 \text{ K}) + (3E4 \text{ kJ/kmol}))$
$\qquad = ? \text{ kJ/h}$

$\dot{H}*(3) = (\dot{n}(3, A) \text{ kmol/h})((50 \text{ kJ/kmol.K})(350 \text{ K} - 298 \text{ K}) + (-2E5 \text{ kJ/kmol}))$
$\qquad + (\dot{n}(3, B) \text{ kmol/h})((70 \text{ kJ/kmol.K})(350 \text{ K} - 298 \text{ K}) + (-1E5 \text{ kJ/kmol}))$
$\qquad + (\dot{n}(3, C) \text{ kmol/h})((90 \text{ kJ/kmol.K})(350 \text{ K} - 298 \text{ K}) + (3E4 \text{ kJ/kmol}))$
$\qquad = ? \text{ kJ/h}$

| Stream Table | | | Mixer solution | |
|---|---|---|---|---|
| Species | M | | Stream | |
| | | 1 | 2 | 3 |
| | kg/kmol | | kmol/h | |
| A (liq) | 20 | 4.0 | 3.0 | 7.0 |
| B (liq) | 30 | 8.0 | 5.0 | 13.0 |
| C (liq) | 40 | 12.0 | 14.9 | 26.9 |
| Total | kg/h | 800 | 806 | 1606 |
| Phase | | L | L | L |
| Temp. | K | 400 | 300 | *350* |
| Press. | kPa(abs) | 150 | 150 | 140 |
| Enthalpy* | kJ/h | *-1.05E+06* | *-6.50E+05* | *-1.70E+06* |
| Mass Balance Check | | | | |
| Mass IN | kg/h | 1606 | Closure % | |
| Mass OUT | kg/h | 1606 | 100.0 | |
| Energy Balance Check | | | | |
| Energy IN | kJ/h | -1.70E+06 | Closure % | |
| Energy OUT | kJ/h | -1.70E+06 | 100.0 | |

*Enthalpy values include heats of formation.*

The combined material and energy balance has four independent equations and four unknowns, i.e. FULLY-SPECIFIED. Solve equations [1 to 4] to get the stream table.

## B.   Separator (adiabatic flash split)

This example shows how a known pressure is used to find an unknown adiabatic temperature and phase split.

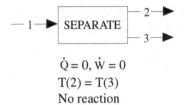

$\dot{Q} = 0,\ \dot{W} = 0$

$T(2) = T(3)$

No reaction

Stream 2 is in thermal and V-L equilibrium with stream 3.

| Stream Table | | Separator Problem | | |
|---|---|---|---|---|
| Species | M | Stream | | |
| | | *1* | *2* | *3* |
| | kg/kmol | kmol/h | | |
| A | 20 | 4 | ? | ? |
| B | 30 | 8 | ? | ? |
| Total | kg/h | 320 | ? | ? |
| Phase | | L | G | L |
| Temp. | K | 600 | ? | ? |
| Press. | kPa(abs) | 25000 | *140* | *140* |
| Enthalpy | kJ/h | ? | ? | ? |

$p*(A) = \exp[17 - 3300/(T - 39)]$ kPa
$p*(B) = \exp[14 - 4000/(T - 55)]$ kPa

This is the classic **adiabatic flash split**.

| Properties | Species | A(liq) | A(gas) | B(liq) | B(gas) |
|---|---|---|---|---|---|
| $C_{p,m}$(ref.298K) | kJ/kmol.K | 50 | 35 | 70 | 45 |
| $h_{f,298K}$ | kJ/kmol | −2.0E+05 | −1.8E+05 | −1.0E+05 | −7.0E+04 |

## Solution: Material balance

Mole balance on A   $0 = 4 - \dot{n}(2, A) - \dot{n}(3, A)$   kmol/h   [1]
Mole balance on B   $0 = 8 - \dot{n}(2, B) - \dot{n}(3, B)$   kmol/h   [2]

Two independent equations. Four unknowns. The material balance alone is UNDER-SPECIFIED.

Equilibrium compositions:

$y(2, A) = k(A)\ x(3, A)$   where  $k(A) = p*(A)/P(2)$   Raoult's law (see *Equation 2.33*) [3]
$y(2, B) = k(B)\ x(3, B)$   where  $k(B) = p*(B)/P(2)$   [4]
$y(i, j) = $ m.f. j in gas i   $1 = y(2, A) + y(2, B)$   [5]
$x(i, j) = $ m.f. j in liquid i   $1 = x(3, A) + x(3, B)$   [6]

Re-arrange equations [1 to 6] by the method of *Example 4.07B* to get:

$1 = 4/[\bar{n}(2)(k(A) - 1) + 12] + 8/[\bar{n}(2)(k(B) - 1) + 12]$   [7]

## Energy balance

Reference condition = *elements* at standard state, 298K
$0 = \dot{H}*(1) - \dot{H}*(2) - \dot{H}*(3) + \dot{Q} - \dot{W}$   kJ/h   [8]

where:
$\dot{H}*(i) = \Sigma[\dot{n}(i, j)(C_{p,m}(j)(T(i) - T_{ref})h^{\circ}_{f,Tref}(j))]$   kJ/h   [Respect the phase]

$\dot{H}^*(1) = (4 \text{ kmol/h})((50 \text{ kJ/kmol.K})(300 \text{ K} - 298 \text{ K}) + (-2E5 \text{ kJ/kmol}))$                kJ/h
$\quad + (8 \text{ kmol/h})((70 \text{ kJ/kmol.K})(300 \text{ K} - 298 \text{ K}) + (-1E5 \text{ kJ/kmol}))$

$\dot{H}^*(2) = (\dot{n}(2, A) \text{ kmol/h})((35 \text{ kJ/kmol.K})(T(2) \text{ K} - 298 \text{ K}) + (-1.8E5 \text{ kJ/kmol}))$    kJ/h
$\quad + (\dot{n}(2, B) \text{ kmol/h})((45 \text{ kJ/kmol.K})(T(2) \text{ K} - 298 \text{ K}) + (-7E4 \text{ kJ/kmol}))$

$\dot{H}^*(3) = (\dot{n}(3, A) \text{ kmol/h})((50 \text{ kJ/kmol.K})(T(3) \text{ K} - 298 \text{ K}) + (-2E5 \text{ kJ/kmol}))$    kJ/h
$\quad + (\dot{n}(3, B) \text{ kmol/h})((70 \text{ kJ/kmol.K})(T(3) \text{ K} - 298 \text{ K}) + (-1E5 \text{ kJ/kmol}))$

RESPECTING THE PHASE

Thermal equilibrium      $T(2) = T(3)$                                                                [9]

The combined material and energy balance has six independent equations and six unknowns, i.e. FULLY-SPECIFIED.

The non-linear equations [7 and 8] must be solved simultaneously to find $T(2) = T(3)$ and $\dot{n}(2)$.

The spreadsheet method is as follows: Fix $T(2) \rightarrow$ Solve equation [7] for $\bar{n}(2)$ by the Solver $\rightarrow$ Calculate the material balance and stream enthalpies.

Bisect $T(2)$ manually to close the energy balance.

| Stream Table | | Separator | | Solution |
|---|---|---|---|---|
| Species | M | | | |
| | | 1 | 2 | 3 |
| | kg/kmol | | kmol/h | |
| A | 20 | 4 | *3.87* | *0.13* |
| B | 30 | 8 | *2.12* | *5.88* |
| Total moles | kmol/h | 12 | *5.99* | *6.01* |
| Total mass | kg/h | 320 | *141.0* | *179.0* |
| Phase | | L | *G* | *L* |
| Temp. | K | 600 | *436* | *436* |
| Press. | kPa(abs) | 25000 | 140 | 140 |
| Enthalpy* | kJ/h | −1.37E+06 | *−8.13E+05* | *−5.56E+05* |
| *Mass Balance Check* | | | | |
| Mass IN | kg/h | 320 | | Closure % |
| Mass OUT | kg/h | 320 | | 100.0 |
| *Energy Balance Check* | | | | |
| Energy IN | kJ/h | −1.37E+06 | | Closure % |
| Energy OUT | kJ/h | −1.37E+06 | | 99.9 |

*Enthalpy values include heats of formation.*

*EXAMPLE:*

| T(2) | $\bar{n}(2)$ | Closure % |
|---|---|---|
| 450 | 7.96 | 95.6 |
| 400 | 3.89 | 105.5 |
| 425 | 5.13 | 102.1 |
| 437 | 6.09 | 99.7 |
| 431 | 5.55 | 101 |
| 433 | 5.72 | 100.6 |
| 435 | 5.89 | 100.2 |
| **436** | **5.99** | **99.9** |

Bisection solution of energy balance
$\quad k(A) = 29.873$
$\quad k(B) = 0.362$
$x(3, A) = 0.022$

## C.   Heat exchanger (indirect contact)

This example shows how the known output temperature is used to find an unknown input and output flow.

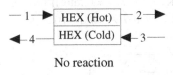

| Stream Table | | Indirect Heat Exchanger | | Problem |
|---|---|---|---|---|
| Species | M | Stream | | | |
| | | 1 | 2 | 3 | 4 |
| | kg/kmol | kmol/h | | | |
| A | 20 | 4 | ? | 3 | ? |
| B | 30 | ? | ? | 7 | ? |
| Total | kg/h | ? | ? | 270 | ? |
| Phase | | G | L | L | G |
| Temp. | K | 500 | 420 | 300 | 440 |
| Press. | kPa(abs) | 600 | 570 | 200 | 170 |
| Enthalpy | kJ/h | ? | ? | ? | ? |

No reaction

$$\dot{Q} = \dot{W} = 0$$
Overall

**Solution: Material balance**

[System = HEX hot side]

Mole balance on A

$$0 = 4 - \dot{n}(2, A) \qquad \text{kmol/h} \qquad [1]$$

Mole balance on B

$$0 = \dot{n}(1, B) - \dot{n}(2, B) \quad \text{kmol/h}$$
$$[2]$$

[System = HEX cold side].

| Properties | Species | A(liq) | A(gas) | B(liq) | B(gas) |
|---|---|---|---|---|---|
| $C_{p,m}$ (ref.298K) | kJ/kmol.K | 50 | 35 | 70 | 45 |
| $h_{f,298K}$ | kJ/kmol | −2.0E+05 | −1.8E+05 | −1.0E+05 | −7.0E+04 |
| $h_{v,298K}$ | kJ/kmol | | 2.00E+04 | | 3.00E+04 |

Mole balance on A      $0 = 3 - \dot{n}(4, A) \qquad$ kmol/h      [3]
Mole balance on B      $0 = 7 - \dot{n}(4, B) \qquad$ kmol/h      [4]

Four independent equations. Five unknowns. The material balance alone is UNDER-SPECIFIED.

**Energy balance** [Overall HEX balance]

Reference condition = *elements* at standard state, 298 K

$$0 = \dot{H}*(1) - \dot{H}*(2) + \dot{H}*(3) - \dot{H}*(4) + 0 - 0 \qquad [5]$$

For a heat exchanger $\dot{Q} = \dot{W} = 0$      [Overall]

where:

$$\dot{H}*(i) = \Sigma[\dot{n}(i, \ j)(C_{p,m} \ (j)(T(i) \ - \ T_{ref}) \ + \ h^{o}_{f,Tref} \ (j))]$$
[Respect the phase]

$\dot{H}*(1) = (4 \text{ kmol/h})((35 \text{ kJ/kmol.K})(500 \text{ K} - 298 \text{ K}) +$
        $(-1.8E5 \text{ kJ/kmol}))$
    $+ (\dot{n}(1, B) \text{ kmol/h})((45 \text{ kJ/kmol.K})(500 \text{ K} - 298 \text{ K}) + (-7E4 \text{ kJ/kmol}))$
    $= ? \text{ kJ/h}$

$\dot{H}*(2) = (\dot{n}(2, A) \text{ kmol/h})((50 \text{ kJ/kmol.K})(400 \text{ K} - 298 \text{ K}) + (-2E5 \text{ kJ/kmol}))$
    $+ (\dot{n}(2, B) \text{kmol/h})((70 \text{ kJ/kmol.K})(400 \text{ K} - 298 \text{ K}) + (-1E5 \text{ kJ/kmol}))$
    $= ? \text{ kJ/h}$

Under-Specified
Problem

$\dot{H}*(3) = (2\ kmol/h)((50\ kJ/kmol.K)(350\ K - 298\ K) + (-2E5\ kJ/kmol))$
$+ (7\ kmol/h)((70\ kJ/kmol.K)(350\ K - 298K) + (-1E5\ kJ/kmol))$
$= ?\ kJ/h$

$\dot{H}*(4) = (\dot{n}(4, A)\ kmol/h)((35\ kJ/kmol.K)(440\ K - 298\ K) + (-1.8E5\ kJ/kmol))$
$+ (\dot{n}(4, B)kmol/h)((45\ kJ/kmol.K)(440\ K - 298\ K) + (-7E4\ kJ/kmol))$
$= ?\ kJ/h$

| Stream Table | | Indirect Heat Exchanger | | | Solution |
|---|---|---|---|---|---|
| Species | M | Stream | | | |
| | | 1 | 2 | 3 | 4 |
| | kg/kmol | kmol/h | | | |
| A | 20 | 4 | 4 | 3 | 3 |
| B | 30 | 8 | 8 | 7 | 7 |
| Total | kg/h | 320 | 320 | 270 | 270 |
| Phase | | G | L | L | G |
| Temp. | K | 500 | 420 | 300 | 440 |
| Press. | kPa(abs) | 600 | 570 | 200 | 170 |
| Enthalpy* | kJ/h | -1.18E+06 | -1.51E+06 | -1.30E+06 | -9.70E+05 |
| *Mass Balance Check* | | | | | |
| Mass IN | kg/h | 5.90E+02 | Closure % | | |
| Mass OUT | kg/h | 5.90E+02 | 100.0 | | |
| *Energy Balance Check* | | | | | |
| Energy IN | kJ/h | -2.48E+06 | Closure % | | |
| Energy OUT | kJ/h | -2.48E+06 | 100.0 | | |

*Enthalpy values include heats of formation.*

The combined M&E balance has five independent equations and five unknowns, i.e. FULLY-SPECIFIED.

Solve equations [1, 3 and 4] for $\dot{n}(2, A)$, $\dot{n}(4, A)$, $\dot{n}(2, B)$; then simultaneous equations [2 + 5] for $\dot{n}(1, B)$ and $\dot{n}(2, B)$. Thermal duty of HEX

$= \dot{H}*(1) - \dot{H}*(2) = \dot{H}*(4) - \dot{H}*(3)$
$= 3.28E + 05\ kJ/h = 91\ kW$

## D. Heat exchanger (direct contact)

This example shows how known output pressure and input flows are used to find an unknown output temperature and flows.

| Stream Table | | Direct Contact HEX | | | Problem |
|---|---|---|---|---|---|
| Species | M | Stream | | | |
| | | 1 | 2 | 3 | 4 |
| | kg/kmol | kmol/h | | | |
| A | 20 | 100 | 100 | 0 | 0 |
| B | 30 | 0 | ? | 800 | ? |
| Total | kg/h | 2000 | ? | 24000 | ? |
| Phase | | G | G | L | L |
| Temp. | K | 300 | ? | 350 | ? |
| Press. | kPa(abs) | 130 | 120 | 130 | 120 |
| Enthalpy | kJ/h | ? | ? | ? | ? |

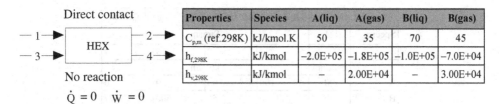

| Direct contact | | Properties | Species | A(liq) | A(gas) | B(liq) | B(gas) |
|---|---|---|---|---|---|---|---|
| | | $C_{p,m}$ (ref.298K) | kJ/kmol.K | 50 | 35 | 70 | 45 |
| HEX | | $h_{f,298K}$ | kJ/kmol | –2.0E+05 | –1.8E+05 | –1.0E+05 | –7.0E+04 |
| No reaction | | $h_{v,298K}$ | kJ/kmol | – | 2.00E+04 | – | 3.00E+04 |

$$\dot{Q} = 0 \quad \dot{W} = 0$$

$p^*(B) = \exp[18 - 4000/(T - 38)]$ kPa

Stream 2 is in thermal and V-L equilibrium with stream 4. [This unit can also function as a humidifier]

**Material balance**
Mole balance on B          $0 = 0 + 800 - \dot{n}(2, B) - \dot{n}(4, B)$                        [1]

One independent equation. Two unknowns. The material balance alone is UNDER-SPECIFIED.

**Energy balance** Reference condition = *elements* at standard state, 298 K

$$0 = \dot{H}^*(1) + \dot{H}^*(3) - \dot{H}^*(2) - \dot{H}^*(4) + \dot{Q} - \dot{W} \qquad \dot{Q} = \dot{W} = 0 \qquad [2]$$

where:

$\dot{H}^*(i) = \Sigma[\dot{n}(i,j)(C_{p,m}(j)(T(i) - T_{ref}) + h^\circ_{f,Tref}(j))]$   [Respect the phase]

$\dot{H}^*(1) = (100$ kmol/h$)$ $((35$ kJ/kmol.K$)$ $(300$ K $- 298$ K$) + (-1.8E5$ kJ/kmol$))$
$\qquad + (0$ kmol/h$)$ $((45$ kJ/kmol.K$)$ $(300$ K $- 298$ K$) + (-7E4$ kJ/kmol$))$
$\qquad = -1.80E+07 \quad$ kJ/h

$\dot{H}^*(2) = (100$ kmol/h$)$ $((35$ kJ/kmol.K$)$ $(T(2)$ K $- 298$ K$) + (-1.8E5$ kJ/kmol$))$
$\qquad + (\dot{n}(2, B)$ kmol/h$)$ $((45$ kJ/kmol.K$)$ $(T(2)$ K $- 298$ K$) + (-7E4$ kJ/kmol$))$
$\qquad = ? \quad$ kJ/h

$\dot{H}^*(3) = (0$ kmol/h$)$ $((50$ kJ/kmol.K$)(350$ K $- 298$ K$) + (-2.0E5$ kJ/kmol$))$
$\qquad + (800$ kmol/h$)((70$ kJ/kmol.K$)(350$ K $- 298$ K$) + (-1.0E5$ kJ/kmol$))$
$\qquad = -7.71E+07 \quad$ kJ/h

$\dot{H}^*(4) = (0$ kmol/h$)((70$ kJ/kmol.K$)(T(4)$ K $- 298$ K$) + (-2.0E5$ kJ/kmol$))$
$\qquad + (\dot{n}(4, B)$ kmol/h$)((50$ kJ/kmol.K$)(T(4)$ K $- 298$ K$) + (-1.0E5$ kJ/kmol$))$
$\qquad = ? \quad$ kJ/h

Equilibria.
Stream temperatures:
$T(2) = T(4)$                                                                                              [3]
Stream compositions: Stream 2

$\dot{n}(2, B)/\bar{n}(2) = p^*(B)/P(2) = f(T(2))/120$                                      [4]

| Stream Table | | Direct Contact HEX | | | Solution |
|---|---|---|---|---|---|
| Species | M | Stream | | | |
| | | 1 | 2 | 3 | 4 |
| | kg/kmol | kmol/h | | | |
| A | 20 | 100 | 100 | 0 | 0 |
| B | 30 | 0 | 56 | 800 | 744 |
| Total | kg/h | 2000 | 3686 | 24000 | 22314 |
| Phase | | G | G | L | L |
| Temp. | K | 300 | 319 | 350 | 319 |
| Press. | kPa(abs) | 130 | 120 | 130 | 120 |
| Enthalpy* | kJ/h | −1.80E+07 | −2.18E+07 | −7.71E+07 | −7.33E+07 |
| *Mass Balance Check* | | | | | |
| Mass IN | kg/h | 26000 | Closure % | | |
| Mass OUT | kg/h | 26000 | 100.0 | | |
| *Energy Balance Check* | | | | | |
| Energy IN | kJ/h | −9.51E+07 | Closure % | | |
| Energy OUT | kJ/h | −9.51E+07 | 100.0 | | |

*Enthalpy values include heats of formation.*

The combined M&E balance has four independent equations and four unknowns: $\dot{n}(2, B)$, $\dot{n}(3, B)$, $T(2)$ and $T(4)$, i.e. FULLY-SPECIFIED.

## E. Pump

This example shows how a known pressure range and work are used to find an unknown flow and output temperature.

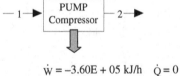

$\dot{W} = -3.60E + 05$ kJ/h     $\dot{Q} = 0$

Adiabatic compression, ideal gas

### Solution: Material balance

Mole balance on A    $0 = \dot{n}(1, A) - \dot{n}(2, A)$  [1]

One independent equation. Two unknowns.
Material balance is UNDER-SPECIFIED.

| Stream Table Compressor | | Problem | |
|---|---|---|---|
| Species | M | Stream | |
| | | 1 | 2 |
| | kg/kmol | kmol/h | |
| A | 20 | ? | ? |
| Total | kg/h | ? | ? |
| Phase | | G | G |
| Temp. | K | 300 | ? |
| Press. | kPa(abs) | 150 | 500 |
| Volume | m3/h | ? | ? |
| Enthalpy | kJ/h | ? | ? |

| Properties | Species | A(liq) | A(gas) |
|---|---|---|---|
| $C_{p,m}$(ref.298K) | kJ/kmol.K | 50 | 35 |
| $h_{f,298K}$ | kJ/kmol | −2.0E+05 | −1.8E+05 |
| Density | − | − | Ideal gas |

### Energy balance

Reference condition = *elements* at standard state, 298 K

$$0 = \dot{H}*(1) - \dot{H}*(2) + \dot{Q} - \dot{W} \qquad\qquad [2]$$

| Stream Table Compressor | | | *Solution* |
|---|---|---|---|
| Species | M | Stream | |
| | | *1* | *2* |
| | kg/kmol | kmol/h | |
| A | 20 | *103.6* | *103.6* |
| Total | kg/h | *2071* | *2071* |
| Phase | | G | G |
| Temp. | K | 300 | *399* |
| Press. | kPa(abs) | 150 | 500 |
| Volume | m3/h | *1721* | *687* |
| Enthalpy* | kJ/h | *-1.86E+07* | *-1.83E+07* |

*Mass Balance Check*

| Mass IN | kg/h | 2071 | Closure % |
|---|---|---|---|
| Mass OUT | kg/h | 2071 | 100.0 |

*Energy Balance Check*

| Energy IN | kJ/h | -1.86E+07 | Closure % |
|---|---|---|---|
| Energy OUT | kJ/h | -1.86E+07 | 100.0 |

*Enthalpy values include heats of formation.

where:
$$H^*(i) = \Sigma[\dot{n}(i,j)(C_{p,m}(j)(T(i) - T_{ref}) + h^\circ_{f,Tref}(j))]$$
[Respect the phase]

$\dot{H}^*(1) = (\dot{n}(1, A) \text{ kmol/h})$
$\quad (35 \text{ kJ/kmol.K}) (300 - 298) \text{ K}$
$\quad + (-1.8E5 \text{ kJ/kmol})$
$\quad = ? \text{ kJ/h}$

$\dot{H}^*(2) = (\dot{n}(2, A) \text{ kmol/h})(35 \text{ kJ/(kmol.K)}$
$\quad (T(2) - 298) \text{ K} + (-1.8E5 \text{ kJ/kmol})$
$\quad = ? \text{ kJ/h}$

$\dot{W} \quad = -\dot{n}(1, A)RT(1)(r/(r - 1))$
$\quad [(P(2)/P(1))^{(r-1)/r} - 1]$
$\quad = -3.60E+05 \text{ kJ/h}$        [3]

$r \quad = C_p/C_v = C_p/(C_p - R) = 35/(35 - 8.314)$
$\quad = 1.312$

The M&E balance has three independent equations and three unknowns $\dot{n}(1, A)$, $\dot{n}(2, A)$ and $T(2)$, i.e. FULLY-SPECIFIED.

$\dot{n}(1, A) = \dot{n}(2, A) = 103.6 \text{ kmol/h}$    $T(2) = 399 \text{ K}$

| Properties | Species | A(liq) | A(gas) | B(liq) | B(gas) |
|---|---|---|---|---|---|
| $C_{p,m}$(ref.298K) | kJ/kmol.K | 50 | 35 | 70 | 45 |
| $h_{f,298K}$ | kJ/kmol | -2.0E+05 | -1.8E+05 | -1.0E+05 | -7.0E+04 |

## F. Reactor

This example shows how a known reactor output temperature is used to find an unknown conversion.

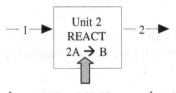

$\dot{Q} = 1.00E + 06 \text{ kJ/h}$      $\dot{W} = 0$

Conversion of A = X(A) unspecified

| Stream Table Reactor | | | *Problem* |
|---|---|---|---|
| Species | M | Stream | |
| | | *1* | *2* |
| | kg/kmol | kmol/h | |
| A | 40 | 10 | ? |
| B | 80 | 0 | ? |
| Total | kg/h | 400 | ? |
| Phase | | L | L |
| Temp. | K | 300 | 350 |
| Press. | kPa(abs) | 180 | 150 |
| Enthalpy | kJ/h | ? | ? |

*Solution*: **Material balance**

Mole balance on A      $0 = 10 - \dot{n}(2, A) + 0 - X(A)\dot{n}(1, A)$                    [1]

Mole balance on B      $0 = 0 - \dot{n}(2, B) + (1/2)X(A)\dot{n}(1, A) - 0$                    [2]

Two independent equations. Three unknowns $\dot{n}(2, A)$, $\dot{n}(2, B)$ and $X(A)$.
Material balance is UNDER-SPECIFIED.

**Energy Balance**

Reference condition = *elements* at standard state, 298 K

$$0 = \dot{H}*(1) - \dot{H}*(2) + \dot{Q} - \dot{W}$$                    [3]

where:

$\dot{H}*(i) = \Sigma[\dot{n}(i, j)(C_{p,m}(j)(T(i) - T_{ref})$
$\qquad + h^{\circ}_{f,Tref}(j))]$   [Respect the phase]

$\dot{H}*(1) = (10 \text{ kmol/h})((50 \text{ kJ/kmol.K})$
$\qquad (300 \text{ K} - 298 \text{ K}) + (-2.0E5 \text{ kJ/kmol}))$
$\qquad + (0 \text{ kmol/h})((70 \text{ kJ/kmol.K})$
$\qquad (300 \text{ K} - 298 \text{ K}) + (-1.0E5 \text{ kJ/kmol}))$
$\qquad = -2.00E+06 \text{ kJ/h}$

$\dot{H}*(2) = (\dot{n}(2, A) \text{ kmol/h})((50 \text{ kJ/kmol.K})$
$\qquad (350 \text{ K} - 298 \text{ K}) + (-2.0E5 \text{ kJ/kmol}))$
$\qquad + (\dot{n}(2, B) \text{ kmol/h})((70 \text{ kJ/kmol.K})$
$\qquad (350 \text{ K} - 298 \text{ K}) + (-1.0E5 \text{ kJ/kmol}))$
$\qquad = ? \text{ kJ/h}$

| Stream Table Reactor | | | *Solution* |
|---|---|---|---|
| Species | M | Stream | |
| | | *1* | *2* |
| | kg/kmol | kmol/h | |
| A | 40 | 10.00 | *4.58* |
| B | 80 | 0.00 | *2.71* |
| Total | kg/h | 400 | 400 |
| Phase | | L | L |
| Temp. | K | 300 | 350 |
| Press. | kPa(abs) | 180 | 150 |
| Enthalpy* | kJ/h | *-2.00E+06* | *-9.99E+05* |
| *Mass balance check* | | | |
| Mass IN | kg/h | 400 | Closure % |
| Mass OUT | kg/h | 400 | 100.0 |
| *Energy balance check* | | | |
| Energy IN | kJ/h | -9.99E+05 | Closure % |
| Energy OUT | kJ/h | -9.99E+05 | 100.0 |

The M&E balance has three independent equations and three unknowns $\dot{n}(2, A)$, $\dot{n}(2, B)$ and $X(A)$, i.e. FULLY-SPECIFIED. Solve for $X(A) = 0.54$.

*Enthalpy values include heats of formation.*

***EXAMPLE 6.04   Simultaneous M&E balance by the overall balance approach (synthesis of ammonia).***

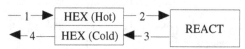

*Continuous adiabatic process at steady-state. Zero work done.*
Stream 1:
Mixture of $N_2$ and $H_2$ in stoichiometric
proportions for *Reaction 1*
Stream 4:
Product mixture of $N_2$, $H_2$ and $NH_3$

This figure shows a flowsheet of an adiabatic process for production of ammonia by the reaction:

$N_2(g) + 3H_2(g) \rightarrow 2NH_3(g)$    *Reaction 1*

GAS  2.2E4 kPa(abs)  298 K

GAS  2.0E4 kPa(abs)  398 K

**Problem:**

Calculate the conversion of nitrogen in this process.

| Data | $N_2(g)$ | $H_2(g)$ | $NH_3(g)$ |
|---|---|---|---|
| $h^\circ_{f,298K}$  kJ/kmol | 0 | 0 | −46E3 |
| $C_{p,m}$  kJ/(kmol.K) | 30 | 31 | 40 |

**Solution:** Basis = 1 kmol/h $N_2$ in stream 1
Check the specifications.

|  | Unit 1 (HEX) | Unit 2 (REACT) | OVERALL |
|---|---|---|---|
| **Material balance** | | | |
| Number of variables [$\dot{n}(i, j)$ + reactions] | IJ = (4)(3) = 12 | IJ + L = (2)(3) + 1 = 7 | IJ + L = (2)(3) + 1 = 7 |
| Number of independent equations* | 6 | 6 | 6 |
| **Energy balance** | | | |
| Number of variables [$P(i), T(i), \dot{Q}(k), \dot{W}(k)$] | 2I + 2 = (2)(4) + 2 = 10 | 2I + 2 = (2)(2) + 2 = 6 | 2I + 2 = (2)(2) + 2 = 6 |
| Number of independent equations* | 7 | 3 | 7 |
| | Net M&E variables = 25, independent equations = 25 | | |

*Includes the basis. Note that the specifications of streams between units are counted at each unit.*

Define the system        = overall process
Component:                 $A = N_2$      $B = H_2$      $C = NH_3$
Specify the quantities  = moles A, B and C        $X(A)$ = conversion of $N_2$

**Material balance**        Rate ACC = Rate IN – Rate OUT + Rate GEN – Rate CON

| | | |
|---|---|---|
| Mole balance on $N_2$ | $0 = 1 - \dot{n}(4, A) + 0 - X(A)$ | [1] |
| Mole balance on $H_2$ | $0 = 3 - \dot{n}(4, B) + 0 - 3X(A)$ | [2] |
| Mole balance on $NH_3$ | $0 = 0 - \dot{n}(4, C) - 2X(A) - 0$ | [3] |

**Energy balance**        $0 = \dot{H}*(1) - \dot{H}*(4) + \dot{Q} - \dot{W}$        $\dot{Q} = 0, \quad \dot{W} = 0$        [4]

$\dot{H}*(1) = 0$        [Elements at 298 K, assumed ideal gases]
$\dot{H}*(4) = \dot{n}(4, A)(30\,(398 - 298) + 0) + \dot{n}(4, B)(31(398 - 298) + 0)$
        $+ \dot{n}(4, C)(40\,(398 - 298) + (-46E3))$
    $= (1 - X(A))(30(398 - 298) + 0) + 3(1 - X(A))(31(398 - 298) + 0)$
        $+ 2X(A)(40\,(398 - 298) + (-46E3))$

Four independent equations, four unknowns, i.e. FULLY-SPECIFIED.

Solve for $X(A) = \underline{\mathbf{0.13}}$

This problem could also be solved by calculating each process unit in sequence to get the complete M&E balance stream table, but that much detail is not needed to find the desired conversion of nitrogen.

### EXAMPLE 6.05  *Simultaneous M&E balance by the iterative sequential modular method.*

The figure shows a generic process in which a feed of "A" is partially converted a product "B". Unconverted "A" is recovered from the reaction product and recycled to the reactor to increase the overall yield of "B" from "A".

| Properties | Species | A (liq) | A (gas) | B (liq) | B (gas) |
|---|---|---|---|---|---|
| $C_{p,m}$ (ref.298K) | kJ/kmol.K | 50 | 35 | 70 | 45 |
| $h_{f,298K}$ | kJ/kmol | -2.0E+05 | -1.8E+05 | -1.0E+05 | -7.0E+04 |
| Density | kg/m³ | 900 | Ideal gas | 900 | Ideal gas |

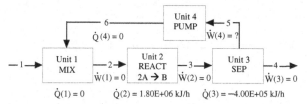

Specifications:

Stream 1:  10 kmol/h pure A                                              2000  kPa(abs) 300 K
Stream 2:  Composition unspecified  Flow unspecified  Liquid  1980  kPa(abs) T(2)
Stream 3:  Composition unspecified  Flow unspecified  Gas      1950  kPa(abs) T(3)
Stream 4:  Composition unspecified  Flow unspecified  Liquid  200   kPa(abs) 320 K
Stream 5:  Composition unspecified  Flow unspecified  Liquid  200   kPa(abs) 400 K
Stream 6:  Composition unspecified  Flow unspecified  Liquid  2000  kPa(abs) T(6)

Unit 2:  Conversion of A = X(A) = exp[–300/T(3)] (Due to the dependence of reaction rate on temperature)

Unit 3:  Separation efficiency of A from stream 3 to stream 5 = 90%

#### Problem:

**A.**  Make a degrees of freedom analysis of the problem.

**B.**  Use a spreadsheet to calculate the stream table for the steady-state material and energy balance.

*Solution*: In this example the material balance and the energy balance are coupled and must be solved together.

**A.**  Check the specifications.

|  | Unit 1 | Unit 2 | Unit 3 | Unit 4 | OVERALL |
|---|---|---|---|---|---|
| **Material balance** | | | | | |
| Number of variables [n(i, j) + reactions] | IJ = 6 | IJ + L = 5 | IJ = 6 | IJ = 4 | IJ + L = 5 |
| Number of independent equations* | 4 | 3 | 3 | 2 | 4 |
| **Energy balance** | | | | | |
| Number of variables [P(i),T(i),$\dot{Q}$(k),$\dot{W}$(k)] | 2I + 2 = 8 | 2I + 2 = 6 | 2I + 2 = 8 | 2I + 2 = 6 | 2I + 2 = 6 |
| Number of independent equations* | 7 | 4 | 7 | 3 | 5 |
| Net M&E variables = 33, independent equations = 33 | | | | | |

*Note that the specifications of streams <u>between</u> units are counted at each unit.*

**B.**   Write the material and energy balances for each process unit in sequence.
Energy balance reference condition = *elements* at standard state, 298K

UNIT 1 (Mixer)

| | | | |
|---|---|---|---|
| Mole balance on A: | $0 = \dot{n}(1, A) + \dot{n}(6, A) - \dot{n}(2, A)$ | kmol/h | [1] |
| Mole balance on B: | $0 = \dot{n}(1, B) + \dot{n}(6, B) - \dot{n}(2, B)$ | kmol/h | [2] |
| Energy balance: | $\dot{H}*(1) + \dot{H}*(6) - \dot{H}*(2) + \dot{Q}(1) - \dot{W}(1)$ | kJ/h | [3] |
| Stream compositions: | $\dot{n}(1, A) = 10$   $\dot{n}(1, B) = 0$ | kmol/h | [4][5] |
| Stream conditions: | $P(1) = 2000$   $T(1) = 300$ | kPa, K | [6][7] |
| | $P(2) = 1980$ | kPa | [8] |
| | $P(6) = 2000$ | kPa | [9] |
| Unit loads: | $\dot{Q}(1) = 0$     $\dot{W}(1) = 0$ | kJ/h | [10][11] |

UNIT 2 (Reactor)

| | | | |
|---|---|---|---|
| Mole balance on A: | $0 = \dot{n}(2, A) - \dot{n}(3, A) + 0 - X(A)\,\dot{n}(2, A)$ | kmol/h | [12] |
| Mole balance on B: | $0 = \dot{n}(2, B) - \dot{n}(3, B) + (1/2)X(A)\,\dot{n}(2, A) - 0$ | kmol/h | [13] |
| Energy balance: | $0 = \dot{H}*(2) - \dot{H}*(3) + \dot{Q}(2) - \dot{W}(2)$ | kJ/h | [14] |
| Conversion: | $X(A) = \exp[-300/T(3)]$ | – | [15] |
| Stream conditions: | $P(3) = 1950$ | kPa | [16] |
| Unit loads: | $\dot{Q}(2) = 1.8E6\ \dot{W}(2) = 0$ | kJ/h | [17][18] |

UNIT 3 (Separator)

| | | | |
|---|---|---|---|
| Mole balance on A: | $0 = \dot{n}(3, A) - \dot{n}(4, A) - \dot{n}(5, A)$ | kmol/h | [19] |
| Mole balance on B: | $0 = \dot{n}(3, B) - \dot{n}(4, B) - \dot{n}(5, B)$ | kmol/h | [20] |
| Energy balance: | $0 = \dot{H}*(3) - \dot{H}*(4) - \dot{H}*(5) + \dot{Q}(3) - \dot{W}(3)$ | kJ/h | [21] |
| Split fraction: | $s(5, A) = \dot{n}(5, A)/\dot{n}(3, A) = 0.9$ | | [22] |
| Stream conditions: | $P(4) = 150$     $T(4) = 320$ | kPa, K | [23][24] |
| | $P(5) = 150$     $T(5) = 400$ | kPa, K | [25][26] |
| Unit loads: | $\dot{Q}(3) = -4.0E5\ \dot{W}(3) = 0$ | kJ/h | [27][28] |

UNIT 4 (Pump)

| | | | |
|---|---|---|---|
| Mole balance on A: | $0 = \dot{n}(5, A) - \dot{n}(6, A)$ | kmol/h | [29] |
| Mole balance on B: | $0 = \dot{n}(5, B) - \dot{n}(6, B)$ | kmol/h | [30] |
| Energy balance: | $0 = \dot{H}*(5) - \dot{H}*(6) + \dot{Q}(4) - \dot{W}(4)$ | kJ/h | [31] |
| Unit loads: | $\dot{Q}(4) = 0$     $\dot{W}(4) = -\dot{V}(5)[P(6) - P(5)]$ | kJ/h | [32][33] |

Here there are 33 independent equations for six streams, two species, one reaction and four process units. The problem is FULLY-SPECIFIED.

The balances are calculated by the iterative sequential modular method, using the spreadsheet "Solver" to solve the non-linear equations for the reactor to get T(3) and the corresponding conversions X(A).

## Everything depends on everything!!

| Stream Table | | Mixer–reactor–separator + recycle | | | | Solution | |
|---|---|---|---|---|---|---|---|
| Species | M | Stream | | | | | |
| | | 1 | 2 | 3 | 4 | 5 | 6 |
| | kg/kmol | kmol/h | | | | | |
| A | 40 | 10.00 | *16.39* | *7.10* | *0.71* | *6.39* | *6.39* |
| B | 80 | 0.00 | *1.63* | *6.27* | *4.64* | *1.63* | *1.63* |
| *Total* | kg/h | 400 | *786* | *786* | *400* | *386* | *386* |
| Phase | | L | L | G | L | L | L |
| Temp. | K | 300 | *347* | *528* | 320 | 400 | *402.0* |
| Press. | kPa(abs) | 2000 | 1980.00 | 1950 | 200 | 200 | 2000 |
| Volume | m³/h | 0.44 | *0.87* | *30* | *0.44* | *0.43* | *0.43* |
| Enthalpy* | kJ/h | −2.00E+06 | *−3.40E+06* | *−1.60E+06* | *−5.99E+05* | *−1.40E+06* | *−1.40E+06* |

| *Mass Balance Check* | | Unit 1 | Unit 2 | Unit 3 | Unit 4 | Overall |
|---|---|---|---|---|---|---|
| Mass IN | kg/h | 786 | 786 | 786 | 386 | 400 |
| Mass OUT | kg/h | 786 | 786 | 786 | 386 | 400 |
| Closure % | | 100.0 | 100.0 | 100.0 | 100.0 | 100.0 |
| *Energy Balance Check* | | | | | | |
| Q | kJ/h | 0.00 | 1.80E+06 | −4.00E+05 | 0.00 | 1.40E+06 |
| W | kJ/h | 0.00 | 0.00 | 0.00 | −774 | −774 |
| Energy IN | kJ/h | −3.40E+06 | −1.60E+06 | −2.00E+06 | −1.40E+06 | −5.99E+05 |
| Energy OUT | kJ/h | −3.40E+06 | −1.60E+06 | −2.00E+06 | −1.40E+06 | −5.99E+05 |
| Closure % | | 100.0 | 100.0 | 100.0 | 100.0 | 100.1 |

*Enthalpy values include heats of formation.*

**EXAMPLE 6.06**    *Simultaneous M&E balance for sustainable development (hydrogen economy).*

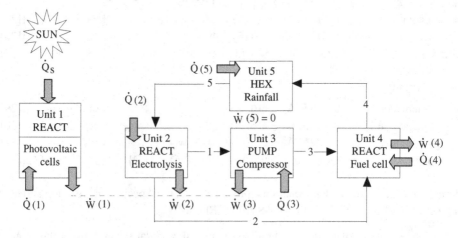

The figure shows a simplified version of the hydrogen economy, in which solar radiation ($\dot{Q}_S$) drives photovoltaic cells (*Unit 1*) to produce DC electricity ($\dot{W}(1)$) which is used to split water (*Stream 5*) to $H_2$ (*Stream 1*) and $O_2$ (*Stream 2*) by electrolysis (*Unit 2, Reaction 1*) and to drive compressors (*Unit 3*) that supply compressed hydrogen (*Stream 3*) to $H_2$/Air fuel cells on board vehicles (*Unit 4*). The water vapour (*Stream 4*) produced by the fuel cell (*Unit 4, Reaction 2*) is dispersed into the atmosphere then condensed as rainfall (*Unit 5*), collected and recycled to the electrolysis step. Oxygen from the electrolysis (*Stream 2*) is discharged into the air and subsequently consumed in the fuel cells.[4]

$2H_2O \rightarrow 2H_2 + O_2$   *Reaction 1*

4F

$2H_2 + O_2 \rightarrow 2H_2O$   *Reaction 2*

| Properties | Species | H₂O(l) | H₂O(g) | H₂(g) | O₂(g) |
|---|---|---|---|---|---|
| $C_{p,m}$(ref.298K) | kJ/kmol.K | 75 | 34 | 29 | 30 |
| $h_{f,298K}$ | kJ/kmol | -2.86E+05 | -2.42E+05 | 0.00 | 0.00 |

The process streams and units are defined as follows:

| | | | | | |
|---|---|---|---|---|---|
| Stream 1. | Pure hydrogen | Flow unspecified | Gas | 110 | kPa(abs) 350 K |
| Stream 2. | Pure oxygen | Flow unspecified | Gas | 110 | kPa(abs) 350 K |
| Stream 3. | Unspecified | Flow unspecified | Gas | 3.00E+04 | kPa(abs) 350 K |
| Stream 4. | Unspecified | Flow unspecified | Gas | 110 | kPa(abs) 350 K |
| Stream 5. | Unspecified | Flow unspecified | Liquid | 110 | kPa(abs) 300 K |

Unit 1.  Photovoltaic cells   Solar flux input  1 kW/m²   6h/day  Energy efficiency = 20%

Unit 2.  Electrochem. reactor  Input voltage   2 Volt/cell

Unit 3.  Compressor                                         Energy efficiency = 70%

Unit 4.  Fuel cells            No. autos      100 kW/  2 h/day  Energy efficiency = 40%
                               700E+06        auto              $H_2$ conversion = 100%

## Problem:

A. Make a degrees of freedom analysis of the problem.
B. Use a spreadsheet to calculate the stream table for the continuous **steady-state** material and energy balance.
C. Calculate the area of photovoltaic cells and the corresponding fraction of the Earth's surface required for the cells.
D. Find the increased rate of rainfall and net thermal load on the Earth required for the energy needs of 700E6 automobiles.

---

[4] A gas compressor (not shown) would be needed in Stream 5 to feed the fuel cell at super-atmospheric pressure.

***Solution:***

**A.** Check the specifications:

|  | Unit 1 | Unit 2 | Unit 3 | Unit 4 | Unit 4 | OVERALL |
|---|---|---|---|---|---|---|
| **Material balance** | | | | | | |
| Number of variables | $IJ = 0$ | $IJ + L = 10$ | $IJ = 6$ | $IJ + L = 10$ | $IJ = 6$ | $IJ + L = 2$ |
| [n(i, j) + reactions] | | | | | | |
| Number of independent equations* | 0 | 7 | 6 | 7 | 3 | 0 |
| **Energy balance** | | | | | | |
| Number of variables | $2 + 3^* = 3$ | $2I + 2 = 8$ | $2I + 2 = 6$ | $2I + 2 = 8$ | $2I + 2 = 6$ | $2I + 3^* = 3$ |
| $[P(i), T(i), \dot{Q}(k), \dot{W}(k)]$ | | | | | | |
| Number of independent equations* | 3 | 8 | 6 | 8 | 6 | 2 |

*Note that the specifications of streams *between* units are counted at each unit. Energy balance unknowns includes $\dot{Q}_s$.

**B.** Write the material and energy balances for each unit in sequence: $H_2O = A$, $H_2 = B$, $O_2 = C$

Energy balance reference condition = *elements* at standard state, 298 K

UNIT 1 (Reactor)     [Zero material flows]

| | | | |
|---|---|---|---|
| Energy balance: | $0 = \dot{Q}_s + \dot{Q}(1) - \dot{W}(1)$ | kW | [1] |
| Efficiency: | $\dot{W}(1) = 0.2\dot{Q}_s$ | kW | [2] |
| Unit loads: | $\dot{W}(1) = -[\dot{W}(2) + \dot{W}(3)]$ | kW | [3] |

UNIT 2 (Reactor)

Mole balance on $H_2O$: $0 = \dot{n}(5, A) - \dot{n}(1, A) - \dot{n}(2, A) + 0 - X(A)\dot{n}(5, A)$     kmol/s     [4]

Mole balance on $H_2$:  $0 = \dot{n}(5, B) - \dot{n}(1, B) - \dot{n}(2, B) + 0 - X(A)\dot{n}(5, A) - 0$  kmol/s     [5]

Mole balance on $O_2$: $0 = \dot{n}(5, C) - \dot{n}(1, C) - \dot{n}(2, C) + (1/2)X(A)\dot{n}(5, B)$
$\qquad\qquad\qquad - \dot{n}(1, B) - \dot{n}(2, B)$     kmol/s     [6]

| | | | |
|---|---|---|---|
| Energy balance: | $0 = \dot{H}^*(5) - \dot{H}^*(1) - \dot{H}^*(2) + \dot{Q}(2) - \dot{W}(2)$ | kW | [7] |
| Stream compositions: | $\dot{n}(1, A) = 0$   $\dot{n}(1, C) = 0$ | kmol/s | [8] [9] |
| | $\dot{n}(2, A) = 0$   $\dot{n}(2, B) = 0$ | kmol/s | [10] [11] |
| Stream conditions: | $P(1) = 110$   $T(1) = 350$ | kPa, K | [12] [13] |
| | $P(2) = 110$   $T(2) = 350$ | kPa, K | [14] [15] |
| | $P(5) = 110$   $T(5) = 300$ | kPa, K | [16] [17] |
| Unit loads: | $\dot{W}(2) = -E_v I' = (2)(2F(\dot{n}(5, A) - \dot{n}(1, A) - \dot{n}(2, A)))$ | kW | [18] |
| Faraday's law: | $F = 96485$ kC/kmol     $I' =$ current     $E_v =$ Voltage | | |

## UNIT 3 (Pump)

| | | | |
|---|---|---|---|
| Mole balance on $H_2O$: | $0 = \dot{n}(1, A) - \dot{n}(3, A)$ | kmol/s | [19] |
| Mole balance on $H_2$: | $0 = \dot{n}(1, B) - \dot{n}(3, B)$ | kmol/s | [20] |
| Mole balance on $O_2$: | $0 = \dot{n}(1, C) - \dot{n}(3, C)$ | kmol/s | [21] |
| Energy balance: | $0 = \dot{H}*(1) - \dot{H}*(3) + \dot{Q}(3) - \dot{W}(3)$ | kW | [22] |
| Stream compositions: | Partially specified at Unit 2 | | |
| Stream conditions: | $P(3) = 30E3 \quad T(3) = 350$ | kPa, K | [23] [24] |
| Unit loads: | $\dot{W}(3) = -P(1)\dot{V}(1)\ln[(P(3)/P(1)]/0.7$ | kW | [25] |

## UNIT 4 (Reactor)

| | | | |
|---|---|---|---|
| Mole balance on $H_2O$: | $0 = \dot{n}(2, A) + \dot{n}(3, A) - \dot{n}(4, A) + X(B)\dot{n}(3, B) - 0 \ X(B) = 1$ | | |
| | | kmol/s | [26] [27] |
| Mole balance on $H_2$: | $0 = \dot{n}(2, B) + \dot{n}(3, B) - \dot{n}(4, B) + 0 - X(B)\dot{n}(3, B)$ | kmol/s | [28] |
| Mole balance on $O_2$: | $0 = \dot{n}(2, C) + \dot{n}(3, C) - \dot{n}(4, C)$ | | |
| | $\quad + 0 - (1/2)X(B)\dot{n}(3, B)$ | kmol/s | [29] |
| Energy balance: | $0 = \dot{H}*(2) + \dot{H}*(3) - \dot{H}*(4) + \dot{Q}(4) - \dot{W}(4)$ | kW | [30] |
| Stream compositions: | Partially specified at Unit 2 | | |
| Stream conditions: | $P(4) = 110 \quad T(4) = 350$ | kPa, K | [31] [32] |
| Unit loads: | $\dot{W}(4) = (700E6 \text{ autos})(100 \text{ kW/auto})(2 \text{ h/24 h})$ | | |
| | $\quad = 5.83E+09$ | kW | [33] |
| Energy efficiency: | $0.4 = \dot{W}(4)/(286E3(\dot{n}(3, B) - \dot{n}(4, B)))$ | | [34] |

## UNIT 5 (Heat exchanger)

| | | | |
|---|---|---|---|
| Mole balance on $H_2O$: | $0 = \dot{n}(4, A) - \dot{n}(5, A)$ | kmol/s | [35] |
| Mole balance on $H_2$: | $0 = \dot{n}(4, B) - \dot{n}(5, B)$ | kmol/s | [36] |
| Mole balance on $O_2$: | $0 = \dot{n}(4, C) - \dot{n}(5, C)$ | kmol/s | [37] |
| Energy balance: | $0 = \dot{H}*(4) - \dot{H}*(5) + \dot{Q}(5) - \dot{W}(5)$ | kW | [38] |
| Stream compositions: | Unspecified | | |
| Unit loads: | $\dot{W}(5) = 0$ | kW | [39] |

*NOTE:* Since $H_2$ and $O_2$ are in the same ratio (2/1) in streams 4 and 5, equations [36 and 37] give only one independent relation.

Here there are 38 independent equations for five streams, three species, two reactions and five process units. 38 stream and unit variables. The problem is FULLY-SPECIFIED.

| Stream Table | | M&E Balance for the Hydrogen Economy | | | | Solution | |
|---|---|---|---|---|---|---|---|
| Species | M | Stream | | | | | |
| | kg/kmol | 1 | 2 | 3 | 4 | 5 | |
| | | kmol/s | | | | | |
| [A] $H_2O$ | 18 | 0.00E+00 | 0.00E+00 | 0.00E+00 | 5.10E+04 | 5.10E+04 | |
| [B] $H_2$ | 2 | 5.10E+04 | 0.00E+00 | 5.10E+04 | 0.00E+00 | 0.00E+00 | |
| [C] $O_2$ | 32 | 0.00E+00 | 2.55E+04 | 0.00E+00 | 0.00E+00 | 0.00E+00 | |
| Total | kg/s | 101981 | 815851 | 101981 | 917832 | 917832 | |
| Phase | | G | G | G | G | L | |
| Temp. | K | 350 | 350 | 350 | 350 | 300 | |
| Press. | kPa | 110 | 110 | 3.00E+04 | 110 | 110 | |
| Volume | m3/s | 1.35E+06 | 6.74E+05 | 4.94E+03 | 1.35E+06 | 918 | |
| Enthalpy* | kW | 7.69E+07 | 3.98E+07 | 7.69E+07 | −1.22E+10 | −1.46E+10 | |
| *Mass Balance Check* | | Unit 1 | Unit 2 | Unit 3 | Unit 4 | Unit 5 | Overall |
| Mass IN | kg/s | 0 | 917832 | 101981 | 917832 | 917832 | 0 |
| Mass OUT | kg/s | 0 | 917832 | 101981 | 917832 | 917832 | 0 |
| Closure % | | 100.0 | 100.0 | 100.0 | 100.0 | 100.0 | 100.0 |
| *Energy Balance Check* | | | | | | | |
| Energy IN | kW | 9.85E+10 | 1.17E+08 | 8.88E+07 | 1.17E+08 | −1.22E+10 | 9.85E+10 |
| Energy OUT | kW | 9.85E+10 | 1.17E+08 | 8.88E+07 | 1.17E+08 | −1.22E+10 | 9.85E+10 |
| Closure % | | 100.0 | 100.0 | 100.0 | 100.0 | 100.0 | 100.0 |

*Enthalpy values include heats of formation.*

$\dot{Q}_s = +9.85E+10$ kW

$\dot{Q}(1) = -7.88E+10$ kW       $\dot{W}(1) = +1.97E+10$ kW

$\dot{Q}(2) = -4.99E+09$ kW       $\dot{W}(2) = -1.97E+10$ kW

$\dot{Q}(3) = -1.19E+07$ kW       $\dot{W}(3) = -1.19E+07$ kW

$\dot{Q}(4) = -6.53E+09$ kW       $\dot{W}(4) = +5.83E+09$ kW

$\dot{Q}(5) = -2.33E+09$ kW       $\dot{W}(5) = 0$ kW

Surface area of Earth = 5.14E+08 km$^2$       Land area of Earth = 1.48E+08 km$^2$

C.  Area of photovoltaic cells = 3.94E+11 m$^2$ = **3.94 E+05 km$^2$ = 0.27% of Earth's land area**[5]

D.  Increased rainfall = 917832 kg/s = **7.93E+07 tonne/day = 0.06 mm/year (average)**

Net thermal load on Earth       = **9.26E+10 kW = 0.04% of net radiant flux from Sun to Earth**

---

[5] This is area of the active surface only. The total area needed would probably be 2 or 3 times this value, approaching 1% of Earth's land area.

## SUMMARY

[1] Simultaneous M&E balances arise for systems in which the material balance is coupled with the energy balance in such a way that the two balances must be solved simultaneously to find the distribution of material and energy across the system.

[2] In simultaneous M&E balances the unknowns are material quantities together with energy variables such as pressure, temperature, heat and work. When the complete simultaneous M&E balance is fully-specified, either the material balance alone is under-specified or the material balance and the energy balance (taken separately) are both under-specified.

[3] Simultaneous M&E balances are more difficult to identify and to specify than separate material and energy balances. In addition, the solution of simultaneous M&E balances is more likely to be complicated by simultaneous non-linear equations.

[4] Simultaneous M&E balances can arise in each of the generic process units: MIX, SEPARATE, HEAT EXCHANGE, PUMP and REACT. A characteristic of these cases is that the outlet stream flows and phase splits are interdependent with the pressure and/or temperature in the unit.

[5] Simultaneous M&E balances for multi-unit processes can be solved by the sequential modular method (outlined in Chapters 4 and 5), with iteration of the recycle loops. The spreadsheet solution may require manual iteration and/or the use of macros to close the simultaneous balances on individual process units.

[6] Simultaneous M&E balances often involve complex non-linear interactions that defy an intuitive knowledge of a system's behaviour. Such interactions are characteristic of chemical processes as well as of natural (ecological) systems that couple the flow of material and energy, and are a main source of the "ingenuity gap" facing society in the 21st century (see *Ref. 7*).

## FURTHER READING

[1]  R. Felder, R. Rousseau and L. Bullard, *Elementary Principles of Chemical Processes*, Wiley, 2018.
[2]  R. Murphy, *Introduction to Chemical Processes*, McGraw-Hill, 2007.
[3]  D. Himmelblau and J. Riggs, *Basic Principles and Calculations in Chemical Engineering,* Pearson, 2022.
[4]  H. Fogler, *Elements of Chemical Reaction Engineering*, Pearson, 2020.
[5]  M. Douglas, *Conceptual Design of Chemical Processes*, McGraw-Hill, 1988.
[6]  S. Sinnott and G. Towler, *Chemical Engineering Design*, Butterworth-Heinemann, 2019.
[7]  T. Homer Dixon, *The Ingenuity Gap*, Vintage Canada 2001.

# CHAPTER SEVEN

# UNSTEADY-STATE MATERIAL AND ENERGY BALANCES

# DIFFERENTIAL MATERIAL AND ENERGY BALANCES WITH ACCUMULATION

Differential material and energy balances can be used to follow the behaviour of systems in which conditions change over time.

When one or more condition in a system changes over time, the rate of accumulation term in the differential material balance and/or energy balance is not zero. Conditions in the system are *transient* and the system is said to be operating in the *unsteady- state* mode. Unsteady-state operation is common in chemical processes and is the rule in natural systems. Examples of unsteady-state systems are as follows:

- Batch chemical process units.
- Continuous chemical process units during "start-up" and "shut-down", or after disturbance to a process variable.
- Biological systems such as living cells, organs and the human body.
- The planet Earth, with respect to its environment, population and resources.

Differential balances are used to predict the temporal behaviour of unsteady-state systems in areas such as process control and bio-reactor engineering, as well as for modelling the Earth's ecology and environment (e.g. predicting the global climate).

Unsteady-state differential balances[1] are powerful tools for process modelling, but they are also relatively difficult in conception and execution.

The calculation of unsteady-state differential M&E balances begins with the general balance equation from Chapter 1, repeated here as *Equation 7.01*, with its corresponding differential form, *Equation 7.02*.

For a defined *system* and a specified *quantity*:

**ACC = IN – OUT + GEN – CON**                                     *Equation 7.01*

**Rate ACC = Rate IN – Rate OUT + Rate GEN – Rate CON**           *Equation 7.02*

where:

Rate ACC      = Rate of accumulation of specified quantity in the system, with respect to time

---

[1] *Also called "dynamic" balances.*

Rate IN        = Rate of input of specified quantity to the system, with respect to time
               (input)

Rate OUT       = Rate of output of specified quantity from the system, with respect to
               time (output)

Rate GEN       = Rate of generation of specified quantity in the system, with respect to
               time (source)

Rate CON       = Rate of consumption of specified quantity in the system, with respect
               to time (sink)

In the case of unsteady-state systems: **Rate ACC ≠ 0**

The "Rate ACC" term in *Equation 7.02* is expressed mathematically as a differential with respect to time.

The differential material balance for an unsteady-state system then appears as *Equation 7.03* or *7.04*.

$$d(m)/dt = (\dot{m})_{in} - (\dot{m})_{out} + (\dot{m})_{gen} - (\dot{m})_{con} \qquad \neq 0 \text{ (mass balance)} \qquad \textit{Equation 7.03}$$
$$d(n)/dt = (\dot{n})_{in} - (\dot{n})_{out} + (\dot{n})_{gen} - (\dot{n})_{con} \qquad \neq 0 \text{ (mole balance)} \qquad \textit{Equation 7.04}$$

where:                                                                        Typical units

| | | |
|---|---|---|
| $(m), (n)$ | = mass, moles of specified material in the system | kg,kmol |
| $(\dot{m})_{in}, (\dot{n})_{in}$ | = mass, mole flow of specified material into the system | kg.s$^{-1}$, kmol.s$^{-1}$ |
| $(\dot{m})_{out}, (\dot{n})_{out}$ | = mass, mole flow of specified material out of the system | kg.s$^{-1}$, kmol.s$^{-1}$ |
| $(\dot{m})_{gen}, (\dot{n})_{gen}$ | = rate of generation of specified material in the system | kg.s$^{-1}$, kmol.s$^{-1}$ |
| $(\dot{m})_{con}, (\dot{n})_{con}$ | = rate of consumption of specified material in the system | kg.s$^{-1}$, kmol.s$^{-1}$ |
| t | = time | s |

In *Equations 7.03* and *7.04* the differential material accumulation terms "d(m)/dt" and "d(n)/dt" are the rates of change of the amount of the specified material in the system (i.e. inside the system envelope) with respect to time. The "in" and "out" terms are the rates of transfer of the specified material across the system envelope, while the "gen" and "con" terms are respectively the rates of generation and consumption of the specified material inside the system envelope. When the quantity being balanced is the total mass, which is conserved in non-nuclear processes, both "gen" and "con" terms are zero. Similarly, if *no chemical reaction* occurs in the system the "gen" and "con" terms are both zero (in non-nuclear processes) when the quantity balanced is either the amount of a component "j" or the total moles. When a chemical reaction occurs in the system, the "gen" and "con" terms for individual reactant and product species are not zero, though the "gen" and "con" terms for total moles may or may not be zero, depending on the reaction stoichiometry.

The differential energy balance on an unsteady-state system appears as *Equation 7.05*. Recall that the generation and consumption terms do not appear in the energy balance because energy is a conserved quantity (in non-nuclear processes).

$$dE/dt = \dot{E}_{in} - \dot{E}_{out} + \dot{Q} - \dot{W} - (P\dot{V}_{out} - P\dot{V}_{in}) \qquad \qquad \textit{Equation 7.05}$$

where:                                                                              Typical units

| | | |
|---|---|---|
| E | = energy content of the system (associated with material)[2] | kJ |
| $\dot{E}_{in}$ | = rate of energy flow into the system (associated with material) | kW |
| $\dot{E}_{out}$ | = rate of energy flow out of the system (associated with material) | kW |
| $(P\dot{V}_{out} - P\dot{V}_{in})$ | = net rate of work to move material in and out of the system | kW |
| $\dot{Q}$ | = net rate of heat transfer into the system | kW |
| $\dot{W}$ | = net rate of work transfer out of the system (a.k.a. shaft power) | kW |
| t | = time | s |

In typical chemical processes where the kinetic, potential and exotic energy terms are considered negligible (see Chapter 5), *Equation 7.05* simplifies to *Equation 7.06*.

$$dU/dt = \ddot{H}_{in} - \ddot{H}_{out} + \dot{Q} - \dot{W} \qquad\qquad\qquad\qquad \textit{Equation 7.06}$$

where:                                                                              Typical units

| | | |
|---|---|---|
| U | = internal energy in the system, | kJ |
| $\ddot{H}_{in}$ | = rate of enthalpy flow into the system, | kW |
| $\ddot{H}_{out}$ | = rate of enthalpy flow out of the system, | kW |
| $\dot{Q}$ | = net rate of heat transfer into the system | kW |
| $\dot{W}$ | = net rate of work transfer out of the system (a.k.a. shaft power) | kW |
| t | = time | s |

In *Equations 7.05* and *7.06* the differential energy accumulation terms "dE/dt" and "dU/dt" are the rates of change of the amount of energy <u>in the system</u> (i.e. inside the system envelope) with respect to time. The $\ddot{H}_{in}$ and $\ddot{H}_{out}$ terms are the rates of enthalpy flow associated with the transfer of material across the system envelope. As outlined in Chapter 5, the $\dot{Q}$ and $\dot{W}$ terms are transfers across the system envelope, respectively in the form of thermal and usually mechanical or electrical energy.

The reference state for internal energy and enthalpy in *Equation 7.06* may be either the *elements* or the *compounds* at specified conditions. As discussed in Chapter 5 the *element* reference state is preferred in processes involving chemical reaction, whereas either the *element* or *compound* reference state may be used, as is convenient, in processes without chemical reaction. Also, the energy reference states can be mixed so long as they each cancel from the energy balance.

*Equations 7.03* to *7.06* are ordinary differential equations (ODEs). In more sophisticated models of unsteady- state systems, you may encounter partial differential equations (PDEs), but PDEs are outside the scope of this text. Differential equations can be solved by analytic calculus or by numerical methods such as *Euler, Runge–Kutta, finite differencing* and *finite element*. Unsteady-state M&E balances generally involve initial-value problems whose numerial solution (e.g. by Runge–Kutta) requires the specification of an initial

---

[2] Remember that the energy terms E, H and U are state functions, that must be defined relative to specified reference conditions.

condition and then proceeds to step through time to obtain a temporal profile of the process variables.

This text contains example problems with ODEs that are solved by simple analytical calculus, plus some examples (*7.03, 7.06* and *7.07*) that are solved by a numerical method.

## CLOSED SYSTEMS (BATCH PROCESSES)

The differential M&E balances for a closed system require the condition:

**Material flow IN = Material flow OUT = 0** [each of $\dot{Q}$ and $\dot{W}$ may or may not be zero]

*Examples 7.01, 7.02* and *7.03* illustrate respectively a differential material balance, a differential energy balance and a simultaneous differential material and energy balance on an unsteady-state closed system.

### EXAMPLE 7.01   *Differential material balance on an unsteady-state closed system (population of Earth).*

The human population of Earth in 2000 AD was 6.035 billion (6.035E9), with birth and death rates respectively 21,000 and 9,000 per million people per year.

***Problem:***   Assuming constant birth and death rates, with zero space travel, calculate the human population of Earth in 2100 AD.

***Solution:***

Define the system = planet Earth (a closed system, i.e. Rate IN = Rate OUT = 0)
Specify the quantity   = number of people = $N_p$

Differential material balance on the number of people:

Rate ACC = Rate IN  −  Rate OUT + Rate GEN     − Rate CON
$dN_p/dt$   = 0       − 0          + (21E3/1E6)$N_p$ − (9E3/1E6)$N_p$ = (12E-3)$N_p$

Separate the variables, integrate and solve for $N_p$, with the boundary (initial) conditions:

$N_p$ = 6.035E9 people         t = 2000 years
$dN_p/N_p$             = (12E-3)dt
$\ln(N_p/6.035E9)$   = (12E-3)(t − 2000)
$N_p$                 = (6.035E9)exp((12E-3)(t − 2000))        Substitute t = 2100 to get:
$N_p$                 = **20.04E9 people**[3] in the year 2100 AD

---

[3] This result probably over-estimates the population because the growth rate is expected to drop as the population increases.

**EXAMPLE 7.02** *Differential energy balance on an unsteady-state closed system (batch heater).*

**Initial condition**

Tank contains 60 kmol pure liquid A at 300 K. Heat is transferred to the contents of the tank at a rate given by:

| HEX | No reaction<br>No phase change |
|-----|--------------------------------|

| Properties | Species | A(liq) |
|------------|---------|--------|
| $C_{p,m}$(ref.298K) | kJ/kmol.K | 50 |
| $h_{f,298K}$ | kJ/kmol | −2.0E + 05 |
| Density | kg/m³ | 900 |
| Assume $C_{p,m} = C_{v,m}$ for liquids. | | |

$\dot{Q} = 3(420 - T)$ kW

$\dot{W} = 0$

$\dot{Q} = 3(420 - T)$                                                                 kW

where:
    $\dot{Q}$ = rate of heat transfer into the contents of the tank,        kW
    T = temperature of liquid in tank,                                            K

**Problem:** Calculate and plot the temperature in the tank as a function of time from t = 0 to t = 2 hours.

**Solution:** Define the system    = contents of the tank
            Specify the quantity = energy
Write the unsteady-state balances, i.e.
    Rate ACC = Rate IN – Rate OUT + Rate GEN – Rate CON

**Energy balance**    Reference condition = *elements* at standard state, 298K

$d[u*n (A)]/dt = \dot{Q} - \dot{W}$                                                    [1]

where:
    u* = specific internal energy of liquid A
        $\approx C_{v,m} (T - 298) + h^{o}_{f,298K}$    kJ/kmol
    n(A) = amount on A in the tank = 60 kmol

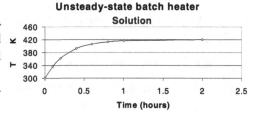

Since $C_{v,m}$ and $h^{o}_{f,298K}$ are constant, equation [1] simplifies to:

$(60)(50)\ dT/dt = (3600)(3)(420 - T)$  kJ/h                                        [2]
$dT/dt\qquad\quad = 3.6(420 - T)$                              K/h                 [3]

Solve the differential equation [3] by separating the variables and integrating: Initial condition T = 300, t = 0

**T = 420 – 120exp(–3.6t)**                    K                    [4]

| t hours | 0 | 0.1 | 0.2 | 0.4 | 0.6 | 0.8 | 1 | 2 |
|---------|-----|-----|-----|-----|-----|-----|-----|-------|
| T K | 300 | 336 | 362 | 392 | 406 | 413 | 417 | 419.9 |

**EXAMPLE 7.03   *Simultaneous differential M&E balance on an unsteady-state***
***closed system (batch reactor).***

REACT
2A(l) → B(l)

$\dot{W} = 0$

| Component | | A(l) | B(l) |
|-----------|-----------|------|------|
| MW | kg/kmol | 45 | 90 |
| $C_{v,m}$ref 298K | kJ/(kmol.K) | 70 | 80 |
| $h^{\circ}_{f,298K}$ | kJ/kmol | –50E3 | –110E3 |
| Density | kg/m³ | 900 | 900 |

A closed batch thermochemical reactor carries out the irreversible liquid phase reaction:

2A(l) → B(l)

Where the reaction rate is a function concentration and temperature, given by the Arrhenius relation:

Reaction rate = $d[A]/dt = -2\exp(-2E3/T)$ [A]                    [1]

| [A] | = concentration of [A] | kmol/m³ |
|-----|------------------------|---------|
| T | = temperature | K |
| t | = time | s |

The reactor initially contains 100 kmol of reactant A and zero B at 298 K, with a volume: $V_R = 5$ m³

The reactor contents are cooled by heat transfer to a water jacket at a rate given by:
$\dot{Q} = -9(T - 298)$ = rate of heat transfer into reactor contents   kW [note negative sign of $\dot{Q}$]   [2]

**Problem:**
Calculate the conversion of A and the temperature in the reactor at 300 seconds after the reaction is initiated.

**Solution:**
Define the system       = contents of the batch reactor
Specify the quantities = mole A, mole B, Energy
Differential material balances: Rate ACC = Rate IN – Rate OUT + Rate GEN – Rate CON
Mole balance on A:  $d(n(A))/dt = 0 – 0 + 0 – V_R(2\exp(-2E3/T)$ [A])
                                    $= -n(A)(2\exp(-2E3/T))$                    [3]
Mole balance on B:  $d(n(B))/dt = 0 – 0 + (1/2)V_R(2\exp(-2E3/T)$ [A]) – 0
                                    $= -(1/2)d(n(A))/dt$                    [4]

Differential energy balance:

| | | | |
|---|---|---|---|
| dE/dt | $= \dot{Q} - \dot{W} = -9(T - 298) - 0$ | $(\dot{W} = 0)$ | [5] |
| E | = energy of reactor contents | | |
| | ≡ internal energy, w.r.t. *elements* at standard state, 298 K | | |
| E | = (n(A) kmol)(70 kJ/(kmol.K)(T − 298) K + (−50E3 kJ/kmol)) | | |
| | + (n(B) kmol)(80 kJ/(kmol.K)(T − 298) K + (−110E3 kJ/kmol)) | | |
| dE /dt | = 70(n(A) (T − 298))/dt − 50E3(n(A))/dt + 80(n(B)(T − 298))/dt | | |
| | − 110E3(n(B))/dt | | |
| | = −9(T − 298) | | [6] |

Initial conditions:     n(A)                = 100 kmol        n(B) = 0 kmol
                 T     = 298 K at t = 0 s    [Assumes u* = h* for liquids]

Equations [1 to 6] form a set of simultaneous differential equations that embodies a complex non-linear interaction between n(A) and T. The formal solution of such simultaneous ODEs is beyond the scope of this text, however, there is a crude method, called Euler's method, that can be used to get an approximate solution by simple spreadsheet calculations. In Euler's method the calculation begins from the initial condition and marches forward in small increments of time to calculate new values of the variables (n(A) and T) using a finite difference equation, e.g.:

$T_{i+1} = T_i + \Delta t \, (dT/dt)_i$

where:

$\Delta t$   = time increment
$T_{i+1}$ = temperature at time $t(i + 1)$
$T_i$   = temperature at time $t(i)$
$(dT/dt)_i$ = differential of T, w.r.t.
        temp. at time $t(i)$

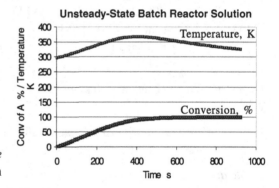

The spreadsheet solution of *Example 7.03* by Euler's method is summarised in *Figure 7.01*.[4]

From the spreadsheet, at t = 300 s

X(A) = **76%**        T = **361 K**

---

[4] *"Closure" in Figure 7.01 indicates dX/dt ≈ 0.*

| Component | | A(l) | B(l) | k | s$^{-1}$ | 2 |
|---|---|---|---|---|---|---|
| M | kg/kmol | 45 | 90 | Exp | K | -2.E+03 |
| $C_{v,m}$ | kJ/kmol.K | 70 | 80 | Vol | m$^3$ | 5 |
| $h_{298K}^o$ | kJ/kmol | -5.00E+04 | -1.10E+05 | $U_{hex}$A | kW/K | 9 |

EULER          Time increment = 10 s          Energy balance converged using "Goal Seek"

| t | n(A) | X(A) | n(B) | Total mass | T | $U_{hex}$A(T–298) | dn(A)/dt | dn(B)/dt | dT/dt | Energy balance |
|---|---|---|---|---|---|---|---|---|---|---|
| s | kmol | % | kmol | kg | K | kW | kmol/s | kmol/s | K/s | closure |
| 0 | 100.0 | 0.0 | 0.0 | 4500 | 298 | 0 | -0.243 | 0.122 | 0.174 | 2.E-12 |
| 100 | 74.2 | 25.8 | 12.9 | 4500 | 317 | 173 | -0.271 | 0.136 | 0.215 | -2.E-12 |
| 200 | 47.0 | 53.0 | 26.5 | 4500 | 340 | 379 | -0.263 | 0.131 | 0.234 | -2.E-12 |
| 300 | 23.7 | 76.3 | 38.2 | 4500 | 361 | 566 | -0.186 | 0.093 | 0.151 | 3.E-12 |
| 400 | 9.8 | 90.2 | 45.1 | 4500 | 369 | 637 | -0.087 | 0.043 | -0.005 | 1.E-12 |
| 500 | 4.0 | 96.0 | 48.0 | 4500 | 364 | 592 | -0.033 | 0.016 | -0.089 | 0.E+00 |
| 600 | 1.8 | 98.2 | 49.1 | 4500 | 354 | 503 | -0.013 | 0.006 | -0.103 | 0.E+00 |
| 700 | 0.9 | 99.1 | 49.5 | 4500 | 344 | 413 | -0.005 | 0.003 | -0.094 | 0.E+00 |
| 800 | 0.5 | 99.5 | 49.7 | 4500 | 335 | 334 | -0.003 | 0.001 | -0.079 | 0.E+00 |
| 900 | 0.3 | 99.7 | 49.8 | 4500 | 328 | 269 | -0.001 | 0.001 | -0.065 | 0.E+00 |

**Figure 7.01.   Spreadsheet solution of Example 7.03.**

## OPEN SYSTEMS (CONTINUOUS PROCESSES)

*Examples 7.04, 7.05* and *7.06* illustrate respectively a material balance, an energy balance and a simultaneous material and energy balance on unsteady-state open systems. All of these examples assume perfect mixing, for which the composition and temperature of the outlet stream are the same as those in the mixing vessel.

### EXAMPLE 7.04    *Differential material balance on an unsteady-state open system (mixer).*

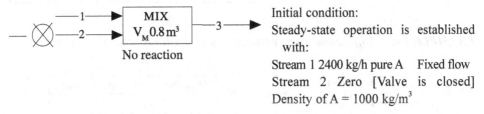

Initial condition:
Steady-state operation is established with:
Stream 1 2400 kg/h pure A    Fixed flow
Stream 2 Zero [Valve is closed]
Density of A = 1000 kg/m$^3$

At time = 0 the valve is opened and stream 2 starts to flow at a fixed rate:
Stream 2    800 kg/h    pure B    Fixed flow    Density of B = 1000 kg/m$^3$
Both stream 1 and stream 2 continue to flow at the fixed rates specified above The volume of material inside the mixing vessel remains fixed at V = 0.8 m$^3$

**Problem:**  Calculate and plot the mass fraction of B in str. 3 as a function of time from t = 0
            to t = 2 hours
**Solution:**  Define the system      = mixing vessel
            Specify the quantities = total mass, mass of B
Let        w(3, B) = mass fraction of B in stream 3 $\rho$ = density of stream 3, kg/m³
Write the unsteady-state balances, i.e.
Rate ACC = Rate IN – Rate OUT + Rate GEN – Rate CON Total mass balance:

$$d(V_{MP}\rho)/dt = 2400 + 800 - \bar{m}(3)$$
$$\text{kg/h [1]}$$

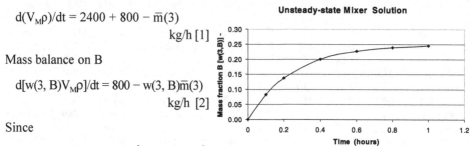

Mass balance on B

$$d[w(3, B)V_{MP}\rho]/dt = 800 - w(3, B)\bar{m}(3)$$
$$\text{kg/h [2]}$$

Since

$$V_{MP}\rho = \text{constant} = (0.8 \text{ m}^3)(1000 \text{ kg/m}^3)$$
$$= 800 \text{ kg}$$

From equation [1]

$$d(V_{MP}\rho)/dt = 0 = 2400 + 800 - \bar{m}(3)$$
$$\bar{m}(3) \quad = 3200 \text{ kg/h}$$

Substitute to equation [2]:

$$800 \, d[w(3, B)]/dt = 800 - 3200w(3, B) \quad \mathbf{d[w(3, B)]/dt = 1 - 4w(3, B)} \quad [3]$$

Solve the differential equation [3] by separating the variables and integrating: Initial
condition:

w(3, B) = 0, t = 0                [4]

**w(3, B) = (1/4)(1 – exp(–4t))**    [5]

| t, hours | 0 | 0.1 | 0.2 | 0.4 | 0.6 | 0.8 | 1 |
|---|---|---|---|---|---|---|---|
| w(3, B) | 0.000 | 0.082 | 0.138 | 0.200 | 0.227 | 0.240 | 0.245 |

**EXAMPLE 7.05**  **Differential energy balance on an unsteady-state open system**
                **(mixer).**

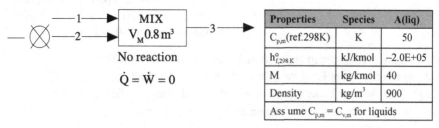

| Properties | Species | A(liq) |
|---|---|---|
| $C_{p,m}$(ref.298K) | K | 50 |
| $h^{\circ}_{f,298 K}$ | kJ/kmol | –2.0E+05 |
| M | kg/kmol | 40 |
| Density | kg/m³ | 900 |
| Ass ume $C_{p,m} = C_{v,m}$ for liquids | | |

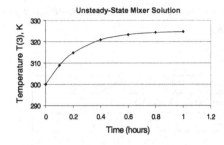

Initial condition:

| Stream 1 | 2400 kg/h pure A | Fixed flow, P and T |
|---|---|---|
| | Liquid | 200 kPa(abs), 300 K |
| Stream 2 | Zero | [Valve is closed] |

At time = 0 the valve is opened and stream 2 starts to flow at a fixed rate:

| Stream 2 | 800 kg/h pure A | Fixed flow, P and T |
|---|---|---|
| | Liquid | 200 kPa(abs), 400 K |

Both *stream 1* and *stream 2* continue to flow at the fixed rates specified above. The volume of material inside the mixing vessel remains fixed at $V_M = 0.8$ m$^3$.

**Problem:** Calculate and plot the temperature of stream 3 as a function of time from t = 0 to t = 2 hours.

*Solution:*
Define the system = mixing vessel
Specify the quantities = total moles, energy

Write the unsteady-state balances, i.e.
**Total mole balance**
$$d(V_M\rho/M)/dt = 2400/40 + 800/40 - \bar{n}(3) \qquad \text{kmol/h} \qquad [1]$$
**Energy balance**
$$d[u^*(3)w(3, B)V_M\rho/M]/dt = \dot{H}^*(1) + \dot{H}^*(2) - \dot{H}^*(3) + 0 - 0 \qquad \text{kJ/h} \qquad [2]$$

Reference condition = *elements* at standard state, 298K

$\dot{H}^*(1) = \Sigma[\dot{n}(i, j)h^*(j)]$ and $h^*(j) = C_{p,m}(T(i) - 298) + h^\circ_{f,298K}$     [Respect the phase]
$u^*(3) \approx C_{v,m}(T(3) - 298) + h^\circ_{f,298K}$             [Assume u* = h* for liquids]
$\dot{H}^*(1) = (60 \text{ kmol/h})((50 \text{ kJ/kmol.K})(300 \text{ K} - 298 \text{ K}) + (-2.0E5 \text{ kJ/kmol}))$
$\dot{H}^*(2) = (20 \text{ kmol/h})((50 \text{ kJ/kmol.K})(400 \text{ K} - 298 \text{ K}) + (-2.0E5 \text{ kJ/kmol}))$
$\dot{H}^*(3) = (\bar{n}(3) \text{ kmol/h})((50 \text{ kJ/kmol.K})(T(3) \text{ K} - 298 \text{ K}) + (-2.0E5 \text{ kJ/kmol}))$
From equation [1] $d(V_M\rho/M)/dt = 0 = 2400/40 + 800/40 - \bar{n}(3) = 80$ then $\bar{n}(3) = 80$   kmol/h

Substitute to equation [2] and collect terms, note that the heat of formation term (–2.0E5 kJ/kmol) cancels out.

$$C_{v, m}(V_M\rho/M)d(T(3))/dt = 6000 + 102000 - 4000(T(3) - 298) \qquad \text{kJ/h} \qquad [3]$$
$$900\, d(T(3))/dt = 1.300E6 - 4000(T(3))$$
$$\mathbf{d(T(3))/dt = 1444 - 4.444(T(3))} \qquad \qquad \text{kJ/h} \qquad [4]$$

Solve the differential equation [4] by separating the variables and integrating:
Initial condition   T(3) = 300, t = 0     $\underline{\mathbf{T(3) = 325 - 5(1 - exp(-4.44t))}}$     K     [5]

| t hours | 0 | 0.1 | 0.2 | 0.4 | 0.6 | 0.8 | 1 |
|---|---|---|---|---|---|---|---|
| T(3) K | 300.0 | 309.0 | 314.7 | 320.8 | 323.3 | 324.3 | 324.7 |

**EXAMPLE 7.06   *Simultaneous differential M&E balance on an unsteady-state open system (mixed and heated tank).***

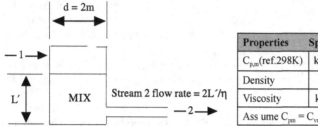

| Properties | Species | A(liq) | |
|---|---|---|---|
| $C_{p,m}$(ref.298K) | kJ/(kg.K) | 3 | |
| Density | kg/m³ | 1200 | |
| Viscosity | kg/(m.s) | (1E-3)exp(2000/T-6) | |
| Assume $C_{pm} = C_{vm}$ for liquids | | | |

Initial condition – steady-state operation with:

Stream 1   2000 kg/h A      Liquid 120 kPa(abs), 300 K
Heat transfer rate:      $\dot{Q} = 0$                                             kJ/h
Liquid level = 1.95 m

$\eta$ = liquid viscosity
At time t = 0 the heater is turned on and delivers heat to the tank at a rate:
$$\dot{Q} = 1.0E + 04 \,(400 - T) \qquad\qquad\qquad\qquad\qquad\qquad\text{kJ/h}$$

Stream 1 flow remains at 2000 kg/h A. Assume no phase change, plus liquid density and heat capacity are independent of temperature.

***Problem***:   Calculate and plot the level (L′) and the temperature (T) in the tank as a function of time for:      t = 0 to t = 10 hours.
***Solution***:   Define the system      = contents of the tank
                Specify the quantities = mass of A, energy

**Material balance on A:**

$$d(V\rho)/dt = \dot{m}(1) - \dot{m}(2) \qquad\qquad\qquad\qquad\qquad\qquad\qquad [1]$$

$V\rho$  = 3.14((2 m)/2)² (L′m)(1200 kg/m³) = 3768L′        kg
$\dot{m}(1)$  = 2000                                          kg/h
$\dot{m}(2)$  = 2 L′/$\eta$                                        kg/h

**Energy balance:**

Reference condition = *liquid compound* A at 298 K
$$dE/dt = \dot{H}(1) - \dot{H}(2) + \dot{Q} \; \dot{W} = 0 \qquad\qquad\qquad\qquad\qquad [2]$$
$\quad$ E = $V\rho$ ($C_{v,m}$(T – 298))
$\qquad$ = (3768L′ kg)(3 kJ/(kg.K))(T – 298) K
$\qquad$ = 11304 L′ (T – 298)                                    kJ
$\dot{H}(1)$ = (2000 kg/h)(3 kJ/(kg.K))(300 – 298) K
$\qquad$ = 12000                                          kJ/h
$\dot{H}(2)$ = $\dot{m}(2)$(3kJ/(kg.K))(T – 298) K
$\qquad$ = 3$\dot{m}$(2)(T – 298)   kJ/h
$\qquad \dot{Q}$ = 10000(400 – T)                                 kJ/h

Initial conditions:     $\dot{m}(2) = 2000$ kg/h          $T = T(2) = 300$ K
                        $L' = 1.95$ m               $t = 0$ hours

Solve the simultaneous differential equations [1 and 2] to obtain $L'$ and $T$ as a function of time. The spreadsheet solution uses Euler's method with a time increment = 0.01 hour.

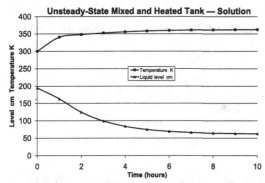

In *Examples 7.02* to *7.06* you can see how an unsteady-state M&E balance model can be useful in process design, process control and safety analysis. For instance, to control the temperature of a batch reactor such as that in *Example 7.03* (or a continuous reactor during start-up) you should know the temperature–time profile and ensure the unit is designed with sufficient heat transfer surface to allow the controller to function properly and prevent a thermal "runaway" of the exothermic reaction. Many modern computer-based process controllers include differential M&E balance models that can anticipate process variations and move to prevent them (see *Ref. 8*).

Differential M&E balances can be used for a large range of problems of which only a small sampling is given above. However, practical unsteady-state problems often involve complexities beyond the scope of this text, including non-linear differential equations that must be solved by numerical methods (for which special software is available), that are more sophisticated than the Euler's method used above in *Example 7.03*.

A final example of the range and versatility of differential material and energy balances is shown in *Example 7.07*. This example presents a highly simplified differential M&E balance model of the atmosphere of planet Earth and shows how such a model can be used to predict aspects of the global climate. In common with most natural systems, the atmosphere of planet Earth is an open system subject to inputs and outputs of material and energy at rates that vary over time. The planet's climate is tied to these material and energy transfers by a set of non-linear interactions that are embodied in the simultaneous differential material and energy balance on the Earth and its atmosphere.

The complexity of *Example 7.06* is magnified many times in *Example 7.07*, but the principles and the crude solution by Euler's method are the same. The material inputs/outputs of carbon dioxide and water vapour are coupled to the input/outputs of radiant energy with non-linear interactions that result in the phenomenon known to climatologists as the *runaway greenhouse effect,*[5] associated with global warming (see *Refs. 6, 10 and 11*) that is looming as the most critical problem yet to face humanity. The kind of complexity shown in *Example 7.07* makes human intuition an unreliable guide in formulating policy for a sustainable future of planet Earth (see *Ref. 9*).

**EXAMPLE 7.07    Simultaneous differential M&E balance on atmosphere of Earth**

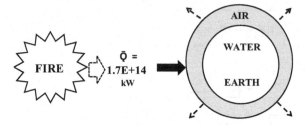

This example presents a highly simplified model of the thermal "greenhouse" effect in the atmosphere of planet Earth. Here the atmosphere is considered as an open-system exchanging material and energy ($CO_2$, $H_2O$ and IR radiation) with the surface of Earth. Carbon dioxide enters the system from combustion of fossil fuel and decomposition of biomass and exits to support photosynthesis on land and sea, and by absorption into the oceans. Water enters the system by evaporation from the oceans, fossil fuel combustion and decomposition of biomass, and exits as precipitation.

Sunlight (radiant energy) falling on Earth is partly absorbed and partly reflected or transmitted back to space as infra-red radiation. At steady-state the Sun/Earth energy balance maintains relatively hospitable conditions on Earth due to the presence in the atmosphere of $CO_2$ and $H_2O$ which absorb infra-red radiation more effectively than $O_2$ and $N_2$. Increasing levels of $CO_2$ in the atmosphere have initiated the dynamic state of "global warming". This complex system can be modeled as an unsteady-state material and energy balance.

---

[5] The *runaway greenhouse effect* is the accelerating upward spiral of the Earth's temperature, initiated by anthropogenic (man-made) $CO_2$ emissions and promoted by the increasing water vapour content of the atmosphere through the coupled M&E balances. The pressure and temperature of the Earth's atmosphere relative to the phase diagram for water (*Figure 2.02*) should prevent Earth from becoming another waterless Venus!

## Specifications:

| | | Formula | Units | Value | Symbol |
|---|---|---|---|---|---|
| y(1), y(2), y(3) = ppm(v) | | r = years from 2022 | | | |
| $CO_2$, $H_2O$, Other* | | $T_e$ = Earth temperature K | | | |
| $CO_2$ sources | Fossil fuel | $a(1+b)^r$ | kmol/y | a = 8E11 b = –0.01 to 0.03 | FOSS |
| | Deforestation, etc. | $c(1+d)^r$ | kmol/y | c = 1E11 d = 0.01 | DEFO |
| | Biomass decomp'n | $e(T_e/288)$ | kmol/y | e = 1E11 | BIOD |
| $CO_2$ sinks | Photosynth' land | $g(y(1)/420)(T_e/288)$ | kmol/y | g = 5E11 | PHLO |
| | Photosynth' ocean | $h(1-j)(y(1)/420)(T_e/288)$ | kmol/y | h = 4E11, j = 0 to 0.001 | PHOO |
| | Absorption ocean | $m(y(1)/420)$ | kmol/y | m = 1E11 | ABSO |
| | Direct Air Capture | X | kmol/y | X = 0 to 1.6E11 | DAC |
| *Other greenhouse gases $CH_4$, $N_2O$, $O_3$ , etc. | | | | | |

| DATA | | Symbol | Value | Units |
|---|---|---|---|---|
| Solar constant | | F | 1.37* | $kW/m^2$ |
| Earth diameter | | D | 12.7E3 | km |
| Earth albedo[1] | | α | 0.30  fixed | — |
| Stephan-Boltzmann constant | | σ | 5.67E-11 | $kW/m^2K^4$ |
| Molar absorption coefficient | k    $CO_2$ | | 2.28 | $m^2/kmol$ |
| | $H_2O$ | | 3.04 | $m^2/kmol$ |
| | Other | | 20 | $m^2/kmol$ |
| Effective absorption height | S    $CO_2$ | | 5E3 | m |
| | $H_2O$ | | 3E3 | m |
| | Other | | 5E3 | m |
| Atmosphere mole fraction | | y(1)  $CO_2$ | 420 initial | ppm(v) |
| | | y (2) $H_2O$ | 6100 initial | ppm(v) |
| | | y(3)  Other | 100 fixed | ppm(v) |
| Atmosphere concentration | | C | Gas law | $kmol/m^3$ |
| Atmosphere IR absorptivity | | f | 1-exp[-Σ(kCS)] | — |
| Atmosphere relative humidity | | RH | 0.25    fixed for $T_e$ | — |
| Atmosphere mean pressure | | P | 75    fixed | kPa(abs) |
| Atmosphere volume | | $V_A$ | 4.0E18 | $Sm^3$ |
| Atmosphere moles | | n(A) | 1.8E17 | kmol |
| Earth land area | | $A_L$ | 1.5E14 | $m^2$ |
| Earth ocean area | | $A_o$ | 3.6E14 | $m^2$ |
| Earth ocean volume | | $V_o$ | 1.4E18 | $m^3$ |
| Earth temperature | | $T_e$ | 288    2022 CE | K |
| Atmosphere temperature | | $T_a$ | 242    2022 CE | K |
| Antoine constants $H_2O$ | | a',b',c' | 16.5362,3985.44,38.9947 | –,K,K |
| Time | | t | From 2022 CE | years |
| *On projected area = $\pi D^2/4$ | | | | |

1 Albedo = Fraction of radiant energy reflected from Earth, assumed constant.

*(Continued)*

Material balances: System = Atmosphere of Earth
Rate ACC = Rate IN – Rate OUT + Rate GEN – Rate CON

$CO_2$     See Specifications.
$d[n(A)y(1)]/dt = (FOSS + DEFO + BIOD) – (PHLO + PHOO + ABSO) + 0 – 0$     *Equation 1*

$H_2O$     $d[n(A)y(2)]/dt = (n(A)/P) \, d[exp(a'- b')/(T_e + c')]/dt – 0 + 0 – 0$     *Equation 2*

Energy balance: System Earth and Atmosphere
Separate radiative energy balances on Earth and on the Atmosphere give the equilibrium surface temperature of Earth (see *Refs. 9, 10*) as:

$$T_e = [F(1-\alpha)/4\sigma(1-f/2)]^{0.25}$$     *Equation 3*

Where:  $f = 1 – exp[-\Sigma(kCS)]$     $\Sigma \equiv$ sum for all greenhouse gases     *Equation 4*

*Note:* The atmospheric absorptivity (f) is a non-linear (exponential) function of $y(1)$, $y(2)$, $T_e$ and t. Escalating absorptivity drives a material and energy recycle "feedback" loop leading to a potential "runaway greenhouse effect".

Numerical methods are needed for simultaneous solution of *Equations 1, 2, 3, 4* for $y(1)$, $y(2)$, f and $T_e$.

The results on the next page were obtained with Euler's method (see *Example 7.03*), stepping forward in 0.25 year increments. Beginning in 2022 CE the graphs show the atmospheric $CO_2$, $H_2O$ and absorptivity levels with consequent increases in Earth temperature ($T_e$) for the following three arbitrary conditions:

  i. "Business as usual" with fossil $CO_2$ emissions increasing at a rate of 3% per year.
 ii. Fossil emissions fixed at the 2021–2022 CE rate, 35 Gigaton $\equiv$ 8E11 kmol $CO_2$ per year.
iii. Fossil emissions fixed at the 2021–2022 CE rate plus direct air capture of $CO_2$ (DAC) beginning in 2027 CE at a rate of 7 Gt/year, equivalent to 10,000 DAC plants in continuous operation each removing 2000 ton/day of $CO_2$ from the atmosphere.

Other conditions and rates were fixed at levels chosen from the Specifications.

Absolute values in the graphs depend on the specification particulars and are not as informative as the trends, which show an alarming trajectory for "business as usual" and a potential reprieve with DAC — if that new technology were viable.

These results have an inherent caveat. They are based on a crude model, using short-cuts, simplifying assumptions and approximations, with no account for many factors affecting climate, such as cloud cover, melting icecaps, aerosols, ocean currents, etc. Such a depiction is far from the sophisticated general circulation models (GCM) used with the IPCC (see *Ref. 12*), but does show the versatility and utility of material and energy balance calculations.

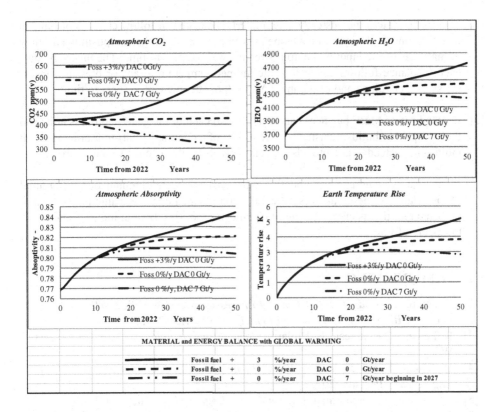

## SUMMARY

**[1]** Unsteady-state material and energy balances are differential balances in which the "rate of accumulation" term in the general balance equation is NOT zero, i.e.:

Rate ACC = Rate IN – Rate OUT + Rate GEN – Rate CON ≠ 0    (NOT ZERO)

The "Rate ACC" term is the rate of accumulation of a specified quantity within the system envelope with respect to time.

Unsteady-state balances take the form of differential equations in time (t), that are solved by analytic or numeric calculus.

**[2]** Unsteady-state M&E balances can follow changes over time and are used in chemical process design for transient conditions, general process control and modelling biological processes, as well as in modelling natural systems such as the environment and climate of Earth.

**[3]** In a closed system the unsteady-state material balance follows changes in composition of the system with respect to time, while the unsteady-state energy balance follows changes in temperature (and/or pressure) in the system with respect to time.

**[4]** In an open system the unsteady-state material balance can follow changes of lumped quantities inside the system envelope as well as changes in the composition and flow of materials of the input and output streams, with respect to time. The unsteady-state energy balance follows changes of the lumped temperature (and/or pressure) inside the system envelope and of the input and output streams with respect to time.

**[5]** Unsteady-state M&E balances can be applied to each of the generic process units and then coupled to build an unsteady-state model of a complete multi-unit process. Such multi-unit unsteady-state models are relatively difficult to conceive and to execute.

**[6]** Many real systems can be modelled with unsteady-state simultaneous differential M&E balances. These balances reflect the complicated non-linear interactive temporal behaviour that is characteristic of chemical[6] processes, both in industry and in the biosphere of Earth.

## FURTHER READING

[1]  R. Murphy, *Introduction to Chemical Processes,* McGraw-Hill, 2007.
[2]  P. Doran, *Bioprocess Engineering Principles*, Elsevier, 2013.
[3]  R. Felder, R. Rousseau, L. Bullard, *Elementary Principles of Chemical Processes,* Wiley, 2018.
[4]  D. Himmelblau, J. Riggs, *Basic Principles and Calculations in Chemical Engineering*, Pearson, 2022.
[5]  J. Mihelcic, J. Zimmerman, *Environmental Engineering*, Wiley, 2021.
[6]  J. Verret, R. Barghout, R. Qiao, *Foundations of Chemical and Biological Engineering*, Victoria, BC Campus, 2020 (Creative Commons).
[7]  H. Fogler, *Elements of Chemical Reaction Engineering*, Pearson, 2020.
[8]  B. Bequette, *Process Control*, Prentice Hall, 2003.
[9]  D. Jacob, *An Introduction to Atmospheric Chemistry*, Princeton University Press, 1999.
[10]  R. Stull, *Meteorology for Scientists and Engineers*, Brooks/Cole, 2000.
[11]  A. Ford, *Modeling the Environment,* Island Press, 2009.
[12]  Intergovernmental Panel on Climate Change, *Sixth Assessment Report — The Physical Science Basis*, IPCC, 2022.
[13]  T. Homer-Dixon, *The Ingenuity Gap*, Vintage Canada, 2001.

---

[6]Throughout this text: "Chemical" process ≡ biochemical, electrochemical, photochemical, physicochemical or thermochemical process.

# APPENDIX

This Appendix has two parts.

**Part 1 "Problems"** presents problem sets matched with Chapters 1 to 7 of the Text.

The problems relate to established chemical processes[1] and environmental issues, along with processes in development or consideration for an environmentally sustainable future.

**Part 2 "Solutions"** provides worked solutions to some problems and un-worked answers to others. The problems with worked solutions are indicated by **"W"**, un-worked by **"U"** and linked to the solutions with a Chapter reference and a problem number, e.g. **2:16 W, 5.10 U.**

Atomic weights of the elements are from the periodic table, rounded to one decimal place.

**Other data not given with the problems can be found in the Text, usually Chapter 2.**

Some problems are best solved in spreadsheets with the help of Goal Seek or Solver.

---

[1] The term "chemical process" embraces biochemical, electrochemical, photochemical, physico-chemical and thermochemical phenomena but excludes nuclear processes.

263

# LIST OF PROBLEMS

# Appendix Part 1

# PROBLEMS

## Chapter 1:   The General Balance Equation

### 1:1 W   Atmospheric $CO_2$

The Earth's atmosphere has a volume of 4.0E18 $m^3$ STP (approximate) and in 2020 CE contained an average 0.0415 vol% $CO_2$. In a "business as usual" scenario, over the period from 2020 to 2050 CE it is estimated that:

- Fossil fuels with the equivalent of 440E12 kg carbon (C) will be burned to $CO_2$
- Changes in land use (e.g. deforestation) will increase atmospheric $CO_2$ by 190E12 kg
- 510E12 kg $CO_2$ will be removed by $CO_2$ sinks on land (e.g. photosynthesis)
- 630E12 kg $CO_2$ will be absorbed by the oceans
- Other sources and sinks of $CO_2$ are relatively insignificant

*Problem*: Calculate the following:
a.   The increase in the mass of $CO_2$ in the atmosphere from 2020 to 2050 CE          [kg]
b.   The "business as usual" concentration of $CO_2$ in the atmosphere in 2050     [ppm(v)]
c.   The concentration of $CO_2$ from "b" expressed as ppm by weight (mass)      [ppm(w)]
*Assume*: Ideal gases with $CO_2$ well mixed in the whole atmosphere.

### 1:2 U   Direct air capture

It is considered feasible to remove $CO_2$ from the air in large scale direct air capture (DAC) absorption or adsorption plants.

*Problem*: If each DAC plant removed 2000 (metric) ton/day $CO_2$ from the air how many such plants, installed in 2025, would be needed to prevent atmospheric $CO_2$ exceeding 450 ppm(v) by 2050 in the "business as usual" case of Problem 1.1?          [–]

*Assume*: Ideal gases with $CO_2$ well mixed in the whole atmosphere. Operation 350 days/ year. DAC feedback does not otherwise affect the carbon cycle.

## 1:3 W  Fossil fuel

Fossil fuels such as coal ($C_{200}H_{90}O_6$), petroleum ($C_{120}H_{220}O$)[2] and gas ($CH_4$) may be characterised in general with a weighted average empirical formula $C_{60}H_{120}O_2$. The combustion of fossil fuels creates $CO_2$ and consumes $O_2$. This $O_2$ is replaced by the natural oxygen cycle and held in the air at 20.95 vol%.

***Problem:***

If oxygen was not replaced by the oxygen cycle by how much would the concentration of $O_2$ in the air be changed by burning the characteristic fossil fuel containing 440E12 kg carbon?

[ppm(vol), up/down]

***Assume:*** Feedback from a change of $O_2$ level does not otherwise affect the oxygen cycle.

## 1:4 W  Gasoline vs. grey hydrogen

A vehicle can be powered by an internal combustion engine (ICE) or by a hydrogen fuel cell. In the ICE octane is thermochemically oxidised in the reaction:

$$2C_8H_{18} + 25O_2 \rightarrow 16CO_2 + 18H_2O$$

In the hydrogen/air fuel cell (FC) hydrogen is electrochemically oxidised in the reaction:

$$2H_2 + O_2 \rightarrow 2H_2O$$

A hydrogen fuel cell would be a "clean, green" power source except that the hydrogen is produced from methane by the combined thermochemical reforming/water-gas shift reaction:

$$CH_4 + 2H_2O \rightarrow CO_2 + 4H_2$$

In a 100 km test drive the ICE engine uses 5 kg octane and the FC uses 0.6 kg hydrogen.

***Problem:*** Considering only the chemical stoichiometry, which power source is responsible for generating the least $CO_2$ for fuelling?

***Assume:*** Zero accumulation of reactants and products in the engines.

## 1:5 U  Green hydrogen and oxygen

"Green" hydrogen is produced by the electrolysis of water, using renewable green energy, according to the reaction:

$$xH_2O \rightarrow yH_2 + zO_2$$

***Problem:*** If 1 kg of $H_2$ has the same fuel value as 10 kg of characteristic fossil fuel ($C_{60}H_{120}O_2$), how much $O_2$ could be released from water electrolysis in place of the carbon dioxide generated from burning fossil fuel with 440E12 kg carbon?

[ppm(v) in atmosphere, up or down]

## 1:6 W  Jet plane

A jet engine consumes kerosene ($C_{14}H_{30}$) at a steady rate of 7920 kg/hour, with complete combustion to $CO_2$ and water.

***Problem:*** Calculate the steady-state rate of $CO_2$ generation by the engine.       [kg/hour]

---

[2] Approximate composition, excluding ca. 1 wt% [S + N].

## 1:7 U Space station $O_2$ and $H_2O$

A space station with total free volume on 100 m³ has 4 occupants, each consuming (breathing) 0.6 kg $O_2$ per day. The minimum recommended level of $O_2$ in air for breathing is 19 vol%.

*Problem:*
a. If the initial $O_2$ level is 20.9 vol% and $O_2$ is not replenished how long would it take for air in the station to reach 19 vol% $O_2$? [days]
b. If $O_2$ in the station is replenished by electrolysing water how much $H_2O$ would be consumed to keep the $O_2$ level at 20.9 vol% for 30 days? [kg]

*Assume:* Total pressure and temperature in station fixed at 100 kPa(a), 25°C. (a) ≡ (abs).

## 1:8 W Space station $CO_2$

A space station with total free volume on 100 m³ has 4 occupants, each generating 1 kg $CO_2$ per day. A maximum recommended level of $CO_2$ in air for long term breathing is 2000 ppm(v).

*Problem:*
a. If the initial $CO_2$ level is 400 ppm(v) and $CO_2$ is not removed how long would it take for air in the station to reach 2000 ppm(v) $CO_2$? [days]
b. If $CO_2$ in the station is removed by absorption into LiOH to form $Li_2CO_3$ how much LiOH would be needed in the station to keep the $CO_2$ level at 400 ppm(v) for 30 days? [kg]

*Assume:* Total pressure and temperature in station fixed at 100 kPa(a), 25°C. (a) ≡ (abs).

## 1:9 U Air balance

A compressed air tank of fixed internal volume initially contains (only) gaseous air at 120 psi(g), 130°F. Over a period of operation, 12000 litres STP of air are discharged from the tank, then the tank is recharged to 900 kPa(a), 67°C by the input of 6.00-pound mass air.

NB. (a) ≡ absolute pressure     (g) ≡ gauge pressure

*Problem:* Calculate the volume of the tank in cubic meters. [m³]
*Assume:* Air is an ideal gas

## 1:10 W Viral infection

For unmasked persons in an enclosed public space with individual virus carriers (e.g. COVID-19) the probability of infection may be roughly estimated by the equation:

$$P = C/S = 1 - \exp(-Igpt/Q)$$

where:
$P$ = probability of individual infection (–)
$C$ = number of infected cases          $S$ = number of susceptible individuals
$I$ = number of infectors               $p$ = individual pulmonary ventilation
                                         (breathing) rate (m³/h)
$Q$ = space ventilation rate (m³ / (h.infector))   $t$ = exposure time (h)
$g$ = quantum generation rate of one infector (?)

**Problem:**
a.  What are the units of "g" in this probability equation?
b.  A university room with a free volume of 500 m$^3$ has 90 students, two of whom are infectious virus carriers. The pressure and temperature in the room are 95 kPa(a), 22°C. The (average) pulmonary ventilation rate is 0.3 m$^3$/h and in this setting: g = 15 (?). Air in the room has the typical ventilation rate of 5 changes per hour. Calculate the probability of at least 1 new infection from a class problem session of 3 hours.          [–]
*Assume:* All in class are susceptible to infection.

## 1:11 U   Strategic materials

Strategic materials such as cobalt, germanium and molybdenum are obtained from their ores in metallurgical processes including flotation, leaching, roasting, solvent extraction and reduction. The production of molybdenum involves separating molybdenum sulphide from copper sulphide ore with froth flotation, where the competitive flotation of copper is "depressed" by added sodium hydrosuphide (NaHS). The NaHS can cause problems in flotation effluents as a source for hydrogen sulphide and sulphuric acid, e.g.

$$NaHS + H^+ \rightarrow Na^+ + H_2S \qquad\qquad H_2S + 2O_2 \rightarrow H_2SO_4$$

The world production of molybdenum is about 300E3 ton per year.

**Problem:** How much sulphuric acid could be produced if this Mo came by 90% recovery of Mo from ore with 0.05 wt% Mo, using 5 kg NaHS per ton of ore for flotation.          [ton/y]

## 1:12 W   Cobalt and lithium

Cobalt is a critical component of lithium-ion batteries, for example with a "cathode" composition: $LiNi_xCo_yAl_zO_2$ where: x = 0.8, y = 0.15 z = 0.05.
In electric vehicles (EVs) these batteries provide 7 kWh per kg lithium.

**Problem:** Calculate the amount of cobalt needed in batteries for production of 30E6 EVs per year, each EV with a battery capacity of 100 kWh.          [ton/y]

## 1:13 U   Germanium

Scarce but critical germanium (Ge) and indium (In) can be obtained from zinc smelter residues. A metallurgical laboratory is planning an experiment to find the effects of solid load and acid concentration in leaching Ge and In from a solid particulate smelter residue.
The 2 × 2 factorial experiment* will be done in an autoclave with fixed volume of 1 litre and requires solid loads of 1 and 2 kg solid per litre of $H_2SO_4$ leachant, matched with initial $H_2SO_4$ concentrations of 30 and 60 wt% in the leachant.
*2 variables × 2 levels.

**Problem:** Use the data below to find the mass of solids (X) and volume of 98 wt% $H_2SO_4$ (Y) to be charged in the autoclave (along with solids and water to make up 1 litre) for each of the four tests indicated in the table.
*Assume:* No change in temperature.

Specific gravity (SG): Solid $= 3.0$   $H_2SO_4 = 1 + (0.0086)w$   $w = wt\% \ H_2SO_4$

| Acid leach | | Solid load kg/L leachate | | | |
|---|---|---|---|---|---|
| Factorial levels | | 1 | | 2 | |
| Total volume 1 litre | | | | | |
| | 30 | X | Y | X | Y |
| Acid conc. | wt% $H_2SO_4$ | kg | L | lg | L |
| | 60 | X | Y | X | Y |

## 1:14 U   Acid dilution

A metallurgical laboratory has stock solutions of 1.0 molar and 0.2 molar hydrochloric acid.

***Problem***: What volume of each should be mixed to obtain 10 litre of 0.70 M HCl(aq)?

[L,L]

***Assume***: Negligible expansion or contraction of the solution(s).

## 1:15 U   Bio-acid

Citric acid is produced in a biochemical process by enzyme catalysed fermentation of molasses according to the reaction:

$$C_xH_yO_z \rightarrow C_6H_8O_7 + 3CO_2 + 2H_2O$$

Where the LHS is an empirical formula for saccharides in the molasses.

***Problem***: Calculate the values of the stoichiometric coefficients x, y and z and the mass of the saccharides converted in this reaction with 100% yield to produce 9600 kg of citric acid.   [kg]

## 1:16 W   Electro-methanol

Carbon dioxide is converted to methanol in an electrochemical process with an imperfectly selective cathode electrocatalyst where the overall process reaction is:

$$3CO_2 + 9H_2O \rightarrow xCH_3OH + yH_2CO_2 + zH_2 + 5O_2$$

The Faradaic efficiency (E) for methanol in the process is given by:

$$E = 6M/[6M + 2F + 2H]$$

where M, F and H are the moles of methanol, formic acid and hydrogen generated.

***Problem***: Find the values of x, y, z and the Faradaic efficiency for methanol.   [–,%]

## 1:17 W   Thermo-fertilizer

The figure below is a conceptual flowsheet for a continuous thermochemical process making ammonium sulphate fertilizer.

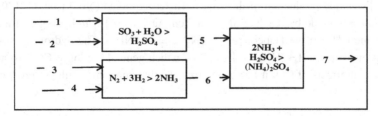

The reactant streams are: **1** $H_2O$, **2** $SO_3$, **3** $N_2$, **4** $H_2$, **5** $H_2SO_4$, **6** $NH_3$, **7** Product $(NH_4)_2SO_4$

***Problem***: Calculate the mass of each reactant required to produce 6600 kg $(NH_4)_2SO_4$
[kg]

***Assume***:  100% efficiency in all process steps with no excess reactants.

### 1:18 U   Solar water-split

The figure below is a conceptual flowsheet of a thermochemical "solar water-splitting" process in which water is decomposed into hydrogen and oxygen by complementary reactions of recycling solid oxides of cerium. Focused heat from the sun supplies the high temperature needed to decompose $CeO_x$.

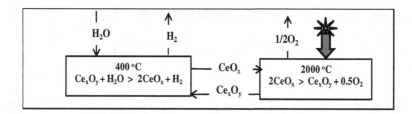

***Problem***: Find the values of stoichiometric coefficients x and y and calculate the mass of hydrogen generated by the reaction of water with 1000 kg $Ce_xO_y$.          [kg]

### 1:19 U   Sodium-ion battery

Due to limitations in the supply of lithium it is proposed to replace lithium-ion with sodium-ion batteries in electric vehicles.

***Problem***: Calculate the mass and volume of lithium and of sodium needed to build 100 million electric vehicles, each with a range of 400 km.       [Li kg, m³], [Na kg, m³]

***Assume***: The energy available from a battery reaction is directly proportional to the standard electrode potential.

| DATA | Redox reaction | Standard electrode potential | Density | Mass |
|---|---|---|---|---|
| Fuel | | Volt SHE at 298 K | kg/m³ | kg/100 km |
| Lithium | Li+ + e− ↔ Li | 3.05 | 530 | 3.0 |
| Sodium | Na+ + e− ↔ Na | 2.71 | 970 | ? |

### 1:20 W   Green ammonia

The world's annual production of the vital fertilizer ammonia ($NH_3$) is about 170E6 tons. This ammonia is made by reacting nitrogen from air ($N_2+O_2$) with hydrogen from natural gas (methane $CH_4$) in the Haber process {$N_2+3H_2 \rightarrow 2NH_3$}. The hydrogen comes from reforming methane {$CH_4+2H_2O \rightarrow 4H_2+CO_2$} which generates a lot of $CO_2$. To avoid the environmental effects of $CO_2$ it has been proposed that ammonia could be produced from air and water, using "green" electricity for the direct electrochemical reduction of nitrogen[3]

$$N_2 + 3H_2O \rightarrow 2NH_3 + (3/2)O_2$$

***Problem***: Calculate the amounts of $CO_2$ averted and $O_2$ generated if the current thermochemical process was fully replaced by direct electrosynthesis of $NH_3$. [ton/year]
***Assume***: 100% efficiency for all process steps.

# Chapter 2:  Process Variables and Their Relationships

## 2:1 W   Water

A.  Specify the phase(s) of pure water ($H_2O$) in its equilibrium state, with zero liquid curvature, under the following conditions:
NB.  (a) ≡ absolute pressure      (g) ≡ gauge pressure
a.    30 kPa(a), 629.9°R
b.    101.3 kPa(a), 20°F
c.    14.696 psi(g), 20°C
d.    1 kPa(a), 200 K
e.    144.3 kPa(g), 260.6°F
f.    0.611 kPa(a), 273.16 K

B.  How many intensive variables are needed to fix:
a.    The vapour pressure of a liquid solution of methanol in water?
b.    The specific volume of a liquid solution of sodium chloride (NaCl) in water?
c.    The specific enthalpy of a liquid solution of sulphuric acid ($H_2SO_4$) in water?

## 2:2 U   Stoichiometry

Reagents in the following reactions are combined in the initial amounts indicated.
(i) $wH_2SO_4 + xNaOH$    →    $yNa_2SO_4 + zH_2O$
350 kg                          240 kg
(ii) $wC_7H_{16} + xO_2$       →    $yCO_2 + zH_2O$
300 kg                          23,000 ft$^3$STP
(iii) $wCO_2 + xH_2$          →    $yCH_3OH + zH_2O$
112 m$^3$ STP                   60 lb

***Problem***: In each case:
a.  Find values for w, x, y, z and identify the limiting reactant.
b.  Assuming the limiting reactant undergoes 100% conversion with no competing reactions find the conversion of the excess reactant.

## 2:3 W   Bio-gas

Biogas is produced in a digester by the anaerobic enzyme catalysed fermentation of an organic waste $C_xH_yO_zN$, according to the representative stoichiometry:
$$C_xH_yO_zN(s) + wH_2O(l) \rightarrow (q+2)CH_4(g) + (q-2)CO_2(g) + NH_3(aq)$$

---

[3] A very slow reaction. The alternative indirect electro-synthesis with green hydrogen is viable.

where x,y,z,w and q are empirical coefficients in the balanced reaction and the gas product
$(CH_4 + CO_2)$ contains 51.69 volume % $CH_4$

*Problem*: Calculate the mass of $CH_4$ produced by conversion of 1000 kg of the waste. [kg]

## 2:4 W   Combustion

A closed vessel of fixed internal volume = 20.0 m³ contains <u>only</u> an unknown mass of
carbon plus 28.0 kg oxygen. The carbon is ignited. Part of the carbon undergoes combustion to carbon dioxide by reaction 1 and the remainder of the carbon is converted to carbon
monoxide by reaction 2 (i.e. overall, 100% conversion of carbon).

$$C(s) + O_2(g) \quad \rightarrow \quad CO_2(g) \quad\quad \text{rxn 1}$$
$$2C(s) + O_2(g) \quad \rightarrow \quad 2CO(g) \quad\quad \text{rxn 2}$$

When the vessel is then cooled to 91°C the pressure is 50 kPa(g).

*Problem*:
a.   Find the original mass of carbon in the vessel.                                        [kg]
b.   Find the <u>weight percent</u> of carbon monoxide in the reaction product gas mixture.   [wt%]
*Assume*: The gas mixture behaves as an ideal gas.

## 2:5 W   Gas/water equilibrium

A closed vessel of fixed internal volume contains only 28.0 kg $N_2$ plus 5300 kg water.
The pressure and temperature in the vessel are 300 kPa(a), 400 K.

*Problem*: Find the internal volume of the vessel.                                          [m³]
*Assume*: Equilibrium conditions, zero solubility of nitrogen in liquid water, ideal gases.

## 2:6 W   Gas/liquid equilibrium

A liquid mixture of hexane ($C_6H_{14}$ SG = 0.659), nonane ($C_9H_{20}$ SG = 0.718) and toluene
($C_7H_8$ SG = 0.866) has a density of 700 kg/m³ at 4°C. 10 kg of this liquid (only) is placed
in an evacuated sealed vessel of fixed volume = 2.00 m³ then heated to vaporize the mixture. When all the liquid is vaporised, the pressure and temperature in the vessel are 250
kPa(a), 580 K.

*Problem*: Find the composition of the mixture in mole %.                              [mol %]
*Assume*: The liquid is an ideal liquid mixture and the vapour behaves as an ideal gas
         (when no liquid is present).

## 2:7 U   Liquid to vapour

An ideal liquid mixture contains three components: X, Y, Z (data below). 5 m³ of this
liquid mixture at 298 K has a mass of 4726 kg. When 106 ft³ of this liquid mixture is
vaporised completely to gas, the volume is 5235 m³ at 900°R, 456 mm Hg vacuum.

| Pure liquids |        | X    | Y    | Z   |
|--------------|--------|------|------|-----|
| Density 298 K | kg/m³ | 800  | 1100 | 900 |
| Molar mass   | kg/kmol | 30  | 90   | 60  |

*Problem*: Find the composition of the gas mixture.                              [mole % X,Y, Z]

## 2:8 W   Mass transfer

When a reaction involves movement of reactants between phases the speed of reaction can be set by the rate of transport of reactant(s) to a phase boundary. Then the process rate would be under "mass transfer control" and, with a fast interfacial reaction, may be modelled with the equation:                     Rate = K[C]                                    where:
K = mass transfer coefficient       [C] = concentration of reactant in its bulk phase kmol/m$^3$
Mass transfer constraints can affect the design of chemical reactors where reactions occur at or via phase boundaries. For example, the design of fixed-bed electrochemical reactors considers liquid-solid mass transfer coefficients, estimated with the correlation:
$$K = (D/d)(1/\varepsilon)Re^{1/3} Sc^{1/3}$$
D = diffusion coefficient   d = bed particle diameter, $\varepsilon$ = bed porosity       dimensionless
   [Re = Reynold number,   Sc = Schmidt number]                   both dimensionless

***Problem***: Calculate the reaction rate, with appropriate units, for:
   D = 1E–9 m$^2$/s, d = 0.1 cm, $\varepsilon$ = 0.70, Re = 10, Sc = 1.0E6, C = 0.02 molar

## 2:9 U   Heat transfer

The overall heat transfer coefficient in a tubular heat exchanger is given by;
   U = 1/(1/h$_i$ + 1/h$_o$ + t/k)             where:       U = overall heat transfer coefficient
   h$_i$ = inside tube film coefficient                     h$_o$= outside tube film coefficient
   k = tube wall thermal conductivity                     t  = tube wall thickness

***Problem***: If the units U and t are respectively [kW/(m$^2$.K)] and [m], what are the consistent units of h$_i$ , h$_o$ and k?

## 2:10 W   Heat exchanger

The thermal load on a shell and tube heat exchanger is given by:
   $\dot{Q}$  = UA($\Delta$T)           where:         $\dot{Q}$ = thermal load        A = heat transfer area
   $\Delta$T = mean* temperature difference between hot and cold streams (*log mean)

A process engineer in a US owned oil refinery is given the design values listed below and asked to trouble-shoot a poorly performing exchanger in which steam is used on the shell side (outside the tubes) to heat a new process liquid.

| | | | | |
|---|---|---|---|---|
| A = 108 | ft$^2$ | | U  = 150 | BTU/(ft$^2$.h.F$°$) |
| t  = 0.0625 | inches | | h$_o$  = 350 | Consistent units |
| k = 100 | Consistent units | | h$_i$  = 260 | Consistent units |

***Problem***: The engineer makes on-site measurements to find: $\dot{Q}$ = 224 kW, $\Delta$T = 60 K.
      Analyse these data and suggest how to recover performance of the exchanger.

## 2:11 W   Biochemical kinetics

The rate of an enzyme catalysed biochemical reaction is represented by the equation:
$$u = 1/\{K/(VS) + 1/V\}$$

where:

u has the same units as V                    ?                    K = constant          ?

S = concentration of substrate          $kmol/m^3$

**Problem**: What are the consistent units of "K" in this equation?

### 2:12 W    Electrochemical kinetics

The rate of an electrochemical reaction under mass transfer constraint is given by:

$$j = nFkKC(\exp(\alpha\eta F/RT)) / [K + k(\exp(\alpha\eta F/RT))]$$

where: k = rate constant                    ?

α = electrochemical charge                          K = mass transfer

transfer coefficient          ?                              coefficient                    ?

C = reactant concentration   $kmol/m^3$          n = electron stoichiometry

coefficient                    −

F = Faraday's number          kC/kmol          R = universal gas constant   kJ/(kmol.K)

j = current density          $kA/m^2$          η = electrode overvoltage          Volt

**Problem**: What are the consistent units of "α", of "k" and of "K"?

### 2:13 U    Photochemical kinetics

The rate of a photochemical reaction can be described by the equation:

$$dC/dt = (-1)kIC$$

where:   C = reactant concentration          $kmol/m^3$          t = time seconds

I = light intensity                    lux                    k = rate constant ?

**Problem**:

a.    What are the units of "k" ?

b.    What is a "lux" and how does it relate to the "Einstein"?

### 2:14 W    Thermochemical kinetics

The rate of a two-phase thermochemical reaction under mass transfer constraint is given by:

$$\text{Rate} = dC/dt = (-1)(k)\exp(-E/RT)KaC / [(k)\exp(-E/RT)+Ka]$$

where:

a = specific interfacial area for mass transfer          $m^2/m^3$

C = reactant concentration                    $kmol/m^3$          E = activation energy          ?

k = reaction frequency factor                    $s^{-1}$          K = mass transfer coefficient   ?

R = gas constant                    kJ/(kmol.K)   T = temperature                    K

t = time                    ?

The reaction is carried out with a reactant MW = 70 kg/kmol in a continuous stirred tank reactor with a steady reaction rate of −5 lb.mol/(ft$^3$.h) and volume of reacting fluid = 3 m$^3$.

**Problem**:

a.    Find the units of E, K and t.

b.    Calculate the mass of reactant consumed per 24-hour day [ton/day]

## 2.15 U  Photo-voltaic array

The current output from a photo-voltaic array is given by the equation:
$$I = N\{p - q[\exp((V/M + IR/N)/nv)] - 1\} - i$$
where:

| | | | |
|---|---|---|---|
| I = current | | A  N = number of PV modules in parallel in the array | – |
| M = number of cells in series in the module | – | R = series resistance | Ohm |
| p = module photo-current | | A  q = saturation current | A |
| V = open circuit voltage | | V  V = diode terminal voltage | V |
| n = diode ideality factor | | ?  i = shunt current | A |

**Problem:** What are the consistent units of n?

## 2:16 W  Solar thermal energy

The solar receiver efficiency in a solar thermal power plant is given by:
$$E = a - be(T_s^4 - T_a^4)/(Cd)$$
where:

| | | | |
|---|---|---|---|
| E = energy efficiency | – | a = absorbance of absorber | ? |
| b = constant | ? | e = emissivity of absorber | dimensionless |
| $T_s, T_a$ = temperature of solar absorber, ambient air | | K | |
| C = geometric solar concentration ratio | | dimensionless | |
| d = normal irradiance from sun | | $W/m^2$ | |

**Problem:** What are the consistent units "a" and of "b" in this equation?

## 2:17 U  Dimensional consistency

In the design of packed tower gas absorbers the following variables are used to calculate the liquid holdup in the packing:

| | | | |
|---|---|---|---|
| h = liquid holdup | dimensionless | u = liquid viscosity | Pa.s |
| V = liquid superficial velocity | m/s | g = gravitational constant | $m/s^2$ |
| p = liquid density | $kg/m^3$ | d = nominal packing diameter | m |

**Problem:** At least one of the following four holdup correlations a, b, c, d is correct.
    Which are incorrect?
a.  $h = 5.7[up/(gV^2)]^{0.5} + 2.3[gd/(uV)]^2$
b.  $h = 3V[up/(gd)]^{0.67}$
c.  $h = 2.2[uV/(gpd^2)]^{0.33} + 1.8[V^2/(gd)]^{0.50}$
d.  $h = 5[gph/(ud^2)]^{0.33} + 2[uV/(gpd)]^{0.5}$

## 2:18 W  Gas flow

Nitrogen gas at 80 psi(g) and 400°C flows at 500 $lb_m$/minute through a 10-inch ID (internal diameter) orifice.

**Problem:** Calculate the average velocity of the gas in the orifice in m/s.    [m/s]
**Assume:** Ideal gas behaviour, negligible change in pressure or temperature.

## 2:19 U   Orifice meter

The figure below depicts a duct with an orifice meter measuring the flow of combustion gas into a regenerative heat exchanger ("stove") for preheating air into an iron blast furnace.

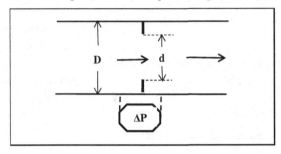

The pressure drop across the orifice is used to calculate the gas flow rate by the equation:

$$F = CA[2\rho\Delta P]^{0.5}$$

where:

| | | | |
|---|---|---|---|
| F | = mass flow rate | C | = dimensionless orifice flow coefficient |
| A | = flow area of duct, corresponding to the larger diameter D | | |
| $\rho$ | = gas density | $\Delta P$ | = pressure drop across the orifice |

***Problem***: Use the data below to calculate:

a.   The volumetric flow and the mean velocity of gas in the duct.          [m³/hour, m/s]
b.   The rate of heat delivery to the stove.                                                    [kW]

***Assume***:   Complete combustion of the gas, zero heat loss, zero condensation.

***Data***:     Gas pressure =1.50 psi(g),                          temperature = 127°C
            D = 2 m      $\Delta P$ = 10 inches of (liquid) water      C = 0.45
            Gas composition, vol%      $N_2$ = 55,      $CO_2$ = 20      CO = 20      $H_2$ = 5
            CO heat of formation = –110 kJ/kmol.                Ideal gases

## 2:20 U   Vaporisation

A liquid fuel mixture is obtained by combining 1.00 m³ liquid methanol ($CH_3OH$) with 2.00 m³ liquid ethanol ($C_2H_5OH$) at 293 K. 7.00 kg of this mixture is placed in a previously evacuated closed vessel of volume = 4.00 m³ and heated to complete vaporisation at 600 K.

***Problem***: Calculate the following.

a.   The weight% of each component in the fuel mixture.                          [wt%]
b.   The pressure in the vessel at 600 K (assume ideal gas phase).          [kPa(a)]
c.   The mole% of methanol in the vapour mixture.                                  [mol%]
d.   The partial volume of ethanol in the vapour at 600 K                          [m³]

***Data***:     Component.                          $CH_3OH(l)$      $C_2H_5OH(l)$
            Density at 293 K      kg/m³          792              789

## 2:21 W   Ignition

Two (2.00) kg of a liquid fuel mixture of methanol and ethanol is injected into a 4.00 m³ closed vessel that initially contains <u>only</u> oxygen gas at 200 kPa(a), 298 K. The vessel is sealed and heated to vaporize the fuel mixture, which is then ignited to cause reaction with oxygen in which the fuel is completely oxidised to carbon dioxide and water. The sealed vessel is then cooled to 350 K, at which point the pressure is 250 kPa(a).    (a) ≡ (abs).

*Problem:*
a.   Find the methanol content (composition) of the original liquid fuel mixture.    [wt%]
b.   Calculate the composition of the final <u>gas-phase</u> in the vessel.    [vol%]
*Assume:* Ideal behaviour of gases and zero solubility of $CO_2$ and $O_2$ in liquid water[4].

## 2:22 W   Steam table

Use the <u>steam table</u> to answer (a) to (f).

*Problem:*
a.   The phase of pure water at:
     (i) 250 kPa(a), 400 K.        (ii) 1 kPa(a), 290 K        (iii) 3000 kPa(a), 400 K
b.   The density of pure *saturated* water vapour at 350 K.    [kg/m³]
c.   The amount of heat input needed to bring 2 kg of pure water from 100 kPa(a), 300 K to 400 kPa(a), 500 K.    [kJ]
d.   The mean heat capacity of pure water at 50,000 kPa(a) over the temperature range 400 K to 500 K.    [kJ/(kg.K)]
e.   The mole percent n-octane in the gas phase in equilibrium with a liquid *emulsion* of n-octane in water at 400 K. *Assume:* n-octane and water are mutually insoluble. [mol%]
f.   The mass of water vapour contained in 7.00 m³ of wet air with 80% relative humidity at 200 kPa(a), 350 K.    [kg]

## 2.23 U   Steam quality

A closed vessel with a total internal volume = 5.00 m³ contains <u>only</u> 10.0 kg of pure water at 400 K.

*Problem:*
a.   Find the pressure in the vessel.    [kPa(a)]
b.   Find the percent of water present in the gas phase (i.e. the "quality").    [%]

## 2:24 W   Methane combustion

Methane ($CH_4$) undergoes complete combustion in stoichiometric oxygen.

*Problem:* Find the dew point temperature of the combustion product gas at 200 kPa(a). [K]
*Assume:* Negligible solubility of $CO_2$ in water.

---

[4] An exact calculation would use Henry's law to account for the small solubility of $CO_2$ and $O_2$ in liquid water.

## 2:25 U   Hydrazine dew and bubble

A vessel with fixed internal volume 10 m$^3$ holds (only) 32 kg hydrazine ($N_2H_4$) and 20 kg $O_2$. The hydrazine is ignited, its combustion produces only $N_2$ and $H_2O$ and consumes all the $O_2$. The temperature of the vessel (contents) is then adjusted to 473 K.

*Problem*: Find —
a.   The equilibrium pressure in the vessel [P(1)] at 473 K.                                   [kPa(a)]
b.   The dew point temperature of gas in the vessel at a pressure 100 kPa above P(1).   [K]
c.   The bubble point pressure of the liquid in the vessel at 373 K.                          [kPa(a)]
*Assume*:
Hydrazine/water form an ideal liquid mixture, ideal gases. Zero solubility of $N_2$ in water. Density of $N_2H_4/H_2O$ liquid mixtures is fixed at 1000 kg/m$^3$
*Data Hydrazine*:
Vapour pressure: $\log_{10}p^* = [5.01105 - 1724.78/(T-41.833)]$   $p^* = $ bar,   T = K   log base 10

## 2.26 W   Boudouard

The standards heats of combustion of carbon [C(s)] and carbon monoxide [CO(g)] are respectively –394E3 and –283E3 kJ/kmol.

*Problem*: Calculate the standard heat of reaction for the Boudouard reaction.   [kJ/kmol C]
$$C(s) + CO_2(g) \rightarrow 2CO(g) \qquad \text{Boudouard}$$
Is this reaction endothermic or exothermic under standard conditions?

## 2:27 U   Fischer–Tropsch

The Fischer–Tropsch process converts CO to hydrocarbons by catalytic thermochemical reduction with $H_2$, according to the stoichiometry:
$$xCO + yH_2 \rightarrow C_nH_{(2n+2)} + zH_2O \qquad \text{Fischer-Tropsch}$$
It is proposed to combine the Boudouard reaction (Problem 2:26) with Fischer–Tropsch, using bio-sourced carbon (C), to convert $CO_2$ to fuels such as octane $C_8H_{18}$.

*Problem*: Given the following (idealised) process specifications calculate the mass of $CO_2$ required in a coupled B-F/T process to produce 1140 ton/day of octane.                   [ton]
Specifications:
Boudouard           C conversion   = 90%,       Selectivity for CO    = 100%.
Fischer–Tropsch    CO conversion = 80%,       Selectivity for octane = 70%.

## 2:28 W   Claus
Elemental sulphur is recovered from "sour gas" in a process based on the Claus reaction.
$$2H_2S(g) + SO_2(g) \rightarrow 3S(s) + 2H_2O(g) \qquad \text{Claus}$$

*Problem*:
a.   Calculate the standard heat of the Claus reaction.                                   [kJ/kmol SO$_2$]
b.   The heats of fusion and of vaporisation of S are respectively 1.7E3 and 9.8E3 kJ/kmol. Calculate the standard heat of the reaction:                               [kJ/kmol SO$_2$]
$$2H_2S(g) + SO_2(g) \rightarrow 3S(g) + 2H_2O(g)$$

**Data:**

| Component | Units | S(s) | H₂(g) | O₂(g) | H₂S(g) |
|---|---|---|---|---|---|
| $h^o_{f\,298K}$ formation | kJ/kmol | 0.0 | 0.0 | 0.0 | Not given |
| $h^o_{c,298K}$ combustion | kJ/kmol | –395E3 | –242E3 (Net) | – | –518E3 (Net) |

Combustion of S: $\quad$ $S(s) + O_2(g)$ $\qquad\rightarrow$ $\quad SO_2(g)$

Combustion of H₂S: $\quad$ $H_2S(g) + (3/2)O_2(g)$ $\quad\rightarrow$ $\quad SO_2(g) + H_2O(g)$

## 2:29 W Raoult and Henry

A closed vessel of total volume = 2.00 m³ contains 10.0 kmol benzene + 2.00 kmol hydrogen all at 350 K.

**Problem:**

a.  Find the equilibrium total pressure in the vessel at 350 K $\hfill$ [kPa(a)]

b.  Find the equilibrium mass of hydrogen dissolved in the liquid phase at 350 K. $\hfill$ [kg]

*Assume:* Liquid volumes are additive. Henry's law applies to the hydrogen and Raoult's law applies to the benzene. This problem requires solution of simultaneous non-linear equations.

**Data:**

Henry's constant for H₂ in liquid benzene (C₆H₆) at 350 K $\quad$ = 0.50E6 kPa

Vapour pressure of benzene at 350 K $\qquad\qquad\qquad\qquad$ = 90.0 kPa(a)

Density of liquid benzene at 350 K $\qquad\qquad\qquad\qquad\quad$ = 830 kg/m³

Molar volume of hydrogen (H₂) in liquid phase at 350 K $\quad$ = 0.050 m³/kmol

NB. Hydrogen dissolves in liquid benzene to give a single-phase liquid.

## 2:30 U Haber-Bosch

A stoichiometric reactant gas mixture of N₂ and H₂ is in a fixed volume sealed vessel at 1.0E4 kPa(a), 300 K. The mixture is heated with a catalyst and undergoes partial conversion to NH₃(g):

$$N_2(g) + xH_2(g) \rightarrow yNH_3(g)$$

to give a pressure of 2.5E4 kPa(a) at 790 K.

**Problem:** Calculate the conversion of nitrogen to ammonia (assume ideal gases). $\hfill$ [%]

## 2:31 W NH₃ to NO

The Haber–Bosch synthesis illustrated in Problem 2:30 produces ammonia used for fertilizer as well as to make nitric acid via catalytic oxidation to nitric oxide, according to:

$$xNH_3 + yO_2 \rightarrow pNO + qH_2O$$

**Problem:** Use the data tabulated below to calculate:

a.  The standard heat of reaction: N₂(g) and H₂(g) to liquid NH₃ (NH₃(l)) [kJ/kmol NH₃]

b.  The dew-point pressure at 300 K of a gas mixture N₂(*30*), H₂(*10*), NH₃(*60*) wt%. $\hfill$ [kPa(a)]

c.  The dew-point temperature under 200 kPa(a) of a product gas mixture from the oxidation of NH₃ with 20% excess (of stoichiometric) dry air and 100% conversion to NO.

| Data Ref. 298 K | | Species > | NH3(g) | NO(g) |
|---|---|---|---|---|
| $h_f$ formation | | kJ/kmol | –46E3 | 90E3 |
| $h_v$ vaporisation | | kJ/kmol | 21E3 | |
| Critical temperature | | K | 406 | 180 |
| Antoine constants for $NH_3$: A = 15.4940 B = 2363.24 | | | | |
| C = –22.6206 ln p* = kPa(a), T = K | | | | |

**Assume:** Ideal behaviour in the gas phase, negligible solubility of NO in liquid water.

### 2:32 U   Biochemical reaction

The rate of an enzyme reaction is given by the Michaelis–Menten equation:
$$r = -(dC/dt) = aC/(K + C)$$

where:

| | | |
|---|---|---|
| r | = reaction rate | $kmol/(m^3.h)$ |
| a | = rate constant | ? |
| C | = reactant ("substrate") concentration | $kmol/m^3$ |
| K | = Michaelis constant | ? |

With reactant concentrations of 0.10 and 0.30 M the reaction rate at 35°C was measured respectively at 0.15 kmol/(m³.h) and 0.18 kmol/(m³.h).

**Problem:** Calculate the values (with units) of "a" and of "K" at 35°C.

### 2:33 U   Azide decomposition

Liquid hydrogen azide ($HN_3$ BP = 309 K) is a toxic primary explosive.

**Problem:**

a.   Write a balanced chemical equation for the decomposition of hydrogen azide to hydrogen and nitrogen.

b.   A mass "Y" of $HN_3$ (only) is sealed in a 5.00 m³ fixed volume vessel and completely decomposed to hydrogen and nitrogen at 200°C. The pressure in the vessel is then 130 psi(g). Calculate the mass "Y".                                                          [kg]

### 2:34 W   Psychrometry

**Problem:**

Use the psychrometric chart in Chapter 2 to find the following.

All at 101.3 kPa(a) total pressure.

a.   Relative humidity of air with dry and wet bulb temperatures respectively 40 and 30°C
                                                                                            [%]

b.   Dew-point temperature of humid air with relative humidity 30% at 40°C.          [°C]

c.   Mass of water condensed when 5 kg of humid air at 40°C dry bulb and 30% relative humidity is cooled to 10°C dry bulb.                                                    [kg]

d.   The amount of heat removed in cooling 5 kg of air with relative humidity of 50% at 40°C from 40°C (dry bulb) to 10°C (dry bulb).                                        [kJ]

## 2:35 W  Gas/liquid, benzene-xylene

A closed vessel of fixed total internal volume $= 3.00$ m$^3$ contains only 1310 kg of a mixture of benzene ($C_6H_6$) and o-xylene ($C_8H_{10}$). At 473 K the pressure in the vessel is 1000 kPa(a).

**Problem:**
a.  Calculate the equilibrium volume of the liquid phase in the vessel at 473 K       [m$^3$]
b.  Find the total mass of benzene in the vessel                                       [kg]
*Assume:* Liquid phase is an ideal liquid mixture and gas phase is an ideal gas mixture.

| Data: | Critical T | Density | Antoine constants | ln, p* = kPa(a), T = K | |
|---|---|---|---|---|---|
| @ 473 K, 1000 kPa(a) | | kg/m$^3$ | A | B | C |
| Benzene | 562 K | 660 | 14.1603 | 2948.78 | –44.5633 |
| o-Xylene | 631 K | 700 | 14.1257 | 3412.02 | –58.6824 |

## 2:36 W  Incomplete combustion

A stream of propane gas ($C_3H_8$) is burned with dry air in an inefficient burner (i.e. poor mixing, with incomplete combustion). The propane undergoes 90% overall conversion in the parallel reactions 1 and 2.

$$C_3H_8 + 5O_2 \rightarrow 3CO_2 + 4H_2O \qquad \text{rxn 1}$$
$$C_3H_8 + 3.5O_2 \rightarrow 3CO + 4H_2O \qquad \text{rxn 2}$$

At steady-state the combustion product gas mixture contains 5.00 vol % oxygen and has a dew point temperature of 340.0 K under 215 kPa(a) total pressure.

**Problem:**
a.  Find the complete analysis of the combustion product gas mixture (wet basis).  [mol%]
b.  Find the selectivity for propane conversion in reaction 1.                       [%]

## 2:37 U  Heat capacity

The heat capacity of methane gas is given by the polynomial:
$$C_p = 19.87 + (5.0E\text{-}2)T + (1.27E\text{-}5)T^2 - (11.00E\text{-}9)T^3 \qquad \text{kJ/(kmol.K)} \qquad T = K$$

**Problem:**
a.  Calculate the mean heat capacity of methane gas for T = 298 to 1000 K   [kJ/(kmol.K)]
b.  The heat of vaporisation of liquid water at its normal boiling point is 40.7E3 kJ/mol
    Calculate the heat of vaporisation of liquid water at its freezing point       [kJ/kmol]
c.  Use data of "b" to estimate the vapour pressure of water at 273 K             [kPa(a)]

## 2:38 W  Heat of mixing

**Problem:** Use the $H_2SO_4$ enthalpy-concentration (h-C) diagram to answer the following:
a.  A 4000 kg batch of sulphuric acid/water mixture with 90 wt% $H_2SO_4$ at 40°C is mixed under <u>adiabatic</u> conditions at 101.3 kPa(a) pressure with 14000 kg liquid water at 20°C. Find the final (equilibrium) temperature of the mixture.                       [K]
b.  When 4000 kg of a sulphuric acid-water mixture at 20°C is mixed with 6000 kg of liquid water initially at 40°C, under adiabatic conditions and 101.3 kPa(a) pressure, the

temperature of the final mixture rises to (equilibrium) 100°C. How much heat would be exchanged with the mixture to maintain <u>isothermal</u> conditions for mixing at 20°C.

[kJ, heating or cooling]

No work is done in either the adiabatic or the isothermal mixing process.

Part "**b**" requires a numerical or a trial-and-error solution of simultaneous equations.

An approximate result (e.g. +/− 5% error) is satisfactory.

### 2:39 U   $H_2/O_2$ conversion

A sealed vessel contains a mixture of $H_2$ and $O_2$ gases at 298 K, 100 kPa(a) with a $H_2/O_2$ mole ratio above 2. The mixture is ignited and the limiting reactant is 100% consumed to form $H_2O$. At 500 K the water is all vaporised and the total pressure in the vessel is 126 kPa(a).

***Problem***: Calculate the $H_2/O_2$ mole ratio in the original gas mixture and the conversion of the non-limiting reactant.                                                                     [–, %]

***Assume***:  Ideal gas behavior of all gases.

### 2:40 W   Nylon and $N_2O$

Electrosynthesis of adiponitrile $(C_4H_6(CN)_2)$ by reaction 1 is a step in the production of Nylon.

$$xC_2H_3CN + yH_2O \rightarrow C_4H_8(CN)_2 + zO_2 \qquad \text{rxn 1}$$

Secondary cathode reactions (e.g. reaction 2) lower selectivity for $C_4H_8(CN)_2$ in reaction 1.

$$pC_2H_3CN + qH_2O \rightarrow C_2H_5CN + wO_2 \qquad \text{rxn 2}$$

After subsequent thermochemical process steps 90 wt% of the $C_4H_8(CN)_2$ from its electrosynthesis is converted to Nylon polymer via the <u>net</u> reaction*:

$$(CH_2)_6(NH_2)_2 + C_4H_8(CN)_2 + 2\{O_2\} \rightarrow H_2O + [C_{12}H_{22}N_2O_2]_n + \{N_2O\} \qquad \text{rxn 3}$$

The conversion of $C_2H_3CN$ in reaction 1 is 95% with 70% selectivity for $C_4H_8(CN)_2$.

**With the intermediate oxidation of adiponitrile to adipic acid.*

### Problem:
a.  Calculate the mass of $C_2H_3CN$ required to make 100E3 tons of Nylon.          [ton]
b.  Find the corresponding mass of nitrous oxide generated with 100% yield in reaction 3.

[ton]

## Chapter 3:   Material and Energy Balances in Process Engineering[5]

### 3:1W   DAC absorption

The figure below is a simplified conceptual flowsheet of a continuous thermochemical process operating at steady-state for the direct air capture of $CO_2$. In this DAC process $CO_2$

---

[5] The data and costs for Chapter 3 are approximate values, used for this exercise. Money is US$ 2022.

is absorbed into a solution of potassium hydroxide to form potassium carbonate by reaction 1, then the hydroxide is regenerated and $CO_2$ recovered in a reaction cycle with the net result of reaction 2:

$$CO_2(g) + 2KOH(aq) \rightarrow K_2CO_3(aq) + H_2O(l) \qquad \text{rxn 1}$$
$$K_2CO_3(aq) + H_2O(l) \rightarrow 2KOH(aq) + CO_2(g) \qquad \text{rxn 2}$$

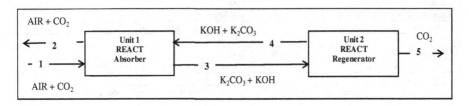

The plant is expected to recover 1000 ton/day $CO_2$ (stream 5) from atmospheric air containing 450 ppm(vol) $CO_2$.

**Problem**: Use the information below to calculate:
a.   The flow rate of air in stream 1                                                    [$Sm^3/h$]
b.   The flow rate of stream 4, i.e. [KOH+$K_2CO_3$](aq), with zero $H_2O$ in stream 5   [$m^3/h$]
c.   The simple ROI for this plant                                                       [%/y]
**Data:**   $CO_2$ in Stream 1 = 450 ppm(v)      $CO_2$ separation efficiency in Unit 1 = 70%
Total potassium [$K^+$] in streams 3,4 = 2 M      [$OH^-$]/[$CO_3^=$] ratio Str. 3 = 1, Str. 4 = 2 M/M
Plant capital cost = 450E6 $ (2022)      Operating cost = 40 $/ton $CO_2$
Value of recovered $CO_2$ (stream 5) = $300 / ton. Interest 5%/y Operating 350, 24 h days/year

### 3:2 U   Syngas

Synthesis gas (CO + $H_2$) can be obtained by thermal treatment of bio-mass or fossil sources of carbon, according to the net reaction {$C_xH_y + xH_2O \rightarrow xCO + (x+y/2)H_2$}, an example of which is steam reforming of methane to produce gas for the synthesis of methanol.

$CH_4 + H_2O \rightarrow CO + 3H_2, T \approx 1200$ K            $(-H_2) >>> CO + 2H_2 \rightarrow CH_3OH$
Heat for the endothermic reforming reaction usually comes from burning methane, with attendant emissions of $CO_2$. Alternatively, it is possible to produce syngas with the correct [2/1] $H_2$/CO ratio by direct electro-reduction of $CO_2$ at T < 400 K, according to:

$CO_2 + 2H_2O$ ---- $6F \rightarrow CO + 2H_2 + (1.5)O_2$      F = Faraday = 96485 Coulomb/mole

**Problem**: Use the data and assumptions given below to calculate and compare the simple return on investment (ROI) for the thermochemical and the electrochemical plants making Syngas to produce 100 ton/day of methanol.            [%/y]

***Data*:**

| THERMOCHEMICAL PLANT | | *All processes 100 % yield, 100 % separation efficiency* |
|---|---|---|
| Carbon tax | | (+) 300 \$/ton $CO_2$ emitted.  Paid by plant |
| Methane consumed for heating | | (1.5) times to match heat of the $CH_4$ reforming reaction |
| CO heat of formation | | −110E3 kJ/kmol |
| Cost of methane | | 500 \$/ton |
| Value of excess hydrogen | | 3000 \$/ton |
| Value of $H_2$ + CO in 2/1 ratio | | 400 \$/ton CO |
| Operating | | 350 day/year, 24 hour/day |
| | | |
| Capital cost | (CAPEX) | 20E6 \$ |
| Labour, maintenance, utilities | (OPEX) | 6% of CAPEX / year  (except carbon tax) |
| Interest rate | | 5 % / year |
| *NB. All values are approximate – for this exercise.* | | Continuous process, steady state |

| ELECTROCHEMICAL PLANT | | *All processes 100 % yield, 100 % separation efficiency* |
|---|---|---|
| Carbon tax | | (–) 300 \$/ton $CO_2$ consumed.  Earned by plant |
| Methane consumed for heating | | Zero |
| Electricity consumed | | 6 Faraday/(mole CO) at 3 Volt |
| Cathode current density (CD) | | 3 kA/m$^2$                               (geometric CD) |
| Electricity cost | | 0.05 \$/kWh |
| Value of $H_2$+CO in 2/1 ratio | | 400 \$/ton CO |
| Value of $O_2$ | | 50 \$/ton $O_2$ |
| Operating | | 350 day/year, 24 hour/day |
| Capital cost | (CAPEX) | 20E3 \$/m$^2$ installed cathode area     (geometric area) |
| Labour, maintenance, utilities | (OPEX) | 6 % of CAPEX / year  (except electricity) |
| Interest rate | | 5 % / year |
| | | CO+$H_2$ from cathode          $O_2$ separate, from anode |
| *NB. All values are approximate values - for this exercise* Continuous process, steady-state | | |

## 3:3 W    Salt, caustic and muriatic acid

The figure below is a simplified conceptual flowsheet and stream table representing a method for producing sodium hydroxide and hydrogen chloride from salt and water. HCl is dissolved in water to give hydrochloric (muriatic) acid.

$$2NaCl + 2H_2O - 2F^* \rightarrow 2NaOH + Cl_2 + H_2 \qquad \text{rxn 1 Unit 1}$$
$$Cl_2 + H_2 \rightarrow 2HCl \qquad \text{rxn 2 Unit 2}$$

*2F indicates electrochemical reaction requiring 2 Faraday of electric charge per mole $H_2$.

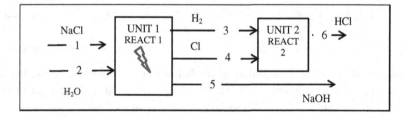

| Stream Table | | Salt, caustic and muriatic acid | | | Continuous process | | Steady state | |
|---|---|---|---|---|---|---|---|---|
| Species | Value | M | | | | Stream | | |
| | $/kg | kg/kmol | 1 | 2 | 3 | 4 | 5 | 6 |
| | | | | | Flow kmol/h | | | |
| NaCl | 0.1 | 58.5 | 114.2 | 0.0 | 0.0 | 0.0 | 0.0 | 0.0 |
| NaOH | 1.0 | 40.0 | 0.0 | 0.0 | 0.0 | 0.0 | ? | 0.0 |
| H2O | 0.01 | 18.0 | 0.0 | ? | 0.0 | 0.0 | 590.2 | 0.0 |
| Cl2 | 0.5 | 71.0 | 0.0 | 0.0 | 0.0 | 57.1 | 0.0 | 0.0 |
| H2 | 4.0 | 2.0 | 0.0 | 0.0 | 57.1 | 0.0 | 0.0 | 0.0 |
| HCl | 0.6 | 36.5 | 0.0 | 0.0 | 0.0 | 0.0 | 0.0 | ? |
| Total | | Kg/h | 6678 | ? | 114 | 4053 | ? | ? |
| Energy kWh/kg NaOH | | | Electrolysis | 4.0 | Pumping, heating, etc. | | 1.0 | |
| Utilities | Electricity | | | | | | | |
| $/kWh | 0.05 | | | | Labour+maintenance: | | 7% of Capital/year | |
| | Capital cost of plant US$2022 = 40E6 | | | | Interest | 6%/y | | |

**Problem**:
a.  Complete the stream table.
b.  Calculate the gross economic potential, net economic potential and simple return on investment (ROI) for this plant.                                    [$/y, $/y, %/y].

## 3:4 U   Green steel

In the production of "green steel" green hydrogen ($H_2$) is used in a vertical shaft furnace replacing the carbon (C) used in a blast furnace to reduce iron ore ($Fe_2O_3$) to iron (Fe). Iron ore and iron are valued respectively at 300 and 1500 $/ton and the prices of hydrogen and carbon (in coke) are 4 and 0.6 $/kg. For a plant producing 1000 ton/day of iron the capital cost of a vertical shaft furnace ($H_2$) and a blast furnace (C) are respectively $300E6 and $200E6. Both plants have the same proportional costs for utilities (4%), labour plus maintenance (5%) and interest (6%) of capital cost (CAPEX) per year, 365 days/year. Water has negligible value. A carbon tax on $CO_2$ produced by the (carbon) blast furnace would give a corresponding cost to its $CO_2$ emissions.

**Problem**: Considering only the basic reaction stoichiometry, all with 100% conversion, at what level of carbon tax would the return on investment for direct reduction by hydrogen equal that for the conventional carbon process?                     [$/t $CO_2$]

## 3:5 W   Peroxide bleach

A paper mill uses hydrogen peroxide ($H_2O_2$) to bleach wood pulp. The mill produces 500 ton/day of bleached pulp with a peroxide "application" of 20 kg of $H_2O_2$ per ton of pulp. The mill has an option of purchasing a 30 wt% solution of $H_2O_2$ in water for $300 per ton or manufacturing peroxide on-site by electrochemical reduction of $O_2$ in the net reaction:

$$O_2 + 2H_2O \rightarrow 2H_2O_2$$

A fully automated on-site electrochemical plant producing 10 ton/day $H_2O_2$ would have an installed cost (CAPEX) of $7.0E6. The combined cost of [operating labour, maintenance (OPEX) and interest on dept] would amount to 9% of CAPEX per year. Electricity could be purchased for 0.05 $/kWh and compressed oxygen (not air) for 50 $/ton. Electricity consumption is 3.0 kWh/kg $H_2O_2$ (100% $H_2O_2$ basis) and oxygen conversion to $H_2O_2$ is 40%.

**Problem:**
a. Compare the cost of on-site $H_2O_2$ to the cost of purchased $H_2O_2$        [$/ton of pulp]
b. How many years operation of an on-site plant would be needed to pay down the original dept of $7.0E6?                                                      [yr]

# Chapter 4:   Material Balances

### 4:1 W   Steam trap

The figure below shows a steam-trap collecting and condensing steam from a heat exchanger.

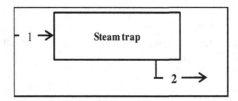

Specifications, continuous operation at steady-state:

| Str. 1 | 70 m³/h | 931.5 kPa(a), 450 K |
|--------|---------|---------------------|
| Str. 2 | 1000 kg/h | 900 kPa(a), 400 K |

**Problem:** Use the steam table to find:
a. The quality of stream 1.                                                      [%]
b. The rate of heat loss from the trap to its surroundings (usually ambient air).   [kW]

### 4:2 U   Alkane distillation

A distillation column in an oil refinery separates products according to their relative volatility. The figure below denotes a column treating a hypothetical crude mixture of alkanes ($C_nH_{2n+2}$) with the following composition, at a rate of 10 ton/hour.

| Stream |  | 1 | 2 | 3 | 4 | 5 | 6 |
|--------|--|---|---|---|---|---|---|
| | | | | | | | |
| n | | 1–30 | 1–3 | 4–10 | 11–16 | ? | 17–30 |
| wt% | | 100 | 10 | 40 | 30 | ? | 20 |
| mol% | | ? | ? | ? | ? | ? | ? |
| M mean | kg/kmol | ? | 30 | 88 | 200 | ? | 330 |
| Phase | | L | G | L | L | G | L |

*(Alkane distillation)*

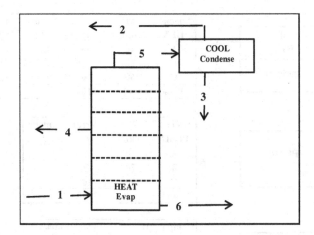

**Problem**: Complete the "Alkane distillation" table above and calculate —
a.  The volumetric flow rate of stream 5 at 7 psi(g), 270°F                                    [m³/h]
b.  The mass of $CO_2$ produced by flaring (a.k.a. burning) stream 2.                     [kg/day]
*Assume*: Continuous operation, steady-state, ideal gases.

### 4:3 U   Generic process units

*Problem*:
For each of the following flowsheets, a, b, c, d, e and f:
Write steady-state mole balance for each species in the process, plus the total mass balance
and any subsidiary relations; solve the balances and complete the stream table.
[NB. 'a to f' are *fully-specified*].

a.

| Species | M | Stream | kmol/h | |
|---|---|---|---|---|
| DIVIDE | kg/kmol | *1* | *2* | *3* |
| A | 40 | 40 | | |
| B | 70 | 60 | | 45 |
| C | 60 | | 5 | |
| Total | kg/h | 7000 | | |

Process unit: **a DIVIDE**, streams 1 → 2, 3

b.

| Species | M | Stream | kmol/h | |
|---|---|---|---|---|
| MIX | kg/kmol | *1* | *2* | *3* |
| A | 40 | 40 | | |
| B | 70 | | 30 | 45 |
| C | 60 | 10 | 15 | |
| Total | kg/h | | 3800 | |

Process unit: **b MIX**, streams 1, 2 → 3

c.

| Species | M | Stream | kmol/h | |
|---|---|---|---|---|
| SEPARATE | kg/kmol | *1* | *2* | *3* 4 |
| A | 40 | | | 4 |
| B | 70 | | 5 | |
| C | 60 | 45 | | 15 |
| Total kg/h | | | | |

Process unit: **c SEPARATE** s(2,A)= 0.8  s(3,B) = 0.9, streams 1 → 2, 3

**d.**

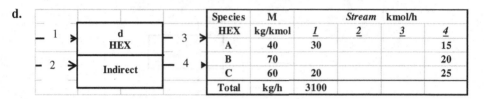

| Species | M | | Stream | kmol/h | |
|---|---|---|---|---|---|
| HEX | kg/kmol | *1* | *2* | *3* | *4* |
| A | 40 | 30 | | | 15 |
| B | 70 | | | | 20 |
| C | 60 | 20 | | | 25 |
| Total | kg/h | 3100 | | | |

**e.**

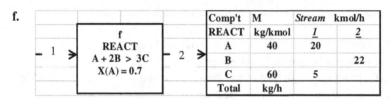

| Species | M | Stream | kmol/h |
|---|---|---|---|
| PUMP | kg/kmol | *1* | *2* |
| A | 40 | 15 | |
| B | 70 | | 20 |
| C | 60 | 10 | |
| Total | kg/h | 2600 | |

**f.**

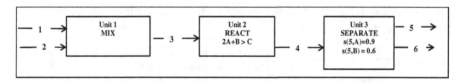

| Comp't | M | Stream | kmol/h |
|---|---|---|---|
| REACT | kg/kmol | *1* | *2* |
| A | 40 | 20 | |
| B | | | 22 |
| C | 60 | 5 | |
| Total | kg/h | | |

## 4:4 W   Generic material balance

The flowsheet and specifications below are for a generic continuous process at steady-state.

| | Unit 1 MIX | | Unit 2 REACT 2A+B > C | | Unit 3 SEPARATE s(5,A)=0.9 s(5,B) = 0.6 | |
|---|---|---|---|---|---|---|
| — 1 → — 2 → | | — 3 → | | — 4 → | | — 5 → — 6 → |

The specifications are:   M(A) = 25, M(B) = 40 kg/kmol.       Conversion of A = 60%
Str. 1 = 50 kmol/h pure A.       Str. 2 = pure B         Str. 4: Total mass flow = 2450 kg/h.
Str. 6 = 50 mol % C.

*Problem*:
a.   Write the material balance equations and subsidiary relations for each process unit.
b.   Solve the balances manually and complete the material balance stream table.

## 4:5 U   Generic balance specification

Each of the following three material balance problems may be under-specified, fully-specified or over-specified. All are continuous processes at steady-state.

*Problem*:
a.   In each case find the degrees of freedom (D of F) of the material balance problem.
b.   Complete the material stream table for any case that is either fully-specified or redundantly over-specified

**a.**

| Species | M kg/kmol | Stream (kmol/h) 1 | 2 | 3 |
|---------|-----------|------|-----|---|
| DIVIDE — 1 → | 2 → / 3 → | | | |
| A | 40 | 40 | 10 | |
| B | 70 | 60 | | |
| C | 60 | | | |
| Total | kg/h | | 2650 | |

**b.**

| Species | M kg/kmol | Stream (kmol/h) 1 | 2 | 3 |
|---------|-----------|------|---|----|
| SEPARATE s(2,A) = 0.8 — 1 → | 2 → / 3 → | | | |
| A | 40 | 50 | | |
| B | 70 | | | 45 |
| C | 60 | | 30 | |
| Total | kg/h | | | 4750 |

**c.**

| Species | M kg/kmol | Stream (kmol/h) 1 | 2 |
|---------|-----------|------|----|
| REACT A + 2B > 3C X(A) = 0.7 — 1 → | 2 → | | |
| A | 40 | 20 | |
| B | 70 | | 30 |
| C | 60 | 5 | |
| Total | kg/h | | 5300 |

## 4:6 W   Recycle material balance

Below is a flowsheet for a generic recycle process in continuous operation at steady-state

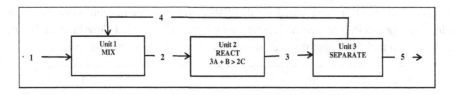

The process specifications are as follows.   M(A) = 30.0      M(B) = 50.0 kg/kmol
Stream 1:   100 kmol/h.      Mixture of [75.0 mol% A + 25.0 mol% B].
Conversion of A per pass in Unit 2 = X(A)
$$X(A) = 0.60[x(2,B)] \times (2,B) = \text{mole fraction B in stream 2}$$
Split fractions in Unit 3:   s(4,A) = 0.99      s(4,B) = 0.98      s(4,C) = 0.10

### Problem:
a.   Write the relevant material balance equations and subsidiary relations needed to complete the material balance stream table for this flowsheet.
b.   Use a spreadsheet (Excel) to set up and solve this material balance problem by the iterative sequential modular method.
   Close the unit and overall mass balances and report the net (overall) yield of C from A.
                                                                             [%]

## 4:7 W   Recycle balance with purge

The figure below is a flowsheet for a continuous process in which an unreactive "non-process" species (N) is caught and accumulates in the reactor recycle loop. The concentration of N in the recycle loop is controlled by the purge stream 6 from the divider Unit 3.

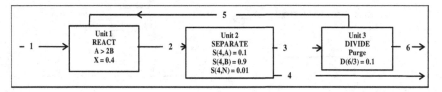

Specifications: Continuous process at steady-state. No thermal effects.

| Reaction A → 2B | Conversion of A per pass = 40% |
|---|---|
| Str. 1 | 100 kmol/h A + 1 kmol/h N |
| Separator fractions | s(4,A) = 0.1, s(4,B) = 0.9, s(4,N) = 0.01 |
| Divider ratio | D(6/3) = 0.1 Purge |

### Problem:
a. Write the relevant material balance equations and subsidiary relations needed to complete the material balance stream table for this flowsheet.
b. Use a spreadsheet (Excel) to set up and solve this material balance problem by the <u>iterative sequential modular method</u>. Ensure closure of all mass balances.
c. Report the concentration of N in stream 5 and net (overall) yield of B from A. [wt%,%]

### 4:8 W   Bio-ethanol

Bio-ethanol ($C_2H_5OH$) is obtained by distillation from fermented bio-mass (e.g. corn, hydrolysed lignocelluloses, etc.). The figure below is a simplified flowsheet for the production of ethanol from glucose in a "wort" of digested and pre-treated corn.

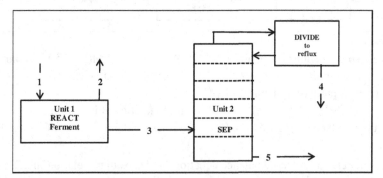

Specifications: Continuous process at steady-state.

| Str. 1 | Wort (glucose 10 wt% solution in water) + enzymes, | liquid | 120 kPa(a), 308 K | 10 ton/hour |
|---|---|---|---|---|
| Str. 2 | $CO_2$, $H_2O$ and $C_2H_5OH$ (EtOH), | gas | 110 kPa(a), 308 K | |
| Str. 3 | Beer — solution of $C_2H_5OH$ in water | liquid | 110 kPa(a), 308 K | |
| Str. 4 | Ethanol 190 proof ≡ 95 vol% $C_2H_5OH$, | liquid | 100 kPa(a), 293 K | |
| Str. 5 | Bottoms, $H_2O$ + $C_6H_{12}O_6$ + $C_2H_5OH$ | liquid | 100 kPa(a), 373 K | |
| Unit 1 Glucose conversion = 95%, EtOH yield = 100% | | | Unit 2 EtOH sep efficiency = 96% | |

Fermentation reaction:     $xC_6H_{12}O_6 \rightarrow yC_2H_5OH + zCO_2$

**Problem:**
a.   Calculate the mass percent (wt%) and mole% of $C_2H_5OH$ in stream 4.
b.   Set up a stream table for the process — streams 1 to 5, with closed mass balances.
*Assume:*  Unit 1 Reactor is well mixed.
           $H_2O$ and $C_2H_5OH$ form an ideal liquid mixture (not true in reality).
*Data:*    SG $C_2H_5OH = 0.789$ at 20°C.
           Antoine equation   $\log_{10}(p^*) = A - (B/(T+C))$   base 10
           $C_2H_5OH$ Antoine constants: A = 5.24687, B = 1598.67, C = –47.424
           $p^* = $ bar, $T = $ K

**4:9 U   Cat cracker**

The figure below is a simplified flowsheet for a catalytic cracker ("cat cracker") in a process used by the oil industry to convert high molecular weight hydrocarbons to fuels such as octane.

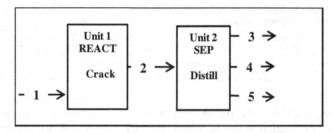

In this process the feed "crude" oil is split with the help of a sold catalyst at temperature around 800 K to produce relatively low molecular weight alkanes and alkenes. For this idealised case the cracker is fed a mixture of $C_{20}H_{42}$ and $C_{18}H_{38}$ (only) that react as follows:

$$C_{20}H_{42} \rightarrow C_8H_{18} + xC_2H_4 \qquad \text{rxn 1}$$
$$C_{18}H_{38} \rightarrow C_8H_{18} + yC_2H_4 \qquad \text{rxn 2}$$

The process specifications are: Continuous process, steady-state.
"Crude" feed rate (stream 1) = 100 metric ton/hour, 60 wt% $C_{20}H_{42}$
Feed conversion:   Reaction 1 = 80%        Reaction 2 = 90%
Unit 2 split fractions:   $[C_{20}H_{42} + C_{18}H_{38}] \equiv A$      $C_8H_{18} \equiv B$      $C_2H_4 \equiv C$
           s(5,A) = 1.0      s(5,B) = s(5,C) = 0      s(4,B) = 0.99      s(3,C) = 0.98

| Str. 1 | "Crude" mixture 60 wt% $C_{20}H_{42}$ | Str. 4 | "Naphtha" medium MW product |
| --- | --- | --- | --- |
| Str. 2 | Cracker product hydrocarbon mixture | Str. 5 | "Heavy" high MW product |
| Str. 3 | "Light" low MW product | | |

**Problem:** Set up a stream table for this process with closed mass balances and calculate the yield of "naphtha" as US gallons octane per ton of feed.   SG octane = 0.70
                                                                          [US gal/ton]

## 4:10 W   Oil burner

The figure below shows a continuous steady-state process where fuel oil undergoes complete combustion in stoichiometric dry air to give a combustion product gas which is compressed, cooled in an indirect heat exchanger and separated into liquid and gas products.

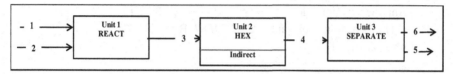

Specifications: NB Assumes zero pressure drop through the system (not true in reality).

| Str. 1 | Fuel oil — Empirical formula $C_7H_8$ | 920 kg/h |
|---|---|---|
| Str. 2 | Stoichiometric dry air | ? |
| Str. 3 | Combustion product $CO_2$, $H_2O$, $N_2$ | ? |
| Str. 4 | Cooled gas-liquid mixture | 400 kPa(a), T(4) = 320 K |
| Str. 5 | Product liquid | 400 kPa(a), T(5) unspecified |
| Str. 6 | Product gas | 400 kPa(a), T(6) unspecified |
| | Zero heat transfer to or from Unit 3 | |

*Problem:*
a. Write the equations needed to find the flows and compositions of streams 1 to 6.
b. Check the specification of this system. Is it under, fully or over specified?
c. Prepare a corresponding stream table showing closure of the mass balances on each process unit and on the overall process, including $N_2$ and $CO_2$ dissolved in the liquid phase.

*Assume:* Material and thermal equilibrium between liquid and gas phases in the separator.
*Data:*    Henry's constant in water at 320 K.        $N_2 = 15E6$ kPa $CO_2 = 4E5$ kPa
          Dry air $N_2/O_2 = 79/21$ by volume.

## 4:11 U   Sulphuric acid

The following is a simplified description of a continuous industrial process for production of sulphuric acid ($H_2SO_4$) from sulphur (S), oxygen ($O_2$) and water ($H_2O$). Fresh feeds of sulphur (Str.1 = 40.0 kmol/h) and excess oxygen (Str.2) pass into a reactor (Unit 1) where the sulphur is burned to sulphur dioxide by reaction 1:

$$S + O_2 \rightarrow SO_2 \qquad\qquad \text{rxn 1}$$

The product from Unit 1 (Str. 3) then goes to a second reactor (Unit 2) where $SO_2$ reacts over a catalyst with the remaining oxygen (from Unit 1) to give sulphur trioxide by reaction 2.

$$2SO_2 + O_2 \rightarrow 2SO_3 \qquad\qquad \text{rxn 2}$$

The product from Unit 2 (Str. 4) goes to an absorber (Unit 3) where the $SO_3$ reacts with water (Str. 5) to give sulphuric acid (Str. 6).

$$SO_3 + H_2O \rightarrow H_2SO_4 \qquad\qquad \text{rxn 3}$$

**Problem:**

a. Draw a conceptual flowsheet of this process with units 1,2 and 3. Name and number the process units and streams.

b. Write material balance equations for each process unit, with components designated: $S \equiv A$, $O_2 \equiv B$, $SO_2 \equiv C$, $SO_3 \equiv D$, $H_2O \equiv E$, $H_2SO_4 \equiv F$

c. Set up a quantitative stream table material balance showing the components in each process stream with their respective mole flow rates       [kmol/h]
   Ensure closure of mass balances on each process unit and on the overall process [kg/h]

*Assume:* Steady-state operation

100% conversion of S, $SO_2$ and $SO_3$, respectively in units 1, 2 and 3

Reactants in <u>stoichiometric proportions</u> in Unit 2 and Unit 3.

## 4.12 W   Antibiotic extraction

The figure below is a simplified version of a flowsheet for a solvent extraction "mixer-settler" process for the commercial production of an antibiotic "A" with MW = 350 kg/kmol.

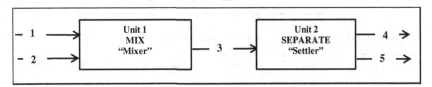

The streams are specified as follows: Continuous operation, steady-state

| Str. 1 | 7000 kg/h | Solution of "A" in water (fermentation "broth") | liquid |
|---|---|---|---|
| Str. 2 | | Pure solvent "S" MW = 120 kg/kmol | liquid |
| Str. 3 | | 2-phase mixture of streams 1 and 2 | liquid/liquid |
| Str. 4 | | Solvent "S" with recovered "A" in solution. Zero water | liquid |
| Str. 5 | | Water with residual "A" in solution. Zero solvent | liquid |

The equilibrium liquid-liquid distribution coefficient D(A) for "A" between solvent "S" and the water "W" is:

$$D(A) = x(A)_S / x(A)_W = 30$$
$$x(A)_S = \text{mole fraction A in solvent} \quad -$$
$$x(A)_W = \text{mole fraction A in water} \quad -$$

**Problem:**

a. Is this material balance problem as defined above: under, fully, or over-specified?

b. Derive an equation that relates the equilibrium mole fraction of A in stream 4 [x(4,A)] to the mass fraction of A in stream 1 [w(1,A)] and the solvent molar feed flow [ṅ(2,S)]. This equation does not need to be explicit in x(4,A).

c. Use a spreadsheet to make a <u>3-D surface plot</u> of the split-fraction (as %) for recovery of A in the overall process [s(4,A) = ṅ(4,A)/ṅ(1,A)] as a function of the concentration of A in stream 1 (in gram/litre) and the flow of S in stream 2 (in kg/h), for the factorial values (i.e. 3 × 3 = 9 points on surface):
   Concentration of A in stream 1:   1         10      20      gram/litre
   Flow of solvent in stream 2:      1000   5000   9000    kg/h

*Assume:* Isothermal conditions. Density of stream 1 is constant at 1000 kg/m³.

Unit 2 behaves as an ideal equilibrium separator.

### 4:13 U   Isothermal flash

The figure below is a flowsheet with specifications for an isothermal flash separation of benzene, toluene and nitrogen. The feed mixture (stream 1) is throttled to drop the pressure and the separator is heated to a controlled temperature $T(2) = T(3) = 380$ K.

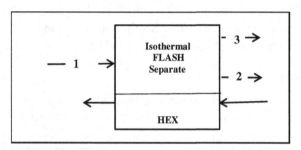

Process specifications: Continuous process, steady-state.

| Str. 1 | 50 kmol/h benzene + 30 kmol/h toluene + 20 kmol/h nitrogen | 300 K, 300 kPa(a) |
|--------|-----------------------------------------------------------|-------------------|
| Str. 2 | Liquid (benzene + toluene + nitrogen) | 380 K, 200 kPa(a) |
| Str. 3 | Gas (benzene + toluene + nitrogen) | 380 K, 200 kPa(a) |

*Problem:*
a.   What are the phases present in stream 1?
b.   Is the system mass balance under-specified, fully specified or over-specified?
c.   Find the composition and flows of streams 2 and 3 with a complete stream table.

*Assume:* Benzene/toluene is an ideal mixture with streams 2, 3 in vapour-liquid equilibrium.

*Data:*

| Antoine constants | A | B | C | ln, kPa(a), K |
|---|---|---|---|---|
| Benzene | 14.1603 | 2948.79 | –44.5633 | |
| Nitrogen | 13.4477 | 658.22 | –2.854 | |

Henry's constant for $N_2$ in benzene/toluene at 380 K = 1E5 kPa

### 4:14 W   Vertical farm

Plant growth in traditional farms depends on the supply of $CO_2$ and $O_2$ from the air, which in 2022 contained 420 ppm(v) $CO_2$ and 20.9 vol% $O_2$. Enclosed "vertical farms" maintain the $CO_2$ level up to 1000 ppm(v) by burning fossil fuel. It is proposed to replace that fossil fuel with $CO_2$ captured directly from the atmosphere. The figure below is a conceptual flowsheet for a process that collects $CO_2$ from air while producing $H_2$, $O_2$ and $CO_2$ (Unit 1) for use in a vertical farm (Unit 2).

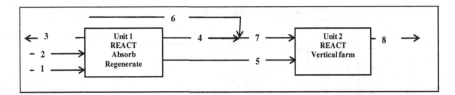

Specifications: Continuous process, steady-state.

| Str. 1 | Air from atmosphere, | gas | 101 kPa(a), 25°C, 420 ppm(v) $CO_2$, 60% RH | |
| Str. 2 | Water | liquid | 101 kPa(a), 25°C | |
| Str. 3 | Air to atmosphere | gas | 101 kPa(a), 25°C, 200 ppm(v) $CO_2$, 60% RH | |
| Str. 4 | $H_2$ | gas | 101 kPa(a), 50°C, 100% RH | $H_2/CO_2 = 3$ v/v |
| Str. 5 | $O_2 + CO_2$ | gas | 101 kPa(a), 50°C, 100% RH | $O_2/CO_2 = 2$ v/v |
| Str. 6 | Air from atmosphere | gas | 101 kPa(a), 25°C, 420 ppm(v) $CO_2$, 60% RH | |
| Str. 7 | Warm air | gas | 101 kPa(a), T(7), 420 ppm(v) $CO_2$ | |
| Str. 8 | Air to atmosphere | gas | 101 kPa(a), 25°C, 1000 ppm(v) $CO_2$ | |

Process Unit 1 takes $CO_2$ from air with water to produce $H_2$ and a mixture of $O_2 + CO_2$ with $H_2/CO_2$ and $O_2/CO_2$ volume ratios respectively 3/1 and 2/1 at a cost of 1 $/kg $CO_2$. The vertical farm with a well-mixed free space volume of 1E4 m$^3$ grows 1000 kg/day of tomatoes. Consumption of $CO_2$ by the tomatoes is 0.01 kg/(kg/day) and the farm is ventilated at 1 volume turnover per hour with fresh air containing 420 ppm(v) $CO_2$ (stream 6).

**Problem:** Write the relevant equations and set up a material balance stream table to find:

a.  The rate and cost of $CO_2$ generation in Unit 1 needed to maintain the (average) $CO_2$ in the farm at 1000 ppm(v).                        [kg/h, $/kg tomatoes]
b.  The change in $O_2$ level in the farm corresponding to (a).              [ppm(v)]
c.  The increase in air temperature entering the farm ($\Delta T$) if preheated by burning all $H_2$ from stream 4 with stream 6 *in situ*.                        [K]

**Assume:** Steady-state operation. Negligible consumption or generation of $O_2$ in farm.

### 4:15 U  DAC adsorption

Carbon dioxide can be recovered from air by temperature swing adsorption (TSA), where $CO_2$ is reversibly adsorbed onto microporous solids with specific surface areas of order 1E3 m$^2$/gram. In this process beds of sorbent are cycled through stages of $CO_2$ adsorption in contact with air (cold) to $CO_2$ desorption in contact with a stripping gas such as steam (hot). Performance of the process depends on the affinity of the sorbent for the adsorbate ($CO_2$), measured by equilibrium adsorption isotherms at the low and high temperatures, represented in the figure and equations below.

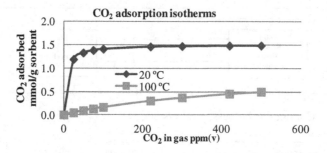

Cold:  q = 0.225y/(1+0.15y)        20°C              q = $CO_2$ adsorbed mmol/gram sorbent
Hot:   q = 0.002y/(1+0.002y)       100°C             y = mole fraction $CO_2$ in gas ppm(v)

This sorbent is to be used in a $CO_2$ direct air capture (DAC) plant to recover 1000 tons per day of $CO_2$. The process will operate 24 hours per day with a 1.5-hour cycle time, consisting of {adsorption 1 hour → desorption 0.5 hour}. Inlet air at 100 kPa(a) 20°C has 420 ppm(v) $CO_2$ and relative humidity of 60%. Air flow will be controlled to maintain a $CO_2$ separation efficiency of 50% throughout each adsorption cycle.

**Problem:**
a.  Find the total mass of air passed to the adsorbent per day.              [ton/day]
b.  Find total mass of water in the air fed to the adsorbent.                [ton/day]
c.  Find the total mass of adsorbent needed for this duty.                   [ton]

**Assume:** Zero effect of water and zero degeneration of adsorbent over time.
          Equilibrium adsorption (i.e. no constraint on the adsorption rate).   mmol ≡ 1E-3 mole

### 4:16 U   Photo-fuel cell

The "photo-fuel cell" illustrated below is a photo-electrochemical device in early stages of development for simultaneous wastewater treatment and clean electricity generation.

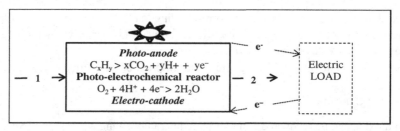

The destruction of xenotoxic chemicals, reduction of BOD and generation of electricity have been reported in laboratory tests with photo-fuel cells. Research data with illumination about 0.5 kW/m² has peak superficial power densities up to 5 W/m² at 0.6 volt/cell and reduction in BOD of wastewater about 80% over several hours of treatment.

**Problem:** Assuming a wastewater BOD is due only to complete oxidation of 5 kg/m³
          sucrose ($C_{12}H_{12}O_{11}$) to $CO_2$ and $H_2O$ use the data above to calculate:
a.  The total (superficial) area of photo-fuel cell needed for an 80% reduction in the BOD of
    1000 m³/day wastewater, with sunlight intensity of 0.5 kW/m² for 10 hours/day.   [m²]

**b.** The electric power output from the photo-fuel cells in sunlight.        [kW]

*Data*: Cathodic reduction of $O_2$ to $H_2O$ requires 4 Faraday per gram mole $O_2$.

## 4:17 W   Artificial leaf

An "artificial leaf" (AL) is a photo or photo-electrochemical device that mimics the function of natural leaves, which use sunlight to convert $H_2O$ and $CO_2$ to $O_2$ and C/H/O compounds (e.g. glucose). It is proposed to use an AL process to convert $CO_2$ to fuels, where the solar-to-fuel (STF) thermodynamic efficiency is calculated as:

$$STF = (F\Delta G)/(IA)$$

| | | | |
|---|---|---|---|
| F = fuel production rate | kmol/s | $\Delta G$ = free energy of reaction | kJ/kmol |
| I = intensity of solar radiation | kW/m$^2$ | A  = area irradiated | m$^2$ |

Methanol may be produced by AL photosynthesis with the idealised stoichiometry:

$$CO_2 + 2H_2O \rightarrow CH_3OH + (3/2)O_2 \qquad \Delta G = 1E3[(-166+0)-(-394-(2)237)] = +702E3 \text{ kJ/kmol}$$

Methanol produced by AL could potentially displace gasoline (e.g. octane) as a transportation fuel, where it may power internal combustion engines (ICE) or fuel cells (FC).

*Problem*: Assuming methanol is produced by the AL with 100% selectivity and harvested with 100% recovery use the data below to find the following:

**a.**  The area irradiated by sunlight 8 hours per day supplying methanol to a methanol fuel cell (FC) equivalent to the gasoline consumed by an ICE car travelling 50 km/day with fuel (octane) consumption of 6 kg/100 km.        [m$^2$]

**b.**  The equivalent $CO_2$ NOT generated per day from burning gasoline in an ICE.   [kg/d]

*Data*:

| | $C_8H_{18}$ | $CH_3OH$ |
|---|---|---|
| Gross heat of combustion kJ/kmol | -5461E3 | -726E3 |

Fuel oxidation (combustion) product water is gas from the ICE, liquid from the FC.

STF = 20%, sunlight intensity = 1 kW/m$^2$   Energy efficiency %:   ICE = 25    FC = 40

## 4:18 U   Solar ferrite cycle

The figure below is a simplified flowsheet of the proposed thermochemical "ferrite--cycle" using focused sunlight to split (decompose) water by the overall (net) reaction:

$$H_2O \rightarrow H_2 + 1/2O_2$$

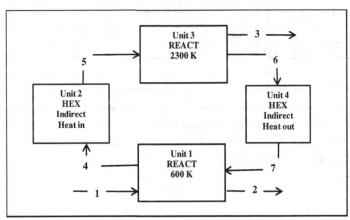

Stream contents

| Str. 1 | Str. 2 | Str. 3 | Str. 4 | Str. 5 | Str. 6 | Str. 7 |
|--------|--------|--------|--------|--------|--------|--------|
| $H_2O$ | $H_2$  | $O_2$  | $Fe_3O_4$ | $Fe_3O_4$ | FeO | FeO |

**Problem:**

Assuming stoichiometric reactants with 100% conversion and yield in all reactions:

a.  Write the balanced reactions for the cycle.
b.  Set up a steady-state material balance stream table showing the quantities of all reactants and products needed to produce 120 tons/day of $H_2$. 24 hours/day.

### 4:19 W   Nitric acid

The figures below are a simplified conceptual flowsheet and partial stream table for a thermochemical process manufacturing nitric acid. The process proceeds through a cascade of reactions with the overall stoichiometry:

$$3NH_3 + xO_2 \rightarrow yHNO_3 + zNO + wH_2O$$

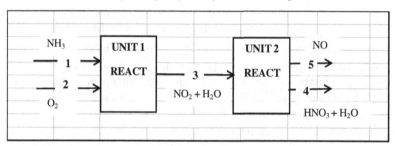

| Stream Table | | Continuous process | | | Steady-state | |
|---|---|---|---|---|---|---|
| Species | M | | Stream | | | |
| | kg/kmol | 1 | 2 | 3 | 4 | 5 |
| | | | Flow kmol/h | | | |
| $NH_3$ | 17.0 | 90.0 | 0.0 | 0.0 | 0.0 | 0.0 |
| NO | 30.0 | 0.0 | 0.0 | 0.0 | 0.0 | ? |
| $NO_2$ | 46.0 | 0.0 | 0.0 | ? | 0.0 | 0.0 |
| $HNO_3$ | 63.0 | 0.0 | 0.0 | 0.0 | ? | 0.0 |
| $H_2O$ | 18.0 | 0.0 | 0.0 | ? | ? | 0.0 |
| $O_2$ | 32.0 | 0.0 | 157.5 | 0.0 | 0.0 | 0.0 |
| Total mass | kg/h | 1530 | 5040 | ? | ? | ? |
| Mass balance | | Unit 1 | | Unit 2 | | Overall |
| In | kg/h | 6570 | | | | |
| Out | kg/h | | | | | |
| Closure % | | | | | | |

**Problem:** Find the values for the stoichiometric coefficients x, y, z, w (they may not be integers) and complete the stream table.

Ensure closure of the Unit 1, Unit 2 and Overall mass balances.

**Assume:** 100% conversion of $NH_3$.

## 4:20 W   Bio-effluent treatment

The figure below is a simplified conceptual flowsheet for a biological activated sludge effluent treatment process used at a pulp and paper mill.

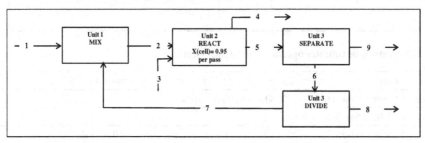

Reaction 1: $C_6H_{10}O_5(s) + \alpha O_2(g) + \beta NH_3(aq) \rightarrow \delta CH_{1.8}O_{0.5}N_{0.2}(s) + 4CO_2(g) + \gamma H_2O(l)$
        Cellulose                              Biomass

Specifications: Continuous operation, steady-state.

| Stream | Specification |
|---|---|
| 1 | 20.0E3 kg/h (total) wastewater mixture containing 2.0 wt% cellulose fibre ($C_6H_{10}O_5$) and stoichiometric $NH_3$ for 95% conversion of cellulose in stream 2 per pass in the reactor (reaction 1). |
| 2 | Mixture of wastewater and recycle biomass sludge. |
| 3 | Wet air at 150 kPa(a), 300 K and 70% relative humidity. 2 times stoichiometric oxygen rate for 95% conversion of cellulose in stream 2 per pass by reaction 1. |
| 4 | Exhaust air and $CO_2$ gas at 100 kPa(a), 310 K, saturated with water vapour (zero $NH_3$). |
| 5 | Reaction product slurry of solids in liquid water (zero $O_2$, $N_2$, $CO_2$). |
| 6 | "sludge" = 15.0 wt% slurry of [biomass + unconverted cellulose] solids in liquid water. |
| 7 | 500 kg/h recycle "sludge". |
| 8 | Excess "sludge" to disposal (probably by drying and burning). |
| 9 | Treated effluent water with zero solids. |

### Problem:
a. Determine values of the stoichiometric coefficients ($\alpha,\beta,\delta,\gamma$) in reaction 1.
b. Write material balance equations for each species in each process unit, together with any relevant subsidiary relations.
c. Set up the flowsheet and material balance stream table in a spreadsheet.
d. Solve the complete material balance and ensure closure of the mass balance over each process unit and the whole process.

*Assume*: Biomass and cellulose are completely insoluble in effluent liquid.

## 4:21 U   Chemical looping combustion

The figure below represents a chemical looping combustion process using circulating fluidised bed reactors. In this process a carbonaceous fuel is oxidised (Unit 1) by a solid oxide "oxygen carrier" which is continuously regenerated with air and recycled from Unit 2. With a manganese oxide oxygen carrier, the reactions in this process are:

$$\text{Unit 1:}\quad 4Mn_3O_4 + CH_4 \rightarrow 12Mn_pO_q + CO_2 + wH_2O$$
$$\text{Unit 2:}\quad xMn_pO_q + O_2 \rightarrow yMn_3O_4$$

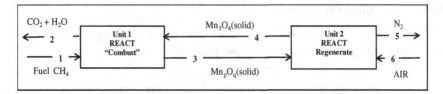

Specifications: Continuous operation at steady-state.

| Unit 1 | Carrier conversion = 90%, selectivity for $CO_2$ = 100%, fuel conversion = 100% |
| --- | --- |
| Unit 2 | Carrier regeneration = 100%, $O_2$ conversion = 70% |
| Str. 1 | 22.40E3 $Sm^3$/hour, 100% $CH_4$ |
| | Mean residence time of Mn oxide: Unit 1 = 30 seconds, Unit 2 = 60 seconds |

**Problem:**
a.  Find the values of p, q, w, x and y.
b.  Find the minimum total mass of Mn in Units 1 and 2 needed to sustain the process.   [kg]
c.  Find the dew-point temperature of stream 2 with total pressure = 90 kPa(a)         [K]
*Assume:* Negligible hold-up of solids in transfer lines, zero solubility of $CO_2$ in water.

# Chapter 5:   Energy Balances

## 5:1 W   Solar energy storage

The following figure is a conceptual flowsheet for energy storage in a solar farm. In this process air heated by focused sunlight transfers heat by direct contact to alternating fluid-ised beds of sand from which the heat is recovered as needed to drive steam turbines.

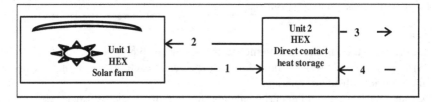

Specifications:

| Str. 1 | Hot air — to sand bed | 150 kPa(a), 1000 K |
| --- | --- | --- |
| Str. 2 | Cool air — from sand bed | 120 kPa(a), 900 K |
| Str. 3 | Hot air — to steam turbine | 140 kPa(a), 600 K |
| Str. 4 | Cool air — from steam turbine | 120 kPa(a), 500 K |

**Problem:**
a.  Find the sand mass in each bed to provide 1E5 kWh to the turbines every 24 hours   [kg]
b.  Find the average flow rate of air in streams 1,2 to charge each sand bed during 8 hours of sunlight every 24-hour day.                                                          [kg/h]
*Assume:* Temperatures are fixed in the specified ranges. Zero heat loss
*Data:*   Heat capacity of sand = 0.9 kJ/(kg.K)

## 5:2 W   Fuel gas combustion

A fuel gas mixture of 20 wt% $H_2$, 70 wt% $CH_4$ and 10 wt% water vapour at 30°C is burned continuously and completely with 20% excess dry air, initially at 25°C.

***Problem***: Use the heat of formation method to calculate the adiabatic flame temperature under 101 kPa(a) then find the dew point temperature of the product gas mixture for complete combustion under a pressure of 190 kPa(a).   [K,K]
Adopt mean heat capacities to simplify the solution.

## 5:3 U   Generic material and energy balances

***Problem***: For each of the flowsheets "a" to "c" below complete the continuous steady-state material and energy balance stream table to find the unknown temperatures and the energy duties ($\dot{Q}$ and $\dot{W}$) of the MIX, SEPARATE and REACT units. In each case ensure closure of the material and the energy balances.

**Flowsheet a:**

- 1 → | a MIX T(2)=T(3) | - 3 →
- 2 →

| STREAM TABLE | | | | MIXER | | | |
|---|---|---|---|---|---|---|---|
| Species | M | $C_{p,m298}$ | $h_{f,298}$ | | Stream | | |
| | kmol/kg | kJ/kmol.K | kJ/kmol | 1 | 2 | 3 | 4 |
| | | | | | Flow kmol/h | | |
| A(liq) | 40 | 70 | -6.0E+04 | | 10 | 20 | NA |
| A(gas) | 40 | 35 | -40+E0.4 | | | | NA |
| B(liq) | 80 | 90 | -1.5E+05 | | 30 | 40 | NA |
| B(gas) | 80 | 45 | -50+E2.1 | | | | NA |
| Total | kg/h | | | | | | NA |
| Phase | | | | L | L | L | |
| Pressure kPa(a) | | | | 500 | 500 | 480 | |
| Temp. K | | | | 290 | ? | ? | |
| Enthalpy kJ/h | | | | | | | |
| Mass balance | IN kg/h | | | Q kW | W kW | | |
| | OUT kg/h | | | 0 | -100 | | |
| | Closure % | | | | | | |
| Energy balance | IN kJ/h | | | | | | |
| | OUT kJ/h | | | | | | |
| | Closure % | | | | | | |

**Flowsheet b:**

- 1 → | b SEPARATE s(2,A)=0.8 s(3,B)=0.9 T(2) = T(3) | - 2 →
- 3 →

| STREAM TABLE | | | | SEPARATOR | | | |
|---|---|---|---|---|---|---|---|
| Species | M | $C_{p,m298}$ | $h_{f,298}$ | | Stream | | |
| | kmol/kg | kJ/kmol.K | kJ/kmol | 1 | 2 | 3 | 4 |
| | | | | | Flow kmol/h | | |
| A(liq) | 40 | 70 | -40+E0.6 | 50 | | | NA |
| A(gas) | 40 | 35 | -40+E0.4 | | | | NA |
| B(liq) | 80 | 90 | -50+E5.1 | 30 | | | NA |
| B(gas) | 80 | 45 | -1.2E+05 | | 3 | | NA |
| Total | kg/h | | | | | 2560 | NA |
| Phase | | | | L | G | L | |
| Pressure kPa(a) | | | | 300 | 280 | 280 | |
| Temp. K | | | | 290 | ? | ? | |
| Enthalpy kJ/h | | | | | | | |
| Mass balance | IN kg/h | | | Q kW | W kW | | |
| | OUT kg/h | | | 200 | 0 | | |
| | Closure % | | | | | | |
| Energy balance | IN kJ/h | | | | | | |
| | OUT kJ/h | | | | | | |
| | Closure % | | | | | | |

**Flowsheet c:**

- 1 → | c REACT 2A > B X(A) = 0.8 | - 2 →

| STREAM TABLE | | | | REACTOR | | | |
|---|---|---|---|---|---|---|---|
| Species | M | $C_{p,m298}$ | $h_{f,298}$ | | Stream | | |
| | kmol/kg | kJ/kmol.K | kJ/kmol | 1 | 2 | 3 | 4 |
| | | | | | Flow kmol/h | | |
| A(liq) | 40 | 70 | -6.0E+04 | 50 | | NA | NA |
| A(gas) | 40 | 35 | -4.0E+04 | 0 | | NA | NA |
| B(liq) | 80 | 90 | -1.5E+05 | 0 | | NA | NA |
| B(gas) | 80 | 45 | -1.2E+05 | 0 | | NA | NA |
| Total | kg/h | | | | | NA | NA |
| Phase | | | | L | G | | |
| Pressure kPa(a) | | | | 200 | 180 | | |
| Temp. K | | | | 350 | 395 | | |
| Enthalpy kJ/h | | | | | | | |
| Mass balance | IN kg/h | | | Q kW | W kW | | |
| | OUT kg/h | | | ? | 0 | | |
| | Closure % | | | | | | |
| Energy balance | IN kJ/h | | | | | | |
| | OUT kJ/h | | | | | | |
| | Closure % | | | | | | |

### 5:4 U   Rocket engine

A rocket is charged with the hypergolic propellant's *dimethyl hydrazine* ($C_2H_8N_2$ — fuel) and *dinitrogen tetroxide* ($N_2O_4$ — oxidant).

**Problem:**
a.   Write a balanced equation for the complete reaction between fuel and oxidant, assuming the only reaction products to be $CO_2$, $H_2O$ and $N_2$.
b.   A rocket with "empty" un-fuelled mass of 5 tons is charged with stoichiometric quantities of the fuel and oxidant (in separate tanks). Assuming that all the chemical energy of reaction, with 100% conversion of reactants, is used to lift the rocket calculate the maximum height reached by the rocket when initially loaded with 8000 kg dimethyl hydrazine.   [m]

*Assume:* Uniform terrestrial gravity ($9.81$ m/s$^2$), zero energy loss to the surroundings and effective mass of [fuel + oxidant] for the whole journey is ½ its initial value.

**Data:**

| | $C_2H_8N_2$ | $N_2O_4$ |
|---|---|---|
| Standard heat of formation kJ/kmol | +48.3E3 | +9.2E3 |

### 5:5 W   Methanol from propane

**Problem:**
Use the data given below to find:
a.   The standard NET heat of combustion of liquid methanol (in air)      [kJ/kmol]
b.   The standard heat of formation of liquid methanol                [kJ/kmol]
c.   The standard heat of reaction for:
  $2C_3H_8(l) + 7O_2(g) \rightarrow 2CH_3OH(l) + 4CO_2(g) + 4H_2O(l)$        [kJ/kmol]
d.   Is reaction (c) exothermic or endothermic?                [Ex,En]

**Data:**

| All values at 298 K | $C_3H_8(l)$ | $CH_3OH(l)$ |
|---|---|---|
| Gross heat of combustion kJ/kmol | –2204E3 | –727E3 |

### 5.6 U   Reactor with heat exchange

The figures below show a flowsheet and partial stream table for a chemical process where liquid reactants are mixed and preheated before passing through a reactor where they are partially converted and vaporised. The reactor outlet temperature is controlled by heat transfer with a process utility and heat is recovered from the reaction product vapour as it condenses to pre-heat the reactor feed in a counter current heat exchanger.

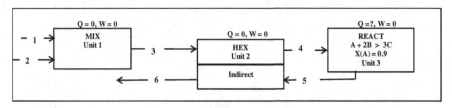

| STREAM TABLE | | REACTOR with HEAT EXCHANGER | | | | | | | | |
|---|---|---|---|---|---|---|---|---|---|---|
| Species | M | $C_{p,m\ 298K}$ | $h^\circ_f$ | | | Stream | | Flow kmol/h | | |
| | kg/kmol | kJ/kmol.K | kJ/kmol | *1* | *2* | *3* | *4* | *5* | *6* | |
| A(l) | 40 | 70 | −5.0E+04 | 40 | 0 | ? | ? | ? | ? | |
| A(g) | | 35 | −40+E5.4 | | | | | | | |
| B(l) | 70 | 90 | −1.0E+05 | 0 | 90 | ? | ? | ? | ? | |
| B(g) | | 45 | −840+E0. | | | | | | | |
| C(l) | 60 | 80 | −2.5E+05 | 0 | 0 | ? | ? | ? | ? | |
| C(g) | | 40 | −50+E4.2 | | | | | | | |
| Total | kg/h | | | 1600 | 6300 | ? | ? | ? | ? | |
| Phase | | | | L | L | L | L | G | L | |
| Pressure | kPa(abs) | | | 300 | 300 | 290 | 260 | 200 | 180 | |
| Temp. | K | | | 300 | 400 | ? | 600 | 700 | ? | |
| Enthalpy* | kJ/h | ref Elements @ 298K | | ? | ? | ? | ? | ? | ? | |
| Mass balance | | | | | Unit 1 | Unit 2 | Unit 3 | Overall | | |
| IN | kg/h | | | | | | | | | |
| OUT | kg/h | | | | | | | | | |
| Closure | % | | | | | | | | | |
| Energy balance | | | | | | | | | | |
| IN | kJ/h | | | | | | | | | |
| OUT | kJ/h | | | | | | | | | |
| Closure | % | | | | | | | | | |

## Problem:

**a.** Write the material and energy balance equations for each process unit.

Use the sequential modular method with Excel Solver or Goal Seek to calculate the complete stream table material and energy balance.

**b.** Report the heat duty of the heat exchanger and the thermal load on the reactor.

[kW, kW heating or cooling?]

## 5:7 W Ammonia vaporisation

Below are a flowsheet and partial stream table for a steam utility heat exchanger vaporising liquid ammonia prior to its use in the production of nitric acid.

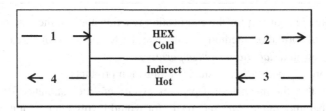

| STREAM TABLE | | VAPORISATION of AMMONIA | | | | | Stream | | |
|---|---|---|---|---|---|---|---|---|---|
| Species | M | Density | $C_{p,m}$ 298K | $h^\circ_f$ | *1* | *2* | *3* | *4* | |
| | kg/kmol | kg/m³ | kJ/kmol.K | kJ/kmol | | Flow kmol/h | | | |
| H2O(l) | 18.0 | | | NA | 0 | ? | ? | ? | |
| H2O(g) | 18.0 | STEAM TABLE | | NA | 0 | ? | ? | ? | |
| NH3(l) | 17.0 | 640 | 71 | −6.6E+04 | 200 | ? | 0 | 0 | |
| NH3(g) | 17.0 | Gas law | 35 | −4.5E+04 | 0 | ? | 0 | 0 | |
| Total | kg/h | | | | ? | ? | ? | ? | |
| Phase | | | | | L | G | L | ? | |
| Pressure | kPa(abs) | | | | 820 | 800 | 240 | 150 | |
| Temp | K | | | | 273 | 350 | 400 | 350 | |
| Volume | m3/h | | | | ? | ? | ? | ? | |
| Enthalpy | ref compounds 298 K | | | | ? | ? | ? | ? | |
| Mass balance | | | | | | | | | |
| IN | kg/h | | | | | | | | |
| OUT | kg/h | | | | | | | | |
| Closure % | | | | | | | | | |
| Energy balance | | | | | | | | | |
| IN | kJ/h | | | | | | | | |
| OUT | kJ/h | | | | | | | | |
| Closure % | | | | | | | | | |

**Problem:**
a.  Write equations needed to solve the M&E balance problem for the complete stream table.
b.  Use the steam table, along with the data supplied above to complete the M&E balance stream table. Base your enthalpy calculations on the <u>compound reference state</u>.

**Assume:**
A continuous process at steady-state
The heat exchanger is insulated to give zero heat exchange with the environment.
Enthalpy of ammonia(liq) and of ammonia(gas) are independent of pressure.

## 5:8 U   Naphthalene combustion

The figure below shows a process in which naphthalene ($C_{10}H_8$) vapour is burned in an insulated furnace.

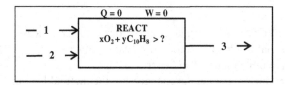

Specifications: Complete combustion, continuous adiabatic process, steady-state.

| Stream 1 | $C_{10}H_8$ vapour 20 kmol/h | 110 kPa(a) | 600 K |
|----------|------------------------------|------------|-------|
| Stream 2 | Wet air 60% RH, 20% excess $O_2$ | 110 kPa(a) | 303 K |
| Stream 3 | Combustion product gases | 100 kPa(a) | T(3) |

**Problem:** Write the equation for the combustion reaction and use the data below with the heat of formation method to calculate the temperature of Stream 3 (a.k.a. the theoretical *adiabatic flame temperature*).                                    [K]

**Assume:** Complete combustion to stable combustion products.
          Note that the mean heat capacities given here are estimates for the expected temperature range. An exact result for would involve integration of the polynomial $C_p$ functions in the energy balance and solution of the resulting quartic equation for T(3).

| Data | | | $C_{10}H_8(s)$ | $C_{10}H_8(l)$ | $C_{10}H_8(g)$ |
|------|------|------|----------------|----------------|----------------|
| $h_{c,298K\ GROSS}$ | Combustion | kJ/kmol | −5.17E+06 | not given | not given |
| $h_{p,298K}$ | Fusion | kJ/kmol | 1.90E+04 | NA | NA |
| $h_{v,298K}$ | Vaporisation | kJ/kmol | NA | 5.50E+04 | NA |
| $C_{p,m298K}$ | | kJ/kmol.K | 170 | 150 | 100 |

## 5:9 W   Gasoline from air

The figure below is a conceptual flowsheet of a proposed process making gasoline from air.

In this process, water and carbon dioxide are recovered from air. Hydrogen is obtained by decomposing the water and combined with the $CO_2$ to make hydrocarbon fuels.

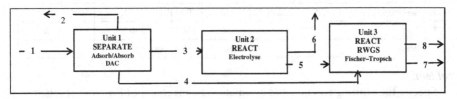

Unit 1:  $H_2O$ and $CO_2$ are obtained by reversible absorption and/or adsorption from air.
Unit 2:  $H_2$ comes from splitting water by electrolysis or solar-thermo/photo/electric methods.
$$H_2O \rightarrow H_2 + 0.5O_2$$
Unit 3:  $CO_2$ is reduced to CO and combined with $H_2$ to give $C_nH_m$      For example:
       $H_2 + CO_2 \rightarrow CO + H_2O$            Reverse water-gas shift (RWGS)
       $17H_2 + 8CO \rightarrow C_8H_{18} + 8H_2O$       Fischer–Tropsch
The streams in flowsheet are as follows:

| Stream | 1 | 2 | 3 | 4 | 5 | 6 | 7 | 8 |
|--------|-----|-----|--------|--------|-------|-------|--------|-----------|
|  | Air | Air | $H_2O$ | $CO_2$ | $H_2$ | $O_2$ | $H_2O$ | $C_8H_{18}$ |

**Problem**: Assuming these operations/reactions could be carried out under practical condition, at useful rates with 100% conversion, yield and separation efficiency in all cases, and without recycle:
a.   Set up a mass balance stream table for the process of the flowsheet, making 11.4-ton octane per hour.
b.   Calculate the minimum energy input for the above three chemical reactions needed to produce 1 litre of octane in this process.                               [kWh]
*Assume*:  Stream 1 is atmospheric air at 20°C with 420 ppm(v) $CO_2$, 100% RH.
*Data*:     $C_8H_{18}$ $h_{f,298K} = -209E3$ kJ/kmol,   SG = 0.70   CO $h_{f,298K} = -110.8E3$ kJ/kmol

### 5:10 U  Catalytic alkylation
The following figure is a simplified flowsheet of a refinery liquid phase catalytic alkylation process making iso-octane, for increasing the "octane-number" of gasoline. The process is at steady-state in a continuous stirred tank with reaction temperature controlled by indirect heat exchange with water.

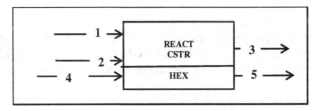

Reaction:      $C_4H_{10} + C_4H_8 \rightarrow C_8H_{18}$                   ($H_2SO_4$ catalyst)
Specifications: 100% conversion of limiting reactant.
                * Yield is based on the limiting reactant.

| Str. 1 | Iso-butane | liquid | 900 kPa(a), 298 K | 8.7E3 kg/h |
|--------|------------|--------|-------------------|------------|
| Str. 2 | Iso-butene | liquid | 900 kPa(a), 298 K | 5.6E3 kg/h |
| Str. 3 | Iso-octane plus unconverted reactant(s) | liquid | 900 kPa(a), 310 K | 100% yield* |
| Str. 4 | Water | phase? | 150 kPa(a), 290 K | 1.2E5 kg/h |
| Str. 5 | Water | phase? | 130 kPa(a), ? K | ? |

**Problem:**
a.  Specify the limiting reactant and calculate the conversion of the other reactant    [%]
b.  Is the alkylation reaction endothermic or exothermic?    [En,Ex]
c.  Find the phase(s) and temperature of stream 5.    [Π,K]
*Assume:* Zero heat transfer with surroundings.

**Data:**

|  |  | $C_4H_{10}(l)$ | $C_4H_8(l)$ | $C_8H_{18}(l)$ |
|---|---|---|---|---|
| Heat of formation $h_{f,298}$ | kJ/kmol | –1.69E5 | –3.9E4 | –2.59E5 |
| Heat capacity $C_{p,m,298}$ | kJ/(kmol.K) | 130 | 121 | 240 |

## 5:11W   Thermochemical water-split

Production of hydrogen by thermal decomposition of $H_2O$ to $H_2$ and $O_2$ requires temperatures over 2000 K, impractical for large scale application. Instead, it is proposed to get hydrogen by splitting $H_2O$ in a thermochemical Solvay cluster employing compounds of sulphur and iodine at temperatures up to 1200 K. Heat for this process would come from very high temperature (VHT) nuclear reactors.
The (<u>unbalanced</u>) reactions, minimum conceptual flowsheet and data are below.

| | | | | |
|---|---|---|---|---|
| $H_2O + SO_2 + I_2 \rightarrow H_2SO_4 + HI$ | 400 K | Irreversible | *Assume all in gas phase* | rxn 1 |
| $HI \leftrightarrow H_2 + I_2$ | 1000 K | Reversible | *Assume all in gas phase* | rxn 2 |
| $H_2SO_4 \leftrightarrow SO_2 + H_2O + O_2$ | 1200 K | Reversible | *Assume all in gas phase* | rxn 3 |

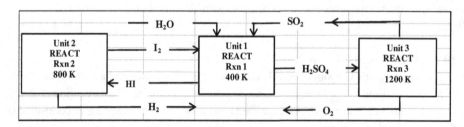

**Problem:**
a.  Balance each reaction 1,2,3 (stoichiometry).
b.  Determine if each reaction is endothermic or exothermic and calculate the (minimum) thermal duty on each process unit in the flowsheet. [Heat IN or OUT, kWh/kg $H_2$]
c.  Calculate the energy efficiency of the process with 100% conversion of all reactants assuming:
   • 100% recovery of the available sensible heat, and
   • Zero recovery of the available sensible heat.

**Data:**

| Species | | H₂SO₄(g) | SO₂(g) | HI(g) | I₂(g) |
|---|---|---|---|---|---|
| $h_f$, 298K | kJ/kmol | −7.44E+05 | −2.71E+05 | −2.60E+05 | 0.00E+00 |
| $C_{p,m}$, 298K | kJ/kmol.K | 100 | 54 | 26 | 53 |

### 5:12 W   Vanadium redox-flow battery

A low-carbon economy relying on intermittent energy sources like sun and wind needs methods of energy storage, such as those available from reversible electrochemical reduction-oxidation (REDOX) cycles. The figure below shows an energy storage system using a vanadium redox-flow battery (VRFB) where energy is reversibly stored and recovered from water solutions of multi-valent vanadium salts separated by an ion-exchange membrane (IEM) in an electrochemical reactor.

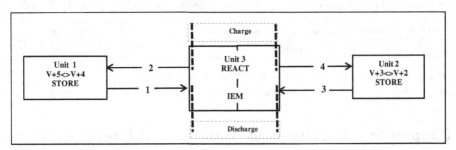

In this system the electrochemical charging (to storage) and discharging (to load) cycles are summarised in the reversible redox reactions 1 and 2, where e⁻ represent electric current (electrons).

| $V^{+3} + e^- \longleftrightarrow V^{+2}$ | charge→    ←discharge | rxn 1 |
|---|---|---|
| $V^{+4} \longleftrightarrow V^{+5} + e^-$ | charge→    ←discharge | rxn 2 |

The thermodynamic open circuit voltage (OCV) of the discharge is about 1.45 Volt/cell but energy losses in practice bring this down around 1 volt/cell ("cell" = one anode/cathode pair).

**Problem:** Using the data tabulated below, calculate:
a. The total volume of reactant solution in Units 1 and 2 needed to provide 2 megaWatt hours per redox cycle. [m³]
b. The total (effective) area of electrochemical cells needed to charge each unit from a solar array over an 8-hour cycle. [m²]
c. The temperature rise of the reactant solution in Units 1 or 2 over each charge cycle [K]
**Assume:** $V^{+5}$ conversion per charge = 90%. Zero heat transfer with surroundings.
**Data:** Values are approximate for this example.

| Voltage/cell | | Current density | Electrolyte | Pressure | Temperature |
|---|---|---|---|---|---|
| Charge | Discharge | Charge/discharge | 2 M vanadium sulphates + 2 M H₂SO₄ | kPa(a) | K |
| 1.6 | 1.2 | 2 kA/m² | SG, Cp = same as 50 wt% H₂SO₄ | 110 | 283–323 |
| Energy converted to heat during charge/discharge ≈ I [ABS(OCV-V)] an approximation. V = operating voltage | | | | | |
| Faraday's law: r = I/(nF) Consistent units | | | | | |
| r = reaction rate, I = current, n = electron stoichiometry coefficient, F = Faraday's number. | | | | | |

### 5:13 U   Thermochemical ammonia

The figure below is a simplified flowsheet for the thermochemical synthesis of ammonia in the Haber–Bosch process, by reaction 1. In this process some heat of reaction from Unit 2 is recovered in Unit 1 by indirect heat exchange with the product stream 3.

$$N_2(g) + 3H_2(g) \rightarrow 2NH_3(g) \qquad \text{rxn 1}$$

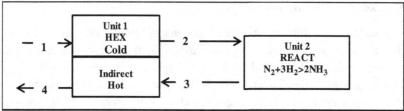

**Specifications:** Continuous adiabatic process at steady-state. Single-pass $\dot{Q} = \dot{W} = 0$

| Str. 1 | $N_2$-$H_2$ mixture in stoichiometric proportions for reaction 1 | G | 3.3E4 kPa(a), 358 K |
|--------|------------------------------------------------------------------|---|---------------------|
| Str. 4 | Product mixture of $N_2$, $H_2$ and $NH_3$ | G | 3.3E4 kPa(a), 500 K |

**Problem:** Calculate the single-pass conversion of $N_2$ in this process.                    [%]
**Data:**    $NH_3(g)$        $C_{p,m,298} = 40$ kJ/(kmol.K),        $h_{f,298} = -46E3$ kJ/kmol.

# Chapter 6:   Simultaneous Material and Energy Balances

### 6:1 W   Partial condenser

The figure below is the flowsheet for a process in which methanol gas undergoes <u>adiabatic</u> compression (Unit 1) followed by cooling in a partial condenser (Unit 2) and separation of the condensed liquid (Unit 3) from the gas.

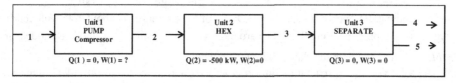

Specifications: Continuous process at steady-state.

| Str. 1 | 2910 m³/h pure $CH_3OH$ gas | G | 100 kPa(a), 350 K |
|--------|-----------------------------|-----|-------------------|
| Str. 2 | Compressed $CH_3OH$ gas. Adiabatic compression | G | 900 kPa(a), T(2) |
| Str. 3 | 2-phase. Gas plus liquid $CH_3OH$ | G+L | 880 kPa(a), T(3) |
| Str. 4 | $CH_3OH$ vapour | G | 870 kPa(a), T(4) |
| Str. 5 | $CH_3OH$ liquid | L | 870 kPa(a), T(5) |

Unit 2 is a utility heat exchanger operating with a thermal duty $= -500$ kW [+ for heat IN] Streams 4 and 5 are in material and thermal equilibrium. T(4)  = T(5)

### Problem:
**a.**   Find the temperature of streams 3, 4 and 5.                                   [K]
**b.**   Find the flow of liquid methanol condensate in Stream 5.                     [kg/h]

| Data | ref 298 K | CH₃OH(g) | CH₃OH(l) |
|------|-----------|----------|----------|
| $C_{p,m}$ | kJ/(kmol.K) | 50.0 | 95.0 |
| $h_v$ | kJ/kmol | | 36.0E3 |
| SG | | | 0.79 |

## 6:2 U   Compressors and intercooler

The figure below represents a 2-stage nitrogen gas compressor with an intercooler between stages.

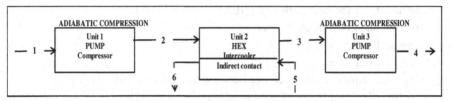

Specifications. Continuous adiabatic operation at steady-state

| Str. 1 | Nitrogen gas | G | 100 kPa(a), 298 K | 2800 kg/h |
|--------|--------------|---|-------------------|-----------|
| Str. 2 | Nitrogen gas | G | 1200 kPa(a), T(2)? | ? |
| Str. 3 | Nitrogen gas | G | 1200 kPa(a), T(3)? | ? |
| Str. 4 | Nitrogen gas | G | 2000 kPa(a),T(4)? | ? |
| Str. 5 | Water | ? | 200 kPa(a), 298 K | ? |
| Str. 6 | Water | ? | 180 kPa(a), T(6) ? | ? |

The rate of heat transfer between the coolant water and nitrogen gas in Unit 2 is given by:
$\dot{Q}$ = UAΔT   U = heat transfer coefficient = 0.1 kW/(m².K),   A = heat transfer area = 40 m²
ΔT = log mean temperature difference   = [ΔTₕ − ΔT_c)]/ln[ΔTₕ/ΔT_c)]       K
ΔTₕ= T(2) − T(6)                        ΔT_c = T(3) − T(5)

### Problem:
Find the water flow (Str. 5) required to hold the N₂ outlet temperature T(4) at 350 K.

[kg/h]

*Assume*: N₂ $C_{p,m}$       fixed = 30 kJ/(kmol.K)

## 6:3 W   Bio-reactor

The following figure is a simplified flowsheet for a bio-reactor used in the production of citric acid by the aerobic fermentation of sucrose. In this process, the mould *Aspergillus niger* is used to direct the conversion of sucrose to citric acid in a water-cooled continuous reactor by the reaction:

$$C_{12}H_{22}O_{11} + xO_2 \rightarrow C_6H_8O_7 + yCO_2 + zH_2O$$

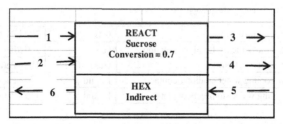

Specifications: Continuous operation at steady-state

| Str. 1 | 3000 kg/h, 20 wt% sucrose in water | Liquid | 120 kPa(a), 293 K |
|--------|-----------------------------------|--------|-------------------|
| Str. 2 | Dry air 100% excess for 100% sucrose conversion | Gas | 150 kPa(a), 350 K |
| Str. 3 | Waste gas $N_2$, excess $O_2$, $CO_2$, saturated with $H_2O$ | Gas | 110 kPa(a), T(3) |
| Str. 4 | Product citric acid, unconverted sucrose, water | Liquid | 110 kPa(a), T(3) |
| Str. 5 | 72E3 kg/h utility water to internal HEX coil | Liquid | 130 kPa(a), 278 K |
| Str. 6 | 72E3 kg/h utility water from internal HEX coil | Liquid | 110 kPa(a), T(6) |

Conversion of sucrose = 70%

The thermal duty of the heat exchange (HEX)     $\dot{Q} = UA\Delta T_{lm}$

$\dot{Q}$    = thermal duty of heat exchanger                               kW

U    = overall heat transfer coefficient          = 0.6           kW/(m².K)

A    = heat transfer area                      = 100            m²

$\Delta T_{lm}$ = log mean temperature difference between hot and cold streams     K

      = [(T(4) – T(5)) – (T(2) – T(6))] / ln[(T(4) – T(5)) / (T(2) – T(6))]     Unspecified

***Problem:***

a.   Find x, y, z then prepare the complete M&E balance stream table for this flowsheet.

b.   Report the temperature of the reactor [i.e. T(3) = T(4)] and the thermal duty of the heat exchanger $\dot{Q}$                                           [K,kW]

***Assume:***   Zero heat exchange with the environment through the walls of the reactor.

             Sucrose, citric acid and water form ideal mixtures (solutions).

| Data | ref 298 K | Sucrose(s) | $C_6H_8O_7$(s) |
|------|-----------|------------|----------------|
| $h_c$ (gross) | kJ/kmol | −5.65E+06 | −1.96E+06 |
| $h_f$ | kJ/kmol | Not given | Not given |
| $C_{pm}$ | kJ/(kmol.K) | 170 | 95 |

## 6:4 W   Hydrogen fuel cell

The following is a simplified version of the flowsheet for a $H_2$/Air PEM (proton exchange membrane) fuel-cell stack powering a car.

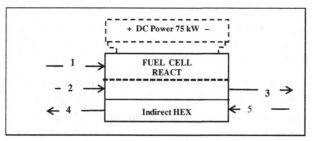

The overall fuel cell reaction is: $H_2 + 0.5O_2$ ----- $2F \rightarrow H_2O$

Specifications: Continuous steady-state.

| Str. 1 | Pure $H_2$ – 100% conversion | Gas | 200 kPa(a). 298 K |
|--------|------------------------------|-----|-------------------|
| Str. 2 | Dry air at 2 x stoichiometric for $H_2$ conversion | Gas | 200 kPa(a). 340 K |
| Str. 3 | Exhaust "air" with $H_2O$ product | Gas or G+L? | 180 kPa(a), T(3) |
| Str. 4 | 4500 kg/hour cooling water (indirect contact) | Liquid | 200 kPa(a), 310 K |
| Str 5 | 4500 kg/h cooling water (indirect contact) | Liquid | 150 kPa(a). [T(3)–10] K |

The "bipolar" fuel cell stack generates 75 kW with 200 cells electrically in series, each operating at 0.6 V, 625 A.

**Problem:**
a. Calculate the complete material and energy balance, including the $H_2O$ gas/liquid "phase split" in Stream 3.
b. Report the temperature of stream 3 and thermal duty of the cooling system. [K,kW]

*Assume:* Overall adiabatic operation (including cooling system), zero parasitic reactions.

Electrochemical stoichiometry $r = I/(2F)$

$r = H_2$ reaction rate kmol/s   I = current kA   F = Faraday's number   Coulomb/mole

## 6.5 U   Direct contact heat exchange

The figure below is a flowsheet for a packed-bed <u>direct contact</u> heat exchanger in which nitrogen gas is cooled and humidified by direct co-current contact with liquid $H_2O$.

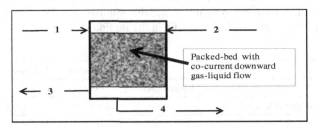

Specifications. Continuous adiabatic process at steady-state

| Str. 1 | Nitrogen 200 kmol/h | Gas | 210 kPa(a) | 600 K |
|--------|---------------------|-----|------------|-------|
| Str. 2 | Water 100 kmol/h | Liquid | 210 kPa(a) | 300 K |
| Str. 3 | Nitrogen + water vapour | Gas | 200 kPa(a) | T(3) |
| Str. 4 | Water | Liquid | 200 kPa(a) | T(4) = T(3) |

**Problem:**
a. Write the appropriate material and energy balance equations for this system.
b. Set up the full stream table material and energy balance with values for T(3),T(4).

*Assume:* Stream 3 and 4 are at equilibrium with T(3) = T(4). Stream 3 saturated with $H_2O$
Zero solubility of nitrogen in liquid water.

## 6:6 W   Ethanol from maltose

The following figure shows a simplified flowsheet for a continuous stirred tank reactor (CSTR) used for production of ethanol by the fermentation of maltose.

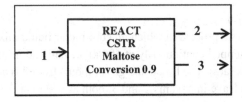

$$C_{12}H_{12}O_{11}(aq) + H_2O(l) \rightarrow 4C_2H_5OH(aq) + 4CO_2(g)$$

Specifications: Continuous adiabatic process, steady-state.

| Str. 1 | 2000 kg/h, 10 wt% maltose in water (+ yeast) | Liq | 110 kPa(a) | 293 K |
|---|---|---|---|---|
| Str. 2 | Exhaust $CO_2$ gas saturated with water and ethanol vapours | Gas | 105 kPa(a) | T(2) |
| Str. 3 | Product ethanol with water and unconverted maltose | Liq | 105 kPa(a) | T(3) |

Maltose conversion = 90%

**Problem:** Write the steady state material and energy balances for this process, plus any useful subsidiary relations. Solve the balances and produce a complete M&E balance stream table.

**Assume:**
- Stream 2 and stream 3 are at vapour/liquid equilibrium, with T(2) = T(3)
- Reactor liquid contents are well mixed (i.e. uniform composition).
- Water/ethanol forms an ideal liquid mixture (not true in reality).
- The amount of yeast in the system is negligible compared to the other components.
- Ideal gases. Negligible solubility of $CO_2$ in the liquid.

**Data:**

| Species | | Maltose(s) | Ethanol(l) | Ethanol(g) |
|---|---|---|---|---|
| $C_{p,m\ 298\ K}$ | kJ/(kmol.K) | 430 | 115 | 71 |
| Heat of combustion (gross) kJ/kmol | | −5.65E+06 | −1.37E+06 | −1.41E+06 |

## 6:7 W   Sulphuric acid dilution

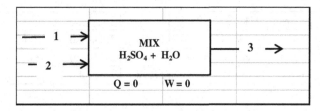

Specifications: Continuous, steady-state, adiabatic operation at 101 kPa(a)

| Str. 1 | 1000 kg/h 90 wt% $H_2SO_4$/water | Liquid | 100°C |
|---|---|---|---|
| Str. 2 | Water | Liquid | 60°C |
| Str. 3 | $H_2SO_4 + H_2O$ mixture | ? | 140°C |

**Problem:**

Use the $H_2SO_4/H_2O$ enthalpy–concentration chart to find the flow of Stream 2.     [kg/h]

## 6:8 W   Adiabatic flash

The figure below shows a process in which an equi-molar liquid mixture of n-hexane and benzene at 400 K is pumped continuously at 600 kPa(a) into an insulated drum where the pressure drops to 120 kPa(a). The liquid and gas (vapour) from this adiabatic "flash" separate under gravity and leave in streams 2 and 3, both at 120 kPa(a), T(2) = T(3).

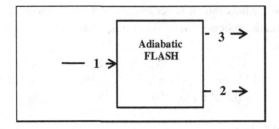

**Problem**: Calculate the continuous steady-state (equilibrium) temperatures and composi-
tions of the liquid and the vapour leaving the drum in stream 2 and in stream 3.
**Assume**: Benzene and n-hexane form an ideal liquid mixture.
Values are independent of pressure and temperature in the present range.

**Data**:

| | | Benzene ($C_6H_6$) | Hexane ($C_6H_{14}$) |
|---|---|---|---|
| Heat of vaporisation | kJ/kmol | 33.8E3 | 31.9E3 |
| Heat capacity (liquid) | kJ/(kmol.K) | 136 | 194 |
| Heat capacity (gas) | kJ/(kmol.K) | 82 | 143 |

Antoine constants for benzene: $A = 14.1603$   $B = 2948.78$   $C = -44.5633$   ln, K, kPa(a)

### 6:9 U   CSTR with heat exchange

The thermochemical reaction $A + B \rightarrow C$ is carried out in the continuous stirred tank reactor (CSTR) of the figure below, in which the temperature is controlled by indirect heat
transfer with water.

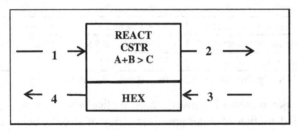

Specifications: Reactor volume = 3 m³ Continuous process, steady-state.

| Str. 1 | Reactant feed stoichiometric mix A and B | L | 120 kPa(a), 298 K | 200 kmol/h, 10 m³/h |
|---|---|---|---|---|
| Str. 2 | Reaction product mix A,B and C | L | 110 kPa(a), T(2) | No phase change 10 m³/h |
| Str. 3 | Water IN | ? | 130 kPa(a), 283 K | ? |
| Str. 4 | Water OUT | ? | 110 kPa(a), T(4) | ? |

The reaction rate is given by: $d[A]/dt = (-1)k[A][B]$     There are no side reactions.
where:
[A], [B] = concentration of A,B kmol/m³,   $k = 2[\exp(-30E3/(RT(2)))]$ m³/(kmol.s)
The rate of heat transfer to/from the reactants is given by: $\dot{Q} = UA\Delta T$
where:
U = heat transfer coefficient = 0.5 kW/(m².K)       A = heat transfer area = 40 m²
$\Delta T = T(2) - T(4)$                                                    K

**Problem:** Calculate the flow rate of water to control the conversion of A at 90%.   [kg/h]

**Assume:** An adiabatic process, ideal mixtures, constant density of reactants and products.

**Data:**

|  |  | A | B | C |
|---|---|---|---|---|
| $h_{f,298K}$ | kJ/kmol | 50E3 | −100E3 | −250E3 |
| $C_{p,m,298K}$ | kJ/(kmol.K) | 90 | 90 | 90 Independent of T |

## 6:10 W   Oxy-coal boiler

Oxygen-enriched combustion of fuels facilitates capture of the product $CO_2$ by increasing its concentration in flue gas. The figure below is a simplified partial flowsheet for an "oxy-coal" electric power generation plant in which coal is burned in air enriched with oxygen.

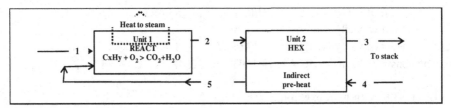

Specifications: Continuous operation, steady-state.

| Str. 1 | Pulverised dry coal $C_{100}H_{50}O_5$* | S | 120 kPa(a), 298 K | 100 ton/hour |
|---|---|---|---|---|
| Str. 2 | Hot boiler combustion products | G | 105 kPa(a), ? K | After heat to steam generation |
| Str. 3 | Cool boiler combustion products | ? | 100 kPa(a), ? K | To stack and atmosphere |
| Str. 4 | Cold $O_2$ enriched air 35 vol% $O_2$ | G | 120 kPa(a), 298 K | 60% RH, 1% excess $O_2$ |
| Str. 5 | Hot $O_2$ enriched air 35 vol% $O_2$ | G | 110 kPa(a), ? K | |

*Impurities, Ash, S, N, Hg, etc. are excluded from these calculations.

Corrosion is a major problem in power plants, particularly in the air pre-heater (APH, Unit 2) where condensation of water promotes attack of steel by acid gases ($SO_2$). Oxy-fuel combustion increases the concentration of $H_2O$ in flue gases, raises their dew-point temperature and complicates process design.

**Problem:** Use the information below to:

a.   Find the temperatures of streams 2, 3 and 5.                                              [K]

b.   Determine whether water will condense in the pre-heater (APH).                 [Yes, No]

**Assume:** 100% conversion of coal $C_{100}H_{50}O_5$, zero heat loss to surroundings.

**Data:**   Coal gross heat of combustion = −30E3 kJ/kg    $C_{p,298K}$ = 1.5 kJ/kg.K

Unit 2 heat transfer rate = UAΔT     U = 0.05 kW/(m².K)      A = 2000 m²

ΔT = {[T(2)+T(3)]/2 − [T(4)+T(5)]/2}             K

Saturated steam generated in Unit 1 = 1200 ton/hour at 450 K from liquid water at 450 K.

## 6.11 U   Solid-oxide fuel cell

The figure below represents a solid-oxide fuel cell (SOFC), used to generate electricity and heat by the electrochemical oxidation of methane. This reactor anode is fed directly with methane but since the electro-oxidation of $CH_4$ is "slow" the catalyst and conditions in the anode chamber are designed for internal reforming to the more reactive $H_2$. In this case the $CH_4$ is converted *in-situ* according to: $CH_4 + 2H_2O \rightarrow 4H_2 + CO_2$, then the net fuel-cell reaction stoichiometry is:

$$CH_4 + 2H_2O + 2O_2 - 4F \rightarrow CO_2 + 4H_2O$$
$$F = \text{Faraday} = 96485 \text{ Coulomb/mole}$$

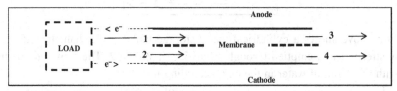

Specifications: Continuous operation at steady-state

| Str. 1 | $CH_4 + H_2O$ in 1 to 2 molar ratio | 150 kPa(a), 298 K | G |
|---|---|---|---|
| Str. 2 | Dry air, excess as needed | 150 kPa(a), 298 K | G |
| Str. 3 | Anode products $CO_2 + H_2O$ | 130 kPa(a), T(3) ? | G |
| Str. 4 | Cathode excess $O_2$ depleted dry air | 130 kPa(s), T(4)? | G |
| Reactor: Single cell with effective anode area A = 0.8 m² Zero heat loss | | | |

Equation for the linearised* reactor performance curve:
$V = 1.5 - 0.1i$     V = reactor output voltage Volt,     i = anode current density   $kA/m^2$
Practical operation of the reactor requires T(3) to be in the range 800°C to 1000°C.
It is proposed to control T(3) by excess air flow to the cathode, with heat transfer at a rate given by:
$\dot{Q} = UA\Delta T$     $U = 0.01 \text{ kW/(m}^2.K)$,     $\Delta T = \{[T(1)+T(3)]/2 - [T(2)+T(4)]/2\}$   K

### Problem:

Find the ranges of <u>current density</u> and <u>air feed rate</u> to keep 800 < T(3) < 1000°C
[kA/m²,kg/h]

### Assume:

Adiabatic conditions with 100% conversion of $CH_4$ and 100% Faradaic efficiency in all cases.
Zero transport of molecular reactants across the separating ion conductive membrane.
Value of T(3) does not affect the performance curve in the given range.
*Real fuel cell performance curves are non-linear and do respond to temperature.

# Chapter 7:   Unsteady-state Material and Energy Balances

### 7:1 W   Storage tank

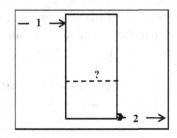

The figure above shows a cylindrical water storage tank, inside diameter 1.5 m, being filled by stream 1 and emptied through an orifice in stream 2. The flow rate in stream 2 varies with the depth of water in the tank according to:

Flow = kL                             $k = 6E\text{-}4 \ m^2/s$          $L$ = depth of water m
Initial depth of water (time zero) = 2 m.     Density of water,     fixed = $1000 \ kg/m^3$
Flow rate of water in stream 1 = 10 m$^3$/hour (fixed)         All at NTP

***Problem*:**
**a.**   Find the depth of water in the tank at 1 hour after start.                [m]
**b.**   Find equilibrium depth of water in the tank.                    [m]

### 7:2 U   Mixed tank dilution

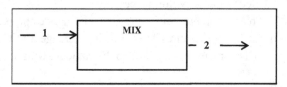

The figure above shows a mixed tank initially containing 2 m$^3$ of 4 molar HCl in water. At zero time:
Stream 1 begins a continuous steady flow of 0.5 m$^3$/h pure water AND
Stream 2 begins an equal steady flow of 0.5 m$^3$/h of HCl solution. All at NTP

***Problem*:** Calculate the concentration of HCl in the tank at 3 hours operation.    [kg/m$^3$]
***Assume*:**   Perfect mixing and density of HCl solution same as that of water.

### 7:3 W   Mixing benzene-xylene

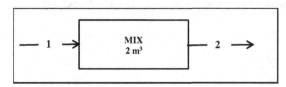

The figure above shows a 2 m$^3$ (fixed volume) mixing tank in a continuous process, initially full of pure liquid o-xylene at 293 K. At zero time an inlet valve opens to allow a steady flow of 1 m$^3$/h liquid benzene at 293 K, while the outlet valve simultaneously opens under level control to keep the tank full of liquid (2 m$^3$).                    All at NTP

**Problem:** Find the wt% benzene in the tank 10 hours after the valves are opened.  [wt%]
**Assume:** SGs equal at 0.880, tank is perfectly mixed, benzene/o-xylene form an ideal
         liquid mixture, no heat effects. SG at 293 K: Benzene = 0.879, o-xylene = 0.880.

### 7:4 W   Vertical pumping

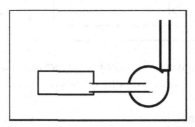

The figure above shows a centrifugal pump delivering a solution of 90 wt% sulphuric acid in water at 293 K from a tank at ground level directly up a vertical 2.90-inch ID (internal diameter) pipe.
The equation for the pump performance curve is:
$P = [25 - F]^{0.5}$    P = pump outlet pressure   Atm(gauge)    F = acid flow rate m$^3$/h
The pump is primed but the pipe is empty.
At zero time the pump is started and acid begins to flow up the pipe.

**Problem:**
**a.**   Calculate the maximum height reached by the acid.                         [m]
**b.**   Find height of the acid 2 hours after start up.                           [m]
**Assume:** Zero suction pressure and friction loss in the pipe. All at NTP.

### 7:5 U   DAC absorbent regeneration

In a process for capture of $CO_2$ directly from the air a KOH absorbent solution is regenerated from $K_2CO_3$ by the causticising reaction:
$$K_2CO_3(aq) + Ca(OH)_2(aq) \rightarrow 2KOH(aq) + CaCO_3(s)$$
The reaction is carried out in a continuous fluidised bed reactor, seeded with 0.25 mm diameter $CaCO_3$ pellets which are grown to 1.0 mm diameter before removal from the reactor.
The pellet growth rate is controlled by mass transfer of Ca cations from the fluidising water slurry of $Ca(OH)_2$ to the pellet surface, according to:
Mass transfer rate = KC
where:
  K = mass transfer coefficient = 2E-4 m/s
  C = $Ca^{++}$ concentration in liquid = 0.04 kmol/m$^3$   (fixed)

**Problem:** Calculate the pellet residence time in the reactor.                    [s]
**Assume:** Spherical pellets with zero porosity and specific gravity = 2.8 (not so in reality).

## 7:6 W   CSTR bio-reactor conversion

The figure below shows a continuous stirred tank 2 m³ volume bio-reactor fed by a substrate solution (stream 1) and enzyme catalyst (stream 2).

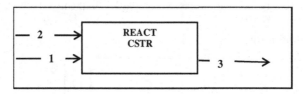

The kinetics of substrate conversion at 310 K are represented by:
$$Rate = dS/dt = (-1)vS/(K + S)$$
where:

| | | | |
|---|---|---|---|
| S | = substrate concentration | | kmol/m³ |
| v | = maximum reaction rate | = 5.0E-2 | kmol/(m³.h) |
| K | = constant | = 3.0E-2 | kmol/m³ |

Prior to startup the reactor is filled with 2 m³ of 0.6 molar solution of substrate.
At start up (time zero) steady flows of stream 1 (0.6 M substrate) and stream 3 are set at 1 m³/h and reaction is initiated and sustained by a negligible volumetric flow of enzyme solution in stream 2. Temperature is constant at 310 K.
**Problem:** Find the conversion of substrate in stream 3 — ten hours after start up.     [%]

## 7:7 U   Heated batch reactor

The steam heated batch mixed tank insulated thermochemical reactor of the figure below is loaded with 2 m³ of solution containing 5 molar reactant A initially at 25°C.

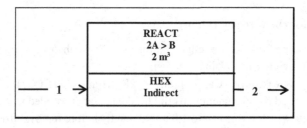

At time zero 245.6 psi(g) saturated steam is passed into stream 1, condensed and expelled as saturated liquid water at 245.6 psi(g) in stream 2. Reaction 1 then proceeds in the fixed reactant volume = 2 m³.
Reaction 1: 2A → B       Reaction rate = $d[A]/dt = (-1)k\{exp(-E/(RT))[A]^2\}$
   [A] = concentration of A kmol/m³
   E  = activation energy = 20E3 kJ/kmol       t = time h       T = reaction temperature K
   k  = rate constant = 100 m³/(kmol.h)       R= gas constant kJ/(kmol.K)

**Problem:** Use the steam table and data below to find —
a.   The reactor temperature at t = 0.6 hour                                                                    [K]
b.   The conversion of A at t = 1.0 hour                                                                        [%]

**Data:**
Adiabatic conditions. Heat of reaction = +100E3 kJ/kmol B (endothermic)
Rate of heat transfer from the steam to reactants = $10(T_s - T)$ kJ/h, $T_s$ = steam temperature K
Heat capacity and density of reactant mixture are fixed at = 4.0 kJ/(kg.K), 1200 kg/m$^3$

## 7:8 W   Mineral waste filtration

A waste mineral/water suspension is to be filtered in a plate and frame press with individual leaf area of 1 m$^2$. The waste has 1 wt% solids and must be treated at a rate of 1,000
m$^3$ per day (24 hours), including 1.5 hours every cycle for washing and discharging the
solids. The incompressible filter cake can accumulate on the filter to a maximum solids
loading of 150 kg/m$^2$ filter area. In the constant pressure process accumulating solids
lower the filtration rate as described in Equation 1.

Filtration rate = $dV/dt = A^2(\Delta P)/(2kV)$                                                              [1]
V             = filtrate volume   m$^3$            t = time            A = filter area   m$^2$
$\Delta P$             = fixed pressure drop through cake (negligible through filter cloth) = 200 kPa
k             = filtration constant, depends on filter cake permeability = 1E7 kg/(m$^3$.s)
Solids density = 3000 kg/m$^3$

**Problem:**
a.   Find the load of solids in the press operated 0.5 hour after the discharge step.   [kg]
b.   Calculate minimum number of filter leafs (1 m$^2$ each) needed for this process.   (–)
**Assume:** Volume fraction of solids in the waste (ca. 0.3 vol%) is negligible.

## 7:9 U   Bio-waste photo-oxidation

A biological wastewater containing xenobiotic materials (e.g. surfactants) with high biochemical oxygen demand (BOD) is to be treated by photochemical oxidation before discharge to a municipal sewage system. The process flowsheet below shows a batch of
wastewater recycled from a perfectly mixed tank through a plug-flow photochemical
reactor (PFR) in which organic contaminants are oxidised to carbon dioxide and water.

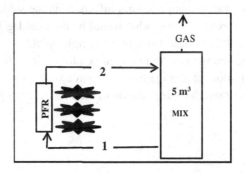

The rate of BOD removal in the photo-reactor is given by: $d[B]dt = (-1)kI[B]$

[B] = BOD "concentration" ppm(w)               $t$ = time

I  = intensity of UV light from mercury vapour lamps     lumen/m$^2$ = lux

k  = reaction rate constant                   1/(lux.s)

Wastewater flow in streams 1 and 2 = 10 m$^3$/h. Fixed volume in tank = 5 m$^3$

In this special case the light intensity is modulated to maintain the conversion of BOD per pass (X) through the plug-flow reactor at 10%.

***Problem:***

a. Calculate the time required to reduce BOD in the 5 m$^3$ tank from 1000 to 20 ppm(w).

b. In practice the BOD conversion per pass (X) could follow the plug-flow reactor profile:

$$X = 1 - \exp(-k'[B]) \text{ where } k' = \text{constant}$$

In that case derive an equation (do not solve it) for the time to reduce tank BOD from 1000 to 20 ppm(w). With an initial conversion 10% per pass and fixed corresponding illumination would the time to reach 20 ppm(w) be longer or shorter than in case "a"?

***Assume:*** Isothermal conditions. Negligible water hold-up in the PFR and piping.

### 7:10 W   Wastewater virus

The spread of a virus (e.g. COVID 19) in a population can be tracked and forecast by analyses of sewage in municipal wastewater treatment plants. The figure below represents the primary and secondary stages of a waste treatment plant, where solids are removed before organics are digested in a bio-reactor (not shown, from stream 5).

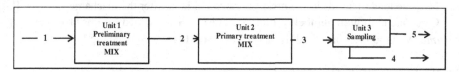

In this system contaminated sewage collected from a population centre flows some kilometers at velocities around 0.8 m/s before entering the treatment plant. The water is sampled (Unit 3, stream 4) for analysis after passing through the preliminary (Unit 1) and secondary (Unit 2) pretreatment stages which may be modelled as ideal mixed tanks. The transport pipe is 5 km long and residence times in Units 1 and 2 are respectively 5 and 10 hours.

***Problem:*** If there were a step change in viral infections in the subject population caused by a "super-spreader" event what would be the time lag before the viral RNA load reached 95% of peak value at the sample point?                [hr]

***Assume:*** Plug flow in sewage pipes and mixed conditions in water pretreatment operations. Steady flow of sewage from its source with a fixed load of virus for the period in question. Zero degradation of viral RNA in the pretreatment system.

## 7:11 U   Heating a pool

A private open air swimming pool 8 m long x 3 m wide x 1.5 m deep is heated by heat exchange from a natural gas ($CH_4$) burner at a fixed rate of 10 kW. Heat is lost from the pool surface at a rate given by the equation: $\dot{Q} = kA(T_p - T_a)$
where:
k = combined coefficient of heat transfer from pool surface by convection, evaporation and radiation = 0.15 kW/(m².K)          A = pool surface area                      m²
$T_p$ = well mixed pool temperature K          $T_a$ = air temperature = 278 K fixed

*Problem*:
a.   Find the time required to heat the pool from 10 to 30°C.                      [hours]
b.   Calculate the mass of $CO_2$ generated by burning natural gas to heat the pool      [kg]
*Assume*: Efficiency of heat transfer from burner to water = 70%.
          Pool is well mixed with mass unchanged by evaporation of water.
          Zero heat loss by conduction through walls, etc.

## 7:12 W   Mixing chlorobenzene-hexane

The figure below shows a tank of fixed volume ($V_T$ = 2.00 m³), which is used for mixing liquid chlorobenzene ($C_6H_5Cl$, Stream 1) with liquid hexane ($C_6H_{14}$, Stream 2)

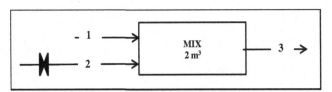

Initially the valve in Stream 2 is closed, the tank is full of liquid chlorobenzene which is fed by Stream 1 at the fixed rate of 1.00 m³/hour. At zero time the valve in stream 2 is opened and liquid hexane is fed to the tank at a fixed rate of 2.00 m³/hour. Stream 1 flow remains at 1.00 m³/hour.

*Problem*: Calculate the mole percent chlorobenzene in Stream 3 one (1) hour after the valve in Stream 2 is opened.                      [%]

*Assume*: The tank remains full of liquid throughout the operation at 298 K, with a fixed fluid volume $V_T$ = 2.00 m³. The liquid in the tank is perfectly mixed. Hexane/chlorobenzene form an ideal liquid mixture. No heat transfer with surroundings.

*Data*:                      Hexane(l)      Chlorobenzene(l)
          Density at 298 K kg/m³      660          1100          **NB. Different densities**

# Appendix Part 2

# SOLUTIONS

The Appendix Part 2 provides the solutions to the problems in Part 1 in Chapter order.
Solutions **1:1** to **7:12** correspond to problems **1:1** to **7:12**.
"W" and "U" designate worked and un-worked solutions.

## Chapter 1:   The General Balance Equation

**1:1 W   Atmospheric $CO_2$**

Define the system = atmosphere of Earth (an open system)
Specify the quantity = kilo-moles (kmol) of $CO_2$
Nomenclature: Component       Name    Molar mass
                  Air           A        28.8   kg/kmol
                  $CO_2$          C        44.0   kg/kmol
Amount of air in Earth's atmosphere = n(A). Ideal gas $PV = n(A)RT$
$n(A) = PV/(RT) = (101.3 \text{ kPa(a)})(4.0E18 \text{ m}^3)/((8.341 \text{ kJ/kmol.K})(273 \text{ K})) = 17.8E16$ kmol
Initial $CO_2$ (2020) = n(i,C) (0.01)(0.0415%)(17.8E16 kmol)          = 73.9E12 kmol
Mass balance on $CO_2$ over the period 2020–2050:
ACC = IN – OUT + GEN – CON                    GEN = CON = 0
ACC = [$CO_2$(fossil) + $CO_2$(land use)] – [$CO_2$(land sink) + $CO_2$(ocean sink)]
a.   $CO_2$ ACC = 440E12(44/12) + 190E12 – (510E12 + 630E12) =      **664E12 kg $CO_2$**
b.   $CO_2$ accumulated = 664E12 kg = 664E12 kg/44 kg/kmol = 15.1E12 kmol
    $CO_2$ Final = n(f,C) = n(i,C) + $CO_2$ ACC = 73.9E12 + 15.1E12 = 89E12 kmol
    n(f,C) = $CO_2$ in atmosphere 2050 = 89.2E12 kmol
    Conc $CO_2$ in 2050 = (1E6)(89E12 kmol)/(17.8E16 kmol) =        **500 ppm(v)**
c.   ppm(w) ≈ ppm(v)($MCO_2$)/(Mair) = 500 ppm(v)[44.0/28.8 ] =       **764 ppm(w)**

NB: Ideal gas: mole fraction = mf = (partial) volume fraction.        ppm(v) = 1E6[mf]

## 1:2 U    Direct air capture

No of 2000 T/d DAC plants =                                                                  **22E3**

## 1:3 W    Fossil fuel

Empirical stoichiometry of fuel combustion:
$C_{60}H_{120}O_2$ molar mass = 872 kg/kmol
$$C_{60}H_{120}O_2 + 89O_2 \rightarrow 60CO_2 + 60H_2O$$
$O_2$ consumed by burning 440E12 kg fuel
= (89 kmol/kmol)(440E12 kg)/(872 kg/kmol) = 4.58E12 kmol $O_2$
Drop in atmosphere $O_2$ = (1E6)(4.58E12 kmol)/(1.78E16 kmol) =    **257 ppm(v), Down**

## 1:4 W    Gasoline vs. grey hydrogen

Molar mass: $C_8H_{18}$ = 8(12.0) + 18(1.0) = 114 kg/kmol. $H_2$ = 2(1.0) = 2.0 kg/kmol
Basis. 100 km test drive. System = ICE
Atom balance on C:
      ACC = 0 = IN – OUT + GEN – CON                                    GEN = CON = 0
      Cin = (5 kg/114 kg/kmol)(8 C atom/mole $C_8H_{18}$) = 0.35 kmol C
      Cout = $CO_2$ out = Cin = 0.35 kmol = (0.35 kmol)(44.0 kg/kmol) = 16.7 kg $CO_2$
      Basis 100 km test drive. System = [$CH_4$ reformer + $H_2$ FC]
      $H_2$ consumed = 0.6 kg/2 kg/kmol = 0.3 kmol
      Mole balance on $CO_2$ (follow the stoichiometry):
      ACC = 0 = IN – OUT + GEN – CON
      $CO_2$ in = 0
      $CO_2$ gen = 1 kmol $CO_2$/4 kmol $H_2$
      $CO_2$ con = 0
      $CO_2$ out = $CO_2$ in + $CO_2$ gen – $CO_2$ con
      $CO_2$ out = 0 + (1/4 kmol $CO_2$/kmol $H_2$)(0.3 kmol $H_2$)(44 kg/kmol) = 3.3 kg $CO_2$
The hydrogen fuel cell generates least $CO_2$.                **3.3 vs. 16.7 kg $CO_2$/100 km**

## 1:5 U    Green hydrogen and oxygen

Change in $O_2$ concentration in atmosphere = $\Delta O_2$ =                **74 ppm(v), UP**

## 1:6 W    Jet plane

System = jet engine        Quantity = $CO_2$
Combustion reaction stoichiometry: $C_{14}H_{30} + (21.5)O_2 \rightarrow 14CO_2 + 15H_2O$
Molar mass $C_{14}H_{30}$ = 198 kg/kmol
Rate of kerosene combustion = (7920 kg/h)/(198 kg/kmol) = 40 kmol/hour
Mole balance on $CO_2$
ACC = 0 = IN – OUT + GEN – CON          steady-state
$CO_2$: 0 = 0 – OUT + (14 kmol $CO_2$/kmol $C_{14}H_{30}$) (40 kmol $C_{14}H_{30}$ /h) – 0
$CO_2$ OUT = (14)(40) = 560 kmol/h ≡ (44 kg/kmol)(560 kmol/h)          = **24.6E3 kg/h**

## 1:7 U   Space station $O_2$ and $H_2O$

Space station. Closed system, batch, reaction.
a.   Time to 19 vol% $O_2$ =                                                                    **1.56 days**
b.   Electrolysis stoichiometry;   $H_2O \rightarrow H_2 + 0.5O_2$ $H_2O$ consumed in 30 days  **= 81 kg**

## 1:8 W   Space station $CO_2$

Space station $CO_2$ balance. Closed system, batch, reaction ACC # 0, IN = OUT = 0
System = air in space station
a.   $CO_2$ balance
   Component = $CO_2$, time = t D = time to reach 2000 ppm(v) $CO_2$ days
   NB. vol% $\equiv$ (100)mole fraction (mf) in ideal gas, and ppm(v) = 1E6(mf)
   Total initial air = PV/RT = (100 kPa(a))(100 $m^3$)/((8.314 kJ/kmol.K)(298K))
                             = 4.04 kmol
   Mole balance on $CO_2$:
   ACC = Final – Initial = IN – OUT + GEN – CON                IN = OUT = CON = 0
   $CO_2$: ACC = 0 – 0 + (4 persons)(D days)(1 kg/d/44 kg/kmol) – 0 = 0.091D kmol $CO_2$
   $CO_2$ increase in D days = (4.04 kmol)(2000E-6 – 400E-6) = 6.46E-3 kmol
   D = 6.46E-3 kmol/0.091 kmol/d                                          **0.071 days**
b.   LiOH stoichiometry;          $2LiOH + CO_2 \rightarrow Li_2CO_3 + H_2O$
   Mole balance on LiOH:       IN = OUT = GEN = 0
   LiOH:       ACC = 0 – 0 + 0 – ((2/1) mol/mol)(0.091 kmol/d)(30 d) = 5.46 kmol
   LiOH consumed in 30 days = (5.46 kmol)(23.9 kg/kmol)                  **= 130.5 kg**

## 1.9 U   Air balance

Tank volume =                                                                   **13 $m^3$**

## 1:10 W   Viral infection

Viral (e.g. COVID) infection. Wells-Riley equation.
$P = C/S = 1 – \exp(–Igpt/Q)$
a.   The exponent (Igpt/Q) is dimensionless.
   Igpt = $m^3$/h      Iqt = (–)($m^3$/h)(h) = $m^3/h^2$                              **g = 1/h**
b.   Calculate the space ventilation rate at the <u>room conditions</u>
   Air flow = (5 changes/h)(500 $m^3$/change) = 2500 $m^3$/h at room conditions
   I = 2 infectors       p = 0.3 $m^3$/h at the room conditions
   Q = 2500/2 = 1250 $m^3$/(h.infector) at room conditions
   g = 15 1/hr, t = 3 hr
   P = 1 – exp[–(2)(15)(0.3)(3)/1250)] = 1 – 0.978
   Probability of infection =                                              **0.021**

## 1:11 U   Strategic materials

Potential $H_2SO_4$ generated =                                  **5.25E6 ton/year $H_2SO_4$**

## 1:12 W   Cobalt and lithium

Stoichiometry: Cathode composition $LiNi_xCo_yAl_zO_2$ x = 0.8, y = 0.15, z = 0.05
Molar mass = 1(6.9) + 0.8(58.7) + 0.15(58.9) + 0.05(27.0) + 2(16) = 96.1 kg/kmol
Co/Li mass ratio in cathode = 0.15(58.9)/1(6.9) = 1.28 Co/Li kg/kg
System: Li batteries for 50E6 EVs., each 100 kW.
Li per EV = 100 kWh/(7 kg/kWh) = 14.3 kg/EV.
Mass balance on Co in EVs: Co ACC = Co IN – 0 + 0
Co IN = Co ACC = (1.28 kg Co/kg Li)(14.3 kg Li/EV) = 18.3 kg Co/EV.
Co per 50E6 EV/year = (30E6 EV/y)(18.3 Co kg/EV) = 549E6 kg/y ≡      **549E3 ton/y**

## 1:13 U   Germanium

The solid and leachate quantities for a 2 × 2 factorial experiment are tabulated below.

| Acid leach | | Solid load kg/L leachate | | | |
|---|---|---|---|---|---|
| Factorial levels | | 1 | | 2 | |
| Total volume 1 litre | | | | | |
| | 30 | 0.75 | 0.157 | 1.2 | 0.125 |
| Acid conc. | wt% $H_2SO_4$ | kg | L | kg | L |
| | 60 | 0.75 | 0.377 | 1.2 | 0.302 |

## 1:14 U   Acid dilution

HCl mixture:                                    **6.25 L 1 M + 3.75 L 0.2 M**

## 1:15 U   Bio-acid

$$C_9H_{12}O_{15} \rightarrow C_6H_8O_7 + 3CO_2 + 2H_2O$$
Mass of saccharides =                             **18.0E3 kg**

## 1:16 W   Electro-methanol

System = electrochemical reactor
$$3CO_2 + 9H_2O \rightarrow xCH_3OH + yH_2CO_2 + zH_2 + 5O_2$$
Atom balances:
   C: 3 = x + y
   H: 18 = 4x + 2y + 2z
   O: 15 = x + 2y + 10
   Solve for x, y, z. x = 1, y = 2, z =5
$$3CO_2 + 9H_2O \rightarrow CH_3OH + 2H_2CO_2 + 5H_2 + 5O_2$$
Faradaic efficiency for methanol = E = 6M/(6M+2F+2H)
E = 6(1)/[6(1)+2(2)+2(5)] = 0.3 i.e.                  **30%**

## 1:17 W   Thermo-fertilizer

This problem can be solved by mass, mole or atom balances. Atom balances on selected
parts of the process give the simplest calculations.

$(NH_4)_2SO_4$, molar mass $= 2(14.0 + 4*1.0) + 32.1 + 4*1.0 = 132.1$ kg/kmol
$(NH_4)_2SO_4 = 6600$ kg/$(132.1$ kg/kmol$) = 50.0 = (n)$ kmol
Designate the species:

| Species | $H_2O$ | $SO_3$ | $N_2$ | $H_2$ | $H_2SO_4$ | $NH_3$ | $(NH_4)_2SO_4$ |
|---------|--------|--------|-------|-------|-----------|--------|----------------|
| kmol | a | b | c | d | e | f | n |

Define the system: Overall process
Atom balances:
  H: $2a + 3d = $ H in stream 7 $= 8(n)$ kmol
  O: $1a + 3b = $ O in stream 7 $= 4(n)$ kmol
  S: $1b$     $= $ S in stream 7 $= 1(n)$ kmol
  N: $2c$     $= $ N in stream 7 $= 2(n)$ kmol
Define the system: Final reactor (Unit 3)
Atom balances:
  H: $2e + 3f = $ H in stream 7     $= 8(n)$ kmol
  O: $4e$     $= $ O in stream 7     $= 4(n)$ kmol
  S: $1e$     $= $ S in stream 7     $= 1(n)$ kmol
  N: $1f$     $= $ N in stream 7     $= 2(n)$ kmol
Solve for a, b, c, d, e, f
$a = n$,  $b = n$,  $c = n$,  $d = 3n$,  $e = n$,  $f = 2n$. Substitute values for mass balances.

| AMMONIUM SULPHATE | | $H_2O$ | $SO_3$ | $N_2$ | $H_2$ | $NH_3$ | $H_2SO_4$ | $(NH_4)_2SO_4$ | Total |
|---|---|---|---|---|---|---|---|---|---|
| M | kg/kmol | 18 | 80 | 28 | 2 | 17 | 98 | 132 | kg |
| Overall | IN  kg | 900 | 4000 | 1400 | 300 | 0 | 0 | 0 | 6600 |
|  | OUT kg |  |  |  |  |  |  | 6600 | 6600 |
| Unit 1 | IN  kg | 900 | 4000 | 0 | 0 | 0 | 0 | 0 | 4900 |
|  | OUT kg | 0 | 0 | 0 | 0 | 0 | 4900 | 0 | 4900 |
| Unit 2 | IN  kg | 0 | 0 | 1400 | 300 | 0 | 0 | 0 | 1700 |
|  | OUT kg | 0 | 0 | 0 | 0 | 1700 | 0 | 0 | 1700 |
| Unit 3 | IN  kg | 0 | 0 | 0 | 0 | 1700 | 4900 | 0 | 6600 |
|  | OUT kg | 0 | 0 | 0 | 0 | 0 | 0 | 6600 | 6600 |
| Mass balances Overall and on each process Unit are "closed" i.e. [MASS OUT / MASS IN] = 1.000 | | | | | | | | | |
| Look at the problem again and see easier solutions e.g. by starting at Units 1 and 2 | | | | | | | | | |

**1:18 U    Solar water-split**

  $x = 2$     $y = 3$     $Ce_2O_3 + H_2O \rightarrow 2CeO_2 + H_2$
In Unit 2 ($O_2$ generator)   $2CeO_2 \rightarrow Ce_2O_3 + 0.5O_2$
$H_2$ OUT =                                                         **6.1 kg**

**1:19 U    Sodium-ion battery**

For 100 million vehicles:
Li Mass =           Mass        =                          **1.2E9 kg**
                    Volume = 1.2E9 kg/530 kg/m$^3$    =      **1.27E6 m$^3$**
Na Mass =           Mass        =                          **4.51E9 kg**
                    Volume = 4.51E9 kg/970 kg/m$^3$   =     **4.65E6 m$^3$**

## 1:20 W   Green ammonia

Reaction stoichiometry                $NH_3$ molar mass = 17.0 kg/kmol
System 1 Reforming and Haber:    $CH_4 + 2H_2O \rightarrow 4H_2 + CO_2$ [1]   $N_2 + 3H_2 \rightarrow 2NH_3$ [2]
System 2 Electrosynthesis:           $N_2 + 3H_2O \rightarrow 2NH_3 + (3/2)O_2$                [3]
Ammonia production 170E6 ton/year. = 170E6 ton/(17 ton/ton mol) = 10E6 ton mol/year
Assuming 100% efficiency in all process steps.
System 1: Overall atom balance on H, with zero accumulation.
0 = H(in) – H(out) = H(in) – (3 H/$NH_3$)(10E6 ton mole/y)        H(in) = 30E6 tmol/year
System 1 Atom stoichiometry C/H = 1/8
$CO_2$ averted = (1/8)30E6 ton mol/y ≡ 44 (ton/ton mol)(1/8)30E6 tmol/y =   **165E6 ton/y**
System 2: Mole stoichiometry      $O_2/NH_3$ = 1.5/2 = 0.75
$O_2$ generated = 0.75(10E6) ≡ (7.5E6 tmol/y)(32 t/tmol) =                    **24E6 ton/y**

# Chapter 2:   Process Variables and Their Relationships

## 2:1 W   Water

A.  Using data from the water phase diagram and steam table:
   a.   629.9°R ≡ 629.9 – 459.6 = 170.3°F ≡ .(170.3 – 32)/1.8 = 76.8°C ≡ 76.8 + 273.15 = 350 K
     vapour pressure $H_2O$ at 350 K = 41.66 kPa(a) at 30 kPa(a), 629.9°R water is a
                                                          **Gas**
   b.   20°F = (20+32)/1.8 = 28.9°C ≡ 302 K   At 302 K v.p. $H_2O$ < 41.6 kPa(a) **Liquid**
   c.   14.696 psi(g) ≡ (14.696 + 14.696) = 29.39 psi(a) ≡ (29.39/14.696)(101.3)
                = 203 kPa(a)                               **Liquid**
   d.   At 1 kPa(a), ln(1) = 0,      200 K on phase diagram           **Solid**
   e.   Pressure = 144.3 kPa(g) + 101.3 ≡ 245.6 kPa(a), 260.3°F ≡ 400 K **Liquid and gas**
   f.   0.611 kPa(abs), 273.16 K is the triple point of water     **Solid, liquid and gas**
B.  Based on the phase rule: F = C – Π + 2.
   a.   C = 2 components, Π = 1 phase     intensive variables = F = 2 – 1 + 2 =     **3**
   b.   In water ($H_2O$) NaCl > $Na^+$ + $Cl^-$ 3 chemical species, but $Na^+$ and $Cl^-$ are not
     independent species. i.e. 2 components  Intensive variables F = 2 – 1 + 2 =    **3**
   c.   In water ($H_2O$) $H_2SO_4$ > $H^+$ + $HSO_4^-$ + $SO_4$ = 5 species, but $H_2SO_4$, $H^+$, $HSO_4^-$ and
     $SO_4$ = tied by chemical equilibria. Not independent species.
     i.e. 2 components        Intensive variables F = 2 – 1 + 2 =                **3**

## 2:2 U   Stoichiometry

a.  (i)   $H_2SO_4 + 2NaOH \rightarrow Na_2SO_4 + H_2O$
                   Limiting reactant =                    **NaOH**
   (ii)  $C_7H_{16} + 11O_2 \rightarrow 7CO_2 + 8H_2O$
                   Limiting reactant =                    **$C_7H_{16}$**
   (iii) $CO_2 + 3H_2 \rightarrow CH_3OH + 2H_2O$
                   Limiting reactant =                    **$H_2$**

**b.** (i)  Excess $H_2SO_4$   Conversion =     **0.84**
   (ii) Excess $O_2$       Conversion =     **0.62**
   (iii) Excess $CO_2$      Conversion =     **0.91**

## 2.3 W  Bio-gas

$$C_xH_yO_zN(s) + wH_2O(l) \rightarrow (q+2)CH_4(g) + (q-2)CO_2(g) + NH_3(aq)$$

Atom balances on biochemical (fermentation) reactor

C: $x = (q+2) + (q-2) = 2q$
H: $y + 2w = 4(q+2) + 3$
O: $z + w = 2(q-2)$

Product gas composition: $(q+2)/[(q+2) + (q-2)] = 0.5169$ gives: $q = 59$
Solve for x,y,w

$x = 2(59) = 118$

Total mass balance:

$$12x + y + 16z + 14 + 18w = 61(16) + 57(44) + 17$$

Solve for: $y = 115$, $z = 68$, $w = 46$

$$C_{118}H_{155}O_{68}N(s) + 46H_2O(l) \rightarrow 61CH_4(g) + 57CO_2(g) + NH_3(aq)$$

Molar mass of solid = 1387 kg/kmol

Mass of $CH_4$ = (61 kmol/kmol waste)(1000 kg/1387 kg/kmol)(16 kg $CH_4$/kmol) =    **704 kg**

## 2:4 W  Combustion

System: Content of vessel. Fixed volume, closed system, reaction.
Let: Initial amount of carbon in vessel = X kmol.
Final mole fraction CO = Y
Initial $O_2$ in vessel = 28 kg/32 kg/kmol = 0.875 kmol
Final amount of gas in vessel = n = PV/(RT)
= $(50 + 101.3)$ kPa$(20\ m^3)/(8.314$ kJ/kmol.K$)(91 + 273)$K$) = 1.00$ kmol

**a.** System = closed vessel (closed system, i.e. IN = OUT = 0)
   Quantity = atoms of C (atoms are conserved)
   Integral balance: ACC = GEN − CON
   Atom balance on C: ACC = 0 = Final C − Initial C
   X = Final C ≡ 1.00 kmol ($CO_2$+CO) ≡ 1 kmol C =    **12 kg**
   Note: Each molecule of $CO_2$ and of CO contains 1 atom of carbon.

**b.** System = closed vessel (closed system, i.e. IN = OUT = 0)
   Quantity = atoms of O (atoms are conserved)
   Integral balance: ACC = GEN − CON
   Atom balance on O: ACC = 0 = Final O − Initial O
   Final O = (1.00 kmol)[Y kmol O/kmol CO) + (1−Y)(2 kmol O/kmol $CO_2$)]
   Initial O = 2(kmol O/kmol $O_2$)(0.875 kmol $O_2$) = 1.75 kmol O
   Balance on O: $(1)[Y + 2(1-Y)] = 1.75$    Y = 0.25
   Wt. frac CO = $(0.25)(28)/[(0.25)(28) + (1-0.25)(44)] = 0.175 \equiv$    **17.5 wt%**

## 2:5 W    Gas/water equilibrium

System: Contents of vessel. Fixed volume, closed system, no reaction ACC=GEN=CON=0

Designate the species:      $A \equiv N_2$   $B \equiv H_2O$

Nitrogen is above its critical temperature and assumed insoluble in any liquid water present, thus all $N_2$ is in gas phase. Water is below its critical temperature and thus may be partly in liquid phase – amount of liquid (if any) is unknown.

$n(A) = 28$ kg/28 kg/kmol = 1 kmol (all gas)

$n(B) = 5300$ kg/18 kg/kmol = 294.4 kmol (gas OR gas + liquid)

**IF** all water were in gas phase, then:

Mole fraction $B = y(B) = n(B)/(n(A)+n(B)) = 294.4/(1 + 294.4) = 0.996$

Partial pressure $B = p(B) = y(B)P = (0.996)(300$ kPa(a)$) = 299$ kPa(a)

Vapour pressure pure $H_2O$ at 400 K = 244 kPa(a) Antoine equation (or steam table)

Since 299 kPa > 244 kPa then part of the water must be in the liquid phase.

At $P = 300$ kPa(a) 380 K is below the dew-point temperature of the gas.

*When pure liquid water is present in equilibrium with the gas then:*

*partial pressure of water vapour* = vapour pressure *of $H_2O$* $\equiv 244$ kPa(a) at 400 K

In the gas phase: $p(B) = p^*(B) = 244$ kPa(a)

Mole fraction $H_2O$ in gas = $y(B) = p^*(B)/P = 244$ kPa/300 kPa(a) = 0.813

$y(B) = n(B)/(n(A)+n(B))$ solve for $n(B) = n(A)y(B)/(1-y(B))$

$n(B)$ in gas = 1 kmol $N_2$ (0.813)/(1 – 0.813) = 4.34 kmol

Mole balance on water in the closed system:

Water in liquid phase = 294.4 – 4.34 = 290.1 kmol

Volume of gas phase = $V_g = (n(A) + n(B))RT/P$ (ideal gas mixture)

$V_g = (1 + 4.34)$ kmol (8.314 kJ/kmol.K)(400 K)/(300 kPa) = 59.2 m$^3$

Volume of liquid = $V_l$

Density liquid $H_2O$ at 300 kPa(a)400 K = 953 m$^3$/kg (steam table)

$V_l = m(B)/\rho = (290.1$ kmol)(18 kg/kmol)/953 kg/m$^3$ = 5.57 m$^3$

Total volume in vessel = $V_g + V_l = 59.2 + 5.6 =$            **64.8 m$^3$**

## 2:6 W    Gas/liquid equilibrium

| Define species   j ≡ | A | B | C |
|---|---|---|---|
| Formula | $C_6H_{14}$ | $C_9H_{20}$ | $C_7H_8$ |
| Molar mass   kg/kmol | 86.0 | 128 | 82 |
| Liquid density at 277 K kg/m$^3$ | 659 | 718 | 866 |

w (j) = mass fraction j

x (j) = mole fraction j in liquid mix

y (j) = mole fraction j in vapour mix when liquid is 100% vaporized.

Liquid mixture density:   $\rho(mix) = 1/\Sigma[w(j)/\rho(j)]$ Ideal liquid mixture 277 K.

     $\rho(mix) = 700$ kg/m$^3$ = $1/[w(A)/659 + w(B)/718 + w(C)/866]$          [1]

After vaporization:               $PV = nRT = [m/(M_m)]RT$

Mean molar mass of vapour:     $M_m = mRT/(PV)$     m = mass, $M_m$ = mean molar mass

$M_m = (10 \text{ kg})(8.314 \text{ kJ/kg.K})(580 \text{ K})/(250 \text{ kPa(abs)}(2 \text{ m}^3)) = 96.4 \text{ kg/kmol}$

and $M_m = 96.4 = \Sigma[M(j)y(j)] = \Sigma[M(j)x(j)]$ (100% vaporization)

$96.4 = 86x(A) + 128x(B) + 92x(C)$ [2]

Also:

$\quad 1 = x(A) + x(B) + x(C)$ [3]

And

$\quad w(A) = x(A)M(A)/\Sigma[M(j)x(j)]$ [4]
$\quad w(B) = x(B)M(B)/\Sigma[M(j)x(j)]$
$\quad w(C) = x[C]M[C]/\Sigma[M(j)x(j)]$

Solve equations 1, 2, 3, 4 for all x(j). Solution by Excel Spreadsheet:

$$x(A) = 62\% \quad x(B) = 23\% \quad x(C) = 15\%$$

## 2:7 U Liquid to vapour

Composition of gas mixture: **X = 37.5, Y = 25.0 Z = 37.5 mol%**

## 2:8 W Mass transfer

Solution: $K = (D/d)(1/\varepsilon)Re^{1/3} Sc^{1/3}$

$K = [(1E\text{-}9 \text{ m}^2/\text{s})/(0.1 \text{ cm}/100 \text{ cm/m})](1/0.7)(10)1/3(1E6)1/3]$

$\quad = (1E\text{-}6 \text{ m})(1.43)(2.15)(100) = 307E\text{-}6 \text{ m/s}$

Rate = K[C] = (307E-6 m/s)(0.02 kmol/m³] = **61E-7 kmol/(m².s)**

## 2:9 U Heat transfer

Units of $h_i$ = Units of $h_o$ = **kW/(m².K)**

Units of k = **kW/(m.K)**

## 2:10 W Heat exchanger

From the test data: $U = Q/(A(\Delta T)) = 224 \text{ kW}/((10 \text{ m}^2)(60 \text{ K})) = 0.373 \text{ kW/(m}^2.\text{K})$

Convert the design specs to SI units:

$U = 150 \text{ [BTU/(ft}^2.\text{h.F}°)](1.054 \text{ kJ/BTU})(3.28^2 \text{ (ft/m)}^2)(1/3600 \text{ s/h})(1.8 \text{ F}°/\text{K})$

$\quad = 0.85 \text{ kW/(m}^2.\text{K})$

t = [0.0625 inch](3.28 ft/m)(1/12 inch/ft) = 0.017 m

$h_o$ = (0.85 kW/m²K)(350 [BTU/(ft².h.R°)](150 BTU/ft²h.R) = 1.98 kW/(m².K)

$h_i$ = (0.85)(250/150) = 1.42 kW/(m².K)

k = (100 BTU/h.ft.F)(1.055 kJ/BTU)(3.28 ft/m)/(1.8 F/K)/3600 s/h = 0.17 kW/(m.K)

Then:

$U = 1/(1/h_i + 1/h_o + t/k) = 1/(1/1.98 + 1/1.42 + 0.017/0.17) = 1/1.21$ = 0.76 kW/(m².K)

But measured U = 0.37 kW/(m²K)

The tube material and wall thickness would not change, with steam on the shell-side the $h_o$ value is probably unchanged — the loss of performance may be caused by the new process liquid, fouling the inside wall of the tubes. Shut down, inspect, and clean the tubes (inside).

## 2:11 W   Biochemical kinetics

Biochemical reaction rate: $u = 1/(K/(VS) + 1/V)$
Units of $S = kmol/m^3$          Units of $u$ = units of $V$
Multiply both sides by $1/V$:
$u/V = 1/(K/S + 1)$
$u/V$ = dimensionless, $K/S$ = dimensionless   Units of $K$ =                                    **kmol/m$^3$**

## 2:12 W   Electrochemical kinetics

Electrochemical reaction rate, current density:
$\qquad j = nFkKC(\exp(\alpha\eta F/RT))/[K + k(\exp(\alpha\eta F/RT))]$          **n = dimensionless**
The term $(\alpha\eta F/RT)$ is an exponent so is dimensionless.
Unit of "$\alpha$" = units of $RT/\eta F$
$= (kJ/kmol.K)(K)/[(Volt)(kC/kmol)] = kJ/(Volt)(kC)$          **$\alpha$ = dimensionless**
[Note: Volt = J/C]
Since $(\exp(a\eta F/RT))$ is dimensionless:
Units of $K$ = Units of $k(\exp(a\eta F/RT)$ = Units of $k$
Units of $K$ = Units of $k = (j/nFC) = (kA/m^2)/[(–)(kC/kmol)(kmol/m^3)]$ =          **m/s**

## 2:13 U   Photochemical kinetics

Photochemical reaction rate.          $dC/dt = (–1)kIC$
    **a.**   Units of $k = [lux.s]^{-1}$
    **b.**   Lux = measure of light intensity = lumen/m$^2$
An "Einstein" is the energy in Avogadro's number = 6.022E23 = 1 mole of photons
The "Einstein" can be used in photochemical processes to specify light intensity, as in
photon flux = x Einstein/(m$^2$.s) a.k.a      x moles of photons/(m$^2$.s)

## 2:14 W   Thermochemical kinetics

Thermochemical reaction kinetics with mass transfer constraint:
$\qquad$ Rate = $dC/dt = – (k)\exp(–E/RT)KaC/[(k)\exp(–E/RT) + Ka]$
First find the unspecified units.
E: E/RT is dimensionless Units of E = units of RT = kJ/(kmol.K)(K) =          **kJ/kmol**
K: Denominator
Units of Ka = units of $(k)\exp(–E/RT)$ = s$^{-1}$ Units of K = $(1/m^2/m^3)(s^{-1})$ =          **m/s**
Units of RHS = $(s^{-1})(–)(m/s)(m^2/m^3)(kmol/m^3)/(s^{-1})(–)$ = kmol m$^{-3}$.s$^{-1}$ = kmol/(m$^3$.s)
Units of LHS = dC/dt = kmol/(m$^3$.s) Units of t kmol/m$^3$/(kmol/(m$^3$.s)) =          **s**
Convert reaction rate from imperial units lbmol/(ft$^3$h) to SI units kmol/(m$^3$.s)
lbmol/(ft$^3$h) = (1/2.205 kmol/lb mole)((3.281)$^3$ft$^3$/m$^3$)(1/3600 hr/s) = 4.4E-3 kmol/(m$^3$.s)
–5 lb.mol/(ft$^3$.h) = (–5 lb.mol/ft$^3$.h)(4.4E-3 kmol/m$^3$.s) = –2.22E-2 kmol/(m$^3$.s)
–5 lb.mol/(ft$^3$.h) = (–3600 s/h)(2.22E-2 kmol/(m$^3$.s) = –80 kmol/(m$^3$.h)
Conversion rate in 3 m$^3$ volume

Differential mole balance on reactant in the open system at steady-state
ACC = 0 = [IN – OUT] + GEN – CON GEN = 0
IN – OUT = CON = (3 m$^3$)(80 kmol/(m$^3$h)(24 hr/d) = 5.77E3 kmol/day
Reactant consumed = (5.77E3 kmol/d)(70 kg/kmol) = 404E3 kg/d =          **404 t/d**

### 2:15 U   Photo-voltaic array

Units of n =                                                             **n = dimensionless**

### 2:16 W   Solar thermal energy

Solar energy receiver efficiency: $E = a - be(T_s^4 - T_a^4)/(Cd)$
Units of a = units of E                                                  **E = dimensionless**
Units of $\{be(T_s^4 - T_a^4)/(Cd)\}$ = units of a                       **a = dimensionless**
Units of $\{be(T_s^4 - T_a^4)\}$ = units of Cd = W/m$^2$
Units of b = (W/m$^2$)/(K$^4$)                                           **b = W/(m$^2$K$^4$)**

### 2:17 U   Dimensional consistency

Correct equations are dimensionally consistent.
For all A,B,C,D LHS = dimensionless
a.   RHS = $[up/(gV^2)]^{0.5} + 2.3[gd/(uV)]^2$
     Dimensionally inconsistent                                          **Incorrect**
b.   RHS = $3V[up/(gd)]^{0.67}$
     Dimensionally inconsistent                                          **Incorrect**
c.   RHS = $[uV/gpd^2]^{0.33} + [V^2/(gd)]^{0.5}$
     Dimensionally consistent, both terms are dimensionless              **OK**
d.   RHS = $[gp/(ud^2]^{0.33} + [uV/gpd]^{0.5}$
     Dimensionally inconsistent                                          **Incorrect**

### 2:18 W   Gas flow

Convert all conditions to SI units.
Pressure = P = (80 psig/14.7 psi/atm)(101.3 kPa/atm) = 551.3 kPa(g)
≡ (551.3 kPa(g) + 101.3 kPa) = 652.67 kPa(a) Assumed unchanged through orifice
Temperature = T = (400°C + 273°C) = 673 K Assumed unchanged through orifice
Orifice diameter = D = (10 inch/12 inch/ft)(1/3.28 ft/m) = 0.254 m
Orifice flow area = 3.14 D$^2$/4 = 3.14 (0.254 m)$^2$/4 = 50.65E-3 m$^2$
Gas mass flow = (500 lb$_m$/min)/((2.2 lb$_m$/kg)(60 s/min)) = m = 3.788 kg/s
Note: lb$_m$ = pounds mass    R = 8.314 kJ/(kmol.K)
Gas volume flow by gas law = V = nRT/P = (m/M)RT/P
V = (3.788 kg/s/28 kg/kmol)(R kJ/(kmol.K)(673 K))/(652.6 kPa(abs)) = 1.16 m$^3$/s
Gas velocity = Volume flow/orifice flow area = (1.16 m$^3$/s)/(50.56E-3 m$^2$) =    **22.9 m/s**

### 2:19 U    Orifice meter

a.   Gas volume flow =                                                                          176E3 m$^3$/h
     Gas mean velocity =                                                                        31 m/s
b.   Heat to stove =                                                                            32 kW

### 2:20 U    Vaporization

a.   Methanol =                                                                                 33.4 wt%
     Ethanol =                                                                                  66.6 wt%
b.   Pressure P =                                                                               217.2 kPa(a)
c.   Methanol in vapour =                                                                       58.3 mol%
d.   Partial volume ethanol =                                                                   2.32 m$^3$

### 2:21 W    Ignition

System: Contents of vessel. Closed system, fixed volume, reaction
Quantities Moles of $CH_3OH$, $C_2H_5OH$, $O_2$, $CO_2$, $H_2O$

a.   Stoichiometry of the oxidation (combustion) reactions:
     $CH_3OH + 1.5O_2 \rightarrow CO_2 + 2H_2O$
     $C_2H_5OH + 3O_2 \rightarrow 2CO_2 + 3H_2O$
     Integral mole balances on each component in vessel (a closed system).
     ACC = Final – Initial = IN – OUT + GEN – CON
     IN = OUT = 0 in a closed system              Final = ACC + Initial
     $CH_3OH$        ACC = GEN – CON = –n(1)            Final = 0
     $C_2H_5OH$      ACC = GEN – CON = –n(2)            Final = 0
     $O_2$   ACC = GEN – CON = –(1.5n(1) + 3n(2))       Final $O_2$ = 0.323 – (1.5(n(1) + 3n(2))
     $CO_2$          ACC = GEN – CON = n(1) + 2n(2)        Final = n(1) + 2n(2)
     $H_2O$          ACC = GEN – CON = 2n(1) + 3n(2)       Final = 2n(1) + 3n(2)
     Total final moles = 0.323 +1.5n(1) + 2n(2)
     Initial mass of liquid mixture = 32n(1) + 46n(2) = 2 kg
     Total final moles = 0.323 + 1.5n(1) + 2(2 – 32n(1))/46
     Total final moles = 0.323 + 1.5n(1) + 0.087 – 1.39n(1) = 0.41 + 0.11n(1)

| Species | $CH_3OH$ | $C_2H_5OH$ | $O_2$ | $CO_2$ | $H_2O$ |
|---------|----------|------------|-------|--------|--------|
| j | 1 | 2 | 3 | 4 | 5 |
| M | 32 | 46 | 32 | 44 | 18 |
| Initial | n(1) | n(2) | 0.323 | 0 | 0 |
| GEN | 0 | 0 | 0 | n(1)+2n(2) | 2n(1)+3n(2) |
| CON | n(1) | n(2) | 1.5n(1)+n(2) | 0 | |
| Final | | | | 0.323 – (1.5n(1)+3n(2)) | |

IF all final components are in the gas phase, then:
P = 250 kPa(a) = nRT/V = (0.41 + 0.11n(1))(R)(350 K)/(4 m$^3$)
Solve for n(1) = – 0.60 kmol NEGATIVE !

There is no real positive value of n(1) to match the "all gas phase" assumption.
Therefore some gas must condense to leave a pressure of 250 kPa(a)
350 K is above the critical temperatures of $O_2$ and $CO_2$ but not of $H_2O$.
Some $H_2O$ condenses at 350 K to leave a total pressure 250 kPa(a)
Vapour pressure $H_2O$ at 350 K = 41.3 kPa (Antoine)
Partial pressure of remaining $[O_2 + CO_2]$ = 250 – 41.3 = 208.7 kPa(a) in gas
Final $[O_2 + CO_2]$ = PV/RT = (208.7 kPa)(4 m³)/(R) 350 K) = 0.286 kmol
Final $[O_2 + CO_2]$ = (0.323 – 1.5n(1) – 3n(2)) + (n(1) + 2n(2))
Final $[O_2 + CO_2]$ = 0.323 – 0.5n(1) – n(2) = 0.323 – 0.5n(1) – (2 – 32n(1)/46)
Solve for: n(1) = 0.0331          n(2) = 0.0204 kmol
Initial wt% MeOH = (100)(32 kg/kmol)(0.0331 kmol)/2 kg =          **53 wt%**

b.   **IF** all $H_2O$ was condensed, then volume of liquid water:
Vol $H_2O$ = ((1/1000) m³/kg)(18((2)(0.033) + (3)(0.0204))kg = 2.3E-3 m³
i.e. Liquid water has negligible effect on gas volume of 4 m³
$H_2O$ in gas = p*V/(RT) = (41.3 kPa)(4 m³)/((R)(350)K) = 0.0568 kmol
$O_2$ in gas = 0.323 – (1.5(0.0331) + 3(0.0204)) = 0.21 kmol
$CO_2$ in gas = 0.0331 + 2(0.0204) = 0.074 kmol
Total gas = 0.343 kmol
Gas phase composition*:
Volume % = mol %          **$H_2O$ = 16.5 vol%  $O_2$ = 61 vol%,  $CO_2$ = 21.5 vol%**
*An exact calculation would use Henry's law to account for the small solubility of $O_2$ and $CO_2$
in liquid water.

## 2:22 W   Steam table

a.   i   Saturated vp at 400 K = 245.6 kPa < 250 kPa          **Liquid**
     ii   Superheated gas water at 300 K 1 kPa < 3.5 kPa          **Gas**
     iii  Subcooled liquid water. 3000 kPa 400 K < 500 K          **Liquid**
b.   Saturated gas water.
     Density = 1/specific volume = 1/0.001027 m³/kg =          **973.7 kg/m³**
c.   Subcooled liquid water. Mean heat capacity:
     $C_{p,m}$ = (h1– h2)/(T1–T2) = (991.1 – 567.3) kJ/kg/(500 – 400) K =   **4.23 kJ/(kg.K)**
d.   Vapour pressure n-octane at 400 K = 104.9 kPa     Antoine equation
     Vapour pressure water at 400 K = 245.6 kPa          Steam table
     Each component of the L/L *emulsion* exerts its independent vapour pressure.
     Mole fraction is proportional to partial pressure
     At equilibrium:
     Mole % n-octane in gas = 100[104.9 kPa/(104.9 + 245.6) kPa] =          **30 mol%**
e.   At 200 kPa(abs), 350 K saturated water vapour pressure: p* = 41.7 kPa(a)
     Relative humidity = p/p* = 0.8
     Partial pressure $H_2O$ = p = 0.8(41.7)kPa(a) = 33.4 kPa(a) (ideal gas)
f.   Water in 7 m³ at 350 K = PV/(RT) = (33.4 kPa(a))(7 m³)/(R)(350 K) = 0.08 kmol
     Mass of water = (18 kg/kmol)(0.08 kmol) =          **1.44 kg**

## 2:23 U   Steam quality

a.   The water must be a two-phase (G/L) system with pressure = P =        **245.6 kPa(a)**
b.   Quality =                                                              **68.4%**

## 2:24 W   Methane combustion

Methane combustion stoichiometry:
$$CH_4 + 2O_2 \rightarrow CO_2 + 2H_2O$$
Mole fraction $H_2O$ in product gas = y = 2/(1 + 2) = 0.667
Partial pressure $H_2O$ in product gas:
(Mole fraction)(total pressure) = yP = (0.667)(200 kPa(abs)) = 133.4 kPa
Dew point occurs when: $p(H_2O) = p^*(H_2O)$ = 133.4 kPa(abs)
In the Antoine equation: $p^* = \exp(A - B/(T+C))$ = 113.4 kPa
Solve for T =                                                               **377 K**
$CO_2$ is above its critical temperature – so cannot condense.

## 2:25 U   Hydrazine dew and bubble

Solution.
a.   Equilibrium pressure   =                                              **490 kPa(a)**
b.   Dew point temperature =                                              **406 K**
c.   Bubble point pressure  =                                              **70 kPa(a)**

## 2:26 W   Boudouard

Combustion reactions 1 and 2:
   1. $C(s) + O_2(g) \rightarrow CO_2(g)$            $h_{c,298K}$ = –394E3 kJ/kmol
   2. $CO(g) + (0.5)O_2(g) \rightarrow CO_2(g)$      $h_{c,298K}$ = –282E3 kJ/kmol
$h_{rxn} = \Sigma[n(j)h_f(j)]$products $- \Sigma[n(j)h_f(j)]$reactants
Reaction 1: $h_{rxn} = h_c$ = –394E3 kJ/kmol – (0 + 0) = –394E3 kJ/kmol
Reaction 1: $h_{rxn}$ = –394E4 = ($h_fCO_2$ – (0 + 0)).   $h_fCO_2$ = –394E3 kJ/kmol
Reaction 2: $h_{rxn}$ = (–394E3) – ($h_fCO$ + 0).
       $h_fCO$ = (–394E3) –(–283E3)           $h_fCO$ = –110E3 kJ/kmol
Reaction 3: $C(s) + CO_2(g) \rightarrow 2CO(g)$          at 298 K
       $h_{rxn}$ = 2(–110E3) – (0 – 394E3) =      **+ 174E3 kJ/kmol Endothermic**

## 2:27 U   Fischer–Tropsch

Stoichiometry                              X = conversion S = selectivity
Boudouard:        $C + CO_2 \rightarrow 2CO$              X1, S1          rxn 1
Fischer-Tropsch: $8CO + 17H_2 \rightarrow C_8H_{18} + 8H_2O$  X2, S2          rxn 2
$CO_2$ = 79.4 ton.mole $CO_2 \equiv$                        **3492 ton/day $CO_2$**

## 2:28 W  Claus

a.  Reactions:

1. Combustion of S:     $S(s) + O_2(g) \rightarrow SO_2(g)$
2. Combustion of $H_2S$: $H_2S(g) + (3/2)O_2(g) \rightarrow SO_2(g) + H_2O(g)$
3. Combustion of $H_2$: $H_2(g) + 0.5O_2(g) \rightarrow H_2O(g)$
4. $2H_2S(g) + SO_2(g) \rightarrow 3S(s) + 2H_2O(g)$

Find the standard heats of formation for $H_2S$, $SO_2$ and $H_2O$.

$h_{c298} = h_{rxn} = \Sigma[n(j)h_f(j)]$products $- \Sigma[n(j)h_f(j)]$reactants

Reaction 1: $h_{rxn} = -297E3 = h_f SO_2 - (h_f S + h_f O_2) = h_f SO_2 - (0 + 0)$
     $h_f SO_2 = -297E3$ kJ/kmol

Reaction 2: $h_{rxn} = -518E3 = (h_f SO_2 + h_f H_2O) - (h_f H_2S + (3/2)h_f O_2)$
$h_f H_2S = (-297E3 + -242E3) - (-518E3) = -21E3$ kJ/kmol

Reaction 3: $h_f H_2O = -242E3 - (0 + 0) = -242E3$ kJ/kmol

Reaction 4: $h_{rxn} = 2h_f H_2O + 3h_f S(s) - (2h_f H_2S + h_f SO_2)$

Reaction 4: $h_{rxn} = 2(-242E3) + 3(0) - (-297E3 + (-2(-21E3))) = $ **–145E3 kJ/kmol SO$_2$**

b.  Adding enthalpies of fusion and vaporization will decrease the (abs) heat of reaction.

$2H_2S(g) + SO_2(g) \rightarrow 3S(l) + 2H_2O(g)$ hrxn $= -145E3 + 1.7E3 = $ **–143.3E3 kJ/kmol SO$_2$**

$2H_2S(g) + SO_2(g) \rightarrow 3S(g) + 2H_2O(g)$ hrxn $= -143.4E3 + 9.8E3 = $ **–133E3 kJ/kmol SO$_2$**

## 2:29 W  Raoult and Henry

System: Contents of vessel. Closed system, no reaction. IN = OUT = ACC = GEN = CON = 0
Quantities: moles of benzene and of hydrogen.
Designate the components:

Benzene ($C_6H_6$)     $j = 1$     $M(1) = 78$ kg/kmol
Hydrogen ($H_2$)     $j = 2$     $M(2) = 2$ kg/kmol

Let: $n_L(j) = $ kmol component "j" in liquid phase at equilibrium.
Total liquid volume
$V_L = (n_L(1)$kmol$)(78$ kg/kmol$)/(850$ kg/m$^3) + (n_L(2)$kmol$)(0.05$ m$^3$/kmol$)$ m$^3$
Total gas volume $= V_G = 2 - V_L$ m$^3$

Mole fraction in liquid:     $x(1) = n_L(1)/(n_L(1) + n_L(2))$
          $x(2) = 1 - x(1)$

Partial pressure in gas:     $p(1) = p*(1)x(1) = 90$ x(1) kPa(a) Raoult's law
          $p(2) = K_H x(2) = 0.5E6$ x(2) kPa(a) Henry's law

Moles in gas:     $n_G(1) = p(1)V_G/RT = p(1)V_G/((R)(350$ K$))$
          $n_G(2) = p(2) V_G/RT = p(2)V_G/(R)(350$ K$))$ Ideal gases

Mole balances:
$0 = 10 - [n_L(1) + n_G(1)]$     $= 10 - f[n_L(1)]$     Benzene     [1]
$0 = 2 - [n_L(2) + n_G(2)]$     $= 2 - f[n_L(2)]$     Hydrogen     [2]

Solve the simultaneous non-linear equations 1 and 2 for:     $n_L(1) = 9.967$ kmol
Use Excel Goal Seek or Solver     $n_L(2) = 0.103$ kmol
$p(1) = 89.1$ kPa(a)

$p(2) = 5110$ kPa(a) Total pressure $= P = p(1) + p(2) =$ **5200 kPa(a)**

Mass of hydrogen in liquid $= (0.103$ kmol$)(2$ kg/kmol$) =$ **0.206 kg**

## 2:30 U   Haber-Bosch

Reaction stoichiometry: $N_2 + 3H_2 \rightarrow 2NH_3$

Conversion of $N_2 = X = 0.365 \equiv$                          **36.5%**

## 2:31 W   NH$_3$ to NO

Reaction stoichiometry: $N_2 + 3H_2 \rightarrow 2NH_3$

Designate: $N_2 \equiv 1$, $H_2 \equiv 2$, $NH_3 \equiv 3$

Heat of reaction $= h_{rxn} = \Sigma(v(j)h_f(j))$ products $- \Sigma(v(j)h_f(j))$ reactants

Standard heat of reaction has $h_{rxn}$ and $h_f$ at 298 K

a.   For $NH_3(g)$

   $h_{rxn} = 2(-46E3)$ kJ/kmol $- (1(0) + 3(0)) = -92E3$ kJ/kmol $= -46E3$ kJ/kmol $NH_3$

   For $NH_3(l)$

   $h_{rxn} = h_{rxn}(gas) - h_{vap} = -46E3 - 21E3 =$              **$-67E3$ kJ/kmol NH$_3$**

b.   Convert gas composition to mol fraction = volume fraction = $y(j)$

   $y(j) = w(j)/M(j)/\Sigma w(j)/M(j)$

   $\Sigma w(j)/M(j) = 0.3/28 + 0.1/2 + 0.6/17 = 0.096$

   $N_2$: $y(1) = 0.3/28/0.096 = 0.12$, $H_2$: $x(2) = 0.1/2/0.096 = 0.52$

   $NH_3$: $x(3) = 0.6/17/0.096 = 0.37$

   $\Sigma y(j) = 0.116 + 0.521 + 0.368 = 1.00$ Check

   Dew point pressure is total pressure to begin condensation of $NH_3$.

   i.e. Partial pressure of $NH_3$ = vapour pressure $NH_3$ at 300 K

   pp $NH_3 = p^*(3) = \exp(15.494 - 236324/(300 - 22.6206)) = 1069$ kPa

   Total pressure $= 1069/0.368 =$                          **2900 kPa(a)**

c.   Reaction (oxidation) stoichiometry: $4NH_3 + 5O_2 \rightarrow 4NO + 6H_2O$

   Define the system: Contents of the oxidation reactor – closed system

   Mole balance: ACC = IN – OUT + GEN – CON          IN = OUT = 0

   With 20% excess $O_2$

   Basis: [4 kmol $NH_3$ + (1.2)5 kmol $O_2$ + (1.2)5(79/21) $N_2$] Initial

   $NH_3$: ACC $= 0 - 0 + 0 - 4 = -4$ kmol

   $O_2$: ACC $= 0 - 0 + 0 - 5 \quad = -5$ kmol

   NO: ACC $= 0 - 0 + 4 - 0 \ = 4$ kmol

   $H_2O$: ACC $= 0 - 0 + 6 - 0 = 6$ kmol

   $N_2$: ACC $= 0 - 0 + 0 - 0 \ = 0$ kmol

Set up a spreadsheet table.

| $4 NH_3 + 5O_2 > 4NO + 6H_2O$ | | | 100 | % conv'n | 20 | % exc $O_2$ |
|---|---|---|---|---|---|---|
| Species | $NH_3$ | $O_2$ | $N_2$ | NO | $H_2O$ | Total |
| Initial | 4 | 6 | 22.6 | 0 | 0 | |
| ACC | -4 | -5 | 0 | 4 | 6 | |
| Final | 0 | 1 | 22.6 | 4 | 6 | 33.6 |
| Mol frac | 0 | 0.03 | 0.67 | 0.12 | 0.18 | |

Critical temperatures of $O_2$ and $N_2$ are 154 and 126 K – will not condense
The dew point temperature is temperature where water begins condensing.
i.e. Where partial pressure of $H_2O$ equals vapour pressure $H_2O$.
p $H_2O$ = p*$H_2O$ = x($H_2O$)(Ptotal) = 0.18(200) kPa = 36 kPa
The Antoine equation:
$$p(H_2O) = p*H_2O = \exp[16.5362 - 3985.44/(T - 38.9974)] = 36 \text{ kPa}$$
Solve for T = $T_D$ (Solution by Excel Goal Seek) =                          **347 K**

### 2:32 U   Biochemical reaction

Solve simultaneous equations:                 **a = 0.20 kmol/(m³.h), K = 0.03 kmol/m³**

### 2:33 U   Azide decomposition

Reaction stoichiometry: $2HN_3 \rightarrow H_2 + 3N_2$
Set up a spreadsheet and solve the mole balances for Y =                    **29.8 kg**

|  | $HN_3$ | $H_2$ | $N_2$ |  |  |
|---|---|---|---|---|---|
| Species | A | B | C | Total | Total mass |
|  | kmole | kmole | kmole | kmole | kg |
| Initial | X | 0 | 0 | X | Y |
| Final | 0 | X/2 | 3X/2 | 2X | 43*X |
| Final total moles n = PV/RT |  |  |  | 1.39 |  |
|  |  |  | X = | 0.693 | 29.81 |

### 2:34 W   Psychrometry

a.   At intersection of vertical $T_{db} = 40°C$ with diagonal $T_{wb} = 30°C$
     Relative humidity = RH =                                              **49%**
b.   Locate the intersection of the $T_{db} = 40°C$ and the rh = 30% lines. Follow horizontally
     back at constant moisture content to meet the saturation curve at $T_{dew}$ =       **19°C**
     Initial condition: $T_{db} = 40°C$ and, rh = 30% meet at: moisture content
     = 13.8 gram/kg dry air = 13.8E-3 kg/kg dry air
c.   Let mass of dry air = m     kg
     Initial mass of water vapour in air = w1     kg
     Then:     m+w1 = 5 kg
     w1/(m+w1) = 13.8E-3 kg/kg     w1 = 0.069 kg     m = 4.931 kg
     Final condition:   $T_{db} = 10°C$
     This condition is below the dew-point ($T_{dew} = 18.8°C$) so the air will be saturated with
     water vapour at 10°C.
     Moisture content saturated at 10°C = 7.6 g/kg dry air
     (from horizontal line on the psychrometric chart).
     Let final mass of water vapour in air = w2 kg
     Mass of dry air is unchanged at       m = 4.931 kg
     Then:       w2/m = 7.6E-3 kg/kg        w2 = 4.931(7.6E-3) = 0.0375 kg
     Mass balance on water: Integral balance. Open system. No reactions

ACC = IN – OUT

OUT = Initial – Final = w1 – w2 = 0.069 kg – 0.038 kg = water condensed = **0.031 kg**

**d.**  Air (wet) at 40°C, 50% rh:

Intersect $T_{db}$ = 40°C with rh = 50%

Follow horizontal to moisture (i.e. water) content = 0.024 kg/kg dry air.

Follow diagonal to Enthalpy = H(1) = 100 kJ/kg dry air

Dry air mass = 5 kg(1/1.024) = 4.88 = m kg (fixed)

Air at 10°C, saturated with water = 0.008 kg/kg dry air

Enthalpy = H(2) = 30 kJ/kg dry air

Heat removed = m(H1-H2) = 4.88 kg dry air (100 – 30) kJ/kg dry air = **341 kJ**

NB. The reference state for these enthalpies is the compounds ($H_2O$) at 273 K.

## 2:35 W   Gas/liquid, benzene-xylene

System: Contents of vessel. Closed system, no reaction    ACC= GEN = CON = 0

Quantities: Moles of benzene and of o-xylene

Designate:   A ≡ benzene,      B ≡ o-xylene      $V_1$ = liquid volume at equilibrium

x(A), x(B) = equilibrium mole fraction A, B in liquid phase.

w(A), w(B) = equilibrium mass fraction A, B in liquid phase

y(A), y(B) = equilibrium mole fraction A, B in gas phase

M(A), M(B) = molar mass A, B

m(A), m(B) = mass A, B        $m_1$ = liquid mass        $\rho_1$ = liquid density

Relations:

| | | |
|---|---|---|
| x(A) + x(B) = 1 | y(A) + y(B) = 1    w(A) + w(B) = 1 | [1] |

Vapour pressure, 473 K: p*(A) = 1448 kPa   p*(B) = 362 kPa(a) Antoine    [2]

Raoult's law:        p(A) = x(A)p*(A)    [3]

p(B) = x(B)p*(B) = [1 – x(A)]p*(B)

Dalton's law:        p(A) + p(B) = P = 1000 kPa(a) at 473 K    [4]

Solve for:        x(A) = 0.59        x(B) = 0.41

Liquid composition:    w(A) = x(A)M(A)/[x(A)M(A) + x(B)M(B)]    [5]

Liquid density:    $\rho_1$ = 1/[w(A)/ρ(A) + w(B)/ρ(B)]    [6]

Liquid mass:    $m_1 = V_1/\rho_1$    [7]

Gas composition    y(A) = p(A)/P      y(B) = 1 – y(A)    [8]

Gas volume:    $V_g = V – V_1$

Gas kmol:    $n_g = PV_g/RT$    [9]

Gas components    A: $n(A)_g = y(A)n_g$

B: $n(B)_g = y(B)n_g$

Liquid components    A: $m(A)_1 = m_1w(A)$

B: $m(B)_1 = m_1w(B)$

Total mass balance:    ACC = 0 = Final – Initial

0 = f(V₁) = M(A)y(A)$n_g$ + M(B)y(A)$n_g$ + $m_1$w(A) + $m_1$w(B) – 1310 kg    [10]

Solve equations for $V_1$ to close the mass balance (Excel Goal seek or Solver).

**a.** Liquid volume =   $V_1$ =                                                   **1.90 m³**
**b.** Total benzene =   m(A) =                                                 **678 kg**

## 2:36 W   Incomplete combustion

System: Continuous reactor. Open system, steady-state, reaction ACC = 0, GEN and CON # 0
Basis:   100 kmol/h propane in stream 1.

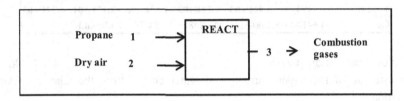

$C_3H_8 + 5O_2 \rightarrow 3CO_2 + 4H_2O$   rxn 1
$C_3H_8 + 3.5O_2 \rightarrow 3CO + 4H_2O$   rxn 2
Let: $\dot{n}(2,B) = O_2$ feed rate kmol/h        S = selectivity for reaction 1 (-)
Designate the species:
$A \equiv C_3H_8$      $B \equiv O_2$      $C \equiv N_2$      $D \equiv CO_2$      $E \equiv CO$      $F \equiv H_2O$
$\dot{n}(j)$ = flow of j  $\bar{n}(i)$ = total flow stream i kmol/hr
Differential mole balances at steady-state:
Rate ACC = 0 = Rate IN – Rate OUT + Rate GEN – Rate CON
Note the stoichiometry, with an arbitrary basis of 100 kmol/h propane feed.

| | |
|---|---|
| A: $0 = 100 - \dot{n}(3,A) - 0.9(100)$ | [1] |
| B: $0 = \dot{n}(2,B) - \dot{n}(3,B) - [5(0.9)(S)(100) + 3.5(0.9)(1-S)(100)]$ | [2] |
| C: $0 = \dot{n}(2,C) - \dot{n}(3,C)$        $N_2$ is un-reactive | [3] |
| D: $0 = 0 - \dot{n}(3,D) + 3(0.9)(S)(100)$ | [4] |
| E: $0 = 0 - \dot{n}(3,E) + 3(0.9)(1-S)(100)$ | [5] |
| F: $0 = 0 - \dot{n}(3,F) + 4(0.9)(100)$ | [6] |
| Composition of dry air:        $79/21 = \dot{n}(2,C)/\dot{n}(2,B)$ | [7] |
| $O_2$ analysis:  $0.050 = \dot{n}(3,B)/\bar{n}(3)$ | [8] |
| Dew point:   $\dot{n}(3,F)/\bar{n}(3) = [p^*(F)$ @ 340 K]/P = 26.96 kPa/215 kPa | [9] |
| Conversion: $0.9 = [\dot{n}(1,A) - \dot{n}(3,A)]/\dot{n}(1,A)$        Where $\dot{n}(1,A) = 100$ | [10] |

Selectivity:
    S = $C_3H_8$ converted to $CO_2$ (rxn 1)/$C_3H_8$ converted to ($CO_2$ + CO) rxns 1 & 2

| | |
|---|---|
| $S = \dot{n}(3,D)/[\dot{n}(3,D) + \dot{n}(3,E)]$ | [11] |
| Total Str. 3:        $\bar{n}(3) = \dot{n}(3,A) + \dot{n}(3,B) + \dot{n}(3,C) + \dot{n}(3,D) + \dot{n}(3,E) + \dot{n}(3,F)$ | [12] |

The above 12 equations are not all independent.
Use a spreadsheet (Excel) to set up the equations. Solve for $\dot{n}(2,B)$ and S.
The rest follows.

**a.**   Gas analysis:

| | $C_3H_8$ | $O_2$ | $N_2$ | $CO_2$ | CO | $H_2O$ |
|---|---|---|---|---|---|---|
| Mole % (wet basis) | 0.35 | 5.00 | 72.7 | 6.7 | 2.7 | 12.5 |

**b.**   Selectivity for reaction 1 = S =                                          **71.3%**

### 2:37 U   Heat capacity

a.   Calculations in Excel

| Heat capacity | | a | b | c | d | |
|---|---|---|---|---|---|---|
| | | 19.9 | 5.00E−02 | 1.27E−05 | −1.10E−08 | |
| Integral | | aT | bT^2/2 | cT^3/3 | dT^4/4 | SUM |
| T K | 1000 | 1.99E+04 | 2.50E+04 | 4.23E+03 | −2.75E+03 | 4.64E+04 |
| T ref K | 298 | 5.92E+03 | 2.22E+03 | 1.12E+02 | −2.17E+01 | 8.23E+03 |
| $C_{p,m}$ | (4.64E04 − 8.23E03)/(1000 − 298) = | | | 54.3 | kJ/kmol.K | |

b.   Heat of vaporization at FP =                                                                 **44.8E3 kJ/kmol**

c.   An estimate of the vapour pressure of water comes from the Clausius–Clapyron equation:

Solve for: $p^*_1$ =                                                                                         **0.67 kPa(a)**

### 2:38 W   Heat of mixing

System = $H_2SO_4$ Water

Consider the mixer as a closed system in which the 4000 kg of 90% acid at 40°C is brought together with the 6000 kg of water at 20°C.

All enthalpies are relative to the compounds in the $H_2SO_4$ h-C diagram at 273 K.

a.   Mass fraction $H_2SO_4$ in initial batch = w(1)

Mass fraction of $H_2SO_4$ in final mixture = w(2)

Mass balance on $H_2SO_4$.

ACC = 0 = IN − OUT + GEN − CON             GEN = CON = 0

0 = 4000w(0.9) − 18000w(2)           w(2) = 0.20 (20 wt%)

Energy balance on system

ACC = 0 = IN − OUT + GEN − CON             GEN = CON = 0

Initial energy (enthalpy): Data read from the $H_2SO_4$ h-C diagram

Initial = H1 = 4E3$h_A$ + 14E3$h_w$ = 4E3 kg(−100 kJ/kg) + 14E3kg(75 kJ/kg) = 650E3 kJ

Final enthalpy = H2 = 650E3 kJ (adiabatic mixing, zero heat transfer).

Energy balance: H1 = H2

Final specific energy = 650E3 kJ/18000 kg = 36.1 kJ/kg of mixture.

Final temperature at 20 wt% acid, by interpolation in h-C diagram = ca. 50°C ≡ **323 K**

b.   Mass fraction $H_2SO_4$ in initial mixture "A" = w(1)

Mass fraction of $H_2SO_4$ in final mixture = w(2)

For the adiabatic process:

Specific enthalpy of initial mixture "A" at 20°C= h(1) kJ/kg

Specific enthalpy of final adiabatic mixture at 100°C = h(2) kJ/kg

Total mass balance:

ACC = IN − OUT + GEN − CON = 0             Final mass − Initial mass

Final mass = Initial mass = (4E3 + 6E3) = 10E3 kg

Mass balance on $H_2SO_4$:              ACC = IN − OUT + GEN − CON = 0

Final mass of $H_2SO_4$ – Initial mass of $H_2SO_4$

i.e. $4E3w(1) = 10E3w(2)$

$w(2) = 0.40w(1)$ [1]

Energy (enthalpy) balance:

$ACC = IN - OUT + GEN - CON = 0$      Adiabatic process, zero work.

Final enthalpy – Initial enthalpy

$0 = (10E3)h(2) - (4E3)h(1) + (6E3 \text{ kg})(75 kJ/kg))$

$$0 = 10h(2) - [4h(1) + 450] \quad [2]$$
$$0 = 2.5h(2) - (h(1) + 113)$$

Also:      $h(1) = f[w(1),T(1)]$ [3]

            $h(2) = f[w(2),T(2)]$ [4]

Values of h(1) and h(2) are in the h-C diagram

Four independent equation with 4 unknowns w(1), w(2), h(1), h(2)

Where $T(1) = 20°C$, $T(2) = 100°C$.

This problem is essentially a simultaneous material and energy balance.

Use bisection of w(1) with the enthalpy concentration diagram to close the energy balance.

| | | | | | | |
|---|---|---|---|---|---|---|
| w(1) 20°C | – | 1.0 | 0.20 | 0.1 | 0.80 | 0.90 |
| w(2) 40°C | – | 0.4 | 0.08 | 0.04 | 0.329 | 0.370 |
| h(1) | kJ/kg | +25 | –75 | 0 | –240 | –150 |
| h(2) | kJ/kg | +160 | +100 | 100 | +60 | +25 |
| RHS of Eqn. 2 | kJ | 1050 | 850 | +371 | +170 | *–4* |

Close enough to zero (i.e. closure)   OR   work through tie lines on the h-C chart.

[Note. This solution is approximate]

For the isothermal process: Compositions are the same as for the adiabatic process

Energy (enthalpy) balance:    $ACC = IN - OUT + GEN - CON = Q$ (heat in)

Isothermal process, zero work.

Final enthalpy – Initial enthalpy $= Q$      all at 20°C. on h-C diagram

i.e. Heat transfer IN to system $= Q =$ Final enthalpy – Initial enthalpy

$Q = (9730 \text{ kg})(-175 \text{ kJ/kg}) - [(4000 \text{ kg})(-150 \text{ kJ/kg}) + (5730 \text{ kg})(90 \text{ kJ/kg})]$

$= -1.703E6 - [-84.3E3] = $            **–1.62E6 kJ (Cooling)**

## 2:39 U   $H_2/O_2$ conversion

Reaction stoichiometry: $2H_2 + O_2 \rightarrow 2H_2O$

Original $H_2/O_2$ mol/mol            **3.03**

Conversion of $H_2$            **66%**

## 2:40 W   Nylon and $N_2O$

a.   First balance the reactions.

$2C_2H_3CN + H_2O \rightarrow C_4H_8(CN)_2 + (1/2)O_2$          rxn 1

$C_2H_3CN + H_2O \rightarrow C_2H_5CN + (1/2)O_2$          rxn 2

Designate: $A \equiv C_2H_3CN$, $B \equiv C_4H_8(CN)_2$, $C \equiv [C_{12}H_{22}N_2O_2]$ ("monomer"), $D \equiv N_2O$
Molar mass: $M(A) = 53$     $M(B) = 108$     $M(C) = [226]_n$     $M(D) = 44$ kg/kmol
Nylon "monomer" production = 1E5 t/y/226 ton/ton mol = 442.5 ton mol
System = Overall electrochemical and thermochemical process. Closed system.
$X$ = conversion of A,     $S$ = selectivity for B in reaction 1.
Integral mole balances:
ACC = IN – OUT + GEN – CON
A:  ACC = Final – Initial = $0 - 0 + 0 - Xn(A)$
B:  ACC = Final – Initial = $0 - 0 + (1/2)(X)(S)n(A) - (1)n(C) = 0$
C:  ACC = Final – Initial = $0 - 0 + 0.9n(B) - 0 = 442.5$ ton mole
$(1/2)(X)(S)n(A) = n(C) = 442.5$ ton mole
$n(A) = (2)442.5/((0.95)(0.7)(0.9)) = 1479$ ton mol
$m(A) = (1479$ ton mole$)(53$ t/tmol$) =$                                  **78.4E3 ton**
b.   Mole balance on $N_2O$.
D: ACC = Final – Initial = $0 - 0 + (1)n(C) = 442.5$-ton mole $\equiv (442.5$ tmol$)(44$ t/tmol$) =$
                                                                        **19.5E3 ton**

NB. The Nylon polymer "n" does not affect the mass ratios of reactants and products.
    $N_2O$ is a potent greenhouse gas.

# Chapter 3:   Material and Energy Balances
# in Process Engineering[1]

### 3:1 W   DAC absorption

Continuous operation, steady-state, reaction        ACC = 0        GEN and CON # 0
Designate: $A \equiv$ Air, $B \equiv CO_2$, $C \equiv KOH$, $D \equiv K_2CO_3$
a.   Flow of air
     System: Overall process. Assume total air flow $\dot{n}(1,A) = \dot{n}(2,A)$ an approximation
     Mole balance on $CO_2$ ACC = IN – OUT + GEN – CON   $y$ = mole fraction in gas
     $0 = y(1,B)\dot{n}(1,A) - y(2,B)\dot{n}(2,A) - \dot{n}(5,B) + 0 - 0$                              [1]
     Then    $\dot{n}(1,A) = \dot{n}(2,A) = \dot{n}(5,B)/[y(1,B) - y(2,B)]$                         [2]
     Substitute values: $n(5,B) = (1000$ t/day/44 kg/kmol$)/24$ h/day)        = 0.947 tmol/h
     $\dot{n}(1,A) = \dot{n}(2,A) = 0.947$ tmol/h/$(0.7(450E-6$ mf$)) = 3.00E3$ tmol/h     = 3E6 kmol/h
     $\dot{n}(1,A) = (3E6$ kmol/h$)(22.4$ Sm$^3$/kmol$) =$      Air flow =        **67E6 Sm$^3$/h**
b.   Flow of stream 4 System Unit 2 Designate $c(i,j)$ = concentration j in stream i
     Absorbent composition: Total $[K+] = 2$ M = $1c(3,C) + 2c(3,D) = 1c(4,C) + 2c(4,D)$
     $[OH^-]/[CO_3=]$ ratios: $c(3,C)/c(3,D) = 1$ and $c(4,C)/c(4.D) = 2$
     Then $c(3,D) = [CO_3=] = 1.00$ M        $c(4,D) = [CO_3=] = 0.5$ M
     Designate:   F = flow of stream 4 = flow stream 3 m$^3$/h
     Atom balance on C        ACC = GEN = CON = 0

---

[1] The data and costs for Chapter 3 are approximate values, used for this exercise.

C:   $0 = (F\ m^3/h)(1\ kmol/m^3) - F(m^3/h)(0.5\ kmol/m^3) - n(5,B) = 947\ kmol/h$

Solve for   $F = (947\ kmol/h)/((1-0.5)\ kmol/m^3) =$          **1.89E3 $m^3/h$**

c.   Value of feed $= 0$    Value of products $= (1000\ t/day)(300\ \$/t)(350\ day/y) = 105E6\ \$/year$

Capital cost (CAPEX) $= 450E6\ \$$

Operating cost (OPEX) $= (40\ \$/t\ CO_2)(1000\ t/d\ CO_2)(360\ d/y) = 14.4E6\ \$/y$

Interest on CAPEX $= (0.05/y)(450E6\ \$) = 22.5E6\ \$/y$

GEP $=$ value of products $-$ value of feeds $= 105E6\ \$/y$

NEP $=$ GEP $-$ (Operating costs $+$ interest) $= 105E6 - (14.4E6 + 22.5E6) = 68.1E6\ \$/y$

ROI $= 68.1E6\ \$/y\ /450E6\ \$ \equiv$              **15%/y**

## 3:2 U   Syngas

A.   Thermochemical stoichiometry:

$CH_4 + H_2O \rightarrow CO + 3H_2\ T \approx 1200\ K$

$CH_4 + 2O_2 \rightarrow CO_2 + 2H_2O$ heating

[$CH_4$ for heating/$CH_4$ to CO] $= 0.256$

$CO + 2H_2 \rightarrow CH_3OH$

B.   Electrochemical stoichiometry:

$CO_2 + 2H_2O$ ---- **6F** $\rightarrow CO + 2H_2 + (1.5)O_2$

$CO + 2H_2 \rightarrow CH_3OH$

Flowsheets.

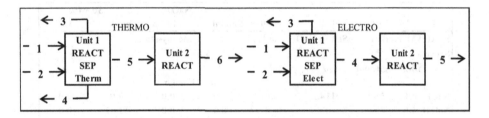

The material balances (not shown) are the basis for the results tabulated below.

ROI: **Thermochemical $= +7\%/y$**       **Electrochemical $= -9\%/y$ (negative)**

| *Economics* THERMO | | Amount | Amount | Amount |
|---|---|---|---|---|
| 350 d/y 24h | | $/t | t/year | $/year |
| Carbon tax | | 300 | 1.06E+04 | 3.17E+06 |
| $CH_4$ | | 500 | 2.42E+04 | 1.21E+07 |
| $H_2$ | | 3000 | 2.19E+03 | 6.56E+06 |
| $2H_2$+CO | syngas | 400 | 3.06E+04 | 1.23E+07 |
| | | $/year | | |
| Value feeds | | 1.53E+07 | | |
| Value prod | | 1.88E+07 | | |
| GEP | | 3.53E+06 | | |
| CAPEX $ | 2.00E+07 | $/year | | |
| OPEX %/y | 6 | 1.20E+06 | | |
| INT %/y | 5 | 1.00E+06 | | |
| NEP | | 1.33E+06 | | |
| **ROI %/y** | **6.7** | | | |

| Economics | ELECTRO | Amount | Amount | Amount |
|---|---|---|---|---|
| 350 d/y 24h | | $/t | t/year | $/year |
| Carbon tax | | −300 | 0.00E+00 | 0.00E+00 |
| $CO_2$ | | −300 | 4.81E+04 | −1.44E+07 |
| $H_2O$ | | 10 | 3.94E+04 | 3.94E+05 |
| $2H_2+CO$ | syngas | 400 | 3.06E+04 | 1.23E+07 |
| $O_2$ | | 50 | 5.25E+04 | 2.63E+06 |
| | | $/year | Elect $/kWh | 5.00E-02 |
| Value feeds $CO_2+H_2O$ | | −1.40E+07 | Elect kWh | 5.28E+08 |
| Elect | | 2.64E+07 | | |
| Value prod | | 1.49E+07 | | |
| GEP | | 2.54E+06 | | |
| CAPEX $ | 1.40E+08 | $/year | | |
| OPEX %/y | 6 | 8.38E+06 | | |
| INT %/y | 5 | 6.98E+06 | | |
| NEP | | −1.28E+07 | | |
| ROI %/y | −9.2 | | | |

### 3:3 W    Salt, caustic and muriatic acid

Stoichiometry:

$$2NaCl + 2H_2O \rightarrow 2NaOH + Cl_2 + H_2$$
$$Cl_2 + H_2 \rightarrow 2HCl$$

| | | | | | | NaOH | | 30 wt% |
|---|---|---|---|---|---|---|---|---|
| HCl | 100 | T/d | 2740 | kmol/d | | 114.16 | kmol/h | 866 |
| STREAM TABLE | | Salt to NaOH and HCl | | | | Steady-state | | |
| Species | M | | | | *Stream* | | | |
| | $/kg kg/kmol | *1* | *2* | *3* | *4* | *5* | | *6* |
| | | | | | Flow kmol/h | | | |
| NaCl  0.1 | 58.5 | 114.2 | 0.0 | 0.0 | 0.0 | 0.0 | | 0.0 |
| NaOH 1.0 | 40.0 | 0.0 | 0.0 | 0.0 | 0.0 | 114.2 | | 0.0 |
| $H_2O$ 0.01 | 18.0 | 0.0 | 704.3 | 0.0 | 0.0 | 590.2 | | 0.0 |
| $Cl_2$   0.5 | 71.0 | 0.0 | 0.0 | 0.0 | 57.1 | 0.0 | | 0.0 |
| $H_2$   4.0 | 2.0 | 0.0 | 0.0 | 57.1 | 0.0 | 0.0 | | 0.0 |
| HCl  0.6 | 36.5 | 0.0 | 0.0 | 0.0 | 0.0 | 0.0 | | 114.2 |
| Total | kg/h | 6678 | 12678 | 114 | 4053 | 15190 | | 4167 |
| Mass balance | | U 1 | | U2 | | Overall | | |
| Mass in | kg/h | 19356 | | 4167 | | 19356 | | |
| Mass out | kg/h | 19356 | | 4167 | | 19356 | | |
| Closure % | | 100.0 | | 100.0 | | 100.0 | | |

Economics:

GEP    = Value of products − value feeds
       = Value (NaOH + HCl) − value (NaCl + $H_2O$)
Products = (1.0 $/kg)(114.2 kmol/h)(40 kg/kmol) + (0.6 $/kg)(4167 kg/h)
Products = 4.56E3 + 2.50 E3 = 7.06E3 $/h
Feeds = (0.1 $/kg)(6678 kg/h) + (0.01$/kg)(12678 kg/h)
Feeds = 6.67E2 + 1.27E2 = 7.95E2 $/h

GEP = ((7.06E3 − 7.95E2) $/h)(8E3 h/y) = 50.1E6 $/y
NEP = GEP − cost(utilities + labour + maintenance)
Utilities: (0.05 $/kWh)(5 kWh/kg NaOH)(12678 kg/h)(8000 h/y) = 25.4E6 $/y
Labour + maintenance = 7% of CAPEX/year = 0.07($100E6) = 7.0E6 $/y
Interest at 6%/y = (0.06)(100E6 $) = 6.0E6 $/y
NEP = 50.1E6 − (25.4E6 + 7.0E6 + 6.0E6) = 11.7E6
ROI = (11.7E6 $/y)/(100E6 $) ≡                                          **11.7%/y**

## 3:4 U   Green steel

Basic stoichiometry of the hydrogen and carbon reactions:

$$3H_2 + Fe_2O_3 \rightarrow Fe + 3H_2O \qquad \text{rxn 1}$$
$$3C + 2Fe_2O_3 \rightarrow 4Fe + 3CO_2 \qquad \text{rxn 2}$$

The material balances and economic estimates are set out below
**The ROI match:**      **At a carbon tax of $700 per ton $CO_2$      ROI = 57%/y**

| MASS BALANCE and $$ | | | Green steel | | | |
|---|---|---|---|---|---|---|
| Species | Value | M | Direct reduction ($H_2$) | | Blast furnace (C) | |
| | | kg/kmol | Feed | Product | Feed | Product |
| | S/T | T/T mol | Ton mol/day | | Ton mol/day | |
| $Fe_2O_3$ | 300 | 159.6 | 6.27 | 0 | 6.27 | 0 |
| Fe | 1500 | 55.8 | 0 | 12.53 | 0 | 12.53 |
| C | 600 | 12.0 | 0 | 0 | 9.40 | 0 |
| $CO_2$ (?) | −700 | 44.0 | 0 | 0 | 0 | 9.40 |
| $H_2$ | 4000 | 2.0 | 18.80 | 0 | 0 | 0 |
| $H_2O$ | 0 | 18.0 | 0 | 18.80 | 0 | 0.00 |
| Total | | T/day | 1.04E+03 | 1.04E+03 | 1.11E+03 | 1.11E+03 |
| Mass closure % | | | 100.0 | | 100.0 | |
| Utili % | L+M % | Interest % | Feeds | Products | Feeds | Products |
| 4 | 5 | 6 | 4.50E+05 | 1.05E+06 | 3.68E+05 | 7.59E+05 |
| GEP $/day | | | 5.98E+05 | | 3.92E+05 | |
| CAPEX $ | | | 3.00E+08 | | 2.00E+08 | |
| Util + Lab &Maint + Interest $/d | | | 1.23E+05 | | 8.22E+04 | |
| NEP $/day | | | 4.75E+05 | | 3.10E+05 | |
| ROI %/y | | | 58 | | 56 | |

## 3:5 W   Peroxide bleach

Amount of $H_2O_2$ required = (500 t/day pulp)(20 kg $H_2O_2$/ton pulp) = 10,000 kg/day
Consider the options:
1.   Purchase 30 wt% $H_2O_2$ Amount = (10 ton/d)/(0.3 t $H_2O_2$/ton) = 33.3 t/day
     Cost = (33.3 t/d)(300 $/t) = 9990 $/day = 3.6E6 $/year (excludes storage, etc.)
     Cost = 3.6E6 $/(500 t/day pulp)(360 day/y) =                    **20 $/ton pulp**
2.   Install on-site plant CAPEX = $7.0E6
     $H_2O_2$ production rate = (33.3 ton/day)/(34 kg./kmol) = 0.98 ton mol/day
                    Reaction stoichiometry: $O_2 + 2H_2O \rightarrow 2H_2O_2$

Expendables:

$O_2$ consumed = (0.98 tmol/d $H_2O_2$)((1/2)$O_2$/$H_2O_2$)/(0.4) = 1.22 tmol/d

Electricity (3 kWh/kg)(0.98 tmol/d)(34000 kg/ton mol) = 100E3 kWh/day

Operating cost:

$O_2$ = (1.22 tmol/d $O_2$)(32 kg/kmol)(50 \$/t) = 1925 \$/day ≡ 705E3 \$/y

Electricity = (100E3 kWh/d)( 0.05 \$/kWh) = 5,000 \$/day ≡ 1.8E6 \$/y

OPEX + interest = 0.09(7E6) = 630E3 \$/y

Total operating cost + interest = 3.14E6 \$/y ≡ 3.14E6 \$/y/(500)(360) ≡   **17.4 \$/ton pulp**

Saving with on-site plant = 3.6E6 – 3.14E6 = 460E3 \$/year

Purchase cost = 7.0E6 \$    Repayment period = (7.0E6 \$)/(460E3E3 \$/y) =   **15 years**

# Chapter 4:   Material Balances

## 4:1 W   Steam trap

Steam trap: Continuous, steady-state, no reaction

**a.**   Overall mass balance for water:

ACC = IN – OUT + GEN = CON      ACC = GEN = CON = 0

$0 = \bar{m}(1) – \bar{m}(2)$        $\bar{m}(1) = \bar{m}(2)$ = 1000 kg/h

Stream 1 at 931.5 kPa(a), 450 K (steam table) is saturated liquid-vapour with quality = $w_q$

$w_q$ = quality = mass fraction vapour in the mixture

Specific mass and volume from the steam table: $V(1)$ =

(1000 kg/h)($w_q$(0.208 $m^3$/kg) + (1 – $w_q$)(0.001123 $m^3$/kg)) = 70 $m^3$/h        $w_q$= **0.333**

**b.**   Enthalpy of streams 1 and 2 from steam table.

$\dot{H}(1)$ = 1000 kg/h (0.333)(2774.9 kJ/kg + (1– 0.333)(749 kJ/kg) = 1.42E6 kJ/h

$\dot{H}(2)$ = 1000 kg/h (0 + (1)523.7 kJ/kg) = 5.33 E5 kJ/h

Heat loss = $\dot{H}(1) – \dot{H}(2)$ = 8.91E5 kJ/h ≡ 8.91E5 kJ/h/3600 s/h =        **247 kW**

## 4:2 U   Alkane distillation

**a.**   Stream 5 volume flow =                                                    **1763 $m^3$/h**

**b.**   Mass of $CO_2$ per day =                                                **70E3 kg/day**

The full table is below

|        |         |      |     | Alkane distillation |       | * = of total |       |
|--------|---------|------|-----|------|------|------|------|
| Stream |         | 1    | 2   | 3    | 4    | 5    | 6    |
| n      |         | 1–30 | 1–3 | 4–10 | 11–16 | 1–10 | 17–30 |
| wt%    |         | 100  | 10  | 40   | 30   | 50*  | 20   |
| mol%   |         | 100  | 33  | 45   | 15   | 79   | 6    |
| M Mean | kg/kmol | 101  | 30  | 88   | 200  | 63   | 330L |
| Phase  |         | L    | G   | L    | L    | G    | L    |

## 4:3 U   Generic process units

### a.   DIVIDE:
Solve mole and mass balance equations to get:

| Species | M | Stream | kmol/h | |
|---|---|---|---|---|
| DIVIDE | kg/kmol | *1* | *2* | *3* |
| A | 40 | 40 | 10 | 30 |
| B | 70 | 60 | 15 | 45 |
| C | 60 | 20 | 5 | 15 |
| Total | kg/h | 7000 | 1750 | 5250 |
| Mass bal | | 7000 | 7000 | |

### b.   MIX:
Solve mole and mass balance equations to get:

| Species | M | Stream | kmol/h | |
|---|---|---|---|---|
| MIX | kg/kmol | *1* | *2* | *3* |
| A | 40 | 40 | 20 | 60 |
| B | 70 | 15 | 30 | 45 |
| C | 60 | 10 | 15 | 25 |
| Total | kg/h | 3250 | 3800 | 7050 |
| Mass bal | kg/h | | 7050 | 7050 |

### c.   SEPARATE:
Solve mole and mass balance equations to get:

| Species | M | Stream | kmol/h | |
|---|---|---|---|---|
| SEPARATE | kg/kmol | *1* | *2* | *3* |
| A | 40 | 20 | 16 | 4 |
| B | 70 | 50 | 5 | 45 |
| C | 60 | 45 | 30 | 15 |
| Total kg/h | | 7000 | 2790 | 4210 |
| Mass bal kg/h | | 7000 | 7000 | |

### d.   HEX:
Solve mole and mass balance equations to get:

| Species | M | Stream | kmol/h | | |
|---|---|---|---|---|---|
| HEX | kg/kmol | *1* | *2* | *3* | *4* |
| A | 40 | 30 | 30 | 15 | 15 |
| B | 70 | 10 | 10 | 20 | 20 |
| C | 60 | 20 | 20 | 25 | 25 |
| Total | kg/h | 3100 | 3100 | 3500 | 3500 |
| Mass bal kg/.h | | 6600 | | 6600 | |

**e. PUMP:**

Solve mole and mass balance equations to get:

| Species | M | Stream | kmol/h |
|---|---|---|---|
| PUMP | kg/kmol | *1* | *2* |
| A | 40 | 15 | 15 |
| B | 70 | 20 | 20 |
| C | 60 | 10 | 10 |
| Total | kg/h | 2600 | 2600 |
| Mass bal kg/h | | 2600 | 2600 |

**f. REACT:**

Solve mole and mass balance equations to get:

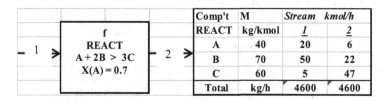

| Comp't | M | Stream | kmol/h |
|---|---|---|---|
| REACT | kg/kmol | *1* | *2* |
| A | 40 | 20 | 6 |
| B | 70 | 50 | 22 |
| C | 60 | 5 | 47 |
| Total | kg/h | 4600 | 4600 |

**4:4 W   Generic material balance**

The specifications are:   M(A) = 25, M(B) = 40 Molar mass kg/kmol
Str. 1 = 50 kmol/h pure A. Str. 2 = pure B Str. 4: Total mass flow = 2450 kg/h
Str. 6 = 50 mol% C.

MANUAL SOLUTION!

a.   For all process units:

$\bar{m}(i) = \Sigma[M(j)\dot{n}(i,j)]$ and $\bar{n}(i) = \Sigma[\dot{n}(i,j)]$                     Eqn E

Unit 1. MIXER. Mole balances at steady-state. GEN = CON = 0
A: $0 = \Sigma\dot{n}(i,A) = 50 + 0 - (3,A)$                     [1]
B: $0 = \Sigma\dot{n}(i,B) = 0 + \dot{n}(2,B) - \dot{n}\ (3,B)$                     [2]
C: $0 = \Sigma\dot{n}(i,C) = 0 + 0 - \dot{n}(3,C)$                     [3]
Total mass balance: $0 = \bar{m}(1) + \bar{m}(2) - \bar{m}(3)$                     [4]

Unit 2. REACTOR. Mole balances at steady-state. GEN and CON # 0
Stoichiometric coefficients $v(A) = 2, v(B) = 1, v(C) = 1$ Signs OK in GBE
$X(A)$ = conversion of A = 0.6
A: $0 = \dot{n}(3,A) - \dot{n}(4,A) - (0.6)\ \dot{n}(3,A)$                     [5]
B: $0 = \dot{n}(3,B) - \dot{n}(4,B) - (1/2)(0.6)\ \dot{n}(3,A)$                     [6]
C: $0 = 0 - \dot{n}(4,C) + (1/2)(0.6)\ \dot{n}(3,A)$                     [7]
Total mass balance: $0 = \bar{m}(3) - \bar{m}(4)$     $= \bar{m}(3) - 2450$                     [8]

Unit 3. SEPARATOR.
Mole balance on A:         $0 = \dot{n}(4,A) - \dot{n}(5,A) - \dot{n}(6,A)$                     [9]

Mole balance on B:      $0 = \dot{n}(4,B) - \dot{n}(5,B) - \dot{n}(6,B)$      [10]

Mole balance on C:      $0 = \dot{n}(4,C) - \dot{n}(5,C) - \dot{n}(6,C)$      [11]

Total mass balance:      $0 = \bar{m}(4) - \bar{m}(5) - \bar{m}(6) = 2450 - \bar{m}(5) - \bar{m}(6)$      [12]

Split fractions:    $0.9 = \dot{n}(5,A)/\dot{n}(4,A)$      [13]

                 $0.6 = \dot{n}(5,B)/\dot{n}(4,B)$      [14]

Stream composition $0.50 = \dot{n}(6,C)/(\dot{n}(6,A) + \dot{n}(6,B) + \dot{n}(6,C))$      [15]

From the reaction stoichiometry (mass is conserved)

$M(C) = 2M(A) + M(B) = (2)(25) + 40 = 90$ kg/kmol

From [4], [8] and eqn E

$\bar{m}(3) = 2450$

$\bar{m}(2) = 2450 - (25)(50) = 1200$ kg/h

$\dot{n}(2,B) = 1200$ kg/h$/40$ kg/kmol $= 30$ kmol/h

From [1],[2],[3] $\dot{n}(3,A) = 50$ $\dot{n}(3,B) = 30$ $\dot{n}(3,C) = 0$    kmol/h

From [5],[6],[7] $\dot{n}(4,A) = 20$ $\dot{n}(4,B) = 15$ $\dot{n}(4,C) = 15$   kmol/h

From [13],[14] $\dot{n}(5,A) = 18$ $\dot{n}(5,B) = 9$                kmol/h

From [9],[10],[15] $\dot{n}(6,A) = 2$, $\dot{n}(6,B) = 6$, $\dot{n}(6,C) = 8$   kmol/h

From [11]          $\dot{n}(5,C) = 7$                      kmol/h

b.    Stream table.        Continuous process at steady-state.

| Species | M | | | Stream | kmol/.h | | |
|---|---|---|---|---|---|---|---|
| GENERIC | kg/kmol | 1 | 2 | 3 | 4 | 5 | 6 |
| A | 25 | 50 | 0 | 50 | 20 | 18 | 2 |
| B | 40 | 0 | 30 | 30 | 15 | 9 | 6 |
| C | 90 | 0 | 0 | 0 | 15 | 7 | 8 |
| Total | kg/h | 1250 | 1200 | 2450 | 2450 | 1440 | 1010 |
| Mass bal | check | Unit 1 | Unit 2 | Unit 3 | Ov'll | | |
| IN | kg/h | 2450 | 2450 | 2450 | 2450 | | |
| OUT | kg/h | 2450 | 2450 | 2450 | 2450 | | |
| Closure | % | 100 | 100 | 100 | 100 | OK! | |

## 4:5 U   Generic balance specification

a.

| Species | M | | Stream | kmol/h | Specification table | |
|---|---|---|---|---|---|---|
| DIVIDE | kg/kmol | 1 | 2 | 3 | Stream variables | (3)(3)=9 |
| A | 40 | 40 | 10 | | Component bals | 3 |
| B | 70 | 60 | | | Specified | 4 |
| C | 60 | | | | Subsidiary | 2 |
| Total | kg/h | | Solution | | Independent | 9 |
| Comp't | M | | Stream | kmol/h | Degrees of freedom | 0 |
| DIVIDE | kg/kmol | 1 | 2 | 3 | Fully specified | |
| A | 40 | 40 | 10 | 30 | | |
| B | 70 | 60 | 15 | 45 | | |
| C | 60 | 80 | 20 | 60 | | |
| Total | kg/h | 10600 | 2650 | 7950 | | |
| Mass bal | kg/h | 10600 | 10600 | OK | | |

**b.**

| Species | M | Stream | kmol/h | | Specification table | |
|---|---|---|---|---|---|---|
| SEP | kg/kmol | 1 | 2 | 3 | Stream variables | (3)(3)=9 |
| A | 40 | 50 | | | Component bals | 3 |
| B | 70 | | | 45 | Specified | 4 |
| C | 60 | | 30 | | Subsidiary | 1 |
| Total | kg/h | | | 4750 | Independent | 8 |
| | | | *No unique solution* | | Degrees of freedom | 1 |
| | | | | | Under-specified | |

**c.**

| Species | M | Stream | kmol/h | | Specification table | |
|---|---|---|---|---|---|---|
| REACT | kg/kmol | 1 | 2 | | Stream variables | (2)(3)=6 |
| A | 40 | 20 | | | Component bals | 3 |
| B | 70 | | 30 | | Specified | 4 |
| C | 60 | 5 | | | Subsidiary | 1 |
| Total | kg/h | | 5300 | | Independent | 8 |
| Comp't | M | Stream | kmol/h | | Degrees of freedom | -2 |
| REACT | kg/kmol | 1 | 2 | | Over-specified | |
| X(A)=0.7 | | | | | *Suspicious. Check for errors* | |
| A | 40 | 20 | 6 | | | |
| B | 70 | 68.6 | 30 | | | |
| C | 60 | 5 | 47 | | Incorrectly over-specified | |
| Total | kg/h | 5300 | *5300* | 5160 | | |

## 4:6 W    Recycle material balance

Flowsheet: MIX > REACT > SEPARATE >> Recycle to MIX
Differential mole balances at steady-state.

Unit 1. MIX

A: $0 = \dot{n}(1,A) + \dot{n}(4,A) - \dot{n}(2,A)$      Feed $\bar{n}(1) = 100$ kmol/h        [1]

B: $0 = \dot{n}(1,B) + \dot{n}(4,B) - \dot{n}(2,B)$      $\dot{n}(1,A) = 0.75\,\bar{n}(1)$        [2]

C: $0 = 0 + \dot{n}(4,C) - \dot{n}(2,C)$        $\dot{n}(1,B) = 0.25\,\bar{n}(1)$        [3]

Unit 2. REACT

A: $0 = \dot{n}(2,A) - \dot{n}(3,A) - X(A)\,(2,A)$        [4]

Conversion: $X(A) = (0.6) \times (2,B) = (0.6)(\dot{n}(2,B)/\bar{n}(2))$        [5]

B: $0 = \dot{n}(2,B) - \dot{n}(3,B) - (1/3) \times (A)\dot{n}(2,A)$        [6]

C: $0 = \dot{n}(2,C) - \dot{n}(3,C) + (2/3) \times (A)\dot{n}(2,A)$        [7]

Unit 3. SEPARATE

A: $0 = \dot{n}(3,A) - \dot{n}(4,A) - \dot{n}(5,A)$        [8]

Split fractions: $s(4,A) = \dot{n}(4,A)/\dot{n}(3,A) = 0.990$        [9]

B: $0 = \dot{n}(3,B) - \dot{n}(4,B) - \dot{n}(5,B)$      $s(4,B) = \dot{n}(4,B)/\dot{n}(3,B) = 0.980$        [10,11]

C: $0 = \dot{n}(3,C) - \dot{n}(4,C) - \dot{n}(5,C)$      $s(4,C) = \dot{n}(4,C)/\dot{n}(3,C) = 0.100$        [12,13]

Tear-stream = stream 4. Set $\dot{n}(4,A) = \dot{n}(4,B) = \dot{n}(4,C) = 0$ for the first pass.

Rearrange the equations for explicit sequential solution in the spreadsheet.

$\dot{n}(2,A) = 75 + \dot{n}(4,A)$        [14]

$\dot{n}(2,B) = 25 + \dot{n}(4,B)$        [15]

$\dot{n}(2,C) = 0 + \dot{n}(4,C)$ [16]

$\dot{n}(3,A) = \dot{n}(2,A) - X(A)\dot{n}(2,A)$    where $X(A) = (0.6)(\dot{n}\ (2,B)/\bar{n}(2))$ [17]

$\dot{n}(3,B) = \dot{n}(2,B) - (1/3)X(A)\dot{n}(2,A)$ [18]

$\dot{n}(3,C) = \dot{n}(2,C) + (2/3)X(A)\dot{n}(2,A)$ [19]

$\dot{n}(4,A) = s(4,A)\dot{n}(3,A)$    Creates a "circular reference" back to the mixer. [20]

$\dot{n}(4,B) = s(4,B)\dot{n}(3,B)$ [21]

$\dot{n}(4,C) = s(4,C)\dot{n}(3,C)$ [22]

$\dot{n}(5,A) = \dot{n}(3,A) - \dot{n}(4,A)$ [23]

$\dot{n}(5,B) = \dot{n}(3,B) - \dot{n}(4,B)$ [24]

$\dot{n}(5,C) = \dot{n}(3,C) - \dot{n}(4,C)$ [25]

Iterate with stream 4. (Excel: Tools > Options > Calculation > Iteration)

Closure = Mass flow OUT/Mass Flow IN.

Convergence is reached when all mass balance closures = 100.0%

The spreadsheet below has the complete solution stream table.

Net yield of C from A =                                                                         **91%**

| STREAM TABLE | | RECYCLE | | Stream | kmol/h | | |
|---|---|---|---|---|---|---|---|
| Species | M kg/kmol | 1 | 2 | 3 | 4 | | 5 |
| | | | | | | | |
| A | 30 | 75 | 743.1 | 674.9 | 668.1 | | 6.7 |
| B | 50 | 25 | 135.2 | 112.5 | 110.2 | | 2.2 |
| C | 70 | 0 | 5.1 | 50.6 | 5.1 | | 45.5 |
| Total | kg/h | 3500 | 2.94E+04 | 2.94E+04 | 2.59E+04 | | 3.50E+03 |
| Mass balance | | Unit 1 | Unit 2 | Unit 3 | Overall | | Yield % |
| IN | kg/h | 2.94E+04 | 2.94E+04 | 2.94E+04 | 3.50E+03 | C from A | |
| OUT | kg/h | 2.94E+04 | 2.94E+04 | 2.94E+04 | 3.50E+03 | | 91.0 |
| Closure | % | 100.0 | 100.0 | 100.0 | 100.0 | | |
| X(A) | 0.6 | x(2,B) | 0.153 | X(A) | 0.092 | | |
| S(4,A) | 0.99 | S(4,B) | 0.98 | S(4,C) | 0.1 | | |

## 4:7 W  Recycle balance with purge

Flowsheet: REACT > SEPARATE > DIVIDE/PURGE $\updownarrow$ Recycle to REACT

Differential mole balances at steady-state.

Unit 1.

A: $0 = \dot{n}(1,A) + \dot{n}(5,A) - \dot{n}(2,A)$    Feed $\dot{n}(1,A) = 100$   $\dot{n}(1,N) = 1$ kmol/h [1]

B: $0 = \dot{n}(1,B) + \dot{n}(5,B) - \dot{n}(2,B)$ [2]

N: $0 = \dot{n}(1,N) + \dot{n}(5,N) - \dot{n}(2,N)$    $+ 0 - 0$   N is unreactive [3]

Conversion: $X(A) = 0.4$    $\dot{n}(2,A) = (1-0.4)[\dot{n}(1,A)+\dot{n}(5,A)]$    Reaction [4]

Stoichiometry: $\dot{n}(2,B) = ((2/1)0.4)[\dot{n}(1,A)+\dot{n}(5,A)]$ [5]

Unit 2.

A: $0 = \dot{n}(2,A) - \dot{n}(3,A) - \dot{n}(4,A)$ [6]

B: $0 = \dot{n}(2,B) - \dot{n}(3,B) - \dot{n}(4,B)$ [7]

N: $0 = \dot{n}(2,N) - \dot{n}(3,N) + \dot{n}(4,N)$ [8]

Separate: $\dot{n}(4,A) = 0.1\dot{n}(2,A)$    $\dot{n}(4,B) = 0.9\dot{n}(2,B)$    $\dot{n}(4,N) = 0.01\dot{n}(2,N)$ [9,10,11]

Unit 3.

A: $0 = \dot{n}(3,A) - \dot{n}(5,A) - \dot{n}(6,A)$                                           [12]

B: $0 = \dot{n}(3,B) - \dot{n}(5,B) - \dot{n}(6,B)$                                           [13]

N: $0 = \dot{n}(3,N) - \dot{n}(5,N) - \dot{n}(6,N)$                                           [14]

Divide: $\dot{n}(6,A)/\dot{n}(3,A) = \dot{n}(6,B)/\dot{n}(3,B) = \dot{n}(6,N)/\dot{n}(3,N) = 0.1$         [15,16,17]

Tear-stream = stream 5. Set $\dot{n}(5,A) = \dot{n}(5,B) = \dot{n}(5,N) = 0$ for the first pass.

Rearrange the equations for explicit sequential solution in the spreadsheet.

NB. *The above singular linear equations are easily made explicit. Solutions with non-linear or simultaneous equations can use the same iterative procedure with the help of macros.*

AS IN PROBLEM 4:7

$\dot{n}(2,A) =, \dot{n}(2,B) =, \dot{n}(2,N) =, \dot{n}(3,A) =, \dot{n}(3,B) =, \dot{n}(3,N) =,$ ETC.

Tear stream:     $\dot{n}(5,A) =, \dot{n}(5,B) =, \dot{n}(5,N) =$ Initially zero, accumulate with recycle.

Creates a circular reference back to the reactor.

Iterate with stream 5. Manual or Automatic. Initial manual interactions to check integrity.

(Excel: Tools > Options > Calculation > Iteration – Manual or Automatic)

$\dot{n}(6,A) =, \dot{n}(6,B) =, \dot{n}(6,N) =$    and $[\dot{n}(4,N) + \dot{n}(6,N)] = \dot{n}(1,N)$ at closure

Closure = Mass flow OUT/Mass Flow IN. Convergence is reached when all mass balance closures = 100.0% The spreadsheet below has the complete solution stream table.

N in $5 = (8.17)(30)/9448 \equiv$ **2.6 wt%** Net yield of B from A = $156/200 = 0.78 \equiv$     **78%**

| STREAM TABLE | | Conv % | 40 | Div 6/3 | 0.1 | | |
|---|---|---|---|---|---|---|---|
| Species | M | RECYCLE + PURGE | | *Stream* | kmol/h | *Tear* | |
| | kg/kmol | *1* | *2* | *3* | *4* | *5* | *6* |
| A | 90 | 100 | 116.7 | 105.1 | 11.7 | 94.6 | 10.5 |
| B | 45 | 0 | 171.0 | 17.1 | 153.9 | 15.4 | 1.7 |
| N  1% | 30 | 1 | 9.2 | 9.1 | 0.1 | 8.2 | 0.9 |
| Total | kg/h | 9030 | 18478 | 10497 | 7980 | 9448 | 1050 |
| Mass balance | | Unit 1 | Unit 2 | Unit 3 | Overall | S(4,A) | 0.1 |
| IN | kg/h | 18478 | 18478 | 10497 | 9030 | S(4,B) | 0.9 |
| OUT | kg/h | 18478 | 18478 | 10497 | 9030 | S(4,N) | 0.01 |
| Closure | % | 100.00 | 100.00 | 100.00 | 100.00 | | |

**4.8 W    Bio-ethanol**

Continuous, steady-state, reaction          ACC = 0, GEN and CON # 0

Reaction stoichiometry: $C_6H_{12}O_6 \rightarrow 2C_2H_5OH + 2CO_2$

Designate: $A \equiv C_6H_{12}O_6$,   $B \equiv C_2H_5OH$,     $C \equiv CO_2$,     $D \equiv H_2O$

a.    The volume of an ideal liquid mixture is the sum of the separate component volumes.

     Let w = mass fraction EtOH in the mixture     $\rho(j)$ = density of j

     Then total volume of mix = $w/\rho(B) + (1-w)/\rho(D)$

     Volume fraction B = $w/\rho(B)/[w/\rho(B) + (1-w)/\rho(D)] = 0.95$ ("proof" by volume)

     Solve for w = 0.94 $\equiv$                      **94 vol% EtOH**

     Mole fraction B = $x(4,B) = 0.94/80/(46)/((0/94/46) + 0.06/18)) =$   **x(4,B) =**   **0.86**

b.    Take the units in sequence

     Unit 1:

     Mole balances, reaction     ACC = 0, GEN and CON # 0

     A: $0 = \dot{n}(1,A) - \dot{n}(2,A) - \dot{n}(3,A) + 0 - 0.95\,\dot{n}(1,A)$                    [1]

B: $0 = \dot{n}(1,B) - \dot{n}(2,B) - \dot{n}(3,B) + 2(0.95 \, \dot{n}(1,A)) - 0$      [2]

C: $0 = \dot{n}(1,C) - \dot{n}(2,C) - \dot{n}(3nC) + 2(0.95 \, \dot{n}(1,A)) - 0$      [3]

D: $0 = \dot{n}(1,D) - \dot{n}(2,D) - \dot{n}(3,D) + 0 - 0$      [4]

Stream 2, gas phase composition.

Pure component vapour pressures $p^*(B)$ and $p^*(D)$ come from the Antoine equation.

Partial pressures from Raoult's Law:    $p(B) = x(3,B)p^*(B)$    $p(D) = (1 - x(3,B))p^*(D)$

     [5][6]

Gas phase composition:

$y(B) = p(B)/P = \dot{n}(2,B)/\Sigma(\dot{n}(2,j))$    $y(D) = p(D)/P = \dot{n}(2,D)/\Sigma \, (\dot{n}(2,j))$      [7][8]

In the gas phase, stream 2:    $\dot{n}(i,j) = \overline{n}(i)y(j)$      [9]

Equations 1 – 9 give the material balance for Unit 1.

Unit 2:

Mole balances, no reactions      ACC = GEN = CON = 0

A: $0 = \dot{n}(3,A) - \dot{n}(4,A) - \dot{n}(5,A) + 0 - 0$      [10]

B: $0 = \dot{n}(3,B) - \dot{n}(4,B) - \dot{n}(5,B) + 0 - 0$      [11]

C: Not present

D: $0 = \dot{n}(3,D) - \dot{n}(4,D) - \dot{n}(5,D) + 0 - 0$      [12]

Separation efficiency: $\dot{n}(4,B)/\dot{n}(3,B) = 0.96$      [13]

Mole fraction B in stream 4: $x(4,B) = 0.86$      [14]

Equations 10–14 give the material balance for Unit 2.

The simultaneous material balances equations are resolved by iteration in Excel. The full stream table is below.

| Species | M | ETHANOL | Stream | | | kmol/h |
|---|---|---|---|---|---|---|
| | kg/kmol | *1* | *2* | *3* | *4* | *5* |
| $C_6H_{12}O_6$ | 180.0 | 5.56 | 0.00E+00 | 0.28 | 0 | 0.28 |
| $C_6H_6O(l)$ | 46.0 | 0 | 0 | 10.35 | 9.94 | 0.41 |
| $C_6H_6O(g)$ | 46.0 | 0 | 0.21 | 0.00 | 0 | 0 |
| $H_2O(l)$ | 18.0 | 500 | 0.00 | 499 | 1.62 | 498 |
| $H_2O(g)$ | 18.0 | 0 | 0.56 | 0 | 0 | 0 |
| $CO_2$ | 44.0 | 0 | 10.56 | 0 | 0 | 0 |
| Total | kmol/h | 505.56 | 11.32 | 510.06 | 11.55 | 498.51 |
| Total | kg/h | 1.00E+04 | 4.84E+02 | 9.52E+03 | 4.86E+02 | 9.03E+03 |
| Pressure | kPa(a) | 120 | 110 | 110 | 100 | 100 |
| T | K | 308 | 308 | 308 | 293 | 373 |
| mf $H_2O$)l) | - | 0.99 | 0 | 0.98 | 0 | 0.999 |
| mf EtOH(l) | - | 0.000 | 0.000 | 0.020 | 0.000 | 0.001 |
| mf $H_2O$(g) | | | 0.050 | | | |
| mf EtOH(g) | | | 0.018 | | 0.86 | |
| pp $H_2O$ | kPa(a) | | 5.471 | | | |
| pp EtOH | kPa(a) | | 1.994 | | | |
| vp$H_2O$ | kPa(a) | | 5.59 | 5.59 | | |
| vpEtOH | kPa(a) | | 98.2 | 98.2 | | |
| Mass balances | | Unit 1 | Unit 2 | Overall | Str. 4 x(B) | 0.86 |
| Mass IN | kg/h | 1.00E+04 | 9.52E+03 | 1.00E+04 | Sep | 0.96 |
| Mass OUT | kg/h | 1.00E+04 | 9.52E+03 | 1.00E+04 | Conv | 0.95 |
| Closure % | | 100.0 | 100.0 | 100.0 | wt % | 10 |
| $H_2O$ Ant | 16.5362 | A    B | 3985.44 | | C   −38.9974 | vol frac |
| EtOH Ant | 5.2468 | log10, bar | 1598.67 | | −46.424 | 0.96 |

## 4:9 U   Cat cracker

Solution                                                                **136 USgal/ton**

The full stream table is below.

| Stream table | | CAT CRACKER | | Crude | | 100 t/h |
|---|---|---|---|---|---|---|
| Species | M | Solution | | *Stream* | kmol/h | |
| | kg/kmol | *1* | *2* | *3* | *4* | *5* |
| $C_{20}H_{42}$ | 282 | 212.8 | 42.6 | 0.0 | 0.0 | 42.6 |
| $C_{18}H_{38}$ | 254 | 157.5 | 15.7 | 0.0 | 0.0 | 15.7 |
| $C_8H_{18}$ | 114 | 0 | 311.9 | 3.1 | 308.8 | 0.0 |
| $C_2H_4$ | 28 | 0 | 1729.9 | 1695.3 | 34.6 | 0.0 |
| Total   kg/h | | 1.00E+05 | 1.00E+05 | 4.78E+04 | 3.62E+04 | 1.60E+04 |
| Mass balance | | Unit 1 | Unit 2 | Overall | 700 | kg/m3 |
| IN | kg/h | 1.00E+05 | 1.00E+05 | 1.00E+05 | 5.17E+01 | m3/hr |
| OUT | kg/h | 1.00E+05 | 1.00E+05 | 1.00E+05 | 1.36E+04 | 264 |
| Closure | % | 100.0 | 100.0 | 100.0 | *136* | *USgal/t* |
| | Conversion | C20H42 | 0.6 | s(5,A) | 1 | |
| 1 | 0.8 | x = | 6 | s(4,B) | 0.99 | s(5,B,C) |
| 2 | 0.9 | y = | 5 | s(3,C) | 0.98 | 0 |

## 4:10 W   Oil burner

Combustion, cooling and phase separation. Assumes zero pressure drop.
Continuous, steady-state, reactions, ACC = 0
System: Process units 1, 2 and 3 succession.

a.  Designate: A ≡ fuel oil, B ≡ $H_2O$, C ≡ $O_2$, D ≡ $N_2$, E ≡ $CO_2$
    Reaction stoichiometry:
    $C_7H_8 + 9O_2 \rightarrow 7CO_2 + 4H_2O$
    Material balances:
    ACC = IN − OUT + GEN − CON
    Unit 1 Reactor = burner.
    A: $0 = \dot{n}(1,A) - \dot{n}(3,A) + 0 - \dot{n}(1,A)$                                    [1]
    B: $0 = \dot{n}(1,B) - \dot{n}(3.B) + 4\dot{n}(1,A) + 0$                                   [2]
    C: $0 = \dot{n}(2,C) - \dot{n}(3,C) + 0 - 9\dot{n}(1,A)$                                   [3]
    D: $0 = \dot{n}(2,D) - \dot{n}(3,C) + 0 - 0$                                              [4]
    E: $0 = 0 - \dot{n}(3,E) + 7\dot{n}\,(1,A) - 0$                                           [5]
    Unit 2 Hex − cooler
    All j (A,B,C,D,E)      $0 = \dot{n}(3,j) - \dot{n}(4,j) + 0 - 0$                      [6,7,8,9,10]
    Unit 3 Separator
    All j (A,B,C,D,E)      $0 = \dot{n}(4,j) - \dot{n}(5,j) - \dot{n}(6,j)$            [11,12,13,14,15]
    Phase equilibria:     x(i,j), y(i,j) = mole fraction j, x in liquid, y in gas
    Raoult's law:   y(6,B) = p*(B)x(5,B)/P                                              [16]
    Henry's law:   y(6,D) = H(D)x(5,D)/P   y(6,E) = H(E)x(5,E)/P                        [17,18]
    Antoine $H_2O$ vapour pressure: p*(B) = exp(14.2515 − 3242.38/(T(4)−47.1806))   [19]

**b.** Specification: 5 components, 6 streams = 30 "unknowns"

19 equations, 3 reaction stoichiometry, 1 feed rate, 1 feed composition, air composition, 3 pressure, 2 temperatures = 30.  **Fully specified**

The full stream table is below.

| Species | M | OIL BURNER | | | Stream kmol/h | | |
|---|---|---|---|---|---|---|---|
| | kg/kmol | *1* | *2* | *3* | *4* | *5* | *6* |
| C7H8(l) | 92.0 | 10 | 0.0 | 0 | 0 | 0 | 0 |
| H2O(l) | 18.0 | 0 | 0.0 | 0.0 | 28.81 | 28.81 | 0.00 |
| H2O(g) | 18.0 | 0 | 0.0 | 40 | 11.19 | 0.00 | 11.19 |
| O2(g) | 32.0 | 0 | 90.0 | 0.0 | 0 | 0 | 0 |
| N2(l) | 28.0 | 0 | 0.0 | 0 | 2.0E-05 | 2.0E-05 | 0.0 |
| N2(g) | 28.0 | 0 | 338.6 | 338.6 | 338.6 | 0 | 338.6 |
| CO2(l) | 44.0 | 0 | 0 | 0 | 1.6E-04 | 1.6E-04 | 0 |
| CO2(g) | 44.0 | 0 | 0 | 70 | 7.0E+01 | 0 | 70.0 |
| Total | kmol/h | 10 | 428.6 | 448.6 | 448.6 | 28.8 | 419.8 |
| Total | kg/h | 920 | 12360 | 13280 | 13280 | 519 | 12761 |
| mf H2O | in gas | 0 | 0.00 | 0.09 | 0.027 | | |
| mf N2 | in gas | 0 | 0.00 | 0.75 | | | |
| mf CO2 | in gas | 0.00 | 0.00 | 0.16 | | | |
| Phase | | L | G | G | L + G | L | G |
| β phase split | G/F | | | | | | |
| Pressure | kPa(abs) | NS | NS | NS | 400 | 400 | 400 |
| Temperature | K | | | | 320.0 | 320.0 | 320.0 |
| Vap. press | H2O kPa | | | | 11 | | |
| Mass balance | | Unit 1 | Unit 2 | Unit 3 | Overall | | |
| IN | kg/h | 13280 | 13280 | 13280 | 13280 | | |
| OUT | kg/h | 13280 | 13280 | 13280 | 13280 | | |
| Closure | % | 100.0 | 100.0 | 100.0 | 100.0 | | |
| Antoine | A | B | C | Henry | kPa | | |
| H2O | 14.2515 | 3242.38 | -47.1806 | N2 | 1.50E+07 | CO2 | 4.00E+05 |

**4:11 U   Sulphuric acid**

**a.** Flowsheet

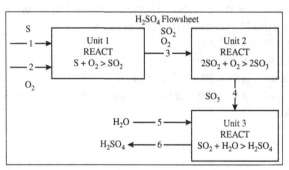

**b.** Material balances:

Unit 1: Mole balances for A, B and C

Unit 2: Mole balances for A, B, C and D

Unit 3: Mole balances for D, E and F

c.

| STREAM TABLE | SULPHURIC ACID | | | Stream | kmol/h | | |
|---|---|---|---|---|---|---|---|
| Species | M kg/kmol | *1* | *2* | *3* | *4* | *5* | *6* |
| | | | | | | | |
| H2SO4 | 98.1 | 0 | 0 | 0 | 0 | 0 | 400 |
| S | 32.1 | 400 | 0 | 0 | 0 | 0 | 0 |
| SO2 | 64.1 | 0 | 0 | 400 | 0 | 0 | 0 |
| SO3 | 80.1 | 0 | 0 | 0 | 400 | 0 | 0 |
| H2O | 18.0 | 0 | 0 | 0 | 0 | 400 | 0 |
| O2 | 32.0 | 0 | 600 | 200 | 0 | 0 | 0 |
| Total | kg/h | 12840 | 19200 | 32040 | 32040 | 7200 | 39240 |
| Mass balance | | Unit 1 | | Unit 2 | | Unit 3 | Overall |
| IN | kg/h | 32040 | | 32040 | | 39240 | 39240 |
| OUT | kg/h | 32040 | | 32040 | | 39240 | 39240 |
| Closure | % | 100.0 | | 100.0 | | 100.0 | 100.0 |

## 4:12 W   Antibiotic extraction

a.   Specification. No reactions occur in this system. Designate components:
A $\equiv$ antibiotic, S $\equiv$ solvent, W $\equiv$ water No of components (A,S,W) = j = 3

| Specification table | Unit 1 | Unit 2 | Unit 3 |
|---|---|---|---|
| Stream variables | IJ=(3)(3)=9 | IJ=(3)(3)=9 | IJ=(3)(4)=12 |
| No. component balances | J=3 | J=3 | J=4 |
| No.specified values | 4 | 2 | 6 |
| No. subsidiary relations | 0 | 1 (D-=30) | 1 (D=30) |
| Total independent relations | 7 | 6 | 10 |
| Degrees of freedom | 9-7=2 | 9-6=3 | 12-10=2 |
| No. of specified values includes those given as zero | | | |
| D of F > 0         Material balance is | | UNDER-SPECIFIED | |

**D of F > 0**   The full 5 stream material balance is **under-specified**.
Stream 3 values are not needed for parts "**b**" and "**c**", just the <u>overall</u> balances.

b.   Overall mole balances: Continuous process, steady-state, no reactions.
$\bar{m}(i)$ = total mass flow stream i, $w(i,j)$ = mass fraction j in stream i
A: $0 = \bar{m}(1)w(1,A)/M(A) - [\bar{n}(4)x(4,A) + \bar{n}(5)x(5,A)]$                    [1]
S: $0 = \dot{n}(2,S) - \bar{n}(4)[1-x(4,A)]$                    [2]
W: $0 = \bar{m}(1)[1-w(1,A)]/M(W)$                    [3]
L-L distribution     : $D = 30 = x(A)_s/x(A)_w$                    [4]
Solve equations 1,2,3 for $\bar{n}(4)$ and $\bar{n}(5)$ and substitute into equation 1 to get:
Balance on A:

0 = m̄(1)w(1,A)/M(A)-[ṅ(2,S)x(4,A)/(1-x(4,A))] + m̄(1)[1-w(1,A](x(4,A)/D/M(W)
   (1-x(4,A)/D]] [5]

This equation [5] is the required relation between x(4,A), w(1,A) and ṅ(2,S)

c.  Arbitrarily fix ranges of values for 2 more variables (w(1,A) & ṅ(2,S)) and fully-specify the problem. Then:

Using Excel with equation 5.

Compile a 3x3 factorial table of the split fraction s(4,A) with:
   w(1,A) = 1,10,20 gram/L, m̄(2) = 1000, 500, 9000 kg/h

| s(4,A) % | S kg/h | | |
|---|---|---|---|
| | 1000 | 5000 | 9000 |
| 1 | 39.2 | 76.3 | 85.3 |
| 10 | 39.7 | 76.5 | 85.4 |
| 20 | 40.1 | 76.8 | 85.7 |

Plot in 3 dimensions:

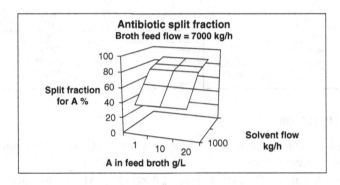

**4:13 U   Isothermal flash**

a.  Stream 1 is partially condensed, i.e.                              **2 phases, liquid+gas**
b.  Check specification: 3 streams, 3 species, 9 "unknowns"          **Fully specified**
c.  Simultaneous equations 1 to 7 can be solved analytically for all = ṅ(i,j)       OR:
    by using the stream table with Excel Solver to close the mass balance.

The complete stream table is below.

| Species | M | ISOTHERMAL FLASH | | | Stream | kmol/h | |
|---|---|---|---|---|---|---|---|
| | kg/kmol | | | *1* | *2* | *3* | |
| A(l) toluene | 78.0 | | | 50 | 30.0 | 0 | |
| A(g) toluene | 78.0 | | | 0 | 0.0 | 20.0 | |
| B(l) benzene | 92.0 | | | 30 | 10.0 | 0 | |
| B(g) benzene | 92.0 | | | 0 | 0.0 | 20.0 | |
| C(l) nitrogen | 28.0 | | | 20 | 10 | 0 | |
| C(g) nitrogen | 28.0 | | | 0 | 0 | 10 | |
| Total | kmol/h | | | 100 | 50.0 | 50.0 | |
| Total | kg/h | | | 7220 | 3540 | 3680 | |
| mf A(l) | | | | 0.5 | 0.60 | 0.40 | |
| mf B(l) | | | | 0.3 | 0.20 | 0.40 | |
| mf C(l) | | | | 0.20 | 0.20 | 0.20 | |
| Phase | | | | L+G | L | G | |
| β phase split G/F | | | 0.50 | | | | |
| Pressure | kPa(abs) | | | 300 | 200 | 200 | |
| Temperature | K | | | 300 | 380.0 | 380.0 | |
| Vap. press | A | | | 4 | 91 | 91 | |
| kPa | B | | | 14 | 215 | 215 | |
| | C | | | 104732 | 167550 | 167550 | |
| K(A) | A | | | 0.01 | 0.27 | 0.18 | |
| K(B) | B | | | 0.01 | 0.21 | 0.43 | |
| K(C) | C | | | 70 | 168 | 168 | |
| Mass balance | | | Antoine | A | B | C | |
| IN | kg/h | 7220 | Toluene | 14.2515 | 3242.38 | -47.1806 | |
| OUT | kg/h | 7220 | Benzene | 14.1603 | 2948.78 | -44.5633 | |
| Closure | % | 100.0 | N2 | 13.7743 | 658.22 | -2.854 | |
| | | | Henry | 1.00E+05 | | | |

## 4:14 W    Vertical farm

Vertical farm $CO_2$. Open system, steady-state, reaction     ACC = 0, GEN and CON # 0

a.    System = contents of farm.

Designate: A ≡ air, C≡$CO_2$, O≡$O_2$, H≡$H_2$ , W≡$H_2O$, TOM≡$CO_2$ to tomatoes kmol/h

$y(i,C)$ = mole fraction $CO_2$ in stream i    NB. ppm(v) = vol/vol = 1E6 (mole fraction)

Air flow in = (farm volume)(turnover) = (1E4 $m^3$)(2/hour) = 20E3 $m^3$/h RTP

               = 810 kmol/h

ACC = IN – OUT + GEN – CON

Mole balances:     Assume zero generation or consumption of air ($N_2+O_2$)

A: $0 = ṅ(4,A) + ṅ(5,A) + ṅ(6,A) – ṅ(8,A)$

C: $0 = [ṅ(4,C) + ṅ(5,C)+ṅ(6,C)] – ṅ(8,C) + 0 – TOM$

O: $0 = [ṅ(4,O) + ṅ(5,O) + ṅ(6,O)] – ṅ(8,O) + 0 – 0.5ṅ(4,H)$ when $H_2$ is burned

H: $0 = [ṅ(4,H)] – ṅ(8,H) + 0 – ṅ(4,H)$             when $H_2$ is burned

W: $0 = [ṅ(4,W) + ṅ(5,W) + ṅ(6,W)] – 0 + ṅ(4,H)$       when $H_2$ is burned

TOM = 0.01 (kg/kg.day)(1000 kg/d)/((24 h/d)(44)) = 9.47E-3 kmol/h $CO_2$ to tomatoes

For water vapour: $y(i,W) = p(i,W)/P = RH(p*(i,W))/P$ p* = vapour pressure $H_2O$ at T(i)

And for all gases:    $n(i,W) = Σ[(ṅ(i,j)-W)(y(i,W)]/(1-y(i,W))$    y = mole fraction in gas

The equations are used in the stream table below, with the following results:

Rate of $CO_2$ generation (stream 5) = 0.25 kmol/h =               **11 kg/h**

Rate of $CO_2$ generation = 11 kg/h/(1000 kg/d/24 h/day) = **0.26 kg/kg**

                         ≡ **\$0.26/kg tomatoes**

**b.** Stream table, extra $O_2$ in stream 8 = 0.12 kmol/h = (1E6)(0.12/407 kmol/h) = **294 ppm(v)**

**c.** Net heat of $H_2$ combustion = $h_{c,gross} - h_{vH2O}$ = 286E3 − 44E3 = 242E3 kJ/kmol
$\Delta T$ = (0.74 kmol/h)(242E3 kJ/kmol)/((408 kmol/h air)(30 kJ/kmol.K)) = **14.6 K**
Complete stream table below.

| Species | M | VERTICAL FARM | | | Stream kmol/h | |
|---|---|---|---|---|---|---|
| | kg/kmol | *4* | *5* | *6* | *7* | *8* |
| H2O(l) | 18.0 | 0 | 0.0 | 0.0 | 0.0 | 0.0 |
| H2O(g) | 18.0 | 0.10 | 0.10 | 7.9 | 8.7 | 9.7 |
| CO2 | 44.0 | 0 | 0.25 | 0.17 | 0.2 | 0.41 |
| AIR | 30.0 | 0.0 | 0.0 | 407.7 | 407.7 | 407.7 |
| H2 | 2.0 | 0.74 | 0 | 0 | 0.0 | 0 |
| O2 | 32.0 | 0 | 0.49 | 0 | 0.12 | 0.12 |
| Total | kg/h | 3.30E+00 | 2.84E+01 | 1.24E+04 | 1.24E+04 | 1.24E+04 |
| mf CO2 | ppm(v) | 250000 | 333000 | 420 | 420 | 1000 |
| Phase | | G | G | G | G | G |
| Press | kPa(a) | 101 | 101 | 101 | 101 | 101 |
| Temp | K | 323 | 323 | 298 | 300 | 300 |
| Volume | m3/h | | | 1.00E+04 | | |
| mf H2O | | 0.12 | 0.12 | 0.03 | 0.04 | 0.04 |
| vpH2O | kPa | 12.2 | 12.2 | 3.2 | 3.5 | 3.5 |
| Mass bal In | kg/h | | 1.24E+04 | TOM | H2/CO2 | O2/CO2 |
| Out | kg/h | | 1.24E+04 | 9.47E-03 | 3 | 2 |
| Closure % | | | 100.0 | | RH | 0.6 |
| Antoine H2O A | 16.5362 | | | B 3985.44 | C -38.9974 | |
| hc H2 | | hc gross - hv = | 286 - 44 | E3 | 2.42E+05 | |
| $\Delta T$ = | 14.6 | K | | | kJ/kmol | |

## 4:15 U  DAC adsorption

**a.** Total mass of air/day = **3.6E6 ton/day**
Assumes negligible effect of $CO_2$ and $H_2O$ on mean molar mass of air.

**b.** Total mass of water = **40E3 ton/day**

**c.** Mass of adsorbent = **29E3 ton**

## 4:16 U  Photo-fuel cell

Photo-fuel cell (PFC) – wastewater treatment.

**a.** Reaction stoichiometry: $C_{12}H_{12}O_{11} + 9.5O_2 \rightarrow 12CO_2 + 6H_2O$
PFC area = **147E3 m$^2$**

**b.** Power from PFC = (5 W/m$^2$)(147E3 m$^2$) = **735 kW**

## 4:17 W  Artificial leaf

Artificial leaf, methanol vs. gasoline
Continuous, steady-state, 8 hour/day, reaction,    ACC = 0, GEN and/or CON # 0

**a.** Combustion of octane in the ICE – stoichiometry
$$C_8H_{18} + (12.5)O_2 \rightarrow 8CO_2 + 9H_2O$$
Fuel consumed for 50 km/day at 6 kg/100 = 6(50/100) = 3 kg = 3 /114 = 0.0263 kmol

The fuel values of octane in an ICE and methanol in a FC are depend respectively on the net and gross heats of combustion.

Octane NET heat of combustion $= -5461E3 + 9(44E3) = -5065E3$ kJ./kmol

Octane/methanol ratio of energy densities $= -5065E3$ kJ/kmol/$(-726E3$ kJ/kmol$) = 7$

Energy efficiency ICE $= 0.25$, FC $= 0.4$

Methanol to match the energy of 3 kg octane $= (0.25/0.4)(7)(0.0263) = 0.115$ kmol

Methanol output from photo-reactor $= 0.115$ kmol/8 hours $= 4.0E-6$ kmol/s

$STF = F\Delta G/(IA) \rightarrow A = F\Delta G/(I(STF))$

Substitute values: With 100% selectivity for and recovery of $CH_3OH$

$A = (4E-6$ kmol/s$)(702E3$ kJ/kmol$)/((1$ kW/m$^2)(0.2)) =$ **14 m$^2$**

**b.**　　$CO_2$ NOT generated $= (8$ mol $CO_2$/mol octane$)(0.0263$ kmol/d$) = 0.21$ kmol/d $\equiv$ **9.3 kg/d**

## 4:18 U　Solar ferrite cycle

Water splitting — "ferrite-cycle"

Continuous, steady state, reactions,　　　　　ACC $= 0$, GEN and/or CON $\# 0$

Reactions:

Unit 1　$3FeO + H_2O \rightarrow Fe_3O_4 + H_2$

Unit 3　$Fe_3O_4 \qquad \rightarrow 3FeO + 0.5O_2$

The complete stream table is below.

| Species | M | FERRITE CYCLE | | | | Stream　kmol/h | | |
|---|---|---|---|---|---|---|---|---|
| | kg/kmol | *1* | *2* | *3* | *4* | *5* | *6* | *7* |
| $Fe_3O_4$ | 231.4 | 0 | 0 | 0 | 2500 | 2500 | 0 | 0 |
| FeO | 71.8 | 0 | 0 | 0 | 0 | 0 | 7500 | 7500 |
| $H_2O$ | 18.0 | 2500 | 0 | 0 | 0 | 0 | 0 | 0 |
| $H_2$ | 2.0 | 0 | 2500 | 0 | 0 | 0 | 0 | 0 |
| $O_2$ | 32.0 | 0 | 0 | 1250 | 0 | 0 | 0 | 0 |
| Total | kg/h | 4.50E+04 | 5.00E+03 | 4.00E+04 | 5.79E+05 | 5.79E+05 | 5.39E+05 | 5.39E+05 |
| | | Unit 1 | Unit 2 | Unit 3 | Unit 4 | Overall | 120 t/d | |
| Mass IN kg/h | | 5.84E+05 | 5.79E+05 | 5.79E+05 | 5.39E+05 | 4.50E+04 | 5.00 t/h | |
| Mass OUT kg/h | | 5.84E+05 | 5.79E+05 | 5.79E+05 | 5.39E+05 | 4.50E+04 | | |
| Closure % | | 100.0 | 100.0 | 100.0 | 100.0 | 100.0 | | |

## 4:19 W　Nitric acid

Overall stoichiometry:

$$3NH_3 + xO_2 \rightarrow yHNO_3 + zNO + wH_2O$$

Find the coefficients by atom balances on N, H and O. IN = OUT

For the overall process (Unit 1 + Unit 2):

N:　　$3 = y + z$　　　H:　　$9 = y + 2w$　　　O:　　$2x = 3y + z + w$

4 unknowns , 3 equations!

For Unit 1 alone $NH_3$ is the only source of N for $NO_2$ and then of $HNO_3$:

Stoichiometry:

$3NH_3 + xO_2 \rightarrow pNO_2 + qH_2O$

N: $3 = p$　　　H: $9 = 2q$　　　O: $2x = 2p + q$

3 unknowns, 3 equations. Solve for:

$x = 5.25$, $p = 3$ $q = 4.5$

Substitute in overall atom balances:

$3NH_3 + 5.25O_2 \rightarrow yHNO_3 + zNO + wH_2O$

N: $3 = y + z$     H: $9 = y + 2w$     O: $10.5 = 3y + z + w$

3 unknowns, 3 equations, solve for:

$y = 2$, $z = 1$, $w = 3.5$

Overall reaction: $3NH_3 + 5.25\ O_2 \rightarrow 2HNO_3 + NO + 3.5H_2O$

The complete stream table is below.

| STREAM TABLE | | NITRIC ACID | | | | |
|---|---|---|---|---|---|---|
| Species | M | | | Stream | | |
| | kg/kmol | 1 | 2 | 3 | 4 | 5 |
| | | | | Flow kmol/h | | |
| NH3 | 17.0 | 90 | 0 | 0 | 0 | 0 |
| NO | 30.0 | 0 | 0 | 0 | 0 | 30 |
| NO2 | 46.0 | 0 | 0 | 90 | 0 | 0 |
| HNO3 | 63.0 | 0 | 0 | 0 | 60 | 0 |
| H2O | 18.0 | 0 | 0 | 135 | 105 | 0 |
| O2 | 32.0 | 0 | 157.5 | 0 | 0 | 0 |
| Total mass | kg/h | 1530 | 5040 | 6570 | 5670 | 900 |
| Mass balance | | Unit 1 | | Unit 2 | | Overall |
| In kg./h | | 6570 | | 6570 | | 6570 |
| Out kg/h | | 6570 | | 6570 | | 6570 |
| Closure % | | 100.0 | | 100.0 | | 100.0 |

## 4:20 W  Bio-effluent treatment

Bio-oxidation reaction:

$$C_6H_{10}O_5 + \alpha O_2 + \beta NH_3 \rightarrow \delta CH_{1.8}O_{0.5}N_{0.2}(s) + 4CO_2(g) + \gamma H_2O(l) \qquad \text{rxn 1}$$

Cellulose                Biomass

Designate component:

$A \equiv C_6H_{10}O_5$ $B \equiv O_2$ $C \equiv NH_3$ $D \equiv CH_{1.8}O_{0.5}N_{0.2}$ $E \equiv CO_2$ $F \equiv H_2O$ $G \equiv N_2$

M kg/kmol 162        32  17        24.6              44     18     28

a.  Atom balances find the stoichiometric coefficients in the balanced reaction.

Atom balances:

C: $6 = \delta + 4$                                                      $\delta = 2$

O: $5 + 2\alpha = 2(0.5) + 8 + \gamma$

N: $\beta = 2(0.2)$                                                    $\beta = 0.4$

H: $10 + 0.4(3) = 2(1.8) + 2\gamma$                        $\gamma = 3.8$ $\alpha = 3.9$

Mole balances on each process unit – in sequence:

b.  Unit 1. MIXER

A: $0 = \dot{n}(1,A) + \dot{n}(7,A) - \dot{n}(2,A)$                                [1]

B, C, D, E, F, G as for A0 $= \dot{n}(1,j) + \dot{n}(7,j) - \dot{n}(2,j)$        [2,3,4,5,6,7]

Total mass balance: $0 = \bar{m}(1) + \bar{m}(7) - \bar{m}(2)$                [8]

Subsidiary:

$\bar{m}(1) = 20.0E3$ kg/h,      $\dot{n}(1,A) = 0.02\bar{m}(1)$, $\dot{n}(1,C) = (0.95)\ \beta\dot{n}(2,A)$          [9,10,11]

$\dot{n}(1,B) = 0$      $\dot{n}(1,D) = 0$      $\dot{n}(1,E) = 0$      $\dot{n}(1,G) = 0$                     [12,13,14,15]

Unit 2. REACTOR.

A: $0 = \dot{n}(2,A) + \dot{n}(3,A) - \dot{n}(4,A) - \dot{n}(5,A) - (1/1)X(A)\dot{n}(2,A)$          [16]

B: $0 = \dot{n}(2,B) + \dot{n}(3,B) - \dot{n}(4,B) - \dot{n}(5,B) - (\alpha/1)X(A)\ \dot{n}(2,A)$          [17]

C: $0 = \dot{n}(2,C) + \dot{n}(3,C) - \dot{n}(4,C) - \dot{n}(5,C) - (\beta/1)X(A)\ \dot{n}(2,A)$          [18]

D: $0 = \dot{n}(2,D) + \dot{n}(3,D) - \dot{n}(4,D) - \dot{n}(5,D) + (\delta/1)X(A)\ \dot{n}(2,A)$          [19]

E: $0 = \dot{n}(2,E) + \dot{n}(3,E) - \dot{n}(4,E) - \dot{n}(5,E) + (4/1)X(A)\ \dot{n}(2,A)$          [20]

F: $0 = \dot{n}(2,F) + \dot{n}(3,F) - \dot{n}(4,F) - \dot{n}(5,F) + (\gamma/1)X(A)\ \dot{n}(2,A)$          [21]

G: $0 = \dot{n}(2,G) + \dot{n}(3,G) - \dot{n}(4,G) - \dot{n}(5,G) + 0 - 0$          [22]

Total mass balance:     $0 = \bar{m}(2) + \bar{m}(3) - \bar{m}(4) - \bar{m}(5)$          [23]

Subsidiary:     $x(A) = 0.95$          [24]

$\dot{n}(3,B) = 2(\alpha)(0.95)\dot{n}(2,A)$      $(3,G) = (79/21)\dot{n}(3,B)$          [25,26]

$\dot{n}(3,F)/\bar{n}(3) = (0.7)(3.536)$ kPa/150 kPa $= 0.0165$ (70% RH 300 K)          [27]

$\dot{n}(4,F)/\bar{n}(4) = (1)(6.23)$ kPa/100 kPa $= 0.0623$ (saturated gas. 310 K)          [28]

$\dot{n}(3,A) = 0\ \dot{n}(3,C) = 0\ (3,D) = 0\ \dot{n}(3,E) = 0$          [29,30,31,32]

$\dot{n}(5,B) = 0\ \dot{n}(5,E) = 0\ \dot{n}(5,G) = 0$          [33,34,35]

Unit 3. SEPARATOR

A: $0 = \dot{n}(5,A) - \dot{n}(6,A) - \dot{n}(9,A)$          [36]

B,C,D,E,F,G as for A     $0 = \dot{n}(5,j) - \dot{n}(6,j) - \dot{n}(9,j)$          [37,38,39,40,41,42]

Total mass balance:     $0 = \bar{m}(5) - \bar{m}(6) - \bar{m}(9)$          [43]

Subsidiary:     $(\dot{n}(6,A) + \dot{n}(6,D))/\bar{m}(6) = 0.15$          [44]

$\dot{n}(9,A) = 0$          [45]

Unit 4. DIVIDER

A: $0 = \dot{n}(6,A) - \dot{n}(7,A) - \dot{n}(8,A)$          [46]

B,C,D,E,F,G as for A  $0 = \dot{n}(6,j) - \dot{n}(7,j) - \dot{n}(8,j)$          [47,48,49,50,51,52]

Total mass balance:     $0 = \bar{m}(6) - \bar{m}(7) - \bar{m}(8)$          [53]

Subsidiary: $\bar{m}(7) = 500$ kg/h          [54]

All streams 6,7 & 8 have the same composition, i.e. for all "j" :

$\dot{n}(6,j)/\bar{n}(6) = \dot{n}(7,j)/\bar{n}(7) = \dot{n}(8,j)/\bar{n}(8)$

Divider flow ratio:   $K_{DIV}(i) = \bar{n}(i)/\bar{n}(6) = \bar{m}(i)/\bar{m}(6)$          [55,56,57,58,59,60]

Stream 7 recycles to the mixer.

c.   The spreadsheet calculation uses the <u>iterative sequential modular method</u>, beginning at Unit 1. The tear stream is stream 7, which is set at zero flow in the first pass of the iterations. The circular reference is resolved in Excel by "tools – options – calculation – iterate"

The Excel spreadsheet with completed stream table is below. $\alpha = 3.9$, $\beta = 0.4$, $\delta = 2$, $\gamma = 3.8$

Solution:

| STREAM TABLE | | BIO-EFFLUENT TREATMENT | | | | Material balance | | | | |
|---|---|---|---|---|---|---|---|---|---|---|
| Species | M | | | *STREAM* | | kmol/h | | | | |
| | kg/kmol | *1* | *2* | *3* | *4* | *5* | *6* | *7* | *8* | *9* |
| A. C6H10O5 | 162.0 | 2.47E+00 | 2.52E+00 | 0.00E+00 | 0.00E+00 | 1.26E-01 | 1.26E-01 | 4.59E-02 | 7.99E-02 | 0.00E+00 |
| B. O2 | 32.0 | 0.00E+00 | 0.00E+00 | 1.86E+01 | 9.32E+00 | 0.00E+00 | 0.00E+00 | 0.00E+00 | 0.00E+00 | 0.00E+00 |
| C. NH3 | 17.0 | 9.56E-01 | 9.56E-01 | 0.00E+00 | 0.00E+00 | 0.00E+00 | 0.00E+00 | 0.00E+00 | 0.00E+00 | 0.00E+00 |
| D. CH1.8O0.5N | 24.6 | 0.00E+00 | 2.75E+00 | 0.00E+00 | 0.00E+00 | 7.53E+00 | 7.53E+00 | 2.75E+00 | 4.78E+00 | 0.00E+00 |
| E. CO2 | 44.0 | 0.00E+00 | 0.00E+00 | 0.00E+00 | 9.56E+00 | 0.00E+00 | 0.00E+00 | 0.00E+00 | 0.00E+00 | 0.00E+00 |
| F. H2O | 18.0 | 1.09E+03 | 1.11E+03 | 1.49E+00 | 5.91E+00 | 1.12E+03 | 6.47E+01 | 2.36E+01 | 4.11E+01 | 1.05E+03 |
| G. N2 | 28.0 | 0.00E+00 | 0.00E+00 | 7.01E+01 | 7.01E+01 | 0.00E+00 | 0.00E+00 | 0.00E+00 | 0.00E+00 | 0.00E+00 |
| Total (not req'd) kmol/h | | 1.09E+03 | 1.12E+03 | 9.02E+01 | 9.49E+01 | 1.12E+03 | 7.23E+01 | 2.64E+01 | 4.59E+01 | 1.05E+03 |
| Total | kg/h | 2.00E+04 | 2.05E+04 | 2.59E+03 | 2.79E+03 | 2.03E+04 | 1.37E+03 | 5.00E+02 | 8.70E+02 | 1.89E+04 |
| Mass balance checks | | | Unit 1 | Unit 2 | Unit 3 | Unit 4 | Overall | | | |
| Mass IN | kg/h | | 2.05E+04 | 2.31E+04 | 2.03E+04 | 1.37E+03 | 2.26E+04 | | | |
| Mass OUT | kg/h | | 2.05E+04 | 2.31E+04 | 2.03E+04 | 1.37E+03 | 2.26E+04 | | | |
| Closure | % | | 100.0 | 100.0 | 100.0 | 100.0 | 100.0 | All OK! | | |

### 4:21 U   Chemical looping combustion

a.  Stoichiometric coefficients   $p = 1$,   $q = 1$,   $w = 2$,   $x = 6$,   $y = 2$

Reactions are:   Unit 1:   $4Mn_3O_4 + CH_4 \rightarrow 12MnO + CO_2 + 2H_2O$

Unit 2:   $12MnO + 2O_2 \rightarrow 4Mn_3O_4$

b.  Mn hold-up in reactors =   **18280 kg**

c.  Stream 2. $T_{dew}$ =   **359 K**

# Chapter 5:   Energy Balances

### 5:1 W   Solar energy storage

Solar energy storage is a cyclic, intermittent process with conditions changing over time. A detailed analysis would invoke differential equations but this problem is solved by global balances for each part of the cycle.

System: Unit 2 sand bed. Cooling – streams 1 and 2 are disconnected.

a.  Quantity: Mass of sand = S kg

Mass balance over 1 cooling cycle:

ACC = IN – OUT + GEN – CON       No reaction       ACC = GEN = CON = 0

$0 = S_{in} – S_{out}$       $S_{in} = S_{out} = S$

Energy balance over 1 cooling cycle:

No reaction, no heat loss, sensible heat only.       Q = 0       *ref. Compounds at 298 K*

$0 = H(4) – H(3) + 0 – W$       W = 1E5 kWh energy out

$0 = SC_{p,sand}[T(4) – T(3)] – W = S(kg)0.9$ kJ/(kg.K)[600 – 500]K – (1E5 kWh)(3600 kJ/kWh)

$S = 1E5(3600)$ kJ/(0.9(100)kJ/kg) =       **4.0E6 kg**

**b.** Quantity: Mass of air
Energy balance over 1 heating cycle:
A = total mass of air passed over the heating cycle kg
$0 = H(1) - H(2) + Q - 0$
$H(2) - H(1) = Q = 1E5 \text{ kWh}$    $H(i) = AC_{p,air}(T(i) - 298)$
$A = Q/(C_{p,air}(T(1) - T(2))) = 1E5 \text{ kWh}(3600 \text{ kJ/kWh})/[(29 \text{ kg/kmol})(31 \text{ kJ/kmol.K})]$
$A = 400E3 \text{ kg}$    Average rate over 8 h = 400E3 kg/8 h =    **50E3 kg/h**

### 5:2 W   Fuel gas combustion

Complete combustion stoichiometry:
$$2H_2 + O_2 \rightarrow 2H_2O$$
$$CH_4 + 2O_2 \rightarrow CO_2 + 2H_2O$$
Designate: $A \equiv H_2$, $B \equiv CH_4$, $C \equiv O_2$, $D \equiv CO_2$, $E \equiv H_2O$, $F \equiv N_2$
Convert all wt% in feed gas to mole fraction y(j) a.k.a mole %
A: $y(A) = (w(A)/M(A)/(\Sigma w(j)/M(j))) = (20/2)/(20/2+70/16+10/18)$    = 67.0 mol%
B: $y(B) = (70/16)/(20/2 + 70/16 +10/18)$    = 29.3 mol%
E: $y(E) = (10/18)/(20/2 + 70/16 + 10/18)$    = 3.7 mol%
Material balances. Continuous steady-state   Dry air (21/79 mol ratio $O_2/N_2$).
To find composition of the combustion product gas.
Basis 100 kmol of fuel gas mixture. Unit time.    $0 = IN - OUT + GEN - CON$
Stoichiometric $O_2 = 67/2 + 2(29.3) = 92.1$   kmol    mol %
A: $0 = 67 - 0 + 0 - 67 = 0$    Aout = 0 kmol    0
B: $0 = 29.3 - 0 + 0 - 29.3$    Bout = 0 kmol    0
C: $0 = 1.2[92.1] - Cout + 0 - 92.1$    Cout = 18.4 kmol    3.1
D: $0 = 0 - Dout + 1(29.3) - 0$    Dout = 29.3 kmol    4.9
E: $0 = 3.7 - Eout + 1(67) + 2(29.3) - 0$    Eout = 129.3 kmol    21.8
F: $0 = (0.79/0.21)(1.2)(92.1) - Fout + 0 - 0$  Fout = 415.8 kmol    70.1
Enthalpy balance over burner. Heat of formation method.
Reference conditions = elements at 298 K
Adiabatic flame temperature = $T_f$
$0 = \dot{H}^*_{in} - \dot{H}^*_{out} + 0 - 0$    Adiabatic conditions $\dot{Q} = 0$, $\dot{W} = 0$
$\dot{H}^*(i) = \Sigma\{\dot{n}(i,j)[(C_{pm,298K}(j)[T(i) - 298] + h^\circ_{f,298K}(j)]\}$ *ref. Elements at 298 K*

| Component | $h_f(298K)$kJ/kmol | $C_{p,m}$ ref 298K kJ/(kmol.K) |
|---|---|---|
| $A = H_2$ | 0 | 28.8 (298 K) |
| $B = CH_4$ | −74.8E3 | 35.8 (298 K) |
| $C = O_2$ | 0 | 34.9 (2070 K) |
| $D = CO_2$ | −393.5E3 | 54.1 (2070 K) |
| $E = H_2O(g)$ | −241.8E3 | 42.8 (2070 K) |
| $F = N_2$ | 0 | 33.2 (2070 K) |

Proceed in Excel. Solve the adiabatic enthalpy balance for $T_{out} = T_f =$     **2173 K**

| FUEL GAS COMBUSTION | | | | | | | | | | |
|---|---|---|---|---|---|---|---|---|---|---|
| Species | All gas | H2 | CH4 | O2 | CO2 | H2O | N2 | T K | H kJ | |
| Cp,m 298 | kJ/kmol.K | 28.8 | 35.8 | 34.9 | 54.1 | 42.8 | 33.2 | Goal-seek | Hout/Hin | |
| h$_{f,298}$ | kJ/kmol | 0 | -7.48E+04 | 0 | -3.94E+05 | -2.41E+05 | 0 | | 1.00E+00 | |
| Gas in | kmol | 29.3 | 29.3 | 110.5 | 0 | 3.7 | 415.8 | 298 | -3.08E+06 | |
| Gas out | kmol | 0 | 0 | 3.1 | 29.3 | 129.3 | 415.8 | 2173 | -3.08E+06 | |

Mole fraction $H_2O$ in gas out = 0.218    pp $H_2O$ = 0.218(190 kPa(a)) = 41.4 kPa(a)
From the steam table vp of $H_2O$ is 41.66 kPa(a) at 350 K. i.e. dew point =     **350 K**

### 5:3 U   Generic material and energy balances

In each case first complete the material balances, which are fully specified and independent of the energy balances. With the material balance defined use Goal-seek or Solver to close the energy balances by changing the unknown temperature, heat or work term.
For each case the material and energy balance is fully specified.
The unknowns are found using Excel Goal-seek to close the energy balances with (energy out/energy in) = 100%. Results are below.

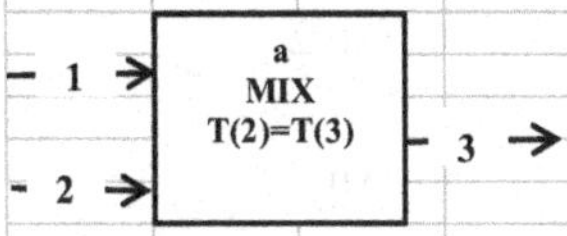

| STREAM TABLE | | | | MIXER | | | |
|---|---|---|---|---|---|---|---|
| Species | M | $C_{p,m,298}$ | $h_{f,298}$ | | Stream | | |
| | kmol/kg | kJ/kmol.K | kJ/kmol | 1 | 2 | 3 | 4 |
| | | | | | Flow kmol/h | | |
| A(liq) | 40 | 70 | -6.0E+04 | 30 | 10 | 20 | NA |
| A(gas) | 40 | 35 | -4.0E+04 | 0 | 0 | 0 | NA |
| B(liq) | 80 | 90 | -1.5E+05 | 70 | 30 | 40 | NA |
| B(gas) | 80 | 45 | -1.2E+05 | 0 | 0 | 0 | NA |
| Total | kg/h | | | 6800 | 2800 | 4000 | NA |
| Phase | | | | L | L | L | |
| Pressure kPa(a) | | | | 500 | 500 | 480 | |
| Temp. K | | | | 290 | 333 | 333 | |
| Enthalpy kJ/h | | | | -1.2E+07 | -5.0E+06 | -7.0E+06 | |
| Mass balance | IN kg/h | 6800 | | Q kW | W kW | | |
| | OUT kg/h | 6800 | | 0 | -100 | | |
| Closure % | | 100.0 | | | | | |
| Energy balance | IN kJ/h | -1.E+07 | | | | | |
| | OUT kJ/h | -1.E+07 | | | | | |
| Closure % | | 100.0 | | | | | |

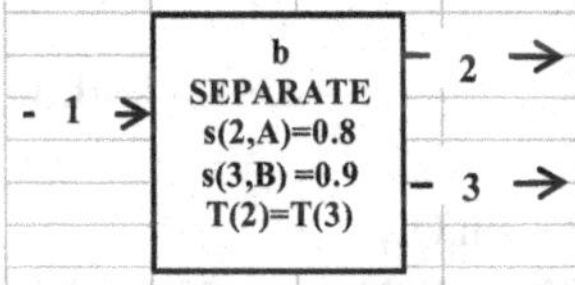

Diagram (left): arrows labelled − 1 → and − 2 → into box **a MIX T(2)=T(3)**, output − 3 →

| STREAM TABLE | | | | SEPARATOR | | | |
|---|---|---|---|---|---|---|---|
| Species | M | $C_{p,m,298}$ | $h_{f,298}$ | | Stream | | |
| | kmol/kg | kJ/kmol.K | kJ/kmol | 1 | 2 | 3 | 4 |
| | | | | | Flow kmol/h | | |
| A(liq) | 40 | 70 | -6.0E+04 | 50 | 0 | 10 | NA |
| A(gas) | 40 | 35 | -4.0E+04 | 0 | 40 | 0 | NA |
| B(liq) | 80 | 90 | -1.5E+05 | 30 | 0 | 27 | NA |
| B(gas) | 80 | 45 | -1.2E+05 | 0 | 3 | 0 | NA |
| Total | kg/h | | | 4400 | 1840 | 2560 | NA |
| Phase | | | | L | G | L | |
| Pressure kPa(a) | | | | 300 | 280 | 280 | |
| Temp. K | | | | 290 | 390 | 390 | |
| Enthalpy kJ/h | | | | -6.9E+06 | -1.8E+06 | -4.3E+06 | |
| Mass balance | IN kg/h | 4400 | | Q kW | W kW | s(2,A) | 0.8 |
| | OUT kg/l | 4400 | | 200 | 0 | s(3,B) | 0.9 |
| Closure % | | 100.0 | | | | | |
| Energy balance | IN kJ/h | -6.1E+06 | | | | | |
| | OUT kJ/l | -6.1E+06 | | | | | |
| Closure % | | 100.0 | | | | | |

Diagram (left): arrow − 1 → into box **b SEPARATE s(2,A)=0.8 s(3,B)=0.9 T(2)=T(3)**, outputs − 2 → and − 3 →

| STREAM TABLE | | | | REACTOR | | | |
|---|---|---|---|---|---|---|---|
| Species | M | $C_{p,m,298}$ | $h_{f,298}$ | Stream | | | |
| | kmol/kg | kJ/kmol.K | kJ/kmol | *1* | *2* | *3* | *4* |
| | | | | Flow kmol/h | | | |
| A(liq) | 40 | 70 | –6.0E+04 | 50 | 0 | NA | NA |
| A(gas) | 40 | 35 | –4.0E+04 | 0 | 10 | NA | NA |
| B(liq) | 80 | 90 | –1.5E+05 | 0 | 0 | NA | NA |
| B(gas) | 80 | 45 | –1.2E+05 | 0 | 20 | NA | NA |
| Total | kg/h | | | 2000 | 2000 | NA | NA |
| Phase | | | | L | G | | |
| Pressure kPa(a) | | | | 200 | 180 | | |
| Temp. K | | | | 350 | 395 | | |
| Enthalpy kJ/h | | | | –2.6E+06 | –2.7E+06 | | |
| Mass balance | IN kg/h | 2000 | | Q kW | W kW | X(A) | 0.8 |
| | OUT kg/h | 2000 | | –10.0 | 0 | 2A > B | |
| Closure % | | 100.0 | | | | | |
| Energy balance | IN kJ/h | –2.7E+06 | | | | | |
| | OUT kJ/h | –2.7E+06 | | | | | |
| Closure % | | 100.0 | | | | | |

Reactor diagram: – 1 → | c REACT 2A > B X(A) = 0.8 | – 2 →

## 5:4 U    Rocket engine

**a.**  Reaction stoichiometry:        $C_2H_8N_2 + 2N_2O_4 \rightarrow 2CO_2 + 4H_2O + 3N_2$

**b.**  For Fuel = 8000 kg        Height =                                    **14.8E3 m**

## 5:5 W    Methanol from propane

First write stoichiometry for the various reactions.

Combustion of methanol: Gross > liquid water $H_2O(l)$

$\qquad CH_3OH(l) + (3/2)O_2(g) \rightarrow CO_2(g) + 2H_2O(l)$        rxn 1

Net heat of combustion: Net > gas water $H_2O(g)$

$\qquad CH_3OH(l) + (3/2)O_2 \rightarrow CO_2(g) + 2H_2O(g)$        rxn 2

Combustion of propane: Gross heat > liquid water $H_2O(l)$

$\qquad C_3H_8(l) + 5O_2(g) \rightarrow 3CO_2(g) + 4H_2O(l)$        rxn 3

Oxidation of propane to methanol:

$\qquad 2C_3H_8(l) + 7O_2(g) \rightarrow 2CH_3OH(l) + 4CO_2(g) + 4H_2O(l)$  rxn 4

Find required heats of formation.

$h_f\ H_2O(l) = h_f\ H_2O(g) + h_v\ H_2O = -242E3 - 44E3$        = –285E3 kJ/kmol

From reaction 3:

$h_{rxn} = (3h_fCO_2(g) + 4h_fH_2O(l)) - (h_fC_3H_8 + 5(0)) = -2204E3$ kJ/kmol

$h_fC_3H_8(l) = 3(-394E3) + 4(-285E3) + 2204E3$        = –118E3 kJ/kmol

**a.**  For reaction 2:

$h_{rxn} = h_{rxn}(1) + 2(h_v(H_2O)) = -727E3 + 2(44E3) =$        **–639E3 kJ/kmol**

**b.**  From reaction 1:

$h_f(CH_3OH(l)) = [h_fCO_2 + 2h_fH_2O(g)] - h_{rxn}CH_3OH(l)$        **–236E3 kJ/kmol**

**c.**  For reaction 4:

$h_{rxn} = [4h_fH_2O(l) + 4h_fCO_2 + 2h_fCH_3OH(l)] - [2h_fC_3H_8(l) + 7h_fO_2]$

$h_{rxn} = [4(-285E3) + 4(-393E3) + 2(-236E3)] - [2(-118E3) + 7(0)] =$    **–2948E3 kJ/ kmol**

**d.**  Negative $h_{rxn}$:                                    **Exothermic**

## 5:6 U   Reactor with heat exchange

a.   The problem is FULLY SPECIFIED.
In Excel the material balances are solved explicitly for outlet flows and the values of T(3), T(6) and Q(3) are found using Goal Seek to close the respective energy balances.
See the full stream table below.

b.   HEX duty = **683 kW**        Reactor thermal load = **−4920 kW** (i.e. Cooling)

| STREAM TABLE | | REACTOR with HEAT EXCHANGER | | | | | | | |
|---|---|---|---|---|---|---|---|---|---|
| Species | M | $C_{p,m\ 298K}$ | $h_f^\circ$ | | | Stream  kmol/h | | | |
| | kg/kmol | kJ/kmol.K | kJ/kmol | *1* | *2* | *3* | *4* | *5* | *6* |
| A(l) | 40 | 70 | −5.0E+04 | 40 | 0 | 40 | 40 | 0 | 4 |
| A(g) | 40 | 35 | −4.5E+04 | 0 | 0 | 0 | 0 | 4 | 0 |
| B(l) | 70 | 90 | −1.0E+05 | 0 | 90 | 90 | 90 | 0 | 18 |
| B(g) | 70 | 45 | −8.0E+04 | 0 | 0 | 0 | 0 | 18 | 0 |
| C(l) | 60 | 80 | −2.5E+05 | 0 | 0 | 0 | 0 | 0 | 108 |
| C(g) | 60 | 40 | −2.4E+05 | 0 | 0 | 0 | 0 | 108 | 0 |
| Total | kg/h | | | 1600 | 6300 | 7900 | 7900 | 7900 | 7900 |
| Phase | | | | L | L | L | L | G | L |
| Pressure | kPa(abs) | | | 300 | 300 | 290 | 260 | 200 | 180 |
| Temp. | K | | | 300 | 400 | *374* | 600 | 700 | *404* |
| Enthalpy* | kJ/h | ref Elements @ 298K | | −2.0E+06 | −8.2E+06 | −1.0E+07 | −7.7E+06 | −2.5E+07 | −2.8E+07 |
| Mass balance | | | | Unit 1 | Unit 2 | Unit 3 | Overall | A + 2B > | |
| IN | kg/h | | | 7900 | 15800 | 7900 | 7900 | X(A) | 0.9 |
| OUT | kg/h | | | 7900 | 15800 | 7900 | 7900 | Q3 | Q2 |
| Closure | % | | | 100.0 | 100.0 | 100.0 | 100.0 | kJ/h | kJ/h |
| Energy | | | | | | | | −1.77E+07 | −2.5E+06 |
| IN | kJ/h | | | −1.0E+07 | −3.6E+07 | −2.5E+07 | −1.0E+07 | kW | kW |
| OUT | kJ/h | | | −1.0E+07 | −3.6E+07 | −2.5E+07 | −1.0E+07 | −4920.3 | −683.3 |
| Closure | % | | | 100.0 | 100.0 | 100.0 | 100.0 | | |

## 5:7 W   Ammonia vaporization

a.   Material and energy balance on an indirect contact heat exchange, no reactions.
Mole balances              $A = H_2O\ B = NH_3$
Cold side
A: $0 = \dot{n}(1,A) - \dot{n}(2,A)$     $n(2,A) = 0$
B: $0 = \dot{n}(3,B) - \dot{n}(4,B)$     $n(2,B) = 200$ kmol/h
Hot side
A: $0 = \dot{n}(3,A) - \dot{n}(4,A)$     $\dot{n}(3,A) = \dot{n}(4,A) = Y$ kmol/h
B: $0 = \dot{n}(3,B) - \dot{n}(4,B)$     $\dot{n}(4,B) - 0$

Overall energy balance     $0 = \dot{H}(1) + \dot{H}(3) - \dot{H}(2) - \dot{H}(4) + \dot{Q} - \dot{W}$       $\dot{Q} = \dot{W} = 0$
Reference = liquid compounds at 273 K. Use the steam table for water.
$\dot{H}(1) = nC_p(T - 273) = (200$ kmol/h$)(71$ kJ/(kmol.K$)(273-273)$K $= 0$ (respecting the phase)
$\dot{H}(2) = (200$ kmol/h$)(35$ kJ.kmol.K$)(350-273)$K $+ h_v = 539E3 + 21E3 = 560E3$ kJ/h
$\dot{H}(3) = (18$ kg/kmol$)(Y$ kmol/h$)[2710$ kJ/kg$] = 48.8E3Y$ kJ/h       Steam table saturated gas
$\dot{H}(4) = (18$ kg/kmol$)(Y$ kmol/h$)[323$ kJ/kg$] = 5.8E3Y$ kJ/h Steam table saturated liquid
The steam table gives the phases and enthalpies of streams 3, 4 G, L, condensing steam

**b.** Use Excel to solve the M&E balances for the water flow rate Y. Complete the stream table – as shown below.

| STREAM TABLE | | VAPORIZATION of AMMONIA | | | | | Stream | | |
|---|---|---|---|---|---|---|---|---|---|
| Species | M | Density | $C_{p,m}$ 298K | $h^\circ_f$ | 1 | 2 | 3 | 4 |
| | kg/kmol | kg/m³ | kJ/kmol.K | kJ/kmol* | | Flow kmol/h | | |
| H2O(l) | 18 | | | NA | 0 | 0 | 0 | 13.0 |
| H2O(g) | 18 | STEAM TABLE | | NA | 0 | 0 | 13.0 | 0 |
| NH3(l) | 17 | 640 | 71 | -6.6E-04 | 200 | 0 | 0 | 0 |
| NH3(g) | 17 | Gas law | 35 | -4.5E-04 | 0 | 200 | 0 | 0 |
| Total | kg/h | | | * ref | 3400 | 3400 | 234 | 234 |
| Phase | | | | elements | L | G | G | L |
| Pressure | kPa(abs) | | | | 820 | 800 | 240 | 150 |
| Temp | K | | | | 273 | 350 | 400 | 350 |
| Volume | m3/h | density, gas law and steam table | | | 5.31 | 7.27E+02 | 171 | 2.34E-01 |
| Enthalpy | kJ/h | ref compounds at 273 K | | | 0 | 5.60E+05 | 6.36E+05 | 7.55E+04 |
| Mass balance | | | | | | h kJ/kg | 4.88E+04 | 5.80E+03 |
| IN | kg/h | 3.63E+03 | | | | specific vol | 0.73 | 1.00E-03 |
| OUT | kg/h | 3.63E+03 | | | | | | |
| Closure % | | 100.0 | | | | | | |
| Energy balance | | | | | | | | |
| IN | kJ/h | 6.36E+05 | | | | | | |
| OUT | kJ/h | 6.36E+05 | | | | | | |
| Closure % | | 100.0 | | | | | | |

## 5:8 U Naphthalene combustion

Combustion of naphthalene <u>vapour</u> in excess air. Reaction stoichiometry:
$C_{10}H_8(g) + 12O_2(g) \rightarrow 10CO_2(g) + 4H_2O(g)$
Use the data to find heat of formation of $C_{10}H_8$ gas.
Use the stream table with Excel Goal Seek to close the energy balance, giving T(3) = **2115 K**

| STREAM TABLE | | | NAPHTHALENE | | Basis | 100 kmol/h C10H8 | |
|---|---|---|---|---|---|---|---|
| Species | M | Cp,m | hf,298K | | Stream kmol/h | | |
| | kg/kmol | kJ/kmol.K | kJ/kmol | 1 | 2 | 3 | |
| C10H8(g) | 128.0 | 220 | 3.80E+03 | 100 | 0 | 0 | |
| H2O(g) | 18.0 | 44 | -2.42E+05 | 0 | 208.7 | 608.7 | |
| CO2(g) | 44.0 | 55 | -3.94E+05 | 0 | 0.0 | 1000.0 | |
| O2(g) | 32.0 | 35 | 0 | 0 | 1440.0 | 240.0 | |
| N2(g) | 28.0 | 34 | 0 | 0 | 5417.1 | 5417.1 | |
| Total | kg/h | | | 1.28E+04 | 2.02E+05 | 2.14E+05 | |
| Phase | | | | G | G | G | |
| Pressure | kPa(a) | | | 110 | 110 | 100 | |
| Temp | K | | | 600 | 303 | *2115* | |
| Enthalpy | kJ/h | ref elements @ 298 K | | 7.02E+06 | -4.93E+07 | -4.22E+07 | |
| Mass | IN kg/h | 2.14E+05 | | | | | |
| balance | OUT kg/h | 2.14E+05 | | | | | |
| | Closure % | 100.0 | | | | | |
| Energy | IN kJ/h | -4.22E+07 | | | | | |
| balance | OUT kJ/h | -4.22E+07 | Xs air % | | 20 | 197760 kg/h | |
| | Closure % | 100.0 | H₂O | | 0.019 kg/kg dry air | | |

## 5:9 W Gasoline from air

Continuous process, steady-state, reactions, 100% conversion and yield.
Designate: $A \equiv Air$, $B \equiv CO_2$, $C \equiv H_2O$, $D \equiv H_2$, $E \equiv O_2$, $F \equiv C_8H_{18}$.

With 100% conversion and yield for all reactions the overall (net) reaction becomes:

$8CO_2 + 9H_2O \rightarrow C_8H_{18} + 12.5O_2$

$C_8H_{18}$ product rate = 11400 kg/hr = 100 kmol/h

$CO_2$ consumed = 800 kmol/h     $H_2O$ consumed = underspecified in the RWGS/FT process.

a.   Mole balances on each process unit in sequence.

   ACC = 0 = IN – OUT + GEN – CON

   Unit 1

   B: $0 = \dot{n}(1,B) - \dot{n}(2,B) - \dot{n}(4,B) + 0 - 0$

   C: $0 = \dot{n}(1,C) - \dot{n}(2,C) - \dot{n}(3,C) + 0 - 0$

   And: $\dot{n}(1,B)/\dot{n}(1,A) = 420E-6$     $\dot{n}(1,B) - \dot{n}(2,B) = 800$ kmol/hr

   $\dot{m}(1,C)/\dot{m}(1,A) = 0.015$ kg/kg

   Unit 2

   C: $0 = \dot{n}(3,C) - \dot{n}(5,C) - \dot{n}(6,C) + 0 - 900$

   D: $0 = \dot{n}(3,D) - \dot{n}(5,D) - \dot{n}(6,D) + 900 - 0$

   E: $0 = \dot{n}(3,E) - \dot{n}(5,E) - \dot{n}(6,R) + 450 - 0$

   Unit 3

   B: $0 = \dot{n}(4,B) - \dot{n}(7,B) - \dot{n}(8,B) + 0 - 800$

   C: $H_2O$ stoichiometry is underspecified in the RWGS/FT process – resolved by using Goal-seek with stream 8 to close the water balance.

   D: $0 = \dot{n}(5,D) - \dot{n}(7,D) - \dot{n}(8.D) + 0 - 900$

   F: $0 = \dot{n}(5,F) - \dot{n}(7,F) - \dot{n}(8.F) + 1000 - 0$

   With 100% conversion and yield:

   $\dot{n}(5,C) = \dot{n}(6,C) = \dot{n}(7,B) = \dot{n}(8,B) = \dot{n}(7,D) = \dot{n}(8,D) = 0$

The full stream table is below.

| Species | M | GASOLINE from AIR | | | | Stream | kmol/h | | |
|---|---|---|---|---|---|---|---|---|---|
| | kg/kmol | 1 | 2 | 3 | 4 | 5 | 6 | 7 | 8 |
| AIR | 28.8 | 1.90E+06 | 1.90E+06 | 0 | 0 | 0 | 0 | 0 | 0 |
| CO2 | 44.0 | 800 | 0 | 0 | 800 | 0 | 0 | 0 | 0 |
| H2O | 18.0 | 4.57E+04 | 4.48E+04 | 900 | 0 | 0 | 0 | 1422 | 0 |
| H2 | 2.0 | 0 | 0 | 0 | 0 | 900 | 0 | 0 | 0 |
| O2 | 32.0 | 0 | 0 | 0 | 0 | 0 | 450 | 0 | 0 |
| C8H18 | 114.0 | 0 | 0 | 0 | 0 | 0 | 0 | 0 | 100 |
| Total | kg/h | 5.57E+07 | 5.57E+07 | 1.62E+04 | 3.52E+04 | 1.80E+03 | 1.44E+04 | 2.56E+04 | 1.14E+04 |
| | | Unit 1 | Unit 2 | Unit 3 | Overall | | | | |
| Mass IN | kg/h | 5.57E+07 | 1.62E+04 | 3.70E+04 | 5.57E+07 | Air | | | |
| Mass OUT | kg/h | 5.57E+07 | 1.62E+04 | 3.70E+04 | 5.57E+07 | 420 | 0.015 | Octane 11400 | |
| Closure % | | 100.0 | 100.0 | 100.0 | 100.0 | ppmCO2 | kg H2O | kg/h | |

b.   The total (minimum) energy needed for the reactions is estimated as the heat of reaction for the overall stoichiometry.

   $8CO_2 + 9H_2O \rightarrow C_8H_{18} + 12.5O_2$

   $h_{rxn} = \Sigma[n(i,j)(h_f(j))]$ products $- \Sigma[n(i,j)(h_f(j))]$ reactants

   $h_{rxn} = 1(-209E3 + 12.5(0)) - (8(-393.5E3) + 9(-241.8E3)) = +5104E3$ kJ/kmol $C_8H_{18}$

   Minimum energy for 1 litre octane:

   $E_{min} = (5104E3$ kJ/kmol$)/(114$ kg/kmol$)(0.70$ kg/L$)/(3600$ kJ/kWh$) =$   **8.6 kWh/L**

## 5:10 U   Catalytic alkylation

a.   Limiting reactant      iso-butene      $C_4H_8$

       Conversion of           iso-butane      $C_4H_{10}$                            **66.7%**

b.   Reaction is                                           **Exothermic**

c.   Stream 5 phase                                **1 phase, Liquid**

       T(5) =                                             **300 K**

The full stream table is below.

| Species | M | $C_{p,}$ 298 | $h_f$ 298 | ALKYLATION | | Stream kmol/h | | |
|---|---|---|---|---|---|---|---|---|
| | kg/kmol | kJ/kmol K | kJ/kmol | 1 | 2 | 3 | 4 | 5 |
| C4H10(l) | 58.0 | 130 | -1.69E-05 | 150.0 | 0.0 | 50.0 | 0.0 | 0 |
| C4H8(l) | 56.0 | 121 | -3.90E+04 | 0.0 | 100.0 | 0.0 | 0.0 | 0 |
| C8H18(l) | 114.0 | 240 | -2.59E-05 | 0.0 | 0.0 | 100.0 | 0.0 | 0 |
| H2O(g) | 18.0 | 33.6 | -2.42E-05 | 0.0 | 0.0 | 0.0 | 0.0 | 0 |
| H2O(l) | 18.0 | 75.3 | -2.86E-05 | 0.0 | 0.0 | 0.0 | 6.6E-03 | 6.6E-03 |
| Total | kg/h | | | 8.7E-03 | 5.6E-03 | 1.4E+04 | 1.20E+05 | 1.20E+05 |
| Phase | | | | L | L | L | L | L |
| Pressure | kPa(abs) | | | 200 | 200 | 180 | 200 | 150 |
| Temp | K | | | 298 | 320 | 310.0 | 290 | 300.0 |
| Mass bal in | | 133821 | | | | | | |
| Out kg/h | | 133821 | | Excess | Limiting | X = 100/150 = | 66.7 % | |
| Closure % | | 100.0 | | | | | | |
| Enthalpy+ kJ/h | refelements 298 K | | | -2.54E+07 | -3.63E+06 | -3.40E+07 | -1.90E+09 | -1.90E+09 |
| Energy bal In kJ/h | | -1.93E+09 | | | | | | |
| Out kJ/h | | -1.93E+09 | | Q kW | W kW | | vp kPa | 3.55 |
| Closure % | | 100.0 | | 0 | 0 | | | |
| Antoine H2O | A | 16.5362 | | B 3985.44 | | C -38.9974 | p+ kPa | |

## 5:11 W   Thermochemical water-split

a.   Stoichiometry: All in gas phase.

     $2H_2O + SO_2 + I_2 \rightarrow H_2SO_4 + 2HI$                                rxn 1

     $2HI$                 $\rightarrow I_2 + H_2$                                    rxn 2

     $H_2SO_4$            $\rightarrow H_2O + SO_2 + (0.5)O_2$                     rxn 3

     $H_2O$             $\rightarrow H_2 + (0.5)O_2$                                [Nett]

b.   Heat of reaction: For each reaction $h_{rxn} = \Sigma[v(j)h_f(j)]$ products $- \Sigma[v(j)h_f(j)]$ reactants

     The heat of formation reference condition is the elements at 298 K.

     Reaction 1: $h_{rxn}$ = (−7.44E5+2(−2.6E4) − (−2.42E5+2.71E5) =      +6.3E4 kJ/kmol

                                                            Endothermic

     Reaction 2: $h_{rxn}$ = (2(0) + 2(0)) − (2(2.6E4))    = −5.2E4 kJ/kmol      Exothermic

     Reaction 3: $h_{rxn}$ = (−242E5 + −271E5) − (−7.44E5) = +2.3E5 kJ/kmol    Endothermic

     Net reaction: $h_{rxn}$ = (1(0) + 0.5(0)) − (−2.42E5) = +2.42E5 kJ/kmol    Endothermic

c.   The minimum energy implies stoichiometric reactants/products with zero heat lost or work done by the system.

     Basis 1 kmol $H_2$ (2 kg) from reaction 2.

     Process Unit 1: Reaction 1 requires one kmol $I_2$ = (1)(+6.3E4) = +6.3E4 kJ

     Process Unit 2: Reaction 2 generates one kmol $H_2$ = (1)( −5.2E4) = −5.2E4 kJ

     Process Unit 3: Reaction 3 requires 1 kmol $H_2SO_4$ = (1)(+2.3E5) = +2.2E5 kJ

     Thermal duty:

     Unit 1 = (6.3E4/2)/3600 =                             8.8 kWh/kg $H_2$ energy IN

     Unit 2 = (−5.2E4/2)/3600 =                       7.2 kWh/kg $H_2$ energy OUT

Unit 3 = (2.2E5/2)/3600 =                                                30.6 kWh/kg $H_2$ energy IN
Overall process minimum energy = 8.8 + 30.6 – 7.2 =     32.2 kWh/kg $H_2$ energy IN
For the net reaction energy = (2.42E5/2)/3600 =            33.6 kWh/kg $H_2$ energy IN
These values should be the same – difference is due to sources of $h_f$ and rounding errors.

Energy efficiency with full recovery of energy =                                      **100%**

If sensible heat is not recovered the energy used to heat streams between process units would be lost. A proper analysis requires a full M&E balance, accounting for incomplete conversion in each reaction, work in to separate and move fluids, heat loss to the surroundings, etc.

The table below has an estimate of sensible heat needed to bring materials from Unit 1 at 400 K to Units 2 and 3 respectively at 800 and 1200 K, assuming 100% conversion of reactants.

| THERMOCHEMICAL WATER - SPLIT | | | | | | | |
|---|---|---|---|---|---|---|---|
| Species | $H_2O$ split | $H_2O$(g) | $H_2SO_4$(g) | $SO_2$(g) | HI(g) | $H_2$(g) | $I_2$(g) |
| $C_{p,m}$ 298K | kJ/kmol K | 38 | 100 | 54 | 26 | 29 | 53 |
| Amount kmol/kmol $H_2$ | | | 1 | | 2 | | |
| Temp | in K | | 400 | | 400 | | |
| | out K | | 1200 | | 800 | | |
| Sens heat | kJ/kmol $H_2$ | | 80000 | | 20800 | | |
| Energy lost kJ/kg $H_2$ | | | Total | 100800 | = 14 kWh/kg $H_2$ | | |

A loss of 14 kWh/kg $H_2$ would drop the overall energy efficiency from 100% to 100(33 – 14)/33 ≡                                                                                     **58%**

More energy loss would be expected from process inefficiencies, fluid movement, heat transfer and separation constraints, etc.

## 5:12 W   Vanadium redox-flow battery

Cycling process, reaction, integral balances.

a.   System: Flow battery through the discharge part of the charge/discharge cycle.
Energy balance:
Output energy per cycle = 2 MWh = 2000 kWh = 2E3(3600) kJ = 7.2E6 kJ
Electrical energy = E = (Voltage)(current)(time) = VIt
E = energy kJ, V = output voltage (Volt), I = current (kA) t = time (seconds)
It = E/V = charge kCoulomb
For E = 7.2E6 kJ, V = 1.2 V on discharge
Reactor discharge stoichiometry: $V^{+5} \rightarrow V^{+4} + e-$
1 electron ≡ 1 Faraday per mole $V^{+5}$
It = 7.2E6 kJ /1.2 V = 6E6 kCoulomb
≡ (1E3 Coulomb/kCoul)(6E6/96485) = 62.1E3 Faraday
Stoichiometric $V^{+5}$ = (62.1E3 Faraday)/(1 Faraday/mole $V^{+5}$ converted) = 62.1 kmol
For 90% $V^{+5}$ conversion $V^{+5}$ required = 62.1/0.9 = 69 kmol per tank.
If fresh electrolyte has 2 M $V^{+5}$ Electrolyte volume = (2)69 kmol/2 kmol/m$^3$ = **69 m$^3$**

b.    Average current needed to charge battery in 8 hours = $I_c$
      $I_c$ = (96485 C/Faraday)(62.1E3 Faraday)/(8hr)(3600 s/h) = 208E3 Amp = 208 kA
      Total cell area = $I_c$/CD = 208 kA/2 kA/m$^2$                                         = **104 m$^2$**
      e.g. 104 cells*, each 1 m$^2$ stacked in parallel and series for required output voltage.

c.    Energy lost to heat during charge = I(ABS(V – OCV)) = 208 kA[ABS(1.2 – 1.5)] V
      Energy to each tank = (208 kA)(0.3 V)(8 hr)(3600 s/h) = 1.8E6 kJ
      Electrolyte properties ρ = 1395 kg/m$^3$, $C_p$ = 100 kJ/kg/40 K = 2.5 kJ/kg (Text HC chart)
      Temperature rise during charge without cooling = ΔT
      ΔT = (1.8E6 kJ)/((69 m$^3$)(1395 kg/m$^3$)(2.5 kJ/kg.K)) =                           **7.5 K**
      *cell ≡ one anode/cathode pair.

### 5:13 U   Thermochemical ammonia

Energy balance equations are solved directly for $N_2$ conversion/pass ≡          **16.5%**

# Chapter 6:   Simultaneous Material and Energy Balances

### 6:1 W   Partial condenser

This problem involves a simultaneous material and energy balance.
The simplest solution is by OVERALL balances (the intermediate conditions are not needed).

a.    If (as indicated in the question) Unit 3 is an equilibrium V/L separator, then the
      streams 4 and 5 must be respectively saturated MeOH vapour and saturated MeOH
      liquid, both at the saturation temperature corresponding the 870 kPa(a).
      i.e. Streams 3, 4 and 5 are at the MeOH dew-point temperature under 870 kPa(a).
      Vapour pressure of $CH_3OH$ = p* = exp[16.4948 – 3593.39/(T – 35.2249)]
      p* = kPa(a), T = K
      Put p*(A) = P(4) = 870 kPa(a) and solve for T(3) = T(4) = T(5) =                   **405 K**

b.    Designate: A ≡ $CH_3OH$
      Feed condition, Stream 1. Assume all gases and vapours behave as ideal gas.
      Total flow = ṅ(1,A) = PV̇/RT = (100 kPa)(2910 m$^3$/h)/((8.314 kJ/kmol.K)(350 K))
                  = 100.0 kmol/h
      Overall material balances:      Continuous process at steady-state
      $CH_3OH$:     0 = 100 – ṅ(4,A) – ṅ(5,A)                                              [1]
      The material balance is under-specified. The energy balance will fulfil the specification.
      For the V/L equilibrium separator (Unit 3):
      Overall energy balance:   0 = Ḣ(1) – Ḣ(4) – Ḣ(5) + Q̇ – Ẇ                          [2]
      Q̇ = –500 kW ≡ (–500 kW)(3600 s/h) = –1.800E6 kJ/h
      Compression work:
      r = $C_p$/$C_v$ = 50/(50 – 8.314) = 1.199
      Ẇ = –$P_1$V̇$_1$ (r/(r–1))[(P$_2$/P$_1$)$^{((r-1)/r)}$ – 1]   [Defined as + for work OUT]
      Ẇ = –(100)(2910)(1.199/(1.199 – 1))[(900/100)$^{((1.199-1)/1.199)}$ – 1] = –7.718E5 kJ/h

Energy balance:

Take the enthalpy reference condition = liquid $CH_3OH$ at 298 K (no reaction)

$\dot{H}(1) = (100)((50)(350 - 298) + 36E3) = 3.86E6$ kJ/h

$\dot{H}(4) = \dot{n}(4,A)((50)(405 - 298)) + 36E3) = (41.35E3)\dot{n}(4,A)$

$\dot{H}(5) = \dot{n}(5,A)((95)(405 - 298) + 0) = (10.165E3)\dot{n}(5,A)$

Substitute eqn. 1 into eqn. 2.

$0 = 3.86E6 - (41.35E3)\ \dot{n}(4,A) - (10.165E3)(100 - \dot{n}(4,A)) + (-1.80E6)$
$\quad - (-7.718E5)$

Solve for: $\dot{n}(4,A) = 1.8153/31.185E3 = 58.2$ kmol/h

$\dot{n}(5,A) = 100 - 58.2$ kmol/h $= 41.8$ kmol/h $\equiv$ **1338 kg/h**

## 6:2 U   Compressors and intercooler

Since there are no reactions only sensible heats are considered (heat of formation not needed).

Work done by compressor $\dot{W} = -(P(1)\dot{V}(1)(r/(r-1))[(P(2)/P(1))^{((r-1)/r)} - 1]$

And $\qquad\qquad\qquad T(2) = T(1)[P(2)/P(1)]^{((r-1)/r)}$

The HEX heat transfer equation is needed to fully specify the problem.

Equations are solved in Excel, using <u>bisection</u> with Goal-seek and/or Solver, to find T(3) and $\dot{m}(5)$ to close the simultaneous mass and energy balances.

Temperature of stream 4 = 350 K, stream 3 = T(3) = 303 K, flow of stream 5 = **6138 kg/h**, The full stream table is below.

| STREAM TABLE | | N2 2-STAGE COMPRESSOR | | | | and INTER-COOLER | | | | |
|---|---|---|---|---|---|---|---|---|---|---|
| Species | M | Cp,m | hf | | | | Stream | kmol/h | | |
| | kg/kmol | kJ/kmol.K | kJ/kmol | *1* | *2* | *3* | *4* | *5* | *6* | |
| N2(g) | 28.0 | 30 | NA | 100.0 | 100.0 | 100.0 | 100.0 | 0.0 | 0 | |
| H2O(l) | 18.0 | 75 | NA | 0.0 | 0.0 | 0.0 | 0.0 | 341.0 | 341.0 | |
| Total mass | kg/h | | | 2800 | 2800 | 2800 | 2800 | 6138 | 6138 | |
| Phase | – | | | G | G | G | G | L | L | |
| Pressure | kPa(a) | | | 100 | 1200 | 1200 | 2000 | 200 | 180 | |
| Temp | K | | | 298 | 593 | 303 | 350 | 293 | 327 | |
| Volume | m3/h | | | 2.48E+03 | 4.11E+02 | 2.10E+02 | 1.45E+02 | 6.14E+00 | 6.14E+00 | |
| Enthalpy | kJ/h | | | 0.00E+00 | 8.86E+05 | 1.64E+04 | 1.55E+05 | -1.28E+05 | 7.42E+05 | |
| Mass balance | | Unit 1 | Unit 2 | Unit 3 | Overall | dlT lm  K | 79 | Q | W(1) | |
| IN | kg/h | 2.80E+03 | 8.94E+03 | 2.80E+03 | 8.94E+03 | Unit | | kW | kW | |
| OUT | kg/h | 2.80E+03 | 8.94E+03 | 2.80E+03 | 8.94E+03 | 1 | | 0 | -246.1 | |
| Closure % | | 100.0 | 100.0 | 100.0 | 100.0 | 2 | 8.70E+05 | 8.70E+05 | W(2) kW | |
| Energy balance | | | | | | 3 | 60.4 | 0 | -38.5 | |
| IN | kg/h | 8.86E+05 | 7.58E+05 | 1.55E+05 | 8.97E+05 | T(2)–T(6) K | 266 | Cv | 21.7 | |
| OUT | kg/h | 8.86E+05 | 7.58E+05 | 1.55E+05 | 8.97E+05 | T(3)–T(5) K | 10 | Cp/Cv | 1.38 | |
| Closure % | | 100.0 | 100.0 | 100.0 | 100.0 | UkW/m2.K | 1.00E-01 | (r–1)/r | 0.277 | |
| | | | | | | A m2 | 40 | delT  K | 79 | |

## 6:3 W   Bio-reactor

a. Reaction stoichiometry, sucrose to citric acid:

$C_{12}H_{22}O_{11} + 7.5O_2 \rightarrow C_6H_8O_7 + 6CO_2 + 7H_2O \qquad$ **x = 7.5, y = 6, z = 7**

Designate: $A \equiv C_{12}H_{22}O_{11}\ B \equiv O_2\ C \equiv N_2\ D \equiv C_6H_8O_7\ E \equiv CO_2\ F \equiv H_2O$

b. Simultaneous material and energy balance with the reactor coupled to the heat exchanger.

System = Reactor (REACT):

Material balances. Continuous process. Steady-state.

A:  $0 = \dot{n}(1,A) - \dot{n}(4,A) - X(A)\dot{n}(1,A)$

B:  $0 = \dot{n}(2,B) - \dot{n}(3,B) - (7.5)X(A)\dot{n}(1,A)$

C:  $0 = \dot{n}(2,C) - \dot{n}(3,C)$

D:  $0 = 0 - \dot{n}(4,D) + (1/1)X(A)\dot{n}(1,A)$

E:  $0 = 0 - \dot{n}(3,D) + (6/1)X(A)\dot{n}(1,A)$

F:  $0 = \dot{n}(1,F) - \dot{n}(3,F) - \dot{n}(4,F) + (7/1)X(A)\dot{n}(1,A)$

Conversion:  $X(A) = 0.7$

Air composition:   $\dot{n}(2,B)/\dot{n}(2,C) = 21/79$

Feed composition:   $w(1,A) = 0.2$

Feed flow:   $\bar{m}(1) = 3000$ kg/h

Feed air stoich:   $2 = \dot{n}(2,B)/(\dot{n}(2,B) - \dot{n}(3,B))$

Water vapour in Str.3: $y(3,F) = p(3,F)/P(3) = x(3,F)p^*(3,F) /P(3)$   Raoult's law.

$p^*(3,F) = \exp[16.5362 - 3985.44/(T(4) - 38.9974)]$   Antoine equation

Thermal equilibrium:   $T(3) = T(4)$

Energy balance:        $0 = \dot{H}^*(1) + \dot{H}^*(2) - \dot{H}^*(3) - \dot{H}^*(4)$        $\dot{Q} = \dot{W} = 0$

Enthalpy reference = elements at standard state, 298 K.

Where:   $\dot{H}^*(i) = \Sigma\{\dot{n}(i,j)[(C_{pm,298K}(j)[T(i) - 298] + h^\circ_{f,298K}(j)]\}$

HEX thermal load:$\rightarrow$ABS$(\dot{Q}) = Q_{HEX} = U_{HEX}A_{HEX}\Delta T_{lm}$

$T_{lm} = [(T(4) - T(5)) - (T(4) - T(6))]/\ln[[(T(4) - T(5))/(T(4) - T(6))]$ Log mean $\Delta T$

System = Heat exchanger (HEX).

Material balance:

F:   $0 = \dot{n}(5,F) - \dot{n}(6,F)$

Given flow   $\dot{n}(5,F) = \dot{n}(6,F) = 72E3$ kg/h

Energy balance:   $0 = \dot{H}(5) - \dot{H}(6) + \dot{Q} - \dot{W}$      $\dot{W} = 0$      $\dot{Q} = -\dot{Q}$

The above equations fully-specify the problem as a simultaneous M&E balance.
The material balance is coupled to the energy balance by vapour pressure of $H_2O$ in Stream 3.
The spreadsheet solution finds the values of $T(4)$ and $T(6)$ that close the simultaneous
REACT and HEX energy balances.  $T(3) = T(4) = \mathbf{306\ K}$      $\dot{Q}_{HEX} = \mathbf{-1192\ kW}$ (cooling)

## 6:4 W   Hydrogen fuel cell

Hydrogen/air fuel cell. Continuous, steady-state, adiabatic, reaction.
Net reaction stoichiometry:

$$2H_2 + O_2 \rightarrow 2H_2O$$

Designate: $A \equiv H_2$, $B \equiv O_2$, $C \equiv H_2O$, $D \equiv N_2$

Mole balances

System = Hot side (fuel cell reactor)

A:   $0 = \dot{n}(1,A) + \dot{n}(2,A) - \dot{n}(3,A) + 0 - \dot{n}(1,A)$      100% conversion

B:   $0 = \dot{n}(1,B) + \dot{n}(2,B) - \dot{n}(3,B) + 0 - (1/2)\dot{n}(1,A)$

C:   $0 = \dot{n}(1,C) + \dot{n}(2,C) - \dot{n}(3,C) + (2/2)\dot{n}(1,A) - 0$
D:   $0 = \dot{n}(1,D) + \dot{n}(2D) - \dot{n}(3,D) + 0 - 0$
System = Cold side – HEX
A:   all 0
B:   all 0
C:   $0 = \dot{n}(4,C) - \dot{n}(5,C) + 0 - 0$
D:   all 0
Energy balance: System = overall fuel cell, reaction and cooling.
Heat of formation method, *ref. elements at 298 K*.
$0 = \dot{H}*(1) + \dot{H}*(2) + \dot{H}*(4) - \dot{H}*(3) - \dot{H}*(5) + \dot{Q} - \dot{W}$          $\dot{Q} = 0$, W = 75 kW
$\dot{H}*(i) = \Sigma\{\dot{n}(i,j)[(C_{pm,298K}(j)[T(i) - 298] + h^{\circ}_{f,298K}(j)]\}$
Set up the M&E balance over the whole system in a stream table
Use an arbitrary (guess) initial value for T(3).
Compute to close the mass balance, then use Excel Goal-seek to find the value of T(3) that
simultaneously closes the energy balance. Str. 3 phase split          G/L = *1.51/0.82*
Make sure the mass balance remains closed. Solution   **T(3) = 337.1 K –89 kW cooling**

| Species | M | $C_{p,m}$ 298 | $h_f$,298 | $H_2$ AIR FUEL CELL | | | Stream kmol/h | |
|---|---|---|---|---|---|---|---|---|
| | kg/kmol | kJ/kmol K | kJ/kmol | 1 | 2 | 3 | 4 | 5 |
| $H_2(g)$ | 2.0 | 29 | 0.00E+00 | 2.33 | 0 | 0 | 0 | 0 |
| $O_2(g)$ | 32.0 | 30 | 0.00E+00 | 0.00 | 2.33 | 1.17 | 0.00 | 0.00 |
| $N_2(g)$ | 29.0 | 29 | 0 | 0.00 | 8.77 | 8.77 | 0.00 | 0.00 |
| $H_2O(g)$ | 18.0 | 34 | -2.42E+05 | 0.00 | 0.00 | 1.51 | 0.00 | 0.00 |
| $H_2O(l)$ | 18.0 | 75 | -2.86E+05 | 0.00 | 0.00 | 0.82 | 250.00 | 250.00 |
| Total | kg/h | | | 4.7 | 329.0 | 333.7 | 4500.0 | 4500.0 |
| Phase | | | | G | G | L+G | L | L |
| Pressure kPa(a) | | | | 200 | 200 | 180 | 200 | 150 |
| Temp   K | | | | 298 | 320 | 337.1 | 310 | 327.1 |
| Mass bal | | 4834 | | | | | O2 stoich | 2 |
| Out kg/h | | 4834 | | | | | | |
| Closure % | | 100.0 | | | | | | |
| Enthalpy* kJ/h | | ref elements 298 K | | 0.00E+00 | 7.14E+03 | -5.85E+05 | -7.13E+07 | -7.10E+07 |
| Energy bal In kJ/h | -7.13E+07 | | | | HEX = | H(4)–H(5) = | -8.90E+01 kW cooling | |
| Out kJ/h | -7.13E+07 | | | Q kW | W kW | Out | Power kW | I kA |
| Closure % | | 100.0 | | 0 | 75 | | 75 | 125 |
| Antoine $H_2O$ | A | 16.5362 | B | 3985.44 | C | -38.9974 | p* kPa(a) | 23.72 |

## 6:5 U   Direct contact heat exchange

Direct contact heat exchange. Continuous, steady-state, adiabatic, no reaction.
a.   Simultaneous material and energy balance.
b.   The full stream table is below, giving T(3) = T(4) = **339.1 K**

| STREAM TABLE | | DIRECT CONTACT HEX | | | | Stream kmol/h | |
|---|---|---|---|---|---|---|---|
| Species | M | $C_{p,m}$ 298K | $h_f^\circ$ | 1 | 2 | 3 | 4 |
| | kg/kmol | kJ/kmol.K | kJ/kmol | | | | |
| $H_2O(l)$ | 18.0 | 75 | -2.86E+05 | 0 | 100 | 0 | 70.12 |
| $H_2O(g)$ | 18.0 | 34 | -2.41E+05 | 0 | 0 | 29.88 | 0 |
| $N_2(g)$ | 28.0 | 29 | 0.0E+00 | 200 | 0 | 200 | 0 |
| Total | kg/h | | | 5600 | 1800 | 6138 | 1262 |
| Phase | | | | L | G | G | L |
| Pressure | kPa(a) | | | 210 | 210 | 200 | 200 |
| Temp | K | | | 600 | 310 | 339.1 | 339.1 |
| Volume | m³/h | | | 4751 | 1.8 | 3241 | 1 |
| vp $H_2O$ | kPa(a) | | | | 6.2 | 26.0 | 26.0 |
| Enthalpy | ref elements 298 K | | | 1.75E+06 | -2.85E+07 | -6.92E+06 | -1.98E+07 |
| Mass balance | | | | | | | |
| IN | kg/h | 7400 | | | | | |
| OUT | kg/h | 7400 | | | | | |
| Closure % | | 100.0 | | | | | |
| Energy balance | | | | | | | |
| IN | kJ/h | -2.68E+07 | | | | | |
| OUT | kJ/h | -2.68E+07 | | | | | |
| Closure % | | 100.0 | | | | | |
| Antoine $H_2O$ | A | 16.5362 | B | 3985.44 | C | -38.9974 | p* kPa(a) |

## 6:6 W    Ethanol from maltose

Maltose fermentation to ethanol.      Continuous, steady-state, adiabatic, reaction.

Reaction:      $C_{12}H_{22}O_{11}(aq) + H_2O(l) \rightarrow 4C_2H_5OH(aq) + 4CO_2(g)$

Designate:   A ≡ maltose      B ≡ Yeast      C ≡ ethanol      D ≡ $CO_2$      T(2) = T(3)

Mole balances. Continuous process at steady-state with reaction.

A:   $0 = \dot{n}(1,A) - \dot{n}(2,A) - \dot{n}(3,A) + 0 - X(A)\,\dot{n}(1,A)$

B:   $0 = \dot{n}(1,B) - \dot{n}(2,B) - \dot{n}(3,B) + 0 - (1/1)X(A)\dot{n}(1,A)$

C:   $0 = \dot{n}(1,C) - \dot{n}(2,C) - \dot{n}(3,C) + (4/1)X(A)\dot{n}(1,A) - 0$

D:   $0 = \dot{n}(1,D) - \dot{n}(2,D) - \dot{n}(3,D) + (4/1)X(A)\dot{n}(1,A) - 0$

Raoult's law   $y(2,C) = p(2,C)/110 = x(3,C)p^*(3,C)/110 = f(T(3))$

Given:         $\dot{n}(1,A) = (0.1)(2000)/342 = 0.585$ kmol/h

               $\dot{n}(1,B) = (0.9)(2000)/18 = 100$ kmol/h

               $\dot{n}(1,C) = 0$      $\dot{n}(1,D) = 0$      $\dot{n}(2,A) = 0$      $\dot{n}(3,C) = 0$      $\dot{n}(3,D) = 0$

               $P(1) = 120$      $P(2) = 110$      $P(3) = 110$      $T(1) = 300$      $T(2) = T(3)$

Energy balance: Ref. *elements* at 298 K      T(2) = T(3)      specified.

$0 = \dot{H}*(1) - \dot{H}*(2) - \dot{H}*(3)$      $\dot{Q} = \dot{W} = 0$

$\dot{H}*(i) = \Sigma\{\dot{n}(i,j)[(C_{pm,298K}(j)[T(i) - 298] + h^\circ_{f,298K}(j)]\}$          Respect the PHASE

The heats of formation for maltose $h_f(A)$ and ethanol $h_f(C)$ come from heats of combustion

Maltose combustion reaction:

$C_{12}H_{22}O_{11} + 17.5O_2 \rightarrow 12CO_2(g) + 11H_2O(l)$ Gross $h_c$.

$h_{rxn} = h_c = 12h_f(D) + 11h_f(B) - [h_f(A) + 17.5h_f(O_2)]$

$h_f(A) = 12h_f(D) + 11h_f(B) - 17.5h_f(O_2)] - h_c(A)$

$h_f(A) = 12(-393.5E3) + 11(-285.8E3) - 17.5(0) - (-5.65E6) = -2.22E6$ kJ/kmol

Ethanol combustion reaction:

$C_2H_5OH + 3.5O_2 \rightarrow 2CO_2 + 3H_2O$

$h_{rxn} = h_c = 2h_f(D) + 3h_f(B) - [h_f(C) + 3.5h_f(O_2)]$

$h_f(C) = 2h_f(D) + 3h_fB) - [3.5h_f(O_2)] - h_c(C)]$

$h_f(C)liq = 2(-393.5E3) + 3(-285.8E3) - 3.5(0) - 1.37E6 = -2.74E5$ kJ/kmol (liquid)

$h_f(C)gas = 2(-393.5E3) + 3(-285.8E3) - 3.5(0) - 1.41E6 = -2.34E5$ kJ/kmol (gas)

The problem is fully specified by simultaneous material and energy balances.
The complete stream table is below.           Giving T(2) = T(3) = **305 K**

| STREAM TABLE | | ETHANOL from MALTOSE | | *Stream* | kmol/h | | |
|---|---|---|---|---|---|---|---|
| Species | M | $C_{p,m}$ 298K | $h_f^o$ | *1* | *2* | *3* | |
| | kg/kmol | kJ/kmol.K | kJ/kmol | | | | |
| $C_{12}H_{22}O_{11}$ (aq) | 342.0 | 300 | -2.22E+06 | 0.58 | 0.00 | 0.058 | |
| $H_2O(l)$ | 18.0 | 75 | -2.86E+05 | 100.00 | 0.00 | 99.40 | |
| EtOH(l) | 46.0 | 115 | -2.74E+05 | 0.00 | 0.00 | 2.10 | |
| $H_2O(g)$ | 18.0 | 37 | -2.42E+05 | 0.00 | 0.13 | 0.00 | |
| EtOH(g) | 46.0 | 71 | -2.34E+05 | 0.00 | 0.01 | 0.00 | |
| $CO_2(g)$ | 44.0 | 37 | -3.94E+05 | 0.00 | 2.11 | 0.00 | |
| Total | kg/h | | | 2000 | 95 | 1906 | |
| Phase | | Hrxn | | L | G | L | |
| Pressure | kPa(a) | -4.00E+03 | | 110 | 105 | 105 | |
| Temp | K | | | 293 | 305 | 305 | T(2)=T(3) |
| Volume | $m^3$/h | Not needed | | | | | |
| vp $H_2O$ | kPa Raoult | kPa | | 2.3E+00 | 4.6E+00 | 4.6E+00 | |
| vp EtOH | kPa Raoult | kPa | | 1.21E-01 | 2.42E-01 | 2.42E-01 | |
| Enthalpy | ref elements 298 K | | | -2.99E+07 | -8.61E+05 | -2.91E+07 | |
| Mass balance | | | | | | | |
| IN | kg/h | 2000 | | | | | |
| OUT | kg/h | 2001 | | | mf EtOH | 0.021 | |
| Closure % | | 100.1 | | | mf $H_2O$ | 0.979 | |
| Energy balance | | | | | | | |
| IN | kJ/h | -2.99E+07 | | Conv | 90 | % | |
| OUT | kJ/h | -2.99E+07 | | Maltose | 10 | wt% | |
| Closure % | | 100.0 | | | | p* kPa | |
| Antoine $H_2O$ | A | 16.5362 | B | 3985.44 | C | -38.9974 | Pure |
| $C_2H_5OH$ | | 16.1952 | | 3424.53 | | -55.7152 | Pure |

## 6:7 W   Sulphuric acid dilution

This problem invokes a simultaneous material and energy balance, based on the $H_2SO_4/H_2O$ enthalpy-concentration (h–C) chart.

Let F = flow of water kg/h, G = flow of product mixture kg/h

w = mass fraction $H_2SO_4$ in mix

System = acid/water mixing tank. Continuous adiabatic operation, steady-state. No reaction.

Overall material balance:

$0 = 1000 + F - G \quad\quad G = 1000 + F$ [1]

Mass balance on $H_2SO_4$:

$0 = 0.9(1000) - wG$ [2]

From equations 1 and 2: w = 900/(1000 + F) [3]

Overall energy balance – reference conditions pure $H_2SO_4$ and $H_2O$ at 273 K, 101 kPa(a).

$0 = \dot{H}(1) + \dot{H}(2) - \dot{H}(3) + 0 - 0 \qquad (\dot{Q} = \dot{W} = 0)$ [4]

Let: h(1), h(2), h(3) = specific enthalpy of streams 1, 2, 3. kJ/kg (from the h-C chart)

$0 = 1000h(1) + Fh(2) - Gh(3) = 1000h(1) + Fh(2) - (1000 + F)h(3)$ [5]

$h(3) = [1000h(1) + Fh(2)]/(1000 + F)$ [6]

From the h–C chart: h(1) = –10 kJ/kg, h(2) = 250 kJ/kg

$h(3) = (-10E3 + 250F)/(1000 + F)$ at 140°C [7]

Draw a tie-line in the h-C chart from 90 wt% $H_2SO_4$/100°C to 0% $H_2SO_4$ (water) at 60°C

At 140°C stream 3 is a 2-phase mixture of water vapour and 60 wt% $H_2SO_4$

From equation 3: F = (900–1000w)/w  **F = 500 kg/h**

Or by trial and error with the h-C chart:

| w | – | 0.2 | 0.4 | 0.5 | 0.6 | 0.7 | |
|---|---|-----|-----|-----|-----|-----|---|
| F | kg/h | 3500 | 1250 | 800 | 500 | 290 | **F = 500 kg/h** |
| h(3) | kJ/kg | na | na | 106 | 77 | –10 | |

## 6:8 W  Adiabatic flash

Continuous, steady-state, adiabatic, no reaction.      ACC = GEN = CON = 0

Designate: A ≡ hexane, B ≡ benzene

Mole fractions: z = stream 1, x = stream 2, y = stream 3.

ACC = IN – OUT + GEN – CON

Mass (mole) balances:

| | | |
|---|---|---|
| Overall: | $0 = \bar{n}(1) - \bar{n}(2) - \bar{n}(3)$   where $\bar{n}(i) = \dot{n}(i,A) + \dot{n}(i,B)$ | [1] |
| A: | $0 = \dot{n}(1,A) - \dot{n}(2,A) - \dot{n}(3,A)$ | [2] |
| B: | $0 = \dot{n}(1,B) - \dot{n}(2,B) - \dot{n}(3,B)$ | [3] |
| Raoult's law: | $y(j) = p*(j)(x(j))/P = K(j)$   Ideal mixture | [4] |
| And | $\Sigma x(j) = 1$ | [5] |
| | $\Sigma y(j) = 1$ | [6] |
| Designate: | $\bar{n}(3)/\bar{n}(1) = \beta$ | [7] |
| From 1,2,3,4,5 | $x(j) = z(j) /\{(K(j) - 1)\beta + 1\}$ | [8] |
| Energy balance: | $0 = \dot{H}(1) - \dot{H}(2) - \dot{H}(3) + 0 - 0$ | [9] |

$\dot{H}(i) = \Sigma\{\dot{n}(i,j)[(C_{pm,298K}(j)[T(i) - 298] + h_{f,298}]\}$ Respect the PHASE [10]

Since there are no reactions the $h°_f$ for vapours is replaced by $h_v$ (heat of vaporization).

$\dot{H}(i) = \Sigma\{\dot{n}(i,j)[(C_{pm,298K}(j)[T(i) - 298] + h_v)]\}$ [11]

T(2) =T(3) [12]

Check specification: 3 stream, 2 component, 3 pressure, 3 temperature = 12 values

Basis, x(1,A) = x(1,B), p(1), p(2) = p(3), T(1), T(2)=T(3) are specified, leaving 6 "unknowns".

Equations 2,3,4,5/6,9 complete the picture. The system is fully specified.

This phase-split problem invokes the difficult solution of simultaneous non-linear equations.

Done using Excel Solver to close the simultaneous mass and energy balances.

The full solution is below.                **T(2) = T(3) = 330.6 K**

| STREAM TABLE | | | ADIABATIC FLASH | | | |
|---|---|---|---|---|---|---|
| Species | M | Cp | hv | Stream | kmol/h | |
| | kg/kmol | kJ/kmol.K | kJ/kmol | *1* | *2* | *3* |
| A (l) hexane | 86 | 194 | 0 | 50 | 44.1 | 0 |
| A(g) hexane | 86 | 143 | 3.19E+04 | 0 | 0.0 | 5.9 |
| B (l) benzene | 78 | 135 | 0 | 50 | 20.0 | 0 |
| B(g) benzene | 78 | 82 | 3.38E+04 | 0 | 0.0 | 30.0 |
| Total | kmol | | kmol | 100 | 64.1 | 35.9 |
| Total | kg | | kg | 8200 | 5351 | 2849 |
| mf A(l) | | | | 0.5 | 0.69 | 0.16 |
| mf B(l) | | | | 0.5 | 0.31 | 0.84 |
| β phase split | | | | 0.36 | | |
| Phase | | | | L | L | G |
| Pressure | kPa(a) | | | 500 | 200 | 200 |
| Temperature | K | | | 400 | *330.6* | *330.6* |
| Vap. press | A | | | 468 | 70 | 70 |
| kPa(a) | B | | | 352 | 240 | 47 |
| K(A) | | | | 0.47 | 0.24 | 0.06 |
| K(B) | | | | 0.35 | 0.37 | 0.20 |
| Enthalpy kJ/h | ref. compounds at 298 K | | | 1.68E+06 | 3.67E+05 | 1.31E+06 |
| Mass balance | | | | | | |
| IN | kg/h | 8200 | | *BASIS* | *100* | *kmol feed* |
| OUT | kg/h | 8200 | | | | |
| Closure | % | 100.0 | | | | |
| Energy balance | | | | | | |
| IN | kJ/h | 1.68E+06 | | | | |
| OUT | kJ/h | 1.68E+06 | | | | |
| Closure | % | 100.0 | | | | |
| Antoine | A | B | C | | | |
| Hexane | 14.0568 | 2825.42 | -42.7089 | | | |

## 6:9 U  CSTR with heat exchange

Solution –                                                    Stream 3 flow = **16369 kg/h**

The full stream table is below.

| STREAM TABLE | | CSTR | REACT and HEX | | | Basis 100 kmol/h A | |
|---|---|---|---|---|---|---|---|
| Species | M | Cp,m | hf | Stream | kmol/h | | |
| | kg/kmol | kJ/kmol.K | kJ/kmol | *1* | *2* | *3* | *4* |
| A | 100.0 | 90 | 5.00E+04 | 100 | 10.3 | 0 | 0 |
| B | 50.0 | 80 | -1.00E+05 | 100 | 10 | 0 | 0 |
| C | 150.0 | 140 | -2.50E+05 | 0.00 | 89.7 | 0 | 0 |
| H₂O (l) | 18.0 | 75 | -2.86E+05 | 0 | 0 | 909 | 909 |
| H₂O(g) | 18.0 | 34 | -2.42E+05 | 0 | 0 | 0 | 0 |
| Total mass | kg/h | | | 15000 | 15000 | *16369* | 16369 |
| Conc A | kmol/m³ | hrxn | | 100.0 | | | |
| Conc B | kmol/m³ | -2.00E+05 | kJ/kmol C | 100.0 | | | |
| k | m³mol.s | | 2.82E+01 | 7.83E-03 | | | |
| Phase | | | m³/kmol.h | L | L | L | ? |
| Pressure | kPa(a) | | | 120 | 110 | 130 | 110 |
| Temperature | K | | | 298 | 651 | 283 | 472 |
| vp H₂O | kPa | | | 3.15E+00 | 2.26E+04 | 1.22E+00 | 1.53E+03 |
| Volume | m³/h | | | 1 | 1 | | |
| Enthalpy | kJ/h | | | -5.00E+06 | -1.79E+07 | -2.61E+08 | -2.48E+08 |
| Mass balance | | REACT | HEX | Q kJ/h | U | 0.5 | kW/m².K |
| IN | kg/h | 15000 | 16369 | -1.29E+07 | VR | 3 | m³ |
| OUT | kg/h | 15000 | 16369 | T(4) | k' | 2 | |
| Closure % | | 100.0 | 100.0 | 1377 | E | 3.00E+04 | kJ/kmol |
| Energy balance | | | | A m² | 40 | Water Str. 3 | 909.4 |
| IN | kg/h | -5.00E+06 | -2.48E+08 | Q1 kW | 1.13E+03 | | kmol/h |
| OUT | kg/h | -5.00E+06 | -2.48E+08 | Q2 kJ/h | 1.29E+07 | Conv A | 0.90 |
| Closure % | | 100.0 | 100.0 | delT K | 179 | F m³/h | 10 |
| Antoine H₂O | A | 16.5362 | B | 3985.44 | C | -38.9974 | p* kPa |

## 6:10 W   Oxy-coal boiler

Continuous, steady state, reaction    ACC = 0, GEN and CON # 0

Systems: 1 Heat exchanger (APH) ≡ HEX    2 Furnace/boiler ≡ REACT

Stoichiometry: $C_{100}H_{50}O_5 + 110O_2 \rightarrow 100CO_2 + 25H_2O$      Mcoal = 1333 kg/kmol

Designate: A ≡ Coal $C_{100}H_{50}O_5$, B ≡ $H_2O$, C ≡ $CO_2$, D ≡ $O_2$, E ≡ $N_2$

HEX Mole balances – no reactions.

| | | |
|---|---|---|
| A: | Not present after 100% conversion in furnace | |
| B: | $0 = \dot{n}(2.B) - \dot{n}(3,B) = \dot{n}(4,B) - \dot{n}(5,B)$ | [1,2] |
| C,D,E: | $0 = \dot{n}(2,j) - \dot{n}(3,j) = \dot{n}(4,j) - \dot{n}(5,j)$ | [3–8] |
| $O_2$ in enriched air: | $\dot{n}(4,D) = 0.35\Sigma[\dot{n}(4,j)]$ | [9] |
| Excess $O_2$: | $\dot{n}(4,D) = 1.01$(stoichiometric $O_2$) | [10] |
| $H_2O$, 60% RH: | $\dot{n}(4,B) = 0.6$($H_2O$ saturation at 298K) | [11] |
| $CO_2$: | $\dot{n}(4,C) = 400E\text{-}6\Sigma[\dot{n}(4,j)]$ optional $CO_2$ in air 400 ppm(v) | [12] |

REACT Mole balances – reactions

| | | | |
|---|---|---|---|
| A: | $0 = \dot{n}(1,A) + \dot{n}(5,A) - \dot{n}(2,A) + 0 - \dot{n}(1,A)$ | 100% conversion | [13] |
| B: | $0 = \dot{n}(1,B) + \dot{n}(5,B) - \dot{n}(2,B) + 25\dot{n}(1,A) - 0$ | | [14] |
| C: | $0 = \dot{n}(1,C) + \dot{n}(5,C) - \dot{n}(2,C) + 100\dot{n}(1,A) - 0$ | | [15] |
| D: | $0 = \dot{n}(1,D) + \dot{n}(5,D) - \dot{n}(2,D) + 0 - 110\dot{n}(1,A)$ | | [16] |
| E: | $0 = \dot{n}(1,E) + \dot{n}(5,E) - \dot{n}(2,E) + 0 - 0$ | | [17] |

HEX    Energy balance.

$$0 = \dot{H}^*(2) + \dot{H}^*(4) - \dot{H}^*(3) - \dot{H}^*(5) + 0 - 0 \qquad [18]$$

REACT Energy balance.

$$0 = \dot{H}^*(1) + \dot{H}^*(5) - \dot{H}^*(2) + \dot{Q}\text{ steam} - 0 \qquad \text{(no work done)} \qquad [19]$$

Where:

$\dot{H}^*(i) = \Sigma\{\dot{n}(i,j)[(C_{pm,298K}(j)[T(i) - 298] + h^\circ_{f,298K}(j)]\}$ Respect the PHASE

$\dot{Q}$ steam = (1E3 kg/T)(1200 T/h)(2026 kJ/kg)      Steam table $h_v$      [20]

Heat transfer in APH:

Q air = $UA\Delta T = UA[(T(2) + T(3))/2 - ((T4) + T(5))/2] = \dot{H}^* (2) - \dot{H}^* (3)$      [21]

**a.**   The set 1-21 are simultaneous equations, solved in Excel by iterative closure of the mass balances with the corresponding energy balances.

Then:                                           **T(2) = 1146 K, T(3) = 893 K, T(5) = 587 K**

**b.**   For stream 3 the $H_2O$ partial pressure = (110 kPa(a))(2.27E3/2.4E4) = 9.8 kPa

This is far below the $H_2O$ vapour pressure at 893 K, which is 1.4E5 kPa(a)

Water will NOT condense in the APV (at least under these idealized conditions)  **No**

The full spreadsheet is below.

| STREAM TABLE | | OXY–COAL | | | 100 | T/h coal | Steam/coal | 12 | T/T |
|---|---|---|---|---|---|---|---|---|---|
| Species | M | Cp,m | hf | | | Stream | kmol/h | | |
| | kg/kmol | kJ/kmol.K | kJ/kmol | 1 | 2 | 3 | 4 | 5 | |
| Coal | 1330.0 | 1.5 | –6.58E+06 | 75.2 | 0.0 | 0 | 0 | 0 | |
| H₂O(l) | 18.0 | 75 | –2.85E+05 | 0 | 0 | 0 | 0 | 0 | |
| H₂O(g) | 18.0 | 34 | –2.42E+05 | 0 | 2.27E+03 | 2.27E+03 | 3.87E+02 | 3.87E+02 | |
| CO₂ | 44.0 | 37 | –3.94E+05 | 0 | 7.53E+03 | 7.53E+03 | 9.55E+00 | 9.55E+00 | |
| O₂ | 32.0 | 30 | 0.00E+00 | 0 | 8.27E+01 | 8.27E+01 | 8.35E+03 | 8.35E+03 | |
| N₂ | 28.0 | 30 | 0.00E+00 | 0 | 1.55E+04 | 1.55E+04 | 1.55E+04 | 1.55E+04 | |
| Total mass | kg/h | | | 1.00E+05 | 8.09E+05 | 8.09E+05 | 7.09E+05 | 7.09E+05 | |
| Phase | | | | S | G | G | G | G | |
| Pressure | kPa(a) | | | 120 | 105 | 100 | 120 | 110 | |
| Temperature | K | | | 298 | 1146 | 893 | 298 | 587 | |
| yH₂O | Str 3 act | 0.09 | | 2.63E–02 | 3.95E+03 | 1.43E+03 | 2.63E–02 | 9.53E+01 | |
| vp H₂O | kPa(a) | | | 3.15E+00 | 4.15E+05 | 1.43E+05 | 3.15E+00 | 1.05E+04 | |
| Volume | m³/h | | | 5.00E+04 | 2.30E+06 | 1.89E+06 | 5.01E+05 | 1.08E+06 | |
| Enthalpy | kJ/h | ref elements 298 K | | –4.94E+08 | –2.81E+09 | –3.02E+09 | –9.72E+07 | 1.13E+08 | |
| Mass balance | | REACT | HEX | O2 % | 35 | U | 0.05 | kW/m².K | |
| IN | kg/h | 8.09E+05 | 1.52E+06 | Qhex kJ/h | 2.08E+08 | A | 2.00E+03 | m2 | |
| OUT | kg/h | 8.09E+05 | 1.52E+06 | (Q/UA) K | 5.77E+02 | Xcs O2 | 1 | % | |
| Closure % | | 100.00 | 100.0 | (T2+T3)/2 | 1020 | RH | 60 | % | |
| Energy balance | | | | T(5) K | 5.87E+02 | Coal | 100 | T/h | |
| IN | kg/h | –3.81E+08 | –2.91E+09 | Q1 kJ/h | 2.43E+09 | Steam | 1.20E+06 | kg/h | |
| OUT | kg/h | –3.81E+08 | –2.91E+09 | Boil Eff | 0.81 | hv | 2026 | kJ/kg | |
| Closure % | | 100.0 | 99.91 | delT K | 577 | Hcomb | –3.00E+04 | kJ/kg | |
| Antoine H₂O | A | 16.5362 | B | 3985.44 | C | –38.9974 | p* kPa(a) | | |

## 6:11 U Solid-oxide fuel cell

Simultaneous equations are used in Excel to close the mass and energy balances for a matrix of currents and air feeds ranging from 1 to 11 Amp with $O_2$ stoichiometric ratios 1.1 to 3.5. The full material and energy balance shown below was used to establish Table 1 which shows a diagonal matrix of conditions (*bold italics*) where air flows and electric currents allow operation from 1073 to 1273 K.

| SOFC RANGES | $O_2$ Stoichiometric/fuel ratio | | | | | |
|---|---|---|---|---|---|---|
| | 1.1 | 1.5 | 2 | 2.5 | 3 | 3.5 |
| | Air rate kg/h | | | | | |
| | 2.82 | 15.37 | 20.5 | 57.7 | 84.6 | 98.6 |
| Current | Viable conditions are in bold italics | | | | | |
| Amp | | | T(3) K | | | |
| 1 | 926 | 849 | | 679 | | |
| 2 | *1097* | 941 | | | | |
| 3 | *1212* | 1032 | | 799 | | |
| 4 | 1328 | *1123* | 963 | | | |
| 5 | | *1215* | 1036 | | | |
| 6 | | 1306 | *1108* | 978 | 887 | |
| 7 | | | | | | |
| 8 | | | *1253* | *1098* | | |
| 9 | | | 1352 | | | |
| 10 | | | | *1218* | *1091* | |
| 11 | | | | 1278 | *1142* | *1041* |

| STREAM TABLE | | SOLID OXIDE FUEL CELL | | | | | |
|---|---|---|---|---|---|---|---|
| Species | M | $C_{p,m}$ | $h_f$ | | *Stream* | kmol/h | |
| | kg/kmol | kJ/kmol.K | kJ/kmol | *1* | *2* | *3* | *4* |
| $CH_4$ | 16.0 | 35.3 | −7.48E+04 | 7.46E−02 | 0 | 0 | 0 |
| $H_2O(g)$ | 18.0 | 33.6 | −2.42E+05 | 1.49E−01 | 0 | 2.99E−01 | 0 |
| $CO_2$ | 44.0 | 37.1 | −3.94E+05 | 0 | 0 | 7.46E−02 | 0 |
| $O_2$ | 32.0 | 29.4 | 0.00E+00 | 0 | 2.99E−01 | 0 | 1.49E−01 |
| $N_2$ | 28.0 | 29 | 0.00E+00 | 0 | 1.E+00 | 0 | 1.E+00 |
| Total mass | kg/h | | | 3.88 | 41.00 | 8.66 | 36.22 |
| Phase | | | | G | G | G | G |
| Pressure | kPa(a) | | | 150 | 150 | 130 | 130 |
| Temperature | K | | | 298 | 298 | *1253* | 1198 |
| Volume | m3/h | | | 3.7 | 23.5 | 29.9 | 97.4 |
| Enthalpy | kJ/h | | | −4.17E+04 | 4.13E+01 | −8.93E+04 | 3.33E+04 |
| Mass balance | | REACT | Volts | 0.50 | | 1.5 | 0.1 |
| IN | kg/h | 45 | kA | 8 | U | 0.01 | kW/m².K |
| OUT | kg/h | 45 | | | A | 8.00E−01 | m2 |
| Closure % | | 100.0 | T(4) K | 1198 | del T | | |
| Energy balance | | | | W | Range C | 800 | 1000 |
| IN | kg/h | −4.16E+04 | | 1.44E+04 | Range K | 1073 | 1273 |
| OUT | kg/h | −4.16E+04 | | kJ/h | O2 stoic | 2 | 2.00 |
| Closure % | | 100.0 | | | Air feed | 41.00 | kg/h |

# Chapter 7:   Unsteady-state Material and Energy Balances

### 7:1 W   Storage tank

Water fed to and draining from a tank. Unsteady state, no reaction. GEN = CON = 0
System = contents of tank (liquid water), fixed density. Open system.
A = tank X-section area $m^2$, L = depth, F = water inlet flow rate $m^3/h$, t = time hr
Differential, unsteady-state mass balance on water in tank.
ACC = IN − OUT + 0 − 0
$d(AL)/dt = A(dL/dt) = F − kL$   $dL/dt = F/A − kL/A = \alpha − \beta L$   $\alpha = F/A, \beta = k/A$
Integrate with limits L = 2 ------ L, t = 0 ------ t
$\ln[\alpha − 2\beta)/(\alpha − L\beta)] = −\beta t$
$[\alpha − 2\beta)/(\alpha − L\beta)] = \exp(−\beta t)$
Solve for $L = [\alpha − (\alpha − 2\beta)\exp(−\beta t)]/\beta$
Substitute for $\alpha$ and $\beta$

a.   At t = 1 hr                                                              L = **3.86 m**
b.   At t => 10 hr                              L reaches "equilibrium" = **4.63 m**

The full time profile is shown below.

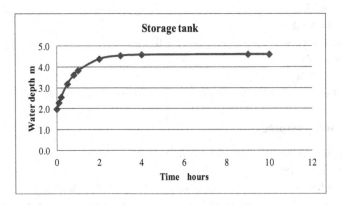

## 7:2 U   Mixed tank dilution

Mixed tank HCl and water. Continuous, unsteady-state, no reaction.
At t = 3 hr          Conc HCl = 1.89 kmol/m³ ≡                                   **69 kg/m³**
HCl concentration vs. time is shown below.

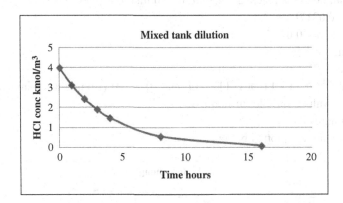

## 7:3 W   Mixing benzene-xylene

Mixing tank: benzene, xylene. Unsteady state, fixed volume, ideal mixture, no reaction
System = contents of tank. Open system. Tank initially full of o-xylene.
Designate: A ≡ o-xylene       B ≡ benzene       w = mass fraction B in tank
ρ = density of mixture = density of benzene = density of o-xylene = 880 kg/m³
v(j) = volume flow stream j m³/h,       V = tank volume = 2 m³
Differential mass balances:
ACC = IN – OUT + GEN – CON          GEN = CON = 0
Total mass: d(ρV)/dt = 0 = ρv(1) – ρv(2) + 0 – 0 → v(1) = v(2) = v          [1]
A: d(ρV(1-w))/dt = ρv(1) – ρ(1-w)v(2) + 0 – 0          [2]
B: d(ρVw))/dt = ρv – ρwv + 0 – 0          [3]
Substitute [1] into [3] and reduce.

$d(\rho V(w))/dt = \rho v - \rho vw = \rho v(1-w)$
$d(w)/dt = v(1-w)/V$                                                               [4]
$d(w)/[1-w] = (v/V)dt$
Integrate with limits: $w = 0$ ------ $w$          $t = 0$ ------- $t$
$\ln(1/(1-w)) = (v/V)t$
$w = 1 - \exp(-(v/V)t)$
Substitute values: $v = 1$ m³/h, $V = 2$ m³          $t = 10$ hr → $w = 0.394$          **39.4 wt%**

## 7:4 W   Vertical pumping

Vertical pumping. Unsteady-state, no reaction.
System: Pump and pipe
Designate: $L$ = height m,          $t$ = time hour          $A$ = pipe flow area m²
$F$ = volume flow m³/h, $\rho$ = density of liquid kg/m³ = 1814 kg/m³          $g$ = gravity m/s²
**a.**   Balance on water in pipe:
    ACC = IN – OUT + GEN – CON          OUT = GEN = CON = 0
    $d(AL/dt) = F = (25 - p^2) = [25 - (1E-5\rho gL)^2] = 25 - 0.0317L^2$ 1 atm ≡ 1E5 Pa
    $dL/dt = [25 - 0.0317L^2]/A$
    Maximum height is reached when $dL/dt = 0$ then $L = (25/0.0317)^{0.5} =$          **28.1 m**
**b.**   $dL/[25 - 0.0317L^2] = Adt$
    Put $a = 25$, $b = 0.0317$
    $dL/(a+bL^2) = Adt$
    Integrate: $L$ 0 ---- $L$          $t = 0$ ------ $t$
    $[1/(2(-ab)^{0.5}\{\ln\{[a + L(-ab)^{0.5}]/[a - L(-ab)^{0.5}]\}] = -At$ (from table of integrals)
    Substitute values and plot to find $L$ as f(t)
    At 10 minutes $L =$                                          **20 m**
    Time vs. height is plotted below

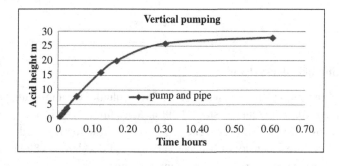

## 7:5 U   DAC absorbent regeneration

Direct air capture of $CO_2$, fluidized pellet bed causticizer.
Pellet residence time =                                          **149 sec**

## 7:6 W   CSTR bio-reactor conversion

Bio-reactor. CSTR   Unsteady-state, material balance, open system, fixed temperature.
Reaction stoichiometry – not given.   Substrate "S", concentration [S] kmol/m$^3$
Reaction kinetics:   Rate = d[S]/dt = (–1)v[S]/(K+[S])
System = contents of reactor
Mole balance on S:
F = feed rate m$^3$/h, V = reactor volume m$^3$, S(j) concentration S in stream j kmol/m$^3$
ACC = IN – OUT + GEN – CON
For a CSTR the concentration in stream 3 = concentration in the reactor, i.e. S(3)
Vd[S(3)]/dt = FS(1) – F(S3) – V{vS(3)/[K+ S(3)]}
dS(3)/dt = (F/V)S(1) – (F/V)(S(3) – vS(3)/[K + S(3)]
d[S(3)] /{(F/V)S(1) – (F/V)(S(3) – vS(3)/[K + S(3)]} = dt                    [1]
Integrate for: S(3) = 0.6 ------- S(3) kmol/m$^3$      t = 0 – t hr
This is a messy analytical integration.
The numerical solution, using Euler's method is shown below.
Conversion of S at 10 hours =                                              **71%**
In this case the time step size was decreased exponentially from 1 to 0.07 hr over 20 hours.

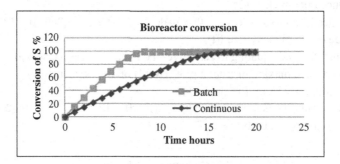

## 7:7 U   Heated batch reactor

Steam heated batch reactor. Unsteady-state, reaction, adiabatic, closed system.
Simultaneous non-linear differential material and energy balance equations.
Solved by numerical integration.                                    **400 K, 85%**

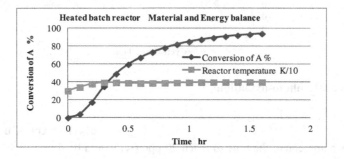

## 7:8 W   Mineral waste filtration

Unsteady-state, semi-continuous, no reaction.

System: Contents of filter.

$V$ = filtrate volume $m^3$, $\rho$ = waste density $kg/m^3$   $W$ = waste mass = $(0.01)\rho V$ kg

Mass balance on solids.

$ACC = IN - OUT + GEN - CON$        $OUT = GEN = CON = 0$

$dW/dt = (0.01)\rho dV/dt - 0 + 0 - 0$                                                                        [1]

$dV/dt = A^2 \Delta P/(2Vk)$

$VdV = [N]dt$              $N = A^2 \Delta P/(2k)$

Integrate      $V = 0$ ----- $V$      $t = 0$ ------- $t$

$[0 - V^2] = -Nt \rightarrow V = (2Nt)^{0.5}$                                                                [2]

Substitute values: $N = (1^2\ m^4)(200E3\ kg/(m.s^2))/((2)(1E7\ kg/m^3s) = 0.01\ m^6/s$

a.    Volume of filtrate in 0.5 hour = $N(1) = (2Nt)^{0.5} = [2(0.01)(0.5)(3600)]^{0.5} = 6.0\ m^3$.
      Load of solids = $6\ m^3$ $(0.01\ kg/kg)(1000\ kg/m^3)$ =                                   **60 kg**

b.    The minimum number of $1\ m^2$ filter leafs will correspond to the maximum filtrate
      produced per 24 hours.

Number of filtration cycles = $(24\ h)/((t + 1.5)\ h)$

Filtrate produced per cycle = $(2Nt)^{0.5}$

Total filtrate per 24 hours = Tot = $((3600)(24))(2Nt)^{0.5})/(t + (1.5)(3600)) = f(t)$

Find the value of "t" to maximize $f(t)$.

Differentiating* $f(t)$ and equating to zero gives a quadratic equation, solved with $t = 1.5$ hr

More easily, pick the maximum from a plot (Excel) of Tot vs. t. (below)          $t = 1.5$ hr

Then filtrate volume = $83\ m^3/(m^2.day)$

Number of $1\ m^2$ filter leafs = $1000\ m^3/d/83\ m^3/(m^2.d) = 12 \rightarrow$                **12 filter leaves**

* $d(u/v)/dx = [vdu/dx - udv/dx]/(v^2)$

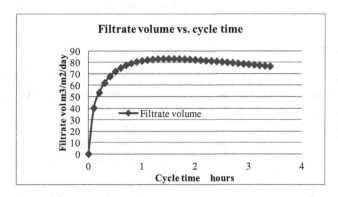

## 7:9 U   Bio-waste photo-oxidation

a.    Time to 20 ppm(w) =                                                                            **19.6 hours**

b.                                                      $dB/[1 - \exp(-k'B)] = (F/V)dt$

Other things being equal, the time to reach 20 ppm(w) would be shorter than in case "a",
because the reaction rate is higher with fixed illumination (I).

## 7:10 W  Wastewater virus

Continuous process, unsteady-state, no reaction. GEN = CON = 0
System: Sequence of sewage transport and treatment vessels.
Plug-flow (pipe) $\rightarrow$ Mixed tank (Unit 1) $\rightarrow$ Mixed tank (Unit 2)
Designate:
$C(0)$ = concentration of COVID RNA at source, $C(i)$ = concentration in stream "i",
$F$ = sewage flow rate $m^3/h$, $V(k)$ = vessel volume $m^3$,
$V(k)/F = \tau(k)$ = sewage residence time in vessel k        h
Mass balances in sequence:
Pipe: $0 = FC(0) - F(C(1) + 0 - 0$       $C(0) = C(1)$                         [1]
Residence time in pipe = distance/flow velocity = (5 km)(1000 m/km)(0.8 m/s) $\equiv$ 1.7 h
Unit 1:  $V(1)(dC(2)/dt = FC(1) - F(C2) + 0 - 0$
$d(C(2)/dt = F/V(1)[C(1) - C(2)] = \tau(1)[C(1) - C(2)]$
$d(C(2))/[C(1) - C(2)] = \tau(1)dt$
Integrate with limits $C(2) = 0$ --------- $C(2)$        $t = 0$ -------- $t(1)$ and solve for $C(2)$
$C(2) = C(1)(1 - \exp[-t(1)/\tau(1)]$                                          [2]
Unit 2:  $V(2)d(C(3)/dt = FC(2) - FC(3) - 0 - 0$
$d(C(3)/[C(2) - C(3)] = \tau(2)[C(2) - C(3)]$
Integrate with limits $C(3) = C(2)$ -------- $C(3)$        $t = t(1)$ ------- $t(2)$ and solve for $C(3)$
$C(3) = C(2)[1 - \exp[-t(2)/\tau(2)]$                                         [3]
Equations 2 and 3 may be combined to give $C(3) = f\,[t(1),t(2),\,\tau(1)],\tau(2)]$        [4]
OR Calculated in series to find the total time for $C(3)$ to reach 95% of its value for $t(2) \rightarrow \infty$
Time to 95% of $\infty$ (see graph below) =                                **33 hours**

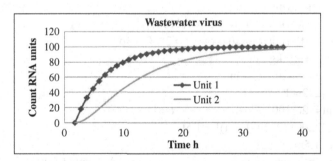

## 7:11 U  Heating a pool

Open air pool.
a.    Time to 30°C =                                                          **88 hours**
       Time vs. $T_p$ plotted below
b.    $CO_2$ generated =                                                       **351 kg $CO_2$**
The temperature vs. time curve is below.

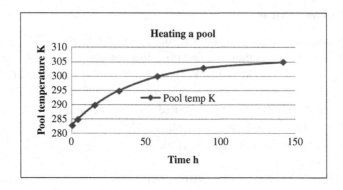

## 7:12 W   Mixing chlorobenzene-hexane

Continuous, unsteady-state, no reaction.                NB. Changing density
System = contents of tank. Open system. Tank in initially full of chlorobenzene.
$A \equiv$ chlorobenzene, $B \equiv$ hexane t = time m = total mass in tank kg V = tank volume $m^3$
Stream i: m(i) = total mass flow kg/h v(i) = volume flow $m^3$/h   w = mass fraction B Str 3
Ideal mixture density = $\rho$ = (1100 kg/$m^3$)w + (660 kg/$m^3$)(1-w) = [440w + 660] kg/$m^3$   [1]
Total volume flow = v(3) = v(1) + v(2) = 1 + 2 = 3 $m^3$/h (ideal mixture)
Differential mass balances:
ACC = IN – OUT + GEN – CON        GEN = CON = 0
Total mass: d(m)/dt = [m(1) + m(2)] – m(3) + 0 – 0                                              [2]
A: d[m(1-w)]/dt = rate in – rate out = 1(1100) – $\rho$v(3)(1 – w)  kg/h                        [3]
B: d[(mw)]/dt = rate in – rate out = 2(660) – $\rho$v(3)(w)            kg/h                      [4]
2 equations, 2 unknowns w and $\rho$
From Eqn 3:      d(m)/dt – d(mw)/dt = 1(1100) – $\rho$v(3)(1 – w)                                [5]
Add [4] to [5]    d(m)/dt = 2420 – $\rho$v(3) = 2420 – 3$\rho$
And      m = (2 $m^3$)($\rho$ kg/$m^3$)
Then     d(m)/dt = 2420 – (3/2)m
            d(m)/[2420-1.5m] = dt                                                               [6]
Integrate for m = 2(1100) ------- m,          t = 0 ----- t
[ln(2420– (1.5)2200)/(2420–1.5m)] = 1.5t
ln(–880/(2420–1.5m)) = 1.5t
–880/(2420–1.5m) = exp(1.5t)          2420 – 1.5m = –880exp(–1.5t)
m = [2420+880exp(–1.5t)]/1.5 = t = 0, m = 2200 kg, t = 1 hr  m = 1744 kg
For m = 1744 kg  $\rho$ = m/2 = 872 kg/$m^3$         w = (872–660)/440 = 0.482 mass fraction B
Mass fraction A = 1 – 0.481 = 0.519 at 1 hour
M chlorobenzene $C_6H_5Cl$ = 112.5 kg/mol, M Hexane $C_6H_{14}$ = 86 kg/kmol
Mole fraction chlorobenzene = x(A) = (0.519/112.5)/(0.519/112.5 + 0.481/86) ≡  **45.2%**

# EPILOGUE

*With apologies to:*
*"Ozymandias" by Percy Bysshe Shelley, 1792–1822*

I met a traveller from a Northern land
who said: Four vast and cab-less rims of chrome
stand in the desert... Near them, on the sand
half sunk, some rusted metal lies, whose frame,
and sparking tip, and gears of cast compound
tell that its owner well those fashions led
which yet survive, stamped on these lifeless things,
by glossy ad extolled and iso-octane fed.
And on the engine block these words appear:
"My name is Suv-Humongous, King of Cars:
Look on my power, ye mighty, and despair!"

Nothing beside remains.
Round the decay of that colossal wreck,
boundless and bare
the lone and level sands stretch
far away...

# INDEX

example calculations, 16, 17, 134, 137,
    149, 201, 210, 222, 225, 234, 236, 251
Thermochemical reactor, 94, 211, 222, 233–234,
    293–294, 299, 301, 304
Thermochemical values (table), 61
Thermochemistry, 60–71,
Thermodynamic chart/diagram, 78–83
    energy balance, reference state, 74,
        174–176,
    example calculations, 80, 83, 197
    non-reactive energy balance, 174–176,
        182, 185
Thermodynamic properties, 74
    calculation, 60–71
    enthalpy-concentration diagram, 78
    example calculations, 70, 77, 80, 83
    tables, diagrams and reference states, 74–81
    water (steam table), 74–78
    water-air (pyschrometric chart) 81–83
Thermodynamic table, 74–77
    steam table (condensed version) 75
    example calculations, 77, 197, 200
Therm-mistor/couple/meter, 32
Threshold effect, 225
Transient conditions, 246
Triple bottom line, 90
Triple point, 35, 74–75
Turbine, 94, 177

**U**

Units (chemical process), 91–94
    generic energy balances, 187–194
    generic material balances, 109–111
Units (physical quantities), 24
    basic and derived (table), 25
    conversion, 28
    dimensional consistency, 26
    example calculations, 26–27, 29
    rules for manipulating, 26
    SI to Imperial/American (table), 29
Unsteady-state, 14, 246–249

example calculations, 249–258
Utility load, 97–98, 173, 176

**V**

Vacuum, 31
Vapour, 35–36
Vapour pressure, 45–50
    Antoine equation, 45
    Clausius–Clapeyron equation, 45
    common substances (table), 46
    example calculations, 36, 47, 50
Vapourization, 274, 278
Vapour-liquid equilibrium, 45
    example calculations, 47, 124, 147, 149 ,
        154, 204, 207, 221, 227, 230
Variable (process), 30–36, 125–126,
    178–180
Vertical farm, 296
Viral infection, 269, 322
Volume, 9, 15, 25, 34, 38, 42, 43
    example calculations, 32, 39, 77

**W**

Water, 9, 19, 75–78, 105, 221, 258
    electrolysis, 268
    splitting, 272, 299, 308
    waste treatment, 321–322
Wet bulb temperature, 81
Work (power)162–164, 168–169, 171, 173,
    176–178
    example calculations, 172, 195, 201, 207,
        236, 238

**X**

Xenobiotic material, 321
Xenotoxic chemical, 298

**Y**

Yield, 55
    example calculations, 56, 130, 134, 136,
        139

Printed in the United States
by Baker & Taylor Publisher Services